Advances in AI for Plant Protection

Editors

Kamel A. Abd-Elsalam

Plant Pathology Research Institute
Agricultural Research Center
Giza, Egypt

and

Aftab Ahmad

Natural and Medical Sciences Research Center (NMSRC)
University of Nizwa, Oman

CRC Press
Taylor & Francis Group
Boca Raton London New York

CRC Press is an imprint of the
Taylor & Francis Group, an **informa** business

A SCIENCE PUBLISHERS BOOK

Cover credit: Image designed and modified using Freepik AI tool for use on the cover.

First edition published 2027
by CRC Press
2385 NW Executive Center Drive, Suite 320, Boca Raton FL 33431

and by CRC Press
4 Park Square, Milton Park, Abingdon, Oxon, OX14 4RN

CRC Press is an imprint of Taylor & Francis Group, LLC

Library of Congress Cataloging-in-Publication Data (applied for)

ISBN: 978-1-032-88567-4 (hbk)
ISBN: 978-1-032-90246-3 (pbk)
ISBN: 978-1-003-54670-2 (ebk)

DOI: 10.1201/9781003546702

Typeset in Times New Roman
by Prime Publishing Services

Preface

In recent years, the agricultural landscape has undergone a remarkable transformation fueled by advancements in technology and data science. Among these developments, artificial intelligence (AI) has emerged as one of the most promising tools in addressing the ever-growing challenges in plant protection, from disease outbreaks and pest infestations to climate-induced stress and resource inefficiencies. The integration of AI into plant health management marks a new era of precision, prediction, and prevention, redefining how we understand and respond to threats in agricultural ecosystems.

Integration of AI with plant protection science represents a landmark achievement for sustainable agriculture and food security. As demand for food security, food safety, and environmental protection grows, modern agricultural practices have transformed conventional approaches to precise, intelligent, and data-driven strategies for managing plant growth and health. This volume, *Advances in AI for Plant Protection*, presents a comprehensive overview of the latest developments in AI-based innovations and solutions in redesigning the future landscape of plant protection to achieve food security and safety.

The introductory chapter establishes the foundation by summarizing the trends of integrating AI with plant protection science, followed by detailed chapters that cover transformative technologies such as remote sensing, image processing, and machine learning. The topics covered in these contributions include insect pest management using satellite and image analysis, automated disease detection systems, and AI-based disease diagnosis and intervention. This volume highlights the use of AI for earlier detection of diseases and insect pest infestations in major crops, including cotton, wheat, tomato, and potato. The earlier detection and diagnosis of diseases and insect pests through AI can significantly contribute to crop productivity and, consequently, to food security and sustainable agriculture.

In addition, the primary focus of this volume is the application of robotic systems and unmanned aerial vehicles (UAVs) for real-time crop monitoring and the precise application of pesticides, aiming to minimize their use while protecting the environment. Several recent studies demonstrate how AI algorithms and machine learning are being utilized for predictive modeling, early disease detection, and targeted prevention measures, with a particular focus on major crops such as wheat rust, a devastating disease at the global level.

The volume also focuses on the broader applications of AI in plant protection, including precise weed control, precision agriculture, disease diagnosis and control, and integrated pest management, while also addressing ethical and data protection challenges. As AI tools become increasingly accessible to everyone, the volume underscores the collaborative efforts between researchers, farmers, policymakers, and consumers to utilize intelligent systems and machine learning approaches to protect crops, reduce input costs, and enhance crop resilience against biotic and abiotic stress, ultimately improving food security and sustainable production.

Finally, *Advances in AI for Plant Protection* serves as a reference book, highlighting interdisciplinary efforts among all stakeholders to inspire innovations, foster collaboration, and ensure the responsible use of AI technologies in agriculture for future goals and developments.

This book, *"Advances in AI for Plant Protection,"* is conceived at the intersection of innovation and necessity. It reflects a global movement towards more innovative, more sustainable agricultural practices that leverage the power of intelligent systems. As the farming sector strives to meet the demands of a growing population under the constraints of shrinking resources and increasing environmental stress, the application of AI offers not only a technical solution but a strategic vision for future farming.

I sincerely thank all the authors who contributed for sharing their expertise and valuable insights, which have been instrumental in shaping this compilation. We are deeply grateful to the entire team, including Raju Primlani, the Editorial Manager, Jyotsna Jangra and Soumya Jain, the Editorial Assistants from CRC Press, Taylor and Francis Books, for their generous cooperation and tireless efforts in bringing this book to fruition. We would also like to extend our sincere appreciation to all the contributing authors for their excellent work and unwavering support in crafting these thought-provoking and scholarly chapters. Furthermore, we want to express our heartfelt gratitude to each reviewer who generously dedicated their valuable time to complete the review process. Lastly, we would like to thank our friends, family, and colleagues for their continuous support and encouragement throughout this endeavor. We extend our sincere gratitude to all the contributing authors, reviewers, and researchers whose work forms the foundation of this volume. Their dedication and vision have made this book a valuable contribution to the field.

Kamel A. Abd-Elsalam
Agricultural Research Center, Giza, Egypt

Aftab Ahmad
NMSRC, University of Nizwa, Oman

Contents

CHAPTER 1

Introductory Chapter

Advances in AI for Plant Protection

Kamel A. Abd-Elsalam,[1,*] *Rawan K. Hassan,*[2] *Toka E. Abdelkhalek,*[2] *Ramadan A. Arafa*[1] and *Hassan Almoammar*[3]

1. Introduction

Agriculture faces significant challenges from pests, diseases, and environmental changes, threatening global food security and crop productivity (Khang et al., 2023). Traditional plant protection methods, often labour-intensive and reliant on extensive chemical use, are increasingly inadequate and unsustainable. Artificial intelligence (AI) offers innovative solutions to address these challenges (Figure 1), transforming plant protection through precision and efficiency (Nahiyoon et al., 2024). AI technologies, including image recognition, machine learning, and deep learning, enable accurate detection of plant diseases and pests, often at early stages (Yue et al., 2025). For example, AI-powered systems can analyse images of leaves or stems to identify symptoms with precision comparable to human experts. Predictive models, using environmental data and real-time monitoring, can forecast disease outbreaks, allowing proactive management and minimizing crop losses (Leybourne et al., 2025). Integrating AI with drones, internet of things (IoT) sensors, and remote sensing has further expanded its application. Drones equipped with high-resolution cameras survey large fields (Mesías-Ruiz et al., 2023), capturing detailed data for crop health analysis, while IoT devices monitor environmental factors like soil moisture and temperature (Upadhyay et al., 2025). These tools drive precision agriculture, ensuring optimal resource utilization and sustainability (Sharada et al.,

[1] Plant Pathology Research Institute, Agricultural Research Center, Giza 12619, Egypt.
[2] Biotechnology English Program, Faculty of Agriculture, Cairo University, Giza 12613, Egypt.
[3] Wellness and Preventive Medicine Institute, Health Sector, King Abdulaziz City for Science and Technology (KACST), Riyadh, Saudi Arabia.
* Corresponding author: kamelabdelsalam@gmail.com

DOI: 10.1201/9781003546702-1

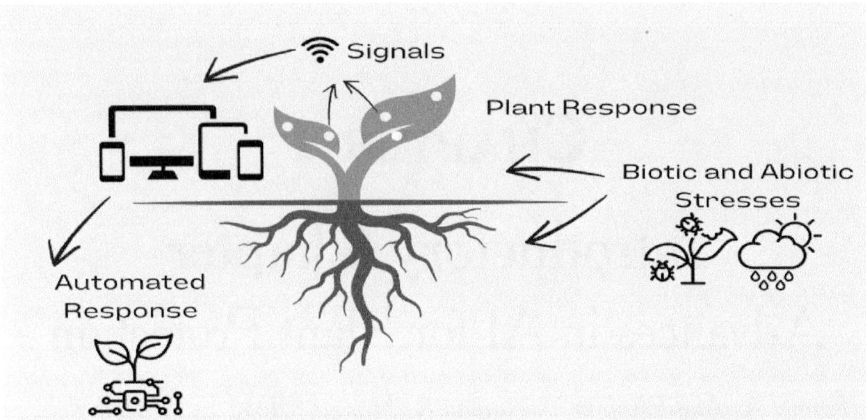

Figure 1. Various artificial intelligence (AI) techniques for the protection of plants have been used. The main technologies are highlighted which enhance the identification of the diseases and pests in plants, image recognition, machine learning and predictive modeling.

2025). Despite these advancements, challenges such as limited data access, high costs, and technical complexity hinder broader adoption, particularly in resource-limited regions AI models must also be refined to address variability across crops, climates, and pathogens (Mahlein et al., 2024a). Nevertheless, the potential benefits of enhanced productivity, reduced environmental impact, and improved resilience highlight AI's transformative role in modern agriculture (Das et al., 2022).

This chapter explores AI-driven advancements in plant protection, focusing on early disease detection, predictive modelling, and resource optimization (Balafoutis et al., 2021). By showcasing real-world applications and addressing challenges, it highlights how AI fosters sustainable agricultural practices and resilience against future threats, paving the way for innovative and efficient plant protection strategies.

2. Challenges in Plant Protection

Plant diseases and invertebrate pests have affected communities throughout history, leading to severe crop losses and famines. At the moment, crops and natural vegetation may suffer minor to major harm as a result of these hazards (Oerke et al., 2012). As population continues to grow, reaching 9 billion by 2045 (US Census Bureau, 2009), it is important to remain vigilant in our efforts to prevent these losses. For instance, coffee leaf rust almost brought Sri Lanka's coffee industry to its knees, and hence led to the economic decline of the country (Kushalappa & Eskes, 1989). Diseases can decrease wheat production by as much as 17% if left unprotected, while pests can decrease it by as much as 11%. These issues and the need to adopt sustainable agriculture in the face of global warming and the growing food demand, cannot be addressed without incorporating cutting edge technologies such as robots and artificial intelligence in plant protection techniques (Cheatham et al., 2009).

2.1 Major Plant Diseases and Pests Affecting Global Agriculture

Plant diseases and pests are major threats to global agriculture, food security and economic stability and caused by some of different biotic reasons (Figure 2). The pathogens from viroid's to more complex organisms—cause symptoms that range from mild to catastrophic, and lead to substantial crop losses. More than 800 million people are hungry today, and these diseases present additional challenges to achieving security (Strange and Scott, 2005).

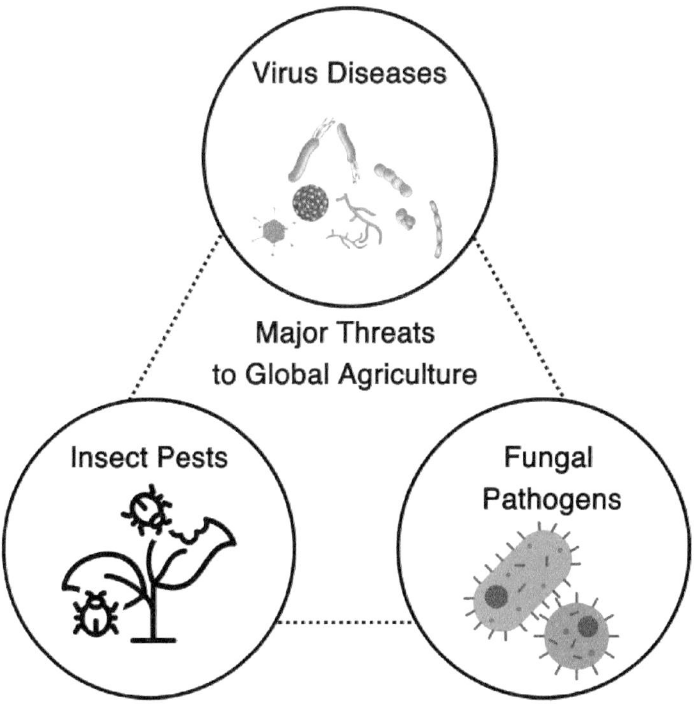

Figure 2. Plant diseases and pests are the world's main agricultural danger, which confirms their impact on both economic stability and food security.

Virus Diseases

Plant viral diseases are a major problem for agricultural food security, especially in the poorer worlds' countries. The history of plant virology can be divided into three main eras, starting with the identification of the virus that caused tobacco mosaic in the late 1800s. The Classical Discovery Period, from 1883 to 1951, made qualitative observations. The Early Molecular Era, from 1952 to 1983, built on this knowledge, advancing the understanding of virus properties. The last period (1983 to the present) is marked by genome engineering and plant genetic engineering for disease resistance (Zaitlin and Palukaitis, 2000). Plant viruses now affect over 200 crop species, including rice, wheat, and bananas, leading to yearly revenue losses of more than thirty billion dollars. The fact that many diseases go undetected complicates yield estimation. Breeding for genetic resistance, RNA interference, CRISPR/Cas9, and integrated pest management (IPM) are all effective methods of controlling virus vectors (Brakke, 1988). These barriers must be removed in order to ensure sustainable farming methods.

Fungal Pathogens

Fungal infections are major hazards to world food supply, affecting a wide range of crops and producing illnesses like rusts, blights, and rots. These infections are highly adaptable and evolve quickly, exacerbated by climate change, posing persistent difficulties to farmers and researchers. Resistant crop types, timely fungicide applications, and integrated pest management (IPM) systems are among the most effective control measures. Recent discoveries in fungal biology, along with technologies such as artificial intelligence and precision agriculture, have increased our ability to prevent, identify, and control these threats to food security (Ferreira et al., 2006). Various species of fungi contribute to these problems, necessitating a diverse approach to their treatment:

Magnaporthe oryzae: Fungal infections are useful not just in agriculture, but also as models for understanding plant-pathogen interactions. The rice blast disease fungus is renowned for its devastating effects on rice, a staple for more than half of the world's population. Research on these fungus improves our understanding of essential biological processes such as cellular respiration, photosynthesis, and genetics (Ou, 1981).

Botrytis cinerea: The second most prevalent pathogen affects a large variety of crops and causes significant pre- and postharvest losses. Moreover, it is also used in wine making thus playing a dual role in agriculture (Patel et al., 2010).

***Puccinia* spp.:** These rust fungi are grouped together according to their shared effect on wheat and are becoming increasingly problematic with the rise of new races like Ug99 that pose a threat to wheat production worldwide (Zadoks, 1985).

Fusarium graminearum: This species is noted for its severe pathology toward cereals especially wheat and maize (Leonard and Szabo, 2005).

Fusarium oxysporum: This fungus is a wide host range pathogen that causes severe disease in crops such as tomatoes, cotton and bananas and is an important problem in many different cropping systems (Oliver, 2024).

Blumeria graminis: This pathogen is famous for causing powdery mildew in cereals and is still a major threat to crop production (Harish et al., 2024).

***Mycosphaerella graminicola*:** This pathogen is especially dangerous to wheat and is currently under study because of its effect on crop production and health (Orton et al., 2011).

***Colletotrichum* spp.:** The present genus comprises several important plant pathogens that are currently undergoing taxonomic revaluation, a process which reveals the complexity of fungal classification and its role in disease management (Prusky, 1996).

Ustilago maydis: This fungus is also known to cause corn smut but has also been studied as a model system for understanding plant immunity and infection processes (Holliday, 2004).

Melampsora lini: This flax rust pathogen is included for scientific importance, helping to understand plant-pathogen interactions and resistance mechanisms (Lawrence, 1988).

Phytophthora infestans: Phytophthora infestans (Mont.) de Bary, a hemibiotrophic oomycete, has been responsible for significant late blight outbreaks in tomato and potato crops globally since the Irish Potato Famine in the 1840s. Developing cultivars that are resistant to late blight is one of the most effective methods to combat this devastating disease. However, P. infestans can overcome host resistance and develop resistance to various fungicides, likely due to its high genetic variability, which arises from both sexual and asexual reproduction. Consequently, understanding the genetic diversity among P. infestans isolates is crucial for effective disease management (Arafa et al. 2020).

Insect Pests

Insect pests are major enemies of agricultural production across the world as they cause direct damage to the crops and are also carriers of diseases which makes pest management difficult. These pests such as aphids, caterpillars and beetles, can cause a severe reduction in yield, impact the quality of the crop and also make the plant more vulnerable to other pathogens (Bala et al., 2018). Integrated pest management, or IPM, is one effective strategy for controlling insect pests. IPM combines chemical, biological, and cultural techniques to lessen insects' effects while using fewer pesticides. Intercropping, crop rotation, and the introduction of natural enemies are examples of cultural controls used for managing pest populations (Gurulingappa et al., 2010). Recent developments in computational intelligence are revolutionizing the management of pests through using environmental data to predict insect activity and suggest the most effective treatments (Strange and Scott, 2005). Farmers can achieve more efficient and ecologically friendly pest management by combining these technologies with traditional farming methods. This comprehensive approach protects crops and promotes environmental health to secure food security despite continuing agricultural challenges.

2.2 Economic and Environmental Impact of Plant Losses

Plant diseases are important in the economic and environmental aspects especially in the recent past due to increased population and different food types needed (He et al., 2016). The agricultural sector is further challenged by factors such as competition for land, degradation of agro-ecosystems and limited resources, which are exacerbated by disease risks that are mostly associated with intensified agriculture and monoculture systems (Palmgren et al., 2015). Plant diseases have been present in the past, affecting food production and the development of societies; large outbreaks have led to well-documented famines, for example the Irish famine of the 1840s which was caused by late blight of potato and the Bengal famine of 1943 which was attributed to rice brown spot. These events show the relationship between plant health and society stability. Although improvement in science and technology has decreased the incidence and spread of plant disease epidemics, post-harvest losses are still put at between 20 and 30 percent (Srilopan et al., 2025). This highlights the need to increase knowledge on the dynamics of disease occurrence and control. However, existing disease control measures have their limitations, which include extinction of useful species, environmental pollution and the development of pathogen resistance. The development of plant disease management has been divided into four stages: The first stage was characterized by minimal intervention, the second stage involved mechanical suppression methods, the third stage was characterized by the application of resistant genes and pesticides and the last stage is ecological pest management (Madhushan et al., 2025). Adaptive management is designed to increase the role of natural controls and prevent pathogen dispersal, based on the knowledge of evolution, not on the return to the primitive agriculture (Krone et al., 2024).

2.3 Potential of AI to Address these Challenges

The agricultural sector is on the verge of transformation with the help of AI to solve the problem of plant losses and food production issues with increasing global population. With conventional farming gradually becoming inadequate, more and more farmers are resorting to AI technologies to enhance crop production, pest control and soil analysis (Javaid et al., 2023). AI systems are able to provide farmers with data driven recommendations for when to plant, what to plant and how to use resources like which type of seed to use and where to apply nutrients to enhance the soil (Ahouandjinou et al., 2017).

Vegetation and crops are managed with precision in order to improve the yield as well as avoid wastage of the resources. AI based health monitoring systems enable the farmer to identify cases of crop assimilate shortage and pest attack early (Jha et al., 2019). Hence, AI improves pest control through better equipment including drones that take high definition pictures that help in identifying pests and determining the health of the crop and therefore recommend proper measures that are least damaging to the environment (Bannerjee et al., 2018).

Machine Intelligence (MI) is a sub discipline of AI that assists in optimizing irrigation by suggesting the right time for watering based on the crop and environmental conditions (Smith, 2018). AI sensors also assist in weed control by using herbicides which are applied only on the areas which have weeds and this reduces the use of chemicals. Although still in the process of being developed, AI is assisting farmers in the process of accessing and analyzing data in order to make better decisions (Waleed et al., 2020a). This transition to data farming increases food production and conservation of the environment and resources.

3. AI Technologies for Plant Protection

Crop threats from pests and diseases are common, along with the ongoing expansion of agricultural output and the growing complexity of environmental change. The swift advancement of artificial intelligence technology offers novel concepts and approaches to address the difficulties associated with managing agricultural pests and diseases. Artificial intelligence technology has the potential to enhance the precision and efficiency of management by prioritizing accurate monitoring as well as disease and pest prediction (Li et al., 2010; Huai et al., 2007).

Image recognition technology, data analysis, and machine learning algorithms provide more rapid and accurate identifications of pests and diseases. Meanwhile, the design and implementation of intelligent early warning systems make management timelier and more flexible. Establishment of a data-driven decision support system architecture, intelligent agricultural management platforms, and the use of automated control technologies provide new approaches and methods for management (Yun et al., 2024; Li and Wang, 2024).

3.1 Image Recognition and Computer Vision

Image recognition technology represents an important branch of AI, mainly trying to allow computers to understand objects, scenes, and features in images as human beings do. It has been done on a large scale, particularly in management of pests and diseases in agricultural production (Xiao et al., 2023).

Not only has deep learning technology been very successful in the field of image recognition, but it has also provided a lot of new ideas and methods for pattern recognition and data analysis in other aspects (Liu et al., 2021). Its prospects are broad in the scenario application. Firstly, it can be used for pest and disease identification. Capture some images of the plant leaves or fruits, and using the trained deep learning model, selected types of diseases and pests can be quickly and precisely identified to help agricultural producers take timely control measures to reduce losses. Second, image recognition technology could be applied for crop growth monitoring (Xiang and Wang, 2023). It therefore needs to be collected in many forms: meteorological, related to soil, crop growth, and pest and disease occurrence data. Real data is collected through sensors, weather stations, and monitoring devices, or it is done through retrospective analysis of historical data so that a comprehensive data foundation is formed (Li et al., 2020). Nevertheless, the use of image recognition technology in agriculture will grow due to ongoing developments in computer vision and hardware equipment modifications (He et al., 2024).

3.1.1 Early detection of plant diseases and pests through image analysis

Photos can be taken to facilitate monitoring and assessment of crop growth, including those on the effect of pests and diseases and status of growth. In view of these factors, agricultural producers may have management references to quickly change management approaches provided (Shahi et al., 2023). In general, the quality of soil, such as the locational information of soil moisture, soil type, and vegetation cover, can be assessed by photographing a drones or sensors' view of the soil surfaces and then combining images of recognition software. This information facilitates farming in terms of soil management and productivity by supporting the producers using science to manage fertilization and irrigation-thus improving efficiency and quality in agricultural production (Jia et al., 2023). Despite recorded success stories in image recognition technologies applied to agriculture, challenges such as lighting condition issues, growth conditions of crops, and quality of images blur the stability and precision of images identified (Fu et al., 2024). Image recognition technology will be used in agriculture more frequently as algorithms, models, and hardware performance continue to evolve. This will allow for more intelligent solutions that boost crop quality, increase productivity, and support the expansion of the agricultural sector (Hernández et al., 2024). Steps of image processing are described in Figure 3.

The huge amount of data regarding agriculture-related activities have been lately stored as an organized database for the creation of the agricultural information system. Data mining technique uses the specific methods and algorithms for the extraction of pattern, weather forecasting, and prediction. Expert systems or knowledge-based systems have been developed using computer programs utilizing artificial intelligence, which tries to emulate the reasoning process of a human expert (Liao et al., 2005). These expert systems use algorithms and data mining approaches to extract information from computer data files that they can use to make decisions (Durkin and Durkin, 1998).

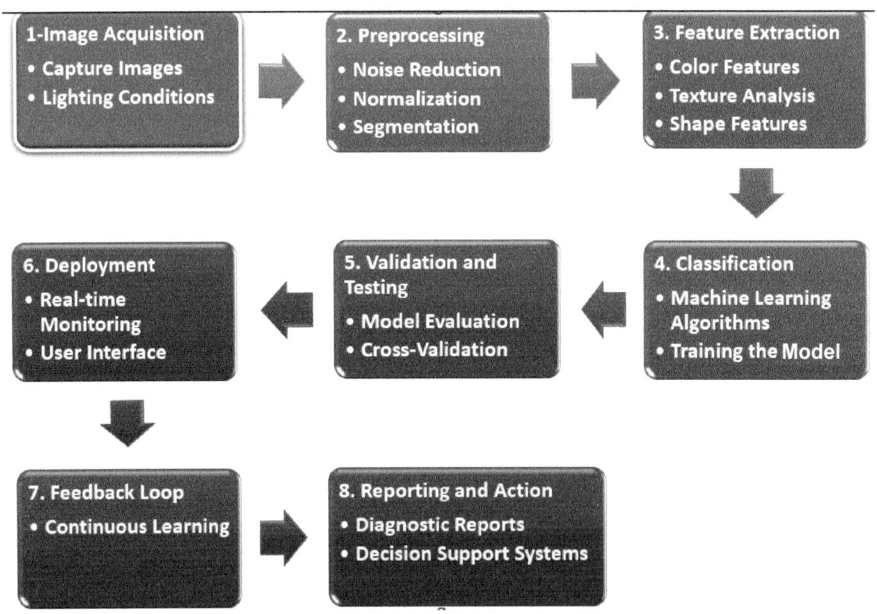

Figure 3. The image processing technique for early recognition of plant diseases and pests involves several important steps. These steps collectively enhance the early recognition of plant diseases and pests, enabling timely intervention and better crop management.

3.1.2 Automated identification of plant species and varieties

Knowing the susceptibility of a given variety to a certain disease, its identification with the aid of this automatic system will allow, for example, a more focused and accurate treatment. Besides the fact that several wine producers need this huge number of grape varieties to produce their most expensive wines—for example, the "*Blend D. Antónia*" wine from the *Quinta do Vallado* has more than 30 grape varieties. For this reason, elaboration on an automatic algorithm for grape variety identification, which would give an automatic split of the different varieties, will have crucial importance for the Douro Demarcated Region (Pereira et al., 2019). In the field of machine learning and artificial intelligence, deep learning (DL) is a very active area of research with several successful applications in many areas, including precision agriculture. Among the strongest advantages of using DL in image classification is a powerful image feature extractor, both from raw or pre-processed data, thus avoiding the traditional and time-consuming process of hand feature extraction. One of the key features of deep learning is its ability to automatically detect low-level features, such as curves and edges, and progressively build higher-level representations through its hierarchical structure (Kamilaris and Prenafeta-Boldú, 2018). The plants represent an ideal taxon on which automated identification accuracy can be tested, implemented in both computational contexts and real-world tests within mobile phone applications, because the taxonomic group is extremely diverse, and it is usually easier to obtain clear photographs of plants under field conditions than is often possible for animals. Since 2013, the Life CLEF challenge has tried to develop and advance the automated identification systems using deep learning frameworks (Goëau et al., 2018). Separate GLMs were developed for each of the five applications of identification to understand both the botanical and image factors controlling accuracy of identification. An ordinal score (0–5) representing the application's accuracy in plant identification served as the dependent variable; a score of 5 meant the first proposal was correct, and a score of 0 meant none of the top five were correctly identified. The existence of flowers, leaves, or fruits as well as the type of plant were independent fixed factors. As variables, the image quality metrics of saliency, focus, and exposure were included. An ordinal or multinomial distribution with a cumulative logit link function was employed in all models (Hart et al., 2023).

3.1.3 Monitoring crop health and growth using drone imagery

Drones are unmanned aerial vehicles that function without the need for a human pilot. These aircraft can be remotely controlled by a human pilot, or they can be programmed with AI software to fly themselves and carry out missions. Using agricultural drones extensively is one of the most revolutionary advances toward "precision agriculture." Drones can investigate dangerous and difficult areas that were previously inaccessible to manned aircraft or satellites because they can carry out flying missions at different heights and viewing angles (Christiansen et al., 2017).

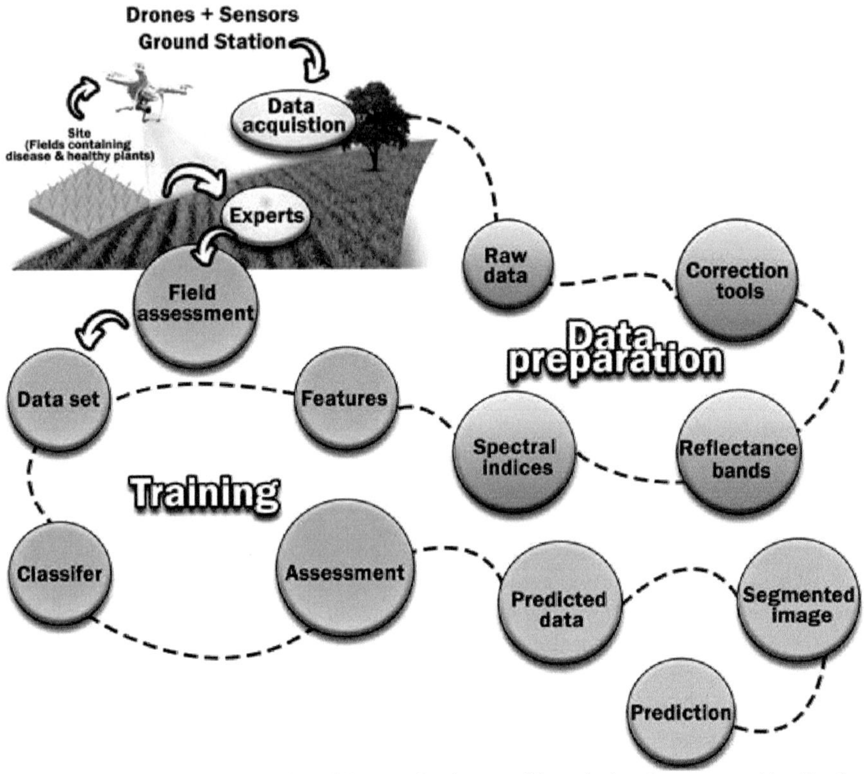

Figure 4. Components and operation of drone technology used in agricultural contexts to identify plant diseases (Sandino et al., 2018).

Recently, a variety of drone types equipped with high-resolution video sensors have found broad usage in many agricultural jobs, such as the detection and treatment of crop diseases. According to how their wings move, drones can be divided into two main categories: hybrid Vertical Take-Off and Landing (VTOL) drones and fixed-wing and rotary-wing drones (Mahlein et al., 2017). An illustration of the composition and functioning of drone technology for the identification of plant diseases is presented in Figure 4 (Sandino et al., 2018).

The model architecture for the detection and classification of plant diseases is a five-step process; image acquisition, image pre-processing, and image segmentation from the feature extraction and classification. Image acquisition is done regarding identification and categorization of crop leaf diseases (Figure 5). A photo dataset is to be gathered in this step that is used afterward in the process. The camera of the drones is used in this regard for capturing. The results can be improved with proper preparation of images. (Martinelli et al., 2015; Mahlein, 2016).

High resolution photos for agricultural monitoring are taken with digital and multispectral cameras, and drones are used in taking a variety of images, in RGB and color infrared (CIR) photos, to help in decision making. Other imaging types like Multispectral (MS), Thermal and Visible and Near Infrared (V-NIR) images are also used in disease identification in plants (Ampatzidis et al., 2017).

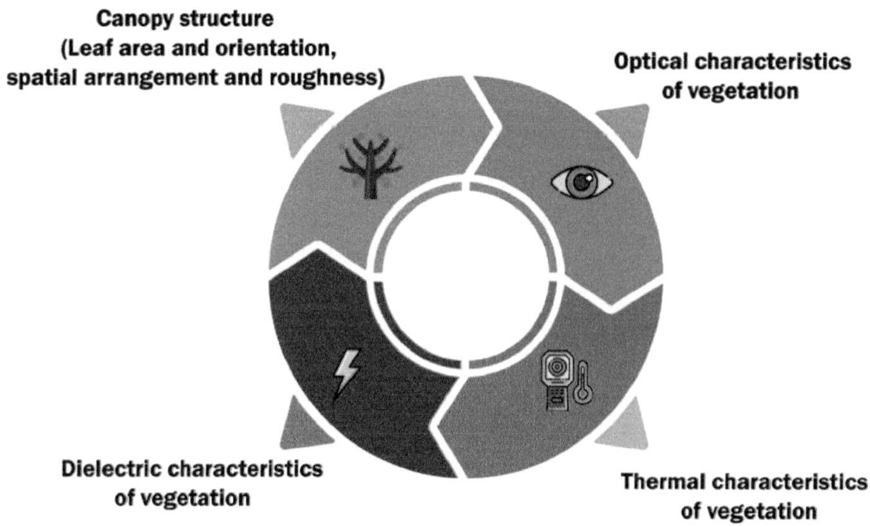

Figure 5. A model architecture of the process of identifying and categorizing plant diseases via or through five steps, which starts from image capture. (Abbas et al., 2023).

Appropriate times for exposure for every camera are chosen based on weather conditions prior to gathering information, and software parameters are established. Drones frequently operate at predetermined altitudes, with one camera sweeping across all experimental plots and another concentrating on particular regions. To ensure efficient image capture, the drones are made to fly over these plots while accounting for wind speed and direction. For these flights to accurately identify plant illnesses, favorable meteorological conditions are necessary (Mirik et al., 2011).

3.2 *Machine Learning and Deep Learning*

By adding more "depth" to the model and transforming the input using a diversity of functions that allow for the representation of hierarchical data with many levels of abstraction, DL extends the capabilities of classical Machine learning (ML) (Schmidhuber, 2015). The ability of deep learning to automatically extract features from raw data en bloc is one of the powerful advantages. The higher-level features of the hierarchy are composed from lower-level ones through a process (Pan and Yang, 2009). Deep learning can solve more difficult problems very well and fast because more complex models are used that allow for large parallelization (LeCun et al., 2015). Efficiency in prediction regarding pest and disease attacks totally depends on the collection and processing of useful data in large amounts (Mingyue et al., 2022).

This would include data from meteorological variables, soil variables, crop growth variables, and incidence of pest and diseases. Data could be acquired in real time through sensors, weather stations, monitoring devices, and so on, or it could retrospectively analyse historic data to form a complete data basis (Zhuoping, 2022). These would require cleaning, processing, and feature extraction for usable inputs on the next steps of machine learning algorithms. Issues to be handled during cleaning and processing include inconsistent qualities of data and non-uniformity of formats while preparing the data for its accuracy and consistency (Ni et al., 2023). Machine learning, including Convolutional Neural Networks (CNNs), provides an automated solution and is likely to have very high accuracy potential. Both require very large and diverse datasets to train the systems. Sistema may experience weaknesses related to rare or unseen pathogens. The relative effectiveness of each approach depends on a series of issues that include scale and budget relevant to diagnosis goals. These approaches integrated together may realize more robust and full plant pest and disease diagnosis (Nigam et al., 2019).

3.2.1 Predictive modeling for disease outbreaks

The technical changes through the digital revolution have been a strong influencer in how predictive modeling has been developing. Large-scale data collection methods, the Internet, and mobile technologies seem to have changed the game and allowed complex datasets to be integrated into models. Modern-day models can now use this real-time information coming from disparate sources, ranging from social media to satellite images and electronic health records, to attain far greater detail and accuracy in predictions (Oldekop et al., 2020).

Data is the foundation for model training and validation. They provide the basis that the models will be built on. The extent of quality, detail, and relevance of the data presented determines how well a model ensures correctness in guaranteeing the outbreaks of diseases. For example, high-resolution mobility data can be integrated into models to predict how a disease spreads geographically (Chumachenko et al., 2023).

Because disease transmission is fundamentally complicated and dynamic, forecasting disease outbreaks is a major issue. Human behaviour, changes in the environment, and pathogen evolution may have a high impact on the spread of diseases; therefore, this is to say that predictions based on various models are like targets that move. Much disease prediction inherently holds uncertainty about future timing and thus requires models that allow for estimates of what might happen with quantification of the uncertainty of predictions. It is here that the role of accuracy and validation becomes critical (Roth et al., 2021).

Predictive modeling of infectious disease outbreaks depends on many data sources, each with different insights into the dynamics of transmission and spread. Indeed, modern epidemiological research without attempts at integrating these streams of data is dead, with work rapidly building from the strengths of one study to another to enhance model accuracy and reliability (Ijeh et al., 2024). Here, we discuss the main sources of data used in predictive modelling, their benefits, and limitations, and the influence of data integration on model performance (International et al., 2020).

3.2.2 Optimizing pesticide application

One of the most important crop management practices that can significantly affect crop output and quality is crop protection. Chemical pesticide treatment is still the most effective way to control weeds, pests, and illnesses, despite the growing use of biological agents (Balsari et al., 2002).

There is a significant danger of exposure and inhalation for nearby livestock and people due to the unavoidable loss of pesticides during crop management, especially spray drift (Feltracco et al., 2022). Pesticide optimization necessitates paying close attention to four fundamental principles: (1) accurately identifying the problem, whether it be physiological, disease-related, or insect-related; (2) choosing the right material for the job; (3) utilizing the chemical in the right equipment at the right time for the target; and (4) applying the chemical at the right time. The public has voiced concerns about the possible risks associated with chemicals being released into the environment (Hall 1984). The use of pesticides for pest control is subject to more regulations. The significance of pest resistance has not decreased. These elements serve as a warning that we need to improve the way pesticides are applied (Guedes et al., 2016). The interaction of numerous intricate levels of activity is presented in Figure 6.

The deployment of intelligent agricultural drones has resulted in important technological advancements in agricultural management. These drones, equipped with high-resolution cameras and sensors, can monitor a variety of vital agriculture characteristics in real time, including crop growth conditions, pest occurrences, and soil moisture levels. This monitoring data provides essential decision assistance, allowing agricultural producers and managers to better understand the state of their farms (Guo, 2023). Intelligent fertilization systems are an important agricultural technology that uses data on soil quality and crop development circumstances to automatically calculate and modify fertilizer amounts and timings. The system may equally distribute nutrients across fields using sprayers or drip irrigation systems, ensuring crops receive adequate nutritional support at various phases of growth (Wang et al., 2021).

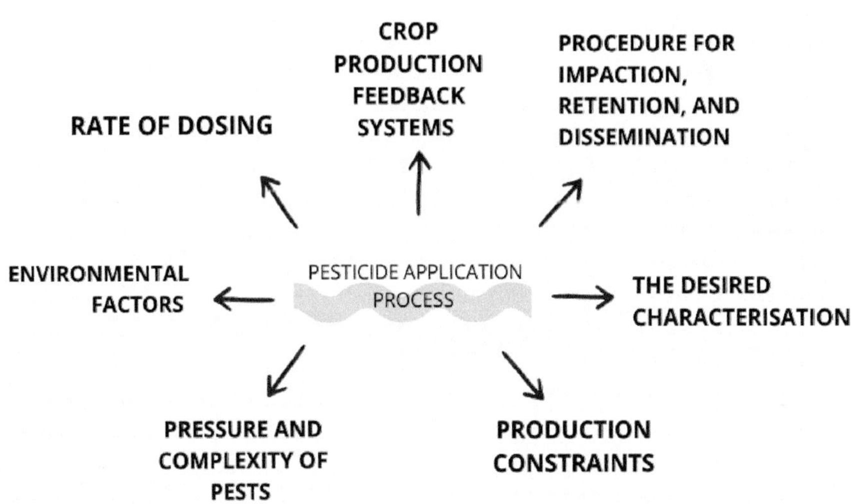

Figure 6. The complexity of relationships between many factors that determine the effectiveness of pesticides application.

3.2.3 Developing personalized plant protection strategies

The term "AI-powered personalized agriculture" describes the use of machine learning and artificial intelligence to offer farmers individualized advice and solutions to maximize their farming operations. To give individualized recommendations for planting, fertilizing, harvesting, and other agricultural operations, this entails gathering and analysing data on weather patterns, soil quality, crop growth, and other characteristics (Tetala, 2023). Integrated pest management (IPM) is critical to sustainable agriculture, and in the following part, I will present a quick overview of the importance of IPM from a sustainable agricultural perspective. Agricultural ecosystems, also known as agroecosystems, are basic spatial and functional units of agricultural activity that comprise both biotic and abiotic components and their interactions (Bonny, 2011).

Sustainability is based on the premise that agricultural systems must meet current demands without jeopardizing future generations' ability to meet their own requirements (Lichtfouse et al., 2009). To use massive datasets as inputs for a specific population model, methods must be developed, which may then be used to create computer programs to simulate pest phenology (Damos and Karabatakis, 2013). Database management system software (DMSS) has typically been used for this purpose. The software is built on a "prediction algorithm," which is a series of logical processes that follow established criteria. The program then selects the appropriate data entry from the available information and stores it in a local host. In the second step, the algorithm feeds this data into the pest model to generate population forecasts and service warnings (Damos, 2015).

3.3 Sensor Networks and IoT

In terms of energy efficiency, the overall energy consumption of sensors and wireless networks must be kept to a minimum when power fluctuates too improve the energy efficiency of IoT sensors, components, and software design (Rao et al., 2019). With IoT, we can control all our household appliances remotely. For example, when we turn on a light, we often forget to turn it off, and when we turn on the TV, we may also forget to turn it off, resulting in energy waste. Also, when we turn on the engine to fill the water tank, we may forget to turn it off, resulting in water waste (Ning et al., 2014). Smart gardening detects moisture levels (dry/wet), and the motor functions accordingly (on/off) (Rao et al., 2019).

3.3.1 Remote sensing for crop health assessment

Crop status estimate requires information input, such as ambient parameters like air temperature, relative humidity, and rainfall, as well as surface factors like soil moisture and temperature (Wang et al., 2009). Remote sensing indices obtained from satellite imagery, such as the Normalized Difference Vegetation Index (NDVI), Land Surface Water Index (LSWI), Temperature-Vegetation Dryness Index (TVDI), Soil Adjusted Vegetation Index (SAVI), Water Deficit Index (WDI), and so on, can be used to determine crop development and/or soil humidity states. Liquid water provides significant luminosity assimilation in the SWIR, and the LSWI is well known to be sensitive to the total amount of liquid water in vegetation and soil (Dong et al., 2014). Furthermore, the Temperature-Vegetation Dryness Index (TVDI) is derived from spatial Land Surface Temperature-NDVI and can be used to estimate soil humidity and hence vegetation water pressure (Rahimzadeh et al., 2012).

The remote sensing technology helps to construct a temporal development profile of plants throughout their growth period. With the recovery of environmental components in addition to remote sensing indices, it is simple to recognize crop development models and their interrelationships, as well as the impact of concerned variables on crop growth (Borengasser et al., 2007). Based on the preceding facts and tests, the remote sensing method is very useful in assessing crop development at both the land and large-scale levels. Assimilation of ecological, surface, and crop state data obtained by remote sensing methods, in addition to soil station data, aids in the advancement of crop state calculation models (Hatfield et al., 2008).

3.3.2 Soil and environmental monitoring

Environment monitoring (EM) includes disaster planning and management, pollution control, and efficiently resolving difficulties that develop because of harmful external conditions. Pollution is caused by a variety of factors, some man-made and others natural, and the duty of EM is to handle the issues to maintain the environment for a healthy society and globe (Ahmed et al., 2019). With recent advances in science and technology, particularly AI and machine learning, EM has evolved into a smart environment monitoring (SEM) system, as technology has enabled EM methods to more precisely monitor environmental factors, resulting in optimal control of pollution and other undesirable effects. Smart city design is replacing traditional ways for creating and planning urban landscapes (Jamil et al., 2015). Wireless networks, also known as wireless sensor networks (WSNs), are made up of sophisticated sensors that use artificial intelligence to monitor and govern themselves. WSNs use IoT devices to manage garbage, mark vehicles, control temperatures, and reduce pollution. Therefore, modern methods of environment monitoring are known as SEM systems, due to the utilization of IoT, AI, and wireless sensors (Bhoomika et al., 2016) Current literature describes the assessment of burned areas using multispectral data captured via satellite imaging and remote sensing, mobile health monitoring systems and IoT-based environment systems, smart marine environment systems using a multimodal sensing network, and many other SME methods. When wireless devices are used over a WSN, specific standards and protocols are required for effective SEM system implementation; accordingly, research have been conducted to define protocols and standards for IoT-based SEM systems (Gaglio et al., 2014; Zhang et al., 2015). The entire world is working to protect the environment for sustainable agriculture, growth, and a healthy society, and thus the primary goal of SEM is to address challenges caused by undesirable environmental effects through smart monitoring so that all key indicators of growth, including societal health, are well regulated. Environmental monitoring methods are used for a variety of applications, including weather forecasting (Kulkarni and Kute, 2016).

4. Applications of AI in Plant Protection

The development of plant protection systems is a significant enhancement in agriculture that is driven by Artificial Intelligence (AI) and leads to increased efficiency in agriculture. These robots apply advanced algorithms, for instance, the ant colony optimization algorithm, in the improvement of accuracy in navigation and performance of operations to attain high efficiency in various farming conditions. As a result, mechanization and automation are becoming more critical in agriculture; hence, AI technologies are being adopted to perform tasks that were formerly done manually or required a lot of labour (Sun et al., 2021). Key innovations include ecological robots that are self-contained and have self-guiding technologies for cultivation and pesticide application. SLAM technology is essential for enabling these robots to navigate autonomously and for precision tasks such as pesticide spraying on specific parts of the plant and monitoring the health of the crop. However, in addition to robotics, AI also assists in real-time monitoring of crop conditions by means of data analysis, and assists farmers in identifying diseases or pests early, and in deciding on the use of resources. This approach enables an effective control of possible losses by acting preventively (An et al., 2022).

4.1 Disease and Pest Management

New and innovative applications and research findings in the application of artificial intelligence (AI) in pest and disease control in agriculture are presented in Figure 1. Intelligent decision support systems are developed and integrated into pest control to help farmers make the right management decisions. New automated technologies such as precision spraying systems and agricultural robots enhance the time and the dosage of pesticide applications, which reduces the amount of chemicals used and their impact on the environment but at the same time controls pests effectively. Combining biological control with AI increases the effectiveness of these approaches (Li and Wang, 2024). In disease resistance breeding, AI analyzes genetic data to identify traits associated with pest and disease resistance, thus facilitating the improvement of crop varieties with better resistance. This integration improves the efficiency of developing disease resistant cultivars and reduces the reliance on chemical controls. The effectiveness of the AI application in pest and disease control is demonstrated through various case studies that highlight its potential. This paper is intended to contribute to the understanding of AI in pest and disease management in order to encourage its application and development in this area (Vidya Madhuri et al., 2025a).

4.1.1 Intelligent pest management strategies

AI techniques such as ML, DL, and computer vision approaches are used in modern technologies for detecting and identifying insects. By identifying patterns and detecting insects, these tools (Figure 7) improve the efficacy of pest treatment. This supports decision-making in pest management by minimizing pesticide use while optimizing the timing of applications (Kariyanna and Sowjanya, 2024). Various AI

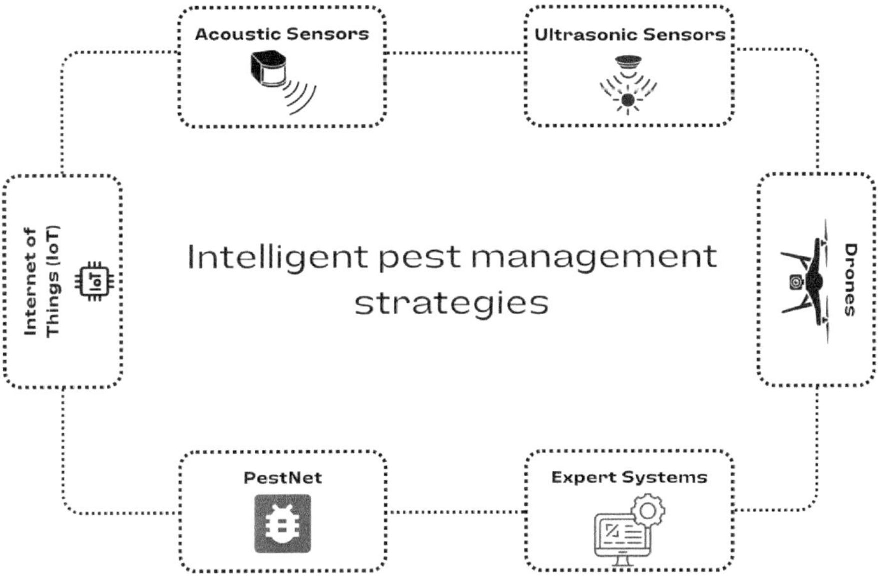

Figure 7. New smart pest control plans that use artificial intelligence methods.

solutions are currently being implemented in pest control methods such as image analysis and optical sensors combined with pheromone traps for pest detection.

Acoustic Sensors

Insect infestations have been identified using sensitive acoustic technology since the early 1900s. According to Balingbing et al. (2024), acoustic sensors can detect insects in soil, grain, and wood from a distance without causing harm. These sensors can map the distribution of cryptic insects, quantify population density, and detect them. Advances in digital signal processing have enhanced the ability to discriminate insect calls from background noise, paving the way for new paths of underground pest monitoring. Furthermore, by recording the low-frequency noises emitted by insects as they move, acoustic technology has been utilized to track insect activity at feeding sites.

Ultrasonic Sensors

Insects that are concealed in wood, seed, and other fibrous materials can be found using ultrasonic sensors. Because there is little background noise at frequencies higher than 20 kHz, these sensors are especially helpful for identifying wood-boring pests because they can detect sounds made during feeding activities (Mankin et al., 2011). An ultrasonic sensor-based pest detection system typically consists of an ultrasonic transducer and a low-noise narrow-band amplifier. Real-time position and movement measurement is made possible by the ultrasonic pulses' ability to bounce back after impacting insects. To guarantee reliable performance in a variety of settings, these systems must be further developed (Ahouandjinou et al., 2017).

Drones (Unmanned Aerial Vehicles)

Drones can help farmers use their inputs of seed, fertilizer, and water more effectively. They can also react quickly to threats such as weeds, pests, and fungi. They also save time in crop scouting and improve variable rate prescriptions in real time. Outfitted with multispectral sensors, they gather data in near infrared and visible spectra to determine crop health and identify pests using image processing methods (Shiva et al., 2020). The data drones gather is analyzed through various software to help farmers know the agricultural conditions and thereby advance precision farming with the help of data interpretation.

Expert Systems

An Expert System (ES) is a type of computer software that has been developed to mimic the problem solving abilities of human experts in a certain field. These systems are important in decision making and in the dissemination of technology which is location specific. Expert systems are used to identify pests, suggest crops, and suggest the best management techniques. They address the issue of staff members' ignorance, particularly in developing nations. For instance, an expert system called TEAPEST was created to recognize insect pests in tea plantations and recommend the best ways to manage them (Vidya Madhuri et al., 2025b).

Pest Net

A deep learning-based network called Pest Net was created in order to precisely identify and detect a variety of pest species in photos. It uses a region suggestion network to find possible pest locations and a new channel-spatial attention module to improve feature extraction. With a mean average precision (mAP) of 75.46%, Pest Net has surpassed conventional techniques in multi-class pest identification by a wide margin (Liu et al., 2019). This model supports its useful uses in agriculture by showcasing a significant improvement in pest identification and categorization.

Internet of Things (IoT)

IoT stands for Internet of Things, a network of physical objects with sensors and software that can connect and share data with one another over the internet. IoT technology is fast and efficient with its systems that are meant for on ground use that can help farmers in identifying the right management decisions that they are likely to make. It enhances the agricultural productivity and sustainability through real time monitoring of the environmental conditions that enable better decision making (Abu et al., 2022). IoT is based on sensor monitoring and automation, which enables optimal use of resources and higher yields in precision farming (Daga and Aruna, 2024).

4.1.2 Early detection and prevention

This is why it is important to detect pests and diseases at an early stage in agriculture for the purpose of control. Some new technologies like image recognition and machine learning algorithms applied in AI help in the detection of pests and diseases from satellite imagery and field sensors and hence play an important role in this effort. These intelligent early warning systems enable farmers to respond proactively to prevent losses in yield (Ristaino et al., 2021).

Hyperspectral and thermal imaging technologies are improving as new plant disease detection methods (Chivasa et al., 2020). High resolution imaging for early disease identification is also enhanced by UAVs equipped with multispectral sensors (Mahlein et al., 2024b). The combination of satellite imagery, UAVs and machine learning makes it possible to respond instantly to management actions (Kamilaris et al., 2017).

Molecular techniques such as PCR, LAMP and NGS are enhancing the speed and sensitivity of the diagnosis (Waleed et al., 2020b). Disease outbreaks can be predicted with the help of historical data and environmental factors with the help of forecasting models (Zhang et al., 2021). However, there is still a problem with the implementation of the precision. It is therefore crucial to combine digital technologies with conventional pest control measures in the development of site-specific control thresholds and to meet the required regulatory standards. More research and education on the proper use of these digital tools in plant disease management are also needed (Mahlein et al., 2024b).

4.3.2 Genetic engineering for improved pest control mechanisms

Genetically modified (GM) crops have emerged as a significant component of modern agriculture, offering potential solutions to pressing global issues such as food security and sustainable agricultural practices. However, their acceptance remains controversial due to concerns about safety, environmental impact, and ethical considerations. As the agricultural sector seeks innovative strategies to enhance the acceptability of GM products, the integration of artificial intelligence (AI) presents a promising avenue (Mmbando, 2024). AI technologies, including algorithms and predictive modeling, can analyse vast datasets encompassing genetic, environmental, and agronomic factors. By leveraging these insights, researchers can expedite risk assessment processes, enhancing the safety evaluations of GM crops. This not only fosters transparency but also builds trust among consumers and stakeholders, potentially increasing public acceptance of GM products (Kjellsson and Simonsen, 2012). Despite the rapid advancements in AI applications within agriculture such as yield optimization, pest management, and resource efficiency, its specific use in GM crop risk prediction has not been extensively explored. Prior research has predominantly focused on conventional risk assessment methods, overlooking the revolutionary potential of AI-driven approaches. Addressing this gap is crucial, as it can provide valuable insights into the safety and efficacy of GM crops (Ghimire et al., 2023). However, challenges remain in implementing AI for GM crop assessments, including data availability, the complexity of AI models, and ethical concerns regarding data privacy and algorithm bias. Overcoming these obstacles will require interdisciplinary collaboration and robust scientific investigation. AI holds transformative potential for enhancing the acceptance of GM crops by providing accurate risk assessments and fostering informed dialogue. Continued research and innovation in this field are essential to ensure that GM technologies can effectively contribute to global food security and sustainable agricultural practices.

5.1 Combining AI with Robotics, Drones, and Sensor Networks

Combining AI with Robotics

This synergy enables precise monitoring of robotic actions and enhances feedback to operators which in turn enhance the surgical precision in agricultural tasks. Thus the need for integration of robotics and AI in agriculture (Gallagher et al., 2022). Plant production ranges from micropropagation in the least to large-scale farming on the other end, and this calls for machinery to lessen labor, increase production, and produce uniform products (Kondo et al., 2003). Although the emphasis is on "Sustainable Crop Protection via Robotics and Artificial Intelligence Solutions" pest control, AI and robotics play their role in other crop protection measures for sustainable food production (Rempelos et al., 2023). These technologies are important in the face of climate change because the advanced algorithms are able to use climatic data to help the farmers in decisions on planting, water application and crop choice (Khan et al., 2023).

Achieving optimal crop nutrition is important in obtaining high yields and good quality of the crop. AI can improve nutrient management by identifying the soil conditions and the needs of the plant and robotics can perform functions such as precise planting and weeding to optimize resources (Rempelos et al., 2023). Moreover, the application of AI can lead to the assessment of the plant's physiological information to support early disease identification and phenological growth stages monitoring for better decision making (Mesías et al., 2023). Since agriculture is labor demanding, the automation of activities such as using autonomous tractors and AI-enabled harvesting robots solves the problem of labor shortage and enhances production (Javaid et al., 2023). In general, the application of AI and robotics is crucial in enhancing the different practices in agriculture in order to increase production and make agriculture more sustainable.

Combining AI with Sensor Networks

The internet has spread to all areas of life and has found its way to the agricultural sector where the IoT is changing the way of work on the farms. With the help of sensors that analyze many factors of the agricultural soil, IoT combines the wireless sensor network to monitor the conditions properly (Rajeswari and Kapoor, 2017). This technology may change the future of agriculture and make farming more effective and efficient. One of the major application of IoT in agriculture is precision farming which uses sensors and other technologies to gather data and take decision on the field and in the right time. For instance, GPS guided machinery leads to better application of fertilizers and pesticides thus reducing on the use of resources and or wastage (Ali et al., 2023). This is because in order to be able to make predictive decisions, data has to be gathered from a number of sources, especially on the weather, which is a key determinant of agricultural production (Aslam et al., 2021).

Sensors are used in every phase of the farming process and can help in improving soil management, nutrients and water management, pest control and yield harvesting (Bochtis et al., 2021). They capture physical measurements of the real world and translate them into digital information that can be processed by Ag-IoT systems to understand the dynamic loading and environmental stress. Farmers can discover technical changes that will lead to more ecologically and economically sustainable farming by monitoring data from sensors in the soil, crops, and livestock, which will encourage a more soil-, plant-, and animal-friendly precision agriculture approach.

6. Future Directions and Challenges

Applying AI and machine learning in agriculture has the potential to increase production, lower costs and improve sustainability. Thus, combining remote sensing with AI can help farmers to enhance yield, quality of the product, and decrease the adverse impact on the environment and production costs which are necessary to meet the challenges of climate change and population increase (Zhang et al., 2021). Nevertheless, security of data issues and problems with data collection, precision, and timing are limiting AI's effectiveness (Chlingaryan et al., 2018; Shaikh et al., 2022b). To realize the full potential of AI, there is a need to have the technologists, the agronomists and the policymakers work together to set up criteria for data collection and exchange (White and Hamilton, 1985).

6.1 Integration of AI with Other Technologies

The integration of artificial intelligence and machine learning in agriculture is about to transform the industry and make it more efficient and productive and sustainable. However, as agricultural systems become more complex, the role of data analysis, predictive modelling, and resource management driven tools will become more critical. For example, optimal yield production can be achieved while minimizing operational costs and environmental impacts through the application of remote sensing technologies paired with AI (Dobrojevic and Bacanin, 2022). The improvements promised to deliver more robust practicing of agriculture that can meet the demands of climate change and a increasing population.

6.1.1 Data Protection in AI-Driven Agriculture

The use of AI in agriculture has brought about great improvements in efficiency and yield but there are some major concerns on the privacy and security of data. Farmers have to share their sensitive operational data with different stakeholders, and this data may be misused resulting in losses (Beloev et al., 2021). There is, therefore, the need to have strong data protection measures in place such as encrypted communications, secure data storage and clear access controls for the AI to be adopted. Thus, farmers, legislators, and technologists can work together to establish uniform data management policies that can address privacy issues without jeopardizing the need to share data. Thus, the solutions of these problems will allow the agricultural industry to fulfill the potential of AI for precision farming, management of resources, and modeling and depend on safe and high-quality data (Beloev et al., 2021).

6.2.1 Nanotechnology in AI-Driven Agriculture

Nanotechnology integrated with AI creates a new and vital domain for sustainable and precise agriculture. Nanotechnology improves nutrient delivery and absorption in crops, while AI manages resources and makes decisions based on real-time crop conditions. This synergy leads to the creation of nanoscale agrochemicals that are safe and effective and can be used effectively in agricultural production systems. By combining AI and nanotechnology with models of crop production, precise nutrient management, assimilation, and soil microbes can be positively improved, which will increase agricultural sustainability and efficiency (Ashique et al., 2025). Nanoscale fertilizers, for example, can be made to release nutrients at particular times, which reduces the potential negative impact on the environment. The sensors and drones can provide data that AI can use to recommend the best time to apply these fertilizers to maximize crop growth. This integration holds a great promise for enhancing agriculture and finding solutions to the problems of sustainability in food production (Rahmadian and Widyartono, 2020). Through greater accuracy in targeting, absorption and distribution of nutrients, farmers can enhance yield while protecting the environment. In the end, AI, nanotechnology, and precision agriculture are integrated, and a powerful system for developing more sustainable and productive farming systems is available.

7. Conclusion

An enormous step toward the development of efficient and sustainable farming methods is the incorporation of artificial intelligence into pest management plans. The threat of pests, disease, and abiotic pressures can be effectively addressed by AI technologies such as machine learning, image recognition, and the Internet of Things. Precision pest control, crop surveillance, and AI-assisted early disease identification optimization enhance decision-making, minimize the use of harmful chemicals, and increase yields. Furthermore, it has been discovered that artificial intelligence technologies, such drones, prediction models, and surveillance systems, can help with proactive and individual farming management. The technologies give actionable insight by combining information from several sources, allowing farmers to respond to risks in a sustainable and efficient manner. Although technology has shown a bright future, obstacles, including the high initial cost, the demand for large databases and the scarcity of technology in resource-poor areas, hinder the application of artificial intelligence in agriculture. However, with the continuous progress of artificial intelligence algorithms, sensing technology and machine learning models, technology will become more extensive and useful with the reduction of obstacles. Its ability to adapt and react to changing environment variables and the emergence of new threats ensures that it will be a part of future methods of sustainable agriculture.

References

Abbas, A., Zhang, Z., Zheng, H., Alami, M. M., Alrefaei, A. F., Abbas, Q. et al. (2023). Drones in plant disease assessment, efficient monitoring, and detection: a way forward to smart agriculture. Agronomy, 13(6): 1524.

Abu, N. S., Bukhari, W. M., Ong, C. H., Kassim, A. M., Izzuddin, T. A., Sukhaimie, M. N. et al. (2022). Internet of things applications in precision agriculture: A review. Journal of Robotics and Control (JRC), 3(3): 338–347.

Ahmed, E. A. S., and Mujtaba, Y. (2019). Elbagir, Internet of things in Smart Environment: Concept, Applications, Challenges, and Future Directions Article Contents. Psjd. Icm. Edu. Pl, 134(1): 1–51.

Ahouandjinou, A. S. R. M., Kiki, P. M. A. F., and Assogba, K. (2017). Smart environment monitoring system by using sensors ultrasonic detection of farm pests. 2017 2nd International Conference on Bio-Engineering for Smart Technologies (BioSMART), 1–5.

Ali, Q., Khan, A. R., Tao, S., Rajer, F. U., Ayaz, M., Abro, M. A. et al. (2023). Broad-spectrum antagonistic potential of *Bacillus* spp. volatiles against *Rhizoctonia solani* and *Xanthomonas oryzae* pv. oryzae. Physiologia Plantarum, 175(6): e14087.

Ampatzidis, Y., De Bellis, L., and Luvisi, A. (2017). iPathology: robotic applications and management of plants and plant diseases. Sustainability, 9(6): 1010.

An, Z., Wang, C., Raj, B., Eswaran, S., Raffik, R., Debnath, S., and Rahin, S. A. (2022). Application of new technology of intelligent robot plant protection in ecological agriculture. Journal of Food Quality, 2022(1): 1257015.

Arafa, R. A., Kamel, S. M., Rakha, M. T., Soliman, N. E. K., Moussa, O. M., and Shirasawa, K. (2020). Analysis of the lineage of Phytophthora infestans isolates using mating type assay, traditional markers, and next generation sequencing technologies. PLoS ONE, 15(1): e0221604.

Ashique, S., Raikar, A., Jamil, S., Lakshminarayana, L., Gajbhiye, S. A., De, S. et al. (2025). Artificial intelligence integration with nanotechnology: a new frontier for sustainable and precision agriculture. Current Nanoscience, 21(2): 242–273.

Aslam, H. M. U., Naveed, K., Hussain, S. I., Shakeel, Q., Ashraf, W., Anwaar, H. A. et al. (2021). First report of brown leaf spot of rice caused by *Bipolaris zeicola* in Pakistan. Plant Disease, 105(1): 212.

Bala, K., Sood, A. K., Pathania, V. S., and Thakur, S. (2018). Effect of plant nutrition in insect pest management: A review. Journal of Pharmacognosy and Phytochemistry, 7(4): 2737–2742.

Balafoutis, A. T., Kavroumatzi, C. K., Moraitis, M., Vaiopoulos, K., Mylonas, N., Tsitsigiannis, I. et al. (2021). Advanced Crop Protection Techniques and Technologies. pp. 112–171. In Modeling for Sustainable Management in Agriculture, Food and the Environment. CRC Press.

Balingbing, C. B., Kirchner, S., Siebald, H., Kaufmann, H.-H., Gummert, M., van Hung, N. et al. (2024). Application of a multi-layer convolutional neural network model to classify major insect pests in stored rice detected by an acoustic device. Computers and Electronics in Agriculture, 225: 109297.

Balsari, P., Marucco, P., and Oggero, G. (2002). Spray applications in Italian apple orchards: Target coverage, ground losses and drift. In 2002 ASAE Annual Meeting (p. 1). American Society of Agricultural and Biological Engineers.

Bannerjee, G., Sarkar, U., Das, S., and Ghosh, I. (2018). Artificial intelligence in agriculture: A literature survey. International Journal of Scientific Research in Computer Science Applications and Management Studies, 7(3): 1–6.

Basu, C. (2010). Editorial [Hot topic: Plant Genomics in the 21st Century (Guest Editor: Chhandak Basu)]. Current Genomics, 11(1), 1.

Beloev, I., Kinaneva, D., Georgiev, G., Hristov, G., and Zahariev, P. (2021). Artificial intelligence-driven autonomous robot for precision agriculture. Acta Technologica Agriculturae, 24(1): 48–54.

Bhoomika, K. N., Deepa, C., Rashmi, R. K., and Srinivasa, R. (2016). Internet of things for environmental monitoring. International Journal of Advanced Networking & Applications, 497–501.

Bonny, S. (2011). L'agriculture écologiquement intensive: nature et défis. Cahiers agricultures, 20(6), 451–462.

Borengasser, M., Hungate, W. S., and Watkins, R. (2007). Hyperspectral remote sensing: principles and applications. CRC press.
Brakke, M. K. (1988). Perspectives on Process in Plant Virology.
Census Bureau. (2009). Statistical abstract of the United States 2010. Government Printing Office.
Cheatham, M. R., Rouse, M. N., Esker, P. D., Ignacio, S., Pradel, W., Raymundo, R. et al. (2009). Beyond yield: plant disease in the context of ecosystem services. Phytopathology, 99(11): 1228–1236.
Chivasa, W., Mutanga, O., and Biradar, C. (2020). UAV-based multispectral phenotyping for disease resistance to accelerate crop improvement under changing climate conditions. Remote Sensing, 12(15): 2445.
Chlingaryan, A., Sukkarieh, S., and Whelan, B. (2018). Machine learning approaches for crop yield prediction and nitrogen status estimation in precision agriculture: A review. Computers and Electronics in Agriculture, 151: 61–69.
Christiansen, M. P., Laursen, M. S., Jørgensen, R. N., Skovsen, S., and Gislum, R. (2017). Designing and testing a UAV mapping system for agricultural field surveying. Sensors, 17(12): 2703.
Chumachenko, D., Dudkina, T., Yakovlev, S., and Chumachenko, T. (2023). Effective utilization of data for predicting COVID-19 dynamics: An Exploration through Machine Learning Models. International Journal of Telemedicine and Applications, 2023(1): 9962100.
Daga, A., and Aruna, M. (2024). IoT based smart agricultural monitoring system. AIP Conference Proceedings, 3075(1).
Damos, P. (2015). Modular structure of web-based decision support systems for integrated pest management. A review. Agronomy for Sustainable Development, 35(4): 1347–1372.
Damos, P., and Karabatakis, S. (2013). Real time pest modeling through the World Wide Web: decision making from theory to praxis. IOBC-WPRS Bulletin, 91: 253–258.
Das, S., Pattanayak, S., and Behera, P. R. (2022a). Application of machine learning: a recent advancement in plant diseases detection. Journal of Plant Protection Research, 122–135.
Das, S., Pattanayak, S., and Behera, P. R. (2022b). Application of machine learning: a recent advancement in plant diseases detection. Journal of Plant Protection Research, 122–135.
Dobrojevic, M., and Bacanin, N. (2022). IoT as a backbone of intelligent homestead automation. Electronics, 11(7): 1004.
Dong, Z., Wang, Z., Liu, D., Song, K., Li, L., Jia, M., and Ding, Z. (2014). Mapping wetland areas using Landsat-derived NDVI and LSWI: A case study of West Songnen plain, Northeast China. Journal of the Indian Society of Remote Sensing, 42: 569–576.
Durkin, J., and Durkin, J. (1998). Expert systems: design and development. Prentice Hall PTR.
Feltracco, M., Barbaro, E., Maule, F., Bortolini, M., Gabrieli, J., De Blasi, F. et al. (2022). Airborne polar pesticides in rural and mountain sites of North-Eastern Italy: An emerging air quality issue. Environmental Pollution, 308,: 119657.
Ferreira, R. B., Monteiro, S., Freitas, R., Santos, C. N., Chen, Z., Batista, L. M. et al. (2006). Fungal pathogens: the battle for plant infection. Critical Reviews in Plant Sciences, 25(6): 505–524.
Fu, X., Ma, Q., Yang, F., Zhang, C., Zhao, X., Chang, F., and Han, L. (2024). Crop pest image recognition based on the improved ViT method. Information Processing in Agriculture, 11(2): 249–259.
Gaglio, S., Re, G. L., Martorella, G., Peri, D., and Vassallo, S. D. (2014, October). Development of an IoT environmental monitoring application with a novel middleware for resource constrained devices. In Proceedings of the 2nd conference on mobile and information technologies in medicine (MobileMed 2014).
Gallagher, R. V., Allen, S. P., Mackenzie, B. D., Keith, D. A., Nolan, R. H., Rumpff, L. et al. (2022). An integrated approach to assessing abiotic and biotic threats to post-fire plant species recovery: Lessons from the 2019–2020 Australian fire season. Global Ecology and Biogeography, 31(10): 2056–2069.
Ghimire, B. K., Yu, C. Y., Kim, W.-R., Moon, H.-S., Lee, J., Kim, S. H. et al. (2023). Assessment of benefits and risk of genetically modified plants and products: current controversies and perspective. Sustainability, 15(2): 1722.
Gill, S. S., Xu, M., Ottaviani, C., Patros, P., Bahsoon, R., Shaghaghi, A. et al. (2022). AI for next generation computing: Emerging trends and future directions. Internet of Things, 19: 100514.

Goëau, H., Bonnet, P., and Joly, A. (2018, September). Overview of expertlifeclef 2018: how far automated identification systems are from the best experts?. In CLEF 2018-Conference and Labs of the Evaluation Forum (Vol. 2125, No. 10).

Guedes, R. N. C., Smagghe, G., Stark, J. D., and Desneux, N. (2016). Pesticide-induced stress in arthropod pests for optimized integrated pest management programs. Annual Review of Entomology, 61(1): 43–62.

Guo, Y. (2023). Intelligent Machine Learning System for Agricultural UAV Based on Artificial Intelligence and Deep Learning [J]. Agricultural Mechanization Research, 45(03): 237–240.

Gurulingappa, P., Sword, G. A., Murdoch, G., and McGee, P. A. (2010). Colonization of crop plants by fungal entomopathogens and their effects on two insect pests when in planta. Biological Control, 55(1): 34–41.

Hall, F. R. (1984). Pesticide application techniques for optimizing efficiency. Arboriculture & Urban Forestry (AUF), 10(4): 104–107.

Harish, M. S., Bhuker, A., and Chauhan, B. S. (2024). Millet production, challenges, and opportunities in the Asia-pacific region: a comprehensive review. Frontiers in Sustainable Food Systems, 8: 1386469.

Hart, A. G., Bosley, H., Hooper, C., Perry, J., Sellors-Moore, J., Moore, O., and Goodenough, A. E. (2023). Assessing the accuracy of free automated plant identification applications. People and Nature, 5(3): 929–937.

Hatfield, J. L., Gitelson, A. A., Schepers, J. S., and Walthall, C. L. (2008). Application of spectral remote sensing for agronomic decisions. Agronomy Journal, 100, S–117.

He, D., Zhan, J., and Xie, L. (2016). Problems, challenges and future of plant disease management: from an ecological point of view. Journal of Integrative Agriculture, 15(4): 705–715.

He, R., Ju, J., Liu, K., Song, J., Zhang, S., Zhang, M., and Liu, H. (2024). Technology of plant factory for vegetable crop speed breeding. Frontiers in Plant Science, 15: 1414860.

Hernández, I., Gutiérrez, S., and Tardaguila, J. (2024). Image analysis with deep learning for early detection of downy mildew in grapevine. Scientia Horticulturae, 331, 113155.

He-Ya, S., Huang, X., Zhou, D., Zhang, J., Bao, G., Tong, S. et al. (2024). Identification of Larch Caterpillar Infestation Severity Based on Unmanned Aerial Vehicle Multispectral and LiDAR Features. Forests, 15(1): 191.

Holliday, R. (2004). Early studies on recombination and DNA repair in Ustilago maydis. DNA Repair, 3(6): 671–682.

Huai-liang, C. H. E. N., Hong, Z. H. A. N. G., and You, L. (2007). Review on research of meteorological conditions and prediction methods of crop disease and insect pest. Chinese Journal of Agrometeorology, 28(02): 212.

Ijeh, S., Okolo, C. A., Arowoogun, J. O., Adeniyi, A. O., and Omotayo, O. (2024). Predictive modeling for disease outbreaks: a review of data sources and accuracy. International Medical Science Research Journal, 4(4): 406–419.

International Olympic Committee Injury and Illness Epidemiology Consensus Group, Bahr, R., Clarsen, B., Derman, W., Dvorak, J., Emery, C. A. et al. (2020). International Olympic Committee consensus statement: methods for recording and reporting of epidemiological data on injury and illness in sports 2020 (including the STROBE extension for sports injury and illness surveillance (STROBE-SIIS)). Orthopaedic journal of sports medicine, 8(2), 2325967120902908.

Jamil, M. S., Jamil, M. A., Mazhar, A., Ikram, A., Ahmed, A., and Munawar, U. (2015). Smart environment monitoring system by employing wireless sensor networks on vehicles for pollution free smart cities. Procedia Engineering, 107: 480–484.

Javaid, M., Haleem, A., Khan, I. H., and Suman, R. (2023). Understanding the potential applications of Artificial Intelligence in Agriculture Sector. Advanced Agrochem, 2(1): 15–30.

Jha, K., Doshi, A., Patel, P., and Shah, M. (2019). A comprehensive review on automation in agriculture using artificial intelligence. Artificial Intelligence in Agriculture, 2: 1–12.

Jia, X., Yin, D., Bai, Y., Yu, X., Song, Y., Cheng, M. et al. (2023). Monitoring Maize Leaf Spot Disease Using Multi-Source UAV Imagery. Drones, 7(11): 650.

Kamilaris, A., and Prenafeta-Boldú, F. X. (2018). Deep learning in agriculture: A survey. Computers and Electronics in Agriculture, 147: 70–90.

Kamilaris, A., Kartakoullis, A., and Prenafeta-Boldú, F. X. (2017). A review on the practice of big data analysis in agriculture. Computers and Electronics in Agriculture, 143: 23–37.

Kariyanna, B., and Sowjanya, M. (2024). Unravelling the use of artificial intelligence in management of insect pests. Smart Agricultural Technology, 100517.

Khan, M., Khan, A. U., Khan, S., and Khan, F. A. (2023). Assessing the impacts of climate change on streamflow dynamics: a machine learning perspective. Water Sci. Technol., 88 (9): 2309–2331.

Khang, A., Rath, K. C., Panda, S., Sree, P. K., and Panda, S. K. (2023). Revolutionizing agriculture: exploring advanced technologies for plant protection in the agriculture sector. In Handbook of Research on AI-Equipped IoT Applications in High-Tech Agriculture (pp. 1–22). IGI Global.

Kjellsson, G., and Simonsen, V. (2012). Methods for risk assessment of transgenic plants: I. Competition, establishment and ecosystem effects. Birkhäuser.

Kondo, H., Maeda, T., and Tamada, T. (2003). Orchid fleck virus: Brevipalpus californicus mite transmission, biological properties and genome structure. Experimental & Applied Acarology, 30(1): 215–223.

Krone, M. J., Dong, Y., and Mideros, S. X. (2024). Effect of Quantitative Wheat Resistance on the Aggressiveness of *Fusarium graminearum*. Phytopathology®, 114(7): 1577–1586.

Kulkarni, P. H., and Kute, P. D. (2016). Internet of things based system for remote monitoring of weather parameters and applications. Int. J. Adv. Electron. Comput. Sci., 3(2): 68–73.

Kushalappa, A. C., and Eskes, A. B. (1989). Coffee rust: epidemiology, resistance and management. CRC Press Boca Raton, Florida.

Lawrence, G. J. (1988). Melampsora lini, rust of flax and linseed. In Advances in plant pathology (Vol. 6, pp. 313–331). Elsevier.

LeCun, Y., Bengio, Y., and Hinton, G. (2015). Deep learning. Nature, 521(7553): 436–444.

Leonard, K. J., and Szabo, L. J. (2005). Stem rust of small grains and grasses caused by *Puccinia graminis*. Molecular Plant Pathology, 6(2): 99–111.

Leybourne, D. J., Musa, N., and Yang, P. (2025). Can artificial intelligence be integrated into pest monitoring schemes to help achieve sustainable agriculture? An entomological, management and computational perspective. Agricultural and Forest Entomology, 27(1): 8–17.

Li, C., and Wang, M. (2024). Pest and disease management in agricultural production with artificial intelligence: Innovative applications and development trends. Advances in Resources Research, 4(3): 381–401.

Li, H., Jin, C. H. E., Song, P., Wang, C., and Bin, T. (2020). Studies on pest prediction based on machine learning and image processing technologies. Journal of Southwest University Natural Science Edition, 42(1): 134–141.

Li, Y., Wang, C., Zhao, B., and Liu, W. (2010). Effects of climate change on agricultural meteorological disaster and crop insects diseases. Transactions of the Chinese Society of Agricultural Engineering, 26(1), 263–271.

Liao, S. H. (2005). Expert system methodologies and applications—a decade review from 1995 to 2004. Expert systems with applications, 28(1), 93–103.

Lichtfouse, E., Navarrete, M., Debaeke, P., Souchère, V., Alberola, C., and Ménassieu, J. (2009). Agronomy for sustainable agriculture: a review. Sustainable Agriculture, 1–7.

Liu, L., Wang, R., Xie, C., Yang, P., Wang, F., Sudirman, S. et al. (2019). PestNet: An end-to-end deep learning approach for large-scale multi-class pest detection and classification. Ieee Access, 7: 45301–45312.

Liu, Z., Xiang, X., Qin, J., Tan, Y., Zhang, Q., and Xiong, N. N. (2021). Image recognition of citrus diseases based on deep learning. Computers, Materials & Continua, 66(1).

Madhushan, A., Weerasingha, D. B., Ilyukhin, E., Taylor, P. W. J., Ratnayake, A. S., Liu, J.-K. et al. (2025). From natural hosts to agricultural threats: The evolutionary journey of phytopathogenic fungi. Journal of Fungi, 11(1): 25.

Mahenge, M. P. J., Mkwazu, H., Sanga, C. A., Madege, R. R., Mwaipopo, B., and Maro, C. (2023). Artificial intelligence and deep learning based technologies for emerging disease recognition and pest prediction in beans (*Phaseolus vulgaris* l.): A systematic review. African Journal of Agricultural Research, 19(3): 260–271.

Mahlein, A. K. (2016). Plant disease detection by imaging sensors–parallels and specific demands for precision agriculture and plant phenotyping. Plant Disease, 100(2): 241–251.

Mahlein, A. K. (2016). Present and future trends in plant disease detection. Plant Dis, 100(2), 1–11.

Mahlein, A. K., Kuska, M. T., Thomas, S., Bohnenkamp, D., Alisaac, E., Behmann, J. et al. (2017). Plant disease detection by hyperspectral imaging: from the lab to the field. Advances in Animal Biosciences, 8(2): 238–243.

Mahlein, A. K., Arnal Barbedo, J. G., Chiang, K.-S., del Ponte, E. M., and Bock, C. H. (2024a). From detection to Protection: the role of Optical Sensors, Robots, and Artificial Intelligence in Modern Plant Disease Management. Phytopathology®, 114(8): 1733–1741.

Mahlein, A.-K., Arnal Barbedo, J. G., Chiang, K.-S., del Ponte, E. M., and Bock, C. H. (2024b). From detection to Protection: the role of Optical Sensors, Robots, and Artificial Intelligence in Modern Plant Disease Management. Phytopathology®, 114(8): 1733–1741.

Mankin, R. W., Hagstrum, D. W., Smith, M. T., Roda, A. L., and Kairo, M. T. K. (2011). Perspective and promise: a century of insect acoustic detection and monitoring. American Entomologist, 57(1): 30–44.

Martinelli, F., Scalenghe, R., Davino, S., Panno, S., Scuderi, G., Ruisi, P. et al. (2015). Advanced methods of plant disease detection. A review. Agronomy for Sustainable Development, 35: 1–25.

Mesías-Ruiz, G. A., Pérez-Ortiz, M., Dorado, J., de Castro, A. I., and Peña, J. M. (2023). Boosting precision crop protection towards agriculture 5.0 via machine learning and emerging technologies: A contextual review. Frontiers in Plant Science, 14: 1143326.

Mingyue, S. H. A. O., Jianhua, Z. H. A. N. G., Quan, F. E. N. G., Xiujuan, C. H. A. I., Ning, Z. H. A. N. G., and Wenrong, Z. H. A. N. G. (2022). Research progress of deep learning in detection and recognition of plant leaf diseases. Smart Agriculture, 4(1): 29.

Mirik, M., Jones, D. C., Price, J. A., Workneh, F., Ansley, R. J., and Rush, C. M. (2011). Satellite remote sensing of wheat infected by wheat streak mosaic virus. Plant Disease, 95(1): 4–12.

Mmbando, G. S. (2024). The use of artificial intelligence in the production of genetically modified (GM) crops: a recent promising strategy for enhancing the acceptability of GM products? Discover Applied Sciences, 6(11): 1–12.

Nabwire, S., Suh, H.-K., Kim, M. S., Baek, I., and Cho, B.-K. (2021). Application of artificial intelligence in phenomics. Sensors, 21(13): 4363.

Nahiyoon, S. A., Ren, Z., Wei, P., Li, X., Li, X., Xu, J. et al. (2024). Recent development trends in plant protection UAVs: A journey from conventional practices to cutting-edge technologies—A comprehensive Review. Drones, 8(9): 457.

Ni, J., Lan, Y., Qiu, B., Zhang, J., and Yuen, P. (Eds.). (2023). Advanced methods, equipment and platforms in precision field crops protection. Frontiers Media SA.

Nigam, A., Garg, S., Agrawal, A., and Agrawal, P. (2019, November). Crop yield prediction using machine learning algorithms. In 2019 Fifth International Conference on Image Information Processing (ICIIP) (pp. 125–130). IEEE.

Ning, H., Liu, H., and Yang, L. T. (2014). Aggregated-proof based hierarchical authentication scheme for the internet of things. IEEE Transactions on Parallel and Distributed Systems, 26(3): 657–667.

Novaković, N., and Rikalović, A. (n.d.). The future of agricultural spraying: ai and uavs for smart and sustainable practices. Proceedings–The First International Conference FUTURE-BME 2024 (Forging the Future: Pioneering Approaches in Business, Management And, 681.

Oerke, E.-C., Dehne, H.-W., Schönbeck, F., and Weber, A. (2012). Crop production and crop protection: estimated losses in major food and cash crops. Elsevier.

Oldekop, J. A., Rasmussen, L. V., Agrawal, A., Bebbington, A. J., Meyfroidt, P., Bengston, D. N. et al. (2020). Forest-linked livelihoods in a globalized world. Nature Plants, 6(12): 1400–1407.

Oliver, R. P. (2024). The history and development of plant pathology. In Agrios' Plant Pathology (pp. 3–61). Elsevier.

Orton, E. S., Deller, S., and Brown, J. K. M. (2011). Mycosphaerella graminicola: from genomics to disease control. Molecular Plant Pathology, 12(5): 413–424.

Ou, S. H. (1981). Pathogen variability and host resistance in rice blast disease.

Palmgren, M. G., Edenbrandt, A. K., Vedel, S. E., Andersen, M. M., Landes, X., Østerberg, J. T. et al. (2015). Are we ready for back-to-nature crop breeding? Trends in Plant Science, 20(3): 155–164.

Pan, S. J., and Yang, Q. (2009). A survey on transfer learning. IEEE Transactions on Knowledge and Data Engineering, 22(10): 1345–1359.

Patel, R. M., van Kan, J. A. L., Bailey, A. M., and Foster, G. D. (2010). Inadvertent gene silencing of argininosuccinate synthase (bcass1) in *Botrytis cinerea* by the pLOB1 vector system. Molecular Plant Pathology, 11(5): 613–624.

Pereira, C. S., Morais, R., and Reis, M. J. (2019). Deep learning techniques for grape plant species identification in natural images. Sensors, 19(22): 4850.

Prusky, D. (1996). Pathogen quiescence in postharvest diseases. Annual Review of Phytopathology, 34(1): 413–434.

Rahimzadeh-Bajgiran, P., Omasa, K., and Shimizu, Y. (2012). Comparative evaluation of the Vegetation Dryness Index (VDI), the Temperature Vegetation Dryness Index (TVDI) and the improved TVDI (iTVDI) for water stress detection in semi-arid regions of Iran. ISPRS Journal of Photogrammetry and Remote Sensing, 68, 1–12.

Rahmadian, R., and Widyartono, M. (2020). Autonomous robotic in agriculture: a review. 2020 Third International Conference on Vocational Education and Electrical Engineering (ICVEE), 1–6.

Rajeswari, P., and Kapoor, R. (2017). Combinatorial efficacy of Trichoderma spp. and Pseudomonas fluorescens to enhance suppression of cell wall degrading enzymes produced by Fusarium wilt of *Arachis hypogaea* L. International Journal of Agricultural Research, Innovation and Technology, 7(2): 36–42.

Rao, P. R., Srinivas, S., and Ramesh, E. (2019). A report on designing of wireless sensor networks for IoT applications. Int. J. Eng. Adv. Technol, 8(6): 2005–2009.

Rempelos, L., Barański, M., Sufar, E. K., Gilroy, J., Shotton, P., Leifert, H. et al. (2023). Effect of Climatic Conditions, and Agronomic Practices Used in Organic and Conventional Crop Production on Yield and Nutritional Composition Parameters in Potato, Cabbage, Lettuce and Onion; Results from the Long-Term NFSC-Trials. Agronomy, 13: 1225.

Ristaino, J. B., Anderson, P. K., Bebber, D. P., Brauman, K. A., Cunniffe, N. J., Fedoroff, N. v et al. (2021). The persistent threat of emerging plant disease pandemics to global food security. Proceedings of the National Academy of Sciences, 118(23): e2022239118.

Roth, J., Chadalawada, J., Jain, R. K., and Miller, C. (2021). Uncertainty matters: Bayesian probabilistic forecasting for residential smart meter prediction, segmentation, and behavioral measurement and verification. Energies, 14(5): 1481.

Rybicki, E. P., and Pietersen, G. (1999). Plant virus disease problems in the developing world. Advances in Virus Research, 53: 127–175.

Sandino, J., Pegg, G., Gonzalez, F., and Smith, G. (2018). Aerial mapping of forests affected by pathogens using UAVs, hyperspectral sensors, and artificial intelligence. Sensors, 18(4): 944.

Schmidhuber, J. (2015). Deep learning in neural networks: An overview. Neural networks, 61: 85–117.

Shahi, T. B., Xu, C. Y., Neupane, A., and Guo, W. (2023). Recent advances in crop disease detection using UAV and deep learning techniques. Remote Sensing, 15(9): 2450.

Shaikh, T. A., Rasool, T., and Lone, F. R. (2022b). Towards leveraging the role of machine learning and artificial intelligence in precision agriculture and smart farming. Computers and Electronics in Agriculture, 198: 107119.

Sharada, K., Choudhary, S. L., Harikrishna, T., Dixit, R. S., Suman, S. K., Ayyappa Chakravarthi, M. et al. (2025). GeoAgriGuard: AI-Driven Pest and Disease Management with Remote Sensing for Global Food Security. Remote Sensing in Earth Systems Sciences, 1–14.

Shiva, R., Vimal, G., and Kaviyarasu, M. (2020). Intelligent farming using delta robot. 2020 International Conference on Power, Energy, Control and Transmission Systems (ICPECTS), 1–5.

Smith, M. J. (2018). Getting value from artificial intelligence in agriculture. Animal Production Science, 60(1): 46–54.

Srilopan, S., Mata, V. A., and Bumrungsri, S. (2025). DNA Metabarcoding Reveals Insights into Diet Partitioning and Pest Suppression by Open Space Bats in Agricultural Landscapes. Acta Chiropterologica, 26(2): 171–187.

Strange, R. N., and Scott, P. R. (2005). Plant disease: a threat to global food security. Annu. Rev. Phytopathol., 43(1): 83–116.

Sun, F., Wang, X., & Zhang, R. (2021). Improved Q-learning algorithm based on approximate state matching in agricultural plant protection environment. Entropy, 23(6): 737.

Tetala, S. S. D. R. (2023). Artificial intelligence powered personalized agriculture (Doctoral dissertation, Colorado State University).

Upadhyay, A., Chandel, N. S., Singh, K. P., Chakraborty, S. K., Nandede, B. M., Kumar, M., Subeesh, A. et al. (2025). Deep learning and computer vision in plant disease detection: a comprehensive review of techniques, models, and trends in precision agriculture. Artificial Intelligence Review, 58(3): 1–64.

Vidya Madhuri, E., Rupali, J. S., Sharan, S. P., Sai Pooja, N., Sujatha, G. S., Singh, D. P. et al. (2025a). Transforming Pest Management with Artificial Intelligence Technologies: The Future of Crop Protection. Journal of Crop Health, 77(2): 48.

Vidya Madhuri, E., Rupali, J. S., Sharan, S. P., Sai Pooja, N., Sujatha, G. S., Singh, D. P. et al. (2025b). Transforming Pest Management with Artificial Intelligence Technologies: The Future of Crop Protection. Journal of Crop Health, 77(2): 48.

Waleed, M., Um, T.-W., Kamal, T., Khan, A., and Iqbal, A. (2020a). Determining the precise work area of agriculture machinery using internet of things and artificial intelligence. Applied Sciences, 10(10): 3365.

Waleed, M., Um, T.-W., Kamal, T., Khan, A., and Iqbal, A. (2020b). Determining the precise work area of agriculture machinery using internet of things and artificial intelligence. Applied Sciences, 10(10): 3365.

Wang, M., Son, S., and Shi, W. (2009). Evaluation of MODIS SWIR and NIR-SWIR atmospheric correction algorithms using SeaBASS data. Remote Sensing of Environment, 113(3): 635–644.

Wang, T., Xu, X., Wang, C., Li, Z., and Li, D. (2021). From smart farming towards unmanned farms: A new mode of agricultural production. Agriculture, 11(2): 145.

White, J. W., and Hamilton, J. H. (1985). Irradiance and plant temperature monitor/controller. Computers and Electronics in Agriculture, 1(1): 95–103.

Xiang, L., and Wang, D. (2023). A review of three-dimensional vision techniques in food and agriculture applications. Smart Agricultural Technology, 5: 100259.

Xiao, F., Wang, H., Li, Y., Cao, Y., Lv, X., and Xu, G. (2023). Object detection and recognition techniques based on digital image processing and traditional machine learning for fruit and vegetable harvesting robots: An overview and review. Agronomy, 13(3): 639.

Yue, X., Qi, K., Na, X. et al. (2025). Deep learning for recognition and detection of plant diseases and pests. Neural Comput & Applic. https://doi.org/10.1007/s00521-025-11125-5.

Yun, D., Qiyao, F., Zhaowen, N., and Yanping, K. (2024). Fine-grained classification of crop diseases and pests based on progressive growing of GANs. Journal of Chinese Agricultural Mechanization, 45(3): 156.

Zadoks, J. C. (1985). Cereal rusts, dogs and stars in antiquity.

Zaitlin, M., and Palukaitis, P. (2000). Advances in understanding plant viruses and virus diseases. Annual Review of Phytopathology, 38(1): 117–143.

Zhan, J., Thrall, P. H., and Burdon, J. J. (2014). Achieving sustainable plant disease management through evolutionary principles. Trends in Plant Science, 19(9): 570–575.

Zhang, D., Eng, B., Prof, S., Connor, N. E. O., and Regan, P. F. (2015). Multi-Modal Smart Sensing Network for School of Electronic Engineering. Dublin City University, Dublin, Ireland.

Zhang, P., Guo, Z., Ullah, S., Melagraki, G., Afantitis, A., and Lynch, I. (2021). Nanotechnology and artificial intelligence to enable sustainable and precision agriculture. Nature Plants, 7(7): 864–876.

Zhuoping, Q. I. (2022). Design of remote intelligent early warning system for agrometeorological disasters. Agricultural Engineering, 12(4): 54–57.

CHAPTER 2

Plant Pest Monitoring Remote Sensing, Image Processing and Artificial Intelligence

Muhammad Atiq,[1,*] *Nasir Ahmad Rajput,*[1] *Azeem Akram,*[1] *Muhammad Jahanzaib Matloob*[1] and *Kamel A. Abd-Elsalam*[2]

1. Introduction

Plant pests constitute one of the most critical challenges in modern agriculture, affecting crop health, productivity, and food security on a global scale (Saleem et al., 2024). These pests, including insects, mites, nematodes, and microbial pathogens, can lead to extensive crop damage and yield losses if not detected and managed in a timely manner (Subedi et al., 2023). The Food and Agriculture Organization (FAO) estimates that pests are responsible for the destruction of up to 40% of global crop production annually. Such significant losses not only impact the economic viability of farming operations but also exacerbate global hunger and disrupt international food supply chains (Cabral et al., 2024). Therefore, the development of reliable and efficient pest monitoring systems is essential for maintaining agricultural sustainability and ensuring food security.

Pest monitoring involves the systematic observation, detection, and recording of pest activity within agricultural ecosystems. The primary objective of this process is to inform decision-making in pest management by enabling early detection, accurate identification, and precise quantification of pest populations (Zhou et al., 2024). Historically, pest monitoring has been conducted through conventional methods such as manual field scouting, visual inspection, use of light or pheromone traps, and farmer observations. These traditional techniques, while foundational to

[1] Department of Plant Pathology, University of Agriculture, Faisalabad, Pakistan.
[2] Plant Pathology Research Institute, Agricultural Research Center, 12619, Giza, Egypt.
* Corresponding author: muhammad.atiq@uaf.edu.pk

DOI: 10.1201/9781003546702-2

pest management practices, possess inherent limitations. They are typically labor-intensive, time-consuming, and spatially constrained (Yones and Ma'moun, 2025). Additionally, these approaches often rely heavily on human expertise, making them prone to subjectivity and inconsistency, particularly in large-scale or remote agricultural landscapes.

In response to these limitations, a paradigm shift has emerged in recent decades toward the integration of advanced technologies in pest surveillance systems. The increasing availability and affordability of remote sensing platforms, high-resolution imaging systems, and computational tools have opened new frontiers in pest monitoring (Azia et al., 2025). These technologies offer the potential for rapid, accurate, and large-scale detection of pest outbreaks, minimizing reliance on manual efforts while improving data accuracy and response times (Shoaib et al., 2025).

Remote sensing technologies, including satellite-based sensors and unmanned aerial vehicles (UAVs), enable the continuous monitoring of vast agricultural areas (Sabir et al., 2024). These platforms can detect crop stress patterns that may indicate pest infestation through multispectral, hyperspectral, and thermal imagery. Image processing techniques further enhance this capability by applying digital algorithms for feature extraction, segmentation, and classification of pest-related visual cues within captured imagery. Moreover, the rise of artificial intelligence (AI) and machine learning (ML) has enabled the automation of complex decision-making processes in pest surveillance (Sharma and Shivandu, 2024). AI models, such as convolutional neural networks (CNNs), are increasingly being trained to identify specific pests and predict infestation patterns based on environmental and visual data.

The integration of these digital technologies is gradually transforming pest monitoring into a more precise, predictive, and data-driven discipline. Advanced systems are now capable of providing near real-time alerts, enabling timely intervention and reducing the need for broad-spectrum pesticide application (Pradeep and Tyagi, 2024). This transition aligns with the principles of precision agriculture and integrated pest management (IPM), promoting sustainability through targeted, environmentally responsible practices (Getahun et al., 2024). The growing body of scientific literature and field implementations has demonstrated the effectiveness of these modern tools in enhancing pest monitoring outcomes (Singh et al., 2024). However, challenges remain in terms of data interpretation, system scalability, and accessibility, particularly in resource-limited settings. Despite these obstacles, the continued evolution of technology holds promise for overcoming current limitations and achieving widespread adoption of advanced pest monitoring solutions (Nahiyoon et al., 2024). This chapter examines modern plant pest monitoring, highlighting the roles of remote sensing, image processing, and artificial intelligence. It outlines their individual and integrated applications, real-world use cases, and current challenges, while also exploring future research directions in this evolving field.

The main objectives of present chapter is (1) to shed light on the integration of remote sensing, image processing and AI in plant pest monitoring (2) to explore the role of satellite imagery and UAVs in large-scale crop health monitoring (3) to highlight the use of image processing techniques for efficient pest detection, feature extraction and classification (4) to discuss the application of AI for real-time pest detection, predictive analysis and precision agriculture.

2. Role of Remote Sensing in Pest Monitoring

Remote sensing can be defined as the discipline and capability to obtain information about an object area, or event through sensors that do not make physical contact with the subject being examined. Remote sensing activities encompass a wide range of processes, including satellite coordination, data acquisition, subsequent data processing, analysis, and the dissemination of the collected data and imagery results (Alsadik et al., 2024). In recent years, advancements in information technologies such as remote sensors, Geographic Information Systems (GIS), and Global Positioning Systems (GPS) have introduced a novel approach to the detection and management of insect pests (Figure 1). This method is fundamentally based on electromagnetic energy, which plays a crucial role in data collection. Technology is employed to measure and record the reflection and emission of electromagnetic radiation from the target area using specialized sensor instruments (Rajabpour and Yarahmadi, 2024).

Remote sensing applications have proven to be highly valuable across various fields, including weather forecasting, agriculture, forest ecosystems, water resource management, insect pest control, and marine sciences. These applications can be categorized into three primary strategies: monitoring and mapping of Earth's resources, developing forecasting and prediction models, and facilitating management and decision-making based on the collected data. The forecasting and prediction models rely on data retrieved from geo- and bio-physical parameters (Aziz et al., 2025).

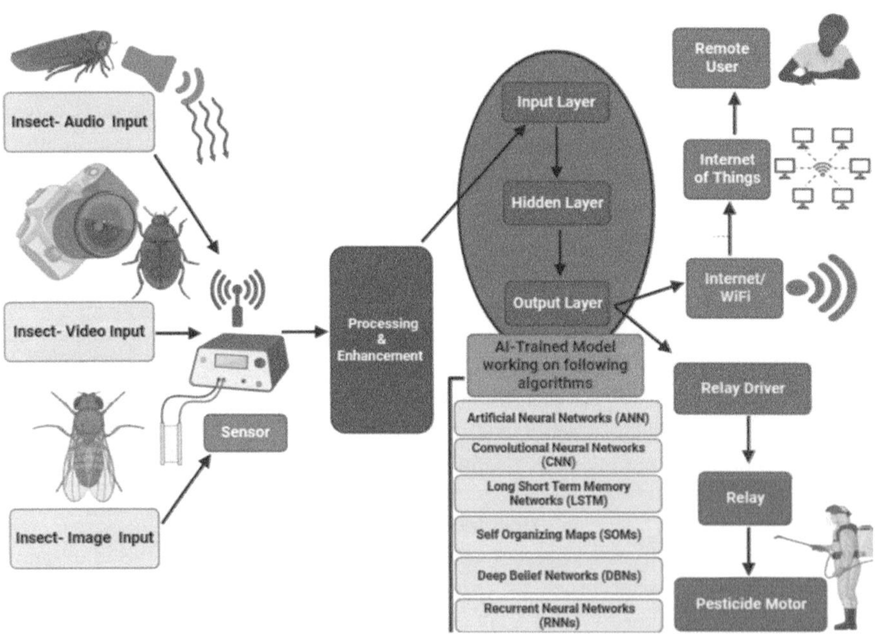

Figure 1. Diagram shows the sequence of the pipeline (data acquisition, data preprocessing & annotation, model training, pest detection & risk prediction, decision support system, feedback & model refinement) adopted in building and deploying an AI system to monitor, identify and regulate insect pests in agricultural systems (reprinted from Aziz et al., 2025).

3. Remote Sensing for Plant Disease Detection

Optical remote sensing utilizes the visible, near infrared, and shortwave infrared regions of the electromagnetic spectrum to capture images of the Earth's surface by detecting the solar radiation reflected from surface features. Different materials reflect and absorb radiation in distinct ways across various wavelengths. The reflectance spectrum of a material represents the proportion of radiation reflected relative to the incident wavelength and serves as a unique signature for that material. In theory, a material can be identified by its spectral reflectance signature if the sensing system possesses sufficient spectral resolution to distinguish its spectrum from those of other materials. This principle underlies many applications of multispectral remote sensing. Consequently, targets can be identified based on their spectral reflectance signatures in remotely sensed images. Indicators of crop stress, such as pest or disease stress and water stress, are often detected through the decline of green pigments and changes in pigment composition (referred to as chlorosis), as well as tissue damage (known as necrosis) or moisture loss in tissues, which may also manifest as an increase in tissue temperature (Zheng et al., 2023).Various data sources can offer extensive remote sensing information across different spatial and temporal scales and resolutions, thereby enhancing the accuracy and speed of data support for monitoring pests and diseases.

3.1 Hyperspectral Technology

Hyperspectral remote sensing data have gained significant traction in the monitoring of crop pests and diseases due to their extensive wavelength information and high spectral resolution. Variations in the physiological and biochemical parameters of vegetation—such as structure, morphology, pigment composition, and water content—lead to notable changes in spectral reflectance. Consequently, capturing these spectral changes in crops affected by pests and diseases has emerged as a crucial tool for remote sensing monitoring in this field (Zhang et al., 2021).

3.2 Multispectral Technology

In optical remote sensing, both multispectral and hyperspectral techniques serve as vital information sources for monitoring crop pests and diseases. In practical applications, the information derived from spectral indices is often more comprehensive and informative than that obtained from raw spectral bands alone. Consequently, there has been a growing body of research focused on developing sensitive vegetation indices specifically for the detection and identification of crop pests and diseases. These studies typically combine pest- and disease-sensitive bands with mathematical functions such as differences, ratios, normalization, or differentiation to enhance detection capabilities (Prabhakar et al., 2022).

3.3 Image Information

In addition to the rich spectral information provided by remote sensing, image data also offers valuable insights for monitoring crop pests and diseases. Image classification is a technique used to identify healthy and diseased leaves by analyzing various morphological features, including shape, texture, color, and pattern. This method allows for the differentiation between healthy and affected vegetation, facilitating timely interventions and management strategies in agricultural practices (Klein et al., 2021).

3.4 Fluorescence Technology

Chlorophyll fluorescence is intricately linked to plant photosynthetic physiology, and changes in fluorescence can occur earlier than alterations in chlorophyll content when plants experience stress. This characteristic makes chlorophyll fluorescence a valuable tool for the early detection of disease-related stress and a promising indicator of how effectively vegetation utilizes light energy. For instance, Jing et al. utilized solar-induced chlorophyll fluorescence in conjunction with green reflectance spectra to detect wheat yellow rust and assess the severity of the disease. This approach highlights the potential of integrating fluorescence data with spectral information for enhanced monitoring and management of crop health (He et al., 2023).

3.5 Thermal Infrared Imaging Technology

Thermal infrared remote sensing captures energy emitted in the thermal infrared band (8–14 μm) that is either reflected or radiated by vegetation and other features from considerable distances. This technology is particularly useful for detecting changes in plant transpiration and water content that may be caused by disease, as well as identifying alterations in plant surface temperature resulting from stress-related physiological changes. Applications of thermal infrared remote sensing include yield estimation, assessment of plant phenotypes, detection of plant water stress, and identification of plant diseases. By monitoring these thermal characteristics, researchers and agricultural practitioners can gain valuable insights into plant health and stress conditions (Zheng et al., 2023).

3.6 Multi-Source Data Fusion

Traditional monitoring methods for crop diseases and pests are often time-consuming and ineffective. However, advancements in remote sensing, meteorology, and plant growth monitoring technologies have paved the way for utilizing multi-source data to monitor these agricultural challenges, marking a new trend and breakthrough in agricultural development. The occurrence of crop diseases and pests—such as the reproduction, transmission, and infection of bacterial spores, as well as the emergence of insect eggs—represents a long-term process that is influenced by appropriate landscape patterns and habitat conditions. Therefore, integrating multi-source temporal remote sensing data with habitat data can enhance the monitoring and early warning capabilities for rice diseases and pests, enabling more proactive and effective management strategies (Yang et al., 2021).

4. Application of Image Processing in Plant Pest Identification

Image processing has become a fundamental component in the field of plant pest identification, offering automated, precise, and scalable solutions for agricultural monitoring (Ngugi et al., 2021). Traditional pest identification relies on visual inspections and manual classification, which are time-consuming and often subject to human error. By contrast, image processing leverages digital imagery and computational algorithms to detect, analyze, and classify pests based on morphological and textural features (Kamilaris and Prenafeta-Boldú, 2018). This technological advancement has significantly improved the efficiency, consistency, and objectivity of pest detection systems in agriculture. One of the initial steps in image-based pest identification is image acquisition, which involves capturing high-quality images of crops using digital cameras, smartphones, or drone-mounted imaging systems (Basri et al., 2023). These images are then subjected to various image enhancement techniques to improve clarity and highlight relevant features. Enhancement methods such as histogram equalization, contrast stretching, and noise reduction are commonly applied to optimize image quality and ensure the accuracy of subsequent analysis steps (Chen, 2012).

Feature extraction follows enhancement and plays a crucial role in isolating meaningful attributes from the images. These attributes may include color, shape, size, edge patterns, and texture, which are essential for differentiating between pest species or distinguishing pests from plant tissue. Techniques such as Gabor filters, gray level co-occurrence matrices (GLCM), and local binary patterns (LBP) are frequently used for extracting texture-based features (Murugappan and Sabeenian, 2019). Accurate feature extraction ensures that classifiers receive informative data for decision-making. Segmentation is another vital stage in the image processing pipeline, where the goal is to isolate regions of interest, typically the pest or affected plant area, from the background. Segmentation techniques range from thresholding and region growing to more sophisticated methods such as active contour models and watershed algorithms. Effective segmentation is crucial for reducing noise and improving classification performance. The final stage involves classification, where machine learning algorithms are employed to categorize the segmented regions into predefined classes, such as specific pest types or pest-free areas. Support vector machines (SVM), decision trees, k-nearest neighbors (k-NN), and deep learning models, particularly convolutional neural networks (CNNs), are widely used in pest classification tasks (Mouelhi et al., 2013). These models are trained on labeled image datasets and can achieve high accuracy in distinguishing between different pests.

Accurate and early identification of tea tree pests is essential to achieve efficient tea tree pest control. The deep learning technique has robust research and practical value to apply to the tea tree pest control area. We introduce a Yolov7-tiny model-based pest detection model into tea plantation. The experiments proved that the Improved Yolov7-tiny model proposed was more efficient in capturing tea tree pest features (Figure 2). The proposed pest detection model has prospective application values and has the ability to decrease the time and financial costs of tea plantation

Figure 2. Photo shows the feature map outputs of the enhanced YOLOv7-tiny model's hidden layers. Panels (a–c) are the original images of three different insect pest species. The corresponding panels (d–f) are the class activation maps (CAMs) of the model highlighting the discriminative parts utilized in pest classification. Through the CAMs, it shows how the model concentrates on particular morphological traits responsible for proper identification and how it improves the model's interpretability and reliability in actual pest identification operations.

pest control. The application of image processing in plant pest identification has enabled the development of automated pest detection systems that can operate in real time and on a large scale. These systems reduce the dependency on human experts, enhance early detection capabilities, and support integrated pest management strategies (Barzman et al., 2015). As imaging technology and computational power continue to advance, image processing will remain a cornerstone of precision agriculture and pest surveillance.

5. Integration of Artificial Intelligence in Pest Surveillance

Agriculture is facing significant challenges. Agricultural manufacturing has evolved into a competitive and globalized industry where farmers and input providers must take into account a wide range of climatic and geographical differences as well as global economic and political factors (Demirel et al., 2024). Therefore, farmers use a variety of different pesticides to improve quality and the storage life of crops.

Smart agriculture has recently been introduced to apply artificial intelligence (AI) techniques, information and wireless communication technologies, such as Internet of Things (IoT) into various aspects of agriculture that are used to achieve precise control over crop pests and disease management as well as fertilization and irrigation in the farm field (Hu et al., 2019). This integration not only increases the effectiveness of pest control but also helps to reduce the costs of material and labour. One of the main applications of smart agriculture is reflected to monitor crop health and assess the farm status with respect to plant pests. However, identifying crop pests remains a difficult challenge for farmers due to their complex structure and high similarities among insect pests. Traditional pest detection method is inefficient, costly and time-consuming. At the early stage of infection, detection of pests is important as it allows farmers to take timely action and apply the appropriate pesticides to stop insects from spreading (Karar et al., 2021). Because of these challenges, AI based tools have become important for helping farmers to effectively monitor, manage and apply agricultural treatments (Azfar et al., 2015). AI techniques play a vital role in modern farming by accurate pest identification, early detection, and predictive modeling, enhancing decision-making for pest control by minimizing indiscriminate pesticide application and optimizing interventions (Leybourne et al., 2025). These tools provide accurate data on weather, soils, and plant development, aiding in crop management and improving agricultural techniques, including crop rotation, fertilization and irrigation (Ben-Lhachemi et al., 2024). AI applications have developed for automatic detection, monitoring and identification of insects, offering potential in enhancing pest control, decreasing excessive pesticide use and improving crop output and quality (Teixeira et al., 2023).

5.1 Machine Learning Model

Machine learning (ML) has appeared together with big data technologies and high-performance computing to create new openings to unravel, quantify and understand data intensive processes in agricultural operational environments. ML is the scientific field that gives machines the ability to learn without being strictly programmed (Samuel, 2000). Year by year, ML applies in more and more scientific fields including, for example, bioinformatics, biochemistry, medicine, meteorology, economic sciences, robotics, aquaculture and food security and climatology (Fang et al., 2017).

5.2 Artificial Neural Networks (ANNs)

ANNs are one of the fundamental elements of machine learning and artificial intelligence (Sebastian et al., 2021). They are inspired by the idea of adjoining computers and man's brain abilities. In a similar way, the main ability of neural networks is to process information simultaneously yet independently. Essentially, the artificial neurons perform similar functions as those found in the brain, opening the door to the development of technological systems like computers that are equipped with a wide range of programs for handling difficult challenges (Kosiński, 2007). NNs are supervised models and are typically used for regression and classification problems. The learning algorithms normally used in ANNs include the radial basis function networks (Broomhead and Lowe, 1988), back-propagation (Linnainma, 1976), perceptron algorithms (Rosenblatt, 1958), and resilient back-propagation (Riedmiller and Braun, 1993). Additionally, numerous ANN-based learning algorithms have been documented, such as counter propagation algorithms (Hecht-Nielsen, 1987), adaptive-neuro fuzzy inference systems (Jang et al., 1993), autoencoder, XY-Fusion, and supervised Kohonen networks (Melsson et al., 2006), as well as Hopfield networks (Hopfield et al., 1982), multilayer perceptron (Pal et al., 1992), self-organising maps (Kohonen et al., 1990), extreme learning machines

(Huang et al., 2006), general regression neural network, ensemble neural networks or ensemble averaging, and self-adaptive evolutionary extreme learning machines (Cao et al., 2012).

5.2.1 Applications for ANNs

They are frequently used to resolve different division and prediction problems. They have also been utilized in the widely defined field of agriculture for a while. They can be included into decision support and precision farming systems. One of the primary substitutes for traditional mathematical models is artificial neural networks, which can take the place of traditional modeling techniques. Artificial neural networks have a very broad range of applications (Kujava et al., 2021). The key benefit of artificial neural networks (ANNs) is that they address problems by producing input signals which foster the network's capacity to learn and identify patterns. For a variety of tasks, ANN can be preferred over intricate algorithms or rule-based programming.

5.3 Real Time Monitoring

Monitoring soil temperature and humidity in real time, together with other factors including weather, wind direction, wind speed, water level, and floods, is one area of study. A wireless farming system can accurately manage agricultural output and increase crop yield by providing monitoring, automated irrigation, and an alarming system that sends a notice to a farmer's phone when a specific circumstance arises. Although traditional cable connections have many applications in the field of real-time monitoring, it also has several drawbacks. The network management platform and wireless sensor network nodes make up the system. Numerous jump routing procedures and a versatile automatic networking monitoring system for soil temperature and humidity are used to accomplish automatic networking. Users can observe the changing soil conditions, such as temperature, humidity, and other elements, from anywhere in the world without physically visiting the field.

6. Successful Case Studies in AI, Remote Sensing, and Image Processing for Plant Pest Monitoring

Several studies revealed a strong relationship between disease and pest emergence/re-emergence and food deficient regions with high population density, where there was a lack of capacity in planning and development (P&D), monitoring and forecasting (Savary et al., 2019; Ristaino et al., 2021). The conventional visual monitoring of P&D has limitations such as lack of real-time data, wide-scale observation is not possible, and it is a time-consuming approach; therefore, crops are at risk of damage (Satapathy, 2020). Meanwhile, remote sensing can detect the abiotic and biotic stresses of plants. Progress in remote sensing sensors, artificial intelligence/geospatial artificial intelligence (GeoAI), machine learning algorithms, internet of things (IoT), high-performance computing, and trained human resources are paving the way for remote sensing-based crop monitoring and reducing the risk of crop damage by P&D (Seetharaman, 2020; Zhang et al., 2019). Recently, the development of information technologies such as remote sensing sensors, GIS

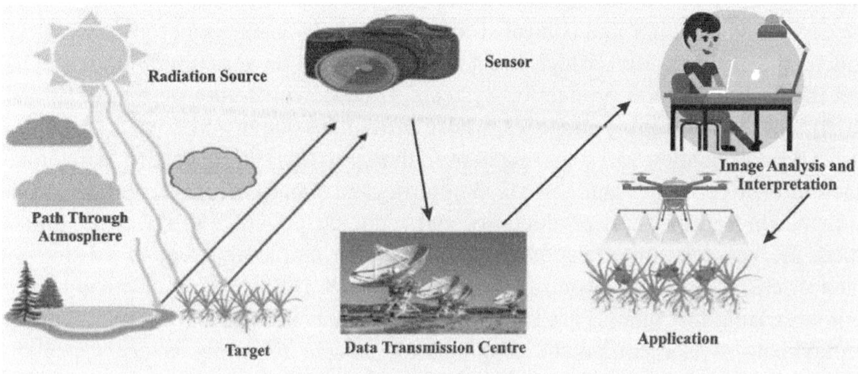

Figure 3. Scientific illustration depicting an integrated pest monitoring framework incorporating remote sensing for pest detection, image processing for infestation mapping and drone-based application targeting identified pest hotspots (reprinted from Aziz et al., 2025).

(geographic information systems), and GPS (global position systems) support a new approach for detecting and controlling insect pests. Remote sensing technologies are mainly depending on receiving and interpretations of information about the earth's surface without any physical contact with it (Abd El-Ghany et al., 2020).

Remote sensing is mainly dependent on the electromagnetic energy which influences the understanding of data collection. This technology is utilized for measuring and recording the reflection and the emission of electromagnetic radiation from the target area using sensor instruments. Highly valuable remote sensing applications reported in various fields include weather, agriculture, forest ecosystems, water resources, insect pest management, and marine sciences. These applications are divided into three main strategies: monitoring and mapping of various earth resources, forecasting and prediction models, and finally management and decision-making depending on the data collected (Figure 3). Forecasting and prediction models depend on the retrieval data from geo-and bio-physical parameters (Oumar and Mutanga 2011; Piekarczyk, 2014).

Today, remote sensing systems are carried on satellites and involve multi-spectral imaging devices characterized by achievement of synoptic view, fast survey, high accuracy, inaccessible coverage area, etc., compared with the conventional methods for monitoring and mapping. For example, remote sensing systems carried on satellites are used to characterize vegetation or calculate water surfaces during floods which supplies ecologists and hydrologists with fast and accurate data on monitoring the water cycle and its influence upon the ecosystem (Malinowski et al. 2015). This review highlights various topics on the principles of remote sensing systems and their applications for agriculture approaches; particularly, management of insect pests and plant diseases.

7. Challenges in Developing Countries

Emerging diseases and pests are becoming increasingly prevalent in developing countries, but their NPPOs (National Plant Protection Organizations) frequently lack the capacity to monitor and control such hazards. As a result, numerous diseases go undetected or unreported, which has a negative impact on agricultural output today and the ability of food production systems to cope with future outbreaks (Flood, 2010). For the developing countries, whose agriculture output and exports frequently form the foundation of their economy, this position is especially dangerous. Therefore, effective pest and disease monitoring and management systems are crucial to increasing agricultural productivity and reducing poverty (Smith et al., 2008). There are several limitations and difficulties with employing remote sensing for plant disease detection, despite all of its advantages. Despite their technical nature, remote sensing instruments are thought to be expensive. Remote sensing equipment deployment, sensor equipment, and training can be relatively costly, especially while more advanced technology is being employed. The greatest number of these are setup expenses, such as purchasing equipment and software, which, depending on the project's goals, might account for 48–78% of the overall cost. The demand for graphics processing units (GPUs) in data mining can cause price fluctuations (Jurczuk et al., 2021). The size and complexity of the data gathered is the second major factor restricting their utility. Most sensors gather vast amounts of remotely sensed data. Sensor-generated data can readily exceed terabytes or petabytes in size and quantity (Wahabzada et al., 2015).

8. Future Prospects

An exceptional platform for the success of precision agriculture may be provided by the future use of robotics and artificial intelligence (AI) in the field of plant disease detection by utilizing all available mechanisms, such as hyperspectral/multispectral imaging and processing images of diseased samples using machine learning or deep learning architecture. Automated pest monitoring started to gain popularity, and its potential was obvious. Based on precise, timely information, crop advisors can more efficiently allocate resources and treatments. FarmSense, Trapview, CropVue, Metos, Semios, RapidAIM, and scoutlabs are among the leading businesses in pest monitoring developments. They provide innovative technologies that combine artificial intelligence (AI), sensors, and imagery to detect pests on their own. An autonomous disease detection system utilizing a hybrid network with superpixel clustering was proposed. Additionally, they classified the photos into three groups using seven different kinds of machine learning technology. To identify plant illness before symptoms, show up, hyperspectral/multispectral imaging should be combined with machine learning (ML) or deep learning (DL) architecture. An exceptional platform for the success of precision agriculture may be provided by the future use of robotics and artificial intelligence (AI) in the field of plant disease detection by utilizing all available mechanisms, such as hyperspectral/multispectral imaging and processing images of diseased samples using machine learning or deep learning architecture Support Vector Machine (SVM):Using pixel patterns, SVMs and the HOG feature extraction technique assist in distinguishing between healthy and pathological features. It is a helpful technique for resolving binary classifiers, such as those focused on classifying plants with diseases and their healthy equivalents.

9. Conclusion

Effective plant pest monitoring is fundamental to reducing crop losses and promoting sustainable agricultural practices. Traditional methods, while foundational, lack the scalability and precision required in modern, large-scale farming. The integration of remote sensing, image processing, and artificial intelligence presents a transformative solution, offering real-time, data-driven insights into pest dynamics. Remote sensing technologies provide extensive spatial coverage and early indicators of pest stress. Image processing enhances detection accuracy by enabling automated analysis of visual data, while AI models bring advanced decision-making capabilities through classification and predictive analytics. Together, these technologies improve the responsiveness and effectiveness of pest management strategies. Despite the promise of these tools, challenges such as limited infrastructure, high implementation costs, and data interpretation complexities persist—particularly in developing regions. Addressing these barriers through targeted research, capacity building, and policy support will be essential for broader adoption. Moving forward, advancements in sensor technology, cloud computing, and open-access datasets are expected to further enhance pest monitoring capabilities. A multidisciplinary approach involving agronomists, data scientists, and policymakers will be crucial for translating technological potential into field-level impact. The continued evolution of smart pest monitoring systems holds great promise for achieving global food security and sustainable agricultural development.

References

Abd El-Ghany, N. M., Abd El-Aziz, S. E., and Marei, S. S. (2020). A review: application of remote sensing as a promising strategy for insect pests and diseases management. Environmental Science and Pollution Research, 27(27): 33503–33515.

Abdullah, H. M., Mohana, N. T., Khan, B. M., Ahmed, S. M., Hossain, M., Islam, K. S. et al. (2023). Present and future scopes and challenges of plant pest and disease (P&D) monitoring: Remote sensing, image processing, and artificial intelligence perspectives. Remote Sensing Applications: Society and Environment, 32: 100996.

Alsadik, B., Ellsäßer, F.J., Awawdeh, M., Al-Rawabdeh, A., Almahasneh, L., Oude Elberink, S. et al. (2024). Remote sensing technologies using UAVs for pest and disease monitoring: A review centered on date palm trees. Remote Sensing, 16(23): 4371.

Azfar, S., Nadeem, A., and Basit, A. (2015). Pest detection and control techniques using wireless sensor network: A review. Journal of Entomology and Zoology Studies, 3(2): 92–99.

Aziz, D., Rafiq, S., Saini, P., Ahad, I., Gonal, B., Rehman, S. A. et al. (2025). Remote sensing and artificial intelligence: revolutionizing pest management in agriculture. Frontiers in Sustainable Food Systems, 9: 1551460.

Barzman, M., Bàrberi, P., Birch, A. N. E., Boonekamp, P., Dachbrodt-Saaydeh, S., Graf, B. et al. (2015). Eight principles of integrated pest management. Agronomy for Sustainable Development, 35(4): 1199–1215.

Basri, B., Indrabayu, I., Achmad, A., and Areni, I. S. (2023). A review of image processing techniques for pest and disease early identification systems on modern cocoa plantation. International Journal of Computing and Digital Systems, 14(1): 10277–10286.

Ben-Lhachemi, N., Benchrifa, M., Nasrdine, S., Mabrouki, J., and Slaoui, M. (2024). Intelligent Temperature and Humidity Control System: Leveraging AI for Precise Climate Management. In Artificial Intelligence Systems in Environmental Engineering (pp. 63–69).

Broomhead, D. S., and Lowe, D. (1988). Multivariable Functional Interpolation and Adaptive Networks. Complex System, 2: 321–355.

Cabral, M. J., Haseeb, M., and Soares, M. A. (2024). Major insect pests of sweet potatoes in brazil and the united states, with information on crop production and regulatory pest management. Insects, 15(10): 823.

Cao, J., Lin, Z., and Huang, G. (2012). Bin Self-adaptive evolutionary extreme learning machine. Neural Processing Letters, 36: 285–305.

Chen, S. D. (2012). A new image quality measure for assessment of histogram equalization-based contrast enhancement techniques. Digital Signal Processing, 22(4): 640–647.

Chouhan, S. S., Singh, U. P., Sharma, U., and Jain, S. (2021). Leaf disease segmentation and classification of *Jatropha Curcas* L. and *Pongamia Pinnata* L. biofuel plants using computer vision based approaches. Measurement, 171: 108796.

Demirel, M., and Kumral, N. A. (2021). Artificial Intelligence in Integrated Pest Management. pp. 289–313. *In*: Tomar, P., and Kaur, G. (Eds.). Artificial Intelligence and IoT-Based Technologies for Sustainable Farming and Smart Agriculture.

Fang, K., Shen, C., Kifer, D., and Yang, X. (2017). Prolongation of SMAP to spatiotemporally seamless coverage of continental US using a deep learning neural network. Geophysical Research Letters, 44(21): 11–030.

Flood, J. (2010). The importance of plant health to food security. Food Secur., 2: 215–231.

Getahun, S., Kefale, H., and Gelaye, Y. (2024). Application of precision agriculture technologies for sustainable crop production and environmental sustainability: A systematic review. The Scientific World Journal, 2024(1): 2126734.

He, J., Chen, K., Pan, X., Zhai, J. and Lin, X. (2023). Advanced biosensing technologies for monitoring of agriculture pests and diseases: A review. Journal of Semiconductors, 44(2): 023104.

Hecht-Nielsen, R. (1987). Counter propagation networks. Applied Optics, 26(23): 4979–4983.

Hopfield, J. J. (1982). Neural networks and physical systems with emergent collective computational abilities. Proceedings of National Academy of Sciences USA, 79(8): 2554–2558.

Hu, Z., Xu, L., Cao, L., Liu, S., Luo, Z., Wang, J., and Wang, L. (2019). Application of non-orthogonal multiple access in wireless sensor networks for smart agriculture. IEEE Access, 7: 87582–87592.

Huang, G. B., Zhu, Q. K., and Siew, C. K. 2006. Extreme learning machine: Theory and applications. Neurocomputing, 70: 489–501.

Jang, J. S. R. (1993). ANFIS: Adaptive-Network-Based Fuzzy Inference System. IEEE Transaction on Systems, Man and Cybernetics, 23(3): 665–685.

Jurczuk, K., Czajkowski, M., and Kretowski, M. (2021). Multi-GPU approach to global induction of classification trees for large-scale data mining. Appl. Intell., 51(8): 5683–5700.

Kamilaris, A., and Prenafeta-Boldú, F. X. (2018). Deep learning in agriculture: A survey. Computers and Electronics in Agriculture, 147: 70–90.

Karar, M. E., F. Alsunaydi, S. Albusaymi, and S. Alotaibi. (2021). A new mobile application of agricultural pests recognition using deep learning in cloud computing system. Alexandria Engineering Journal, 60(5): 4423–4432.

Khan, A., Vibhute, A. D., Mali, S., and Patil, C. H. (2022). A systematic review on hyperspectral imaging technology with a machine and deep learning methodology for agricultural applications. Ecological Informatics, 69: 101678.

Klein, I., Oppelt, N. and Kuenzer, C. (2021). Application of remote sensing data for locust research and management review. Insects, 12(3): 233.

Kohonen, T. (1990). The Self-Organizing Map. Proceedings of IEEE, 78(9): 1464–1480.

Kosinski, R. A., and Zaremba, M. (2007). Dynamics of the model of the Caenorhabditis Elegans neural network. Acta Physica Polonica B, 38(6).

Kujawa, S., and Niedbała, G. (2021). Artificial neural networks in agriculture. Agriculture, 11(6): 497.

Leybourne, D. J., Musa, N., and Yang, P. (2025). Can artificial intelligence be integrated into pest monitoring schemes to help achieve sustainable agriculture? An entomological, management and computational perspective. Agricultural and Forest Entomology, 27(1): 8–17.

Linnainmaa, S. (1976). Taylor expansion of the accumulated rounding error. BIT, 16: 146–160.

Malinowski, R., Groom, G., Schwanghart, W., and Heckrath, G. (2015). Detection and delineation of localized flooding from WorldView-2 multispectral data. Remote Sensing, 7(11): 14853–14875.

Melssen, W., Wehrens, R., and Buydens, L. (2006). Supervised Kohonen networks for classification problems. Chemometrics and Intelligent Laboratory System, 83(2): 99–113.

Mouelhi, A., Sayadi, M., Fnaiech, F., Mrad, K., and Romdhane, K. B. (2013). Automatic image segmentation of nuclear stained breast tissue sections using color active contour model and an improved watershed method. Biomedical Signal Processing and Control, 8(5): 421–436.

Murugappan, V., and Sabeenian, R. S. (2019). Texture based medical image classification by using multi-scale Gabor rotation-invariant local binary pattern (MGRLBP). Cluster Computing, 22(Suppl 5), 10979–10992.

Ngugi, L. C., Abelwahab, M., and Abo-Zahhad, M. (2021). Recent advances in image processing techniques for automated leaf pest and disease recognition–A review. Information processing in agriculture, 8(1): 27–51.

Oumar, Z., and Mutanga, O. (2011). The potential of remote sensing technology for the detection and mapping of *Thaumastocoris peregrinus* in plantation forests. Southern Forests: A Journal of Forest Science, 73(1): 23–31.

Pal, S. K., and Mitra, S. (1992). Multilayer Perceptron, Fuzzy Sets, and Classification. IEEE Transaction of Neural Network, 3(5): 683–697.

Piekarczyk, J. (2014). Application of remote sensing in agriculture. Geoinformatica Polonica, 13(1): 69–75.

Pradeep, M., and Tyagi, A. K. (2024). Smart sensor-based smart agriculture for better crop production in this smart era. In AI Applications for Business, Medical, and Agricultural Sustainability IGI Global Scientific Publishing, pp. 236–266.

Prabhakar, M., Thirupathi, M., and Mani, M. (2022). Principles and application of remote sensing in crop pest management. Trends in Horticultural Entomology, pp. 157–183.

Pusatkar, A. C., and Gulhane, V. C. (2016). Implementation of wireless sensor network for real time monitoring of agriculture. International Research Journal of Engineering and Technology (IRJET), 3(05): 5.

Rajabpour, A., and Yarahmadi, F. (2024). Remote Sensing, Geographic Information System (GIS), and Machine Learning in the Pest Status Monitoring. In Decision System in Agricultural Pest Management (pp. 247–353). Singapore: Springer Nature Singapore.

Riedmiller, M., and Braun, H. (1993). A direct adaptive method for faster back propagation learning: The RPROP algorithm. In Proceedings of the IEEE International Conference on Neural Networks, San Francisco, CA, USA, pp. 586–591.

Ristaino, J. B., Anderson, P. K., Bebber, D. P., Brauman, K. A., Cunniffe, N. J., Fedoroff, N. V. et al. (2021). The persistent threat of emerging plant disease pandemics to global food security. Proceedings of the National Academy of Sciences, 118(23): e2022239118.

Rosenblatt, F. (1958). The perceptron: A probabilistic model for information storage and organization in the brain. Psychol. Rev., 65: 386–408.

Sabir, Rehan Mehmood, Kashif Mehmood, Abid Sarwar, Muhammad Safdar, Nalain E. Muhammad, Nimra Gul, Fareeha Athar et al. (2024). Remote sensing and precision agriculture: A sustainable future. In Transforming Agricultural Management for a Sustainable Future: Climate Change and Machine Learning Perspectives, pp. 75–103. Cham: Springer Nature Switzerland.

Saleem, S., Sharif, M. I., Sharif, M. I., Sajid, M. Z., and Marinello, F. (2024). Comparison of deep learning models for multi-crop leaf disease detection with enhanced vegetative feature isolation and definition of a new hybrid architecture. *Agronomy*, 14(10): 2230.

Samuel, A. L. (2000). Some studies in machine learning using the game of checkers. IBM Journal of research and development, 44(1.2): 206–226.

Satapathy, R. R. (2020). Remote sensing in plant disease management. J. Pharmacogn. Phytochem, 9: 1813–1820.

Savary, S., Willocquet, L., Pethybridge, S. J., Esker, P., McRoberts, N., and Nelson, A. (2019). The global burden of pathogens and pests on major food crops. Nature Ecology & Evolution, 3(3): 430–439.

Sebastian, A. M., and Peter, D. (2022). Artificial intelligence in cancer research: trends, challenges and future directions. Life, 12(12): 1991.

Seetharaman, K. (2021). A fully automated crop disease monitoring and management system based on IoT: IoT-Based disease identification for banana leaf. In Deep Learning Applications and Intelligent Decision Making in Engineering (pp. 192–211).

Sharma, K., and Shivandu, S. K. (2024). Integrating artificial intelligence and internet of things (IoT) for enhanced crop monitoring and management in precision agriculture. Sensors International, p. 100292.

Shoaib, M., Sadeghi-Niaraki, A., Ali, F., Hussain, I., and Khalid, S. (2025). Leveraging deep learning for plant disease and pest detection: a comprehensive review and future directions. Frontiers in Plant Science, 16: 1538163.

Singh, S., Singh, S., Pawar, R., and Kulhar, K. S. (2024). Artificial intelligence approaches for predictive power consumption modeling in machining-Short review. In E3S Web of Conferences EDP Sciences, 540: 06015.

Smith, J. J., Waage, J., Woodhall, J W., Bishop, S. J., and Spence, N J. (2008). The challenge of providing plant pest diagnostic services for Africa. Eur. J. Plant Pathol., 121: 365–375.

Specht, D. F. (1991). A general regression neural network. IEEE Transaction on Neural Network, 2: 568–576.

Subedi, B., Poudel, A., and Aryal, S. (2023). The impact of climate change on insect pest biology and ecology: Implications for pest management strategies, crop production, and food security. Journal of Agriculture and Food Research, 14: 100733.

Subeesh, A., and Mehta, C. R. (2021). Automation and digitization of agriculture using artificial intelligence and internet of things. Artificial Intelligence in Agriculture, 5: 278–291.

Teixeira, A. C., Ribeiro, J., Morais, R., Sousa, J. J., and Cunha, A. (2023). A systematic review on automatic insect detection using deep learning. Agriculture, 13(3): 713.

Upadhyay, A., Chandel, N. S., Singh, K. P., Chakraborty, S. K., Nandede, B. M., Kumar, M. et al. (2025). Deep learning and computer vision in plant disease detection: a comprehensive review of techniques, models, and trends in precision agriculture. Artificial Intelligence Review, 58(3): 1–64.

Wahabzada, M., Mahlein, A. K., Bauckhage, C., Steiner, U., Oerke, E. C., and Kersting, K. (2015). Metro maps of plant disease dynamics—automated mining of differences using hyperspectral images. Plos one, 10(1): e0116902.

Yang, R., Lu, X., Huang, J., Zhou, J., Jiao, J., Liu, Y. et al. (2021). A multi-source data fusion decision-making method for disease and pest detection of grape foliage based on ShuffleNet V2. Remote Sensing, 13(24): 5102.

Yones, M., and Ma'moun, S. (2025). Earth-observation data and simulation modeling for integrated pest management programs: monitoring distribution dimensions of insect pests along coastal cities. In Modelling and Advanced Earth Observation Technologies for Coastal Zone Management (pp. 229–259). Cham: Springer Nature Switzerland.

Zhang, J., Huang, Y., Pu, R., Gonzalez-Moreno, P., Yuan, L., Wu, K. et al. (2019). Monitoring plant diseases and pests through remote sensing technology: A review. Computers and Electronics in Agriculture, 165: 104943.

Zhang, N., Guijun, Y., Chunjiang, Z., Zhang, J., Xiaodong, Y. A. N. G., Yuchun, P. A. N. et al. (2021). Progress and prospects of hyperspectral remote sensing technology for crop diseases and pests. National Remote Sensing Bulletin, 25(1): 403–422.

Zheng, Q., Huang, W., Xia, Q., Dong, Y., Ye, H., Jiang, H. et al. (2023). Remote sensing monitoring of rice diseases and pests from different data sources: A review. Agronomy, 13(7): 1851.

Zhou, W., Arcot, Y., Medina, R. F., Bernal, J., Cisneros-Zevallos, L., and Akbulut, M. E. (2024). Integrated pest management: an update on the sustainability approach to crop protection. ACS Omega, 9(40): 41130–41147.

Chapter 3

Robotic Systems and Drones for Precision Spraying and Monitoring

Hadjer Saifi,[1] *Rayane Saifi,*[2,*] *Bekri Xhemali,*[3] *Hysen Kokiçi,*[4] *Messaouda Benabdelkader,*[5] *Ali Kaan Askin*[6] and *Izzet Akça*[6]

1. Introduction

Modern plant health protection, pollinator insects, farmers, and entire ecosystems face increasing challenges related to effectiveness, sustainability, and cost-efficiency (Bazargani and Deemyad, 2024; Jeger et al., 2021; Song et al., 2020). These challenges are particularly significant due to the rapidly growing global population and the continuous rise in food demand, further intensified by extreme weather fluctuations (Rajak et al., 2023; Jeger et al., 2021). Researchers are increasingly focused on improving and automating agricultural processes through the use of robotic devices and drones, as this topic has garnered significant interest (An et al., 2022). Amongst the most promising areas where these advanced technologies are

[1] Biogeochemistry of Desert Environments Laboratory, Scientific and Technical Research Center in Physicochemical Analysis, Department of Biological Sciences, Faculty of Natural and Life Sciences, Kasdi Merbah University, Ouargla, Algeria.
[2] Amine Elokkal El Hadj Moussa Egakhamouk University of Tamanghasset, Research Laboratory of Sciences and Environment: Bioresources, Geochemical-Physics, Legislation and socio-economic development (C1810200), 11039, Sersouf, 11000, Tamanghasset, Algeria.
[3] Laboratory of Plant Protection, Kosovo Institute of Agriculture, Str. Adem Jashari, no. 244, 30000 Pejë, Kosovo.
[4] Institute for Sustainable Plant Protection, National Research Council (CNR), Via Amendola 122/D, Bari 70126, Italy.
[5] Laboratory of Applied Microbiology, Setif University, Department of Environment and Agronomy, Mohamed Seddik Ben Yahia University, B.P. 98, Ouled Aissa, 18000 Jijel, Algeria.
[6] Department of Plant Protection, Faculty of Agriculture, University of Ondokuz Mayıs, Samsun, Turkiye.
* Corresponding author: rayanesaifi2015@gmail.com, rayanesaifi@univ-tam.dz

DOI: 10.1201/9781003546702-3

being applied are spraying and crop monitoring (Bazargani and Deemyad, 2024; Wei et al., 2024). However, the broader adaptation of these innovations in agriculture remains limited due to inadequate promotion and their relatively high costs (Balaska et al., 2023).

Traditionally, crop management practices have relied on manual, auxiliary, or towed, or self-propelled sprayers to rectify issues identified by visual inspections (Jin and Han, 2024; Patel et al., 2020). However, these methods are often time-consuming, labor intensive, and lead to inconsistent chemical application, potentially harming both humans and the environment (Misra and Ghosh, 2024; Balaska et al., 2023).

The integration of innovative technology into robotic systems and drones had made it possible to apply crop spraying with exceptional accuracy, efficiency, and uniformity (Papadopoulos et al., 2024; Oberti and Schmilovitch, 2021). This advancement significantly reduces the amount of fertilizers and plant protection treatments needed, especially in large or hard-to-reach agricultural fields, while also helping to preserve the environment (Fan et al., 2023). Agricultural robotics has made significant strides, now enabling precise surveillance operations to detect and analyze factors such as vegetation density, plant nutrition, soil moisture, diseases, pests, all of which can impact crop production (Papadopoulos et al., 2024; Sparrow and Howard, 2021). This technology facilitates interventions based on data collected by onboard sensors and high-resolution cameras (Barua et al., 2024; Basri et al., 2021; Subramanian et al., 2021).

Numerous studies have praised the incorporation of robots and drones in agricultural practices, particularly in crop spraying and precision control, for their ability to reduce production costs related to labor and chemical usage (Barua et al., 2024; Fountas et al., 2020). This integration also contributes to safeguarding biodiversity by minimizing ecosystem pollution thereby improving the well-being of farmers and consumers (Papadipoulos et al., 2024; Basri et al., 2021). However, some scholars have emphasized the need to consider the high costs associated with obtaining, maintaining, updating these technologies, as well as the need for farmer training and ensuring the safety and confidentiality of the collected data (Balaska et al., 2023; Hafeez et al., 2023).

This chapter has thoroughly explored the latest scientific advancements in crop field spraying and monitoring technology, with a particular focus on the use of robotic systems and drones. It underscores the critical importance of plant protection systems, detailing the various components and mechanisms that enhance their functionality. The chapter offers an in-depth description of these systems' spraying and monitoring capabilities, along with a balanced discussion of their advantages, disadvantages, and limitations in practical applications.

2. Plant Protection and Spraying and Monitoring Requirements

Agriculture is a crucial pillar of global economy, playing a key role in employment, exports, and food sovereignty (Barua et al., 2024; Faizan et al., 2024). A major concern for nations is the decline in agricultural yields, often beginning with changed in the physiological condition of plants (Papadipoulos et al., 2024; Mitra, 2021). These alterations make plants more vulnerable to infections by bacteria, viruses, and fungal pathogens, as well as to attacks by predators, weed infestations, and adverse growth and weather conditions (Reddy et al., 2024).

The most common plant protection strategies today include plowing to remove plant debris and destroy parasites, rotating different crops on the same field to preserve soil fertility by preventing the depletion of specific nutrients, and cultivating varieties resistant to epidemics. Additionally, the use of predatory insects, such as ladybugs and parasitoid wasps, offers a natural method to control certain pests. Manual removal of unwanted weeds, which could compete with crops for light, nutrients and water is also widely practiced. Finally, the application of phytosanitary products to eradicate phytopathogenic agents and fertilizers to improve the nutritional status of plants and their fruits is a key component of modern plant protection (He et al., 2024; Misra and Ghosh, 2024; Upadhyay et al., 2024; Deguine et al., 2023).

Despite their affordability, the traditional agricultural methods present significant challenges (He et al., 2024; Jin and Han, 2024). They require a large workforce, substantial time investment, physical effort, and careful attention from farmers. Moreover, these methods often struggle to maintain soil health in large-scale farming, leading to long-term productivity declines and jeopardizing global food security. Beyond agricultural implications, these practices can also harm human health, biodiversity, and overall ecosystem balance (Figure 1) (Bilal et al., 2024; Moshayedi et al., 2024; Deguine et al., 2023; Lykogianni et al., 2021).

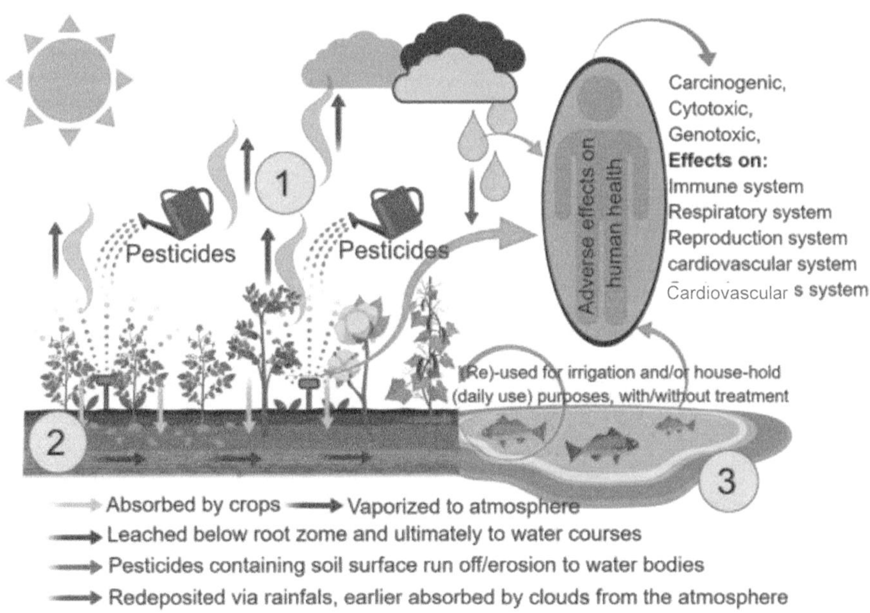

Figure 1. Impact of incompatible use of chemical pesticides/fertilizers on the environment and human health (Bilal et al., 2019).

The current challenge for countries is to optimize agricultural production while maintaining crop quality, protecting human health, and conserve agricultural and financial resources for future generations (Bazargani and deemyad, 2024; Faizan et al., 2024). This requires the adoption of emerging, sustainable, and innovative technologies that are precise, efficient, energy-saving and environmentally friendly, while also meeting ethical standards. Among these techniques are autonomous robotics for spraying and precise surveillance (Zheng et al., 2024). These systems are equipped with high-resolution cameras, extremely sensitive connected sensors, chemical product tanks, and software for data modelling powered by artificial intelligence (Figure 2) (Balaska et al., 2023; Hafeez et al., 2023; Basri et al., 2021).

Spraying is a crucial agricultural practice used to manage phytopathologies, pests and harmful weeds that can negatively affect crop quality and productivity (Baghel et al., 2022). Monitoring, on the other hand, is a systematic procedure designed to assess and track crop and soil conditions to identify specific areas needing additional care (Abbas et al., 2023). The effectiveness of both spraying and monitoring is influenced by several factors, including the type of equipment used, the dimensions and terrain of the agricultural field, the specific plant species, prevailing weather conditions, and the regulatory guidelines governing these activities (Abbas et al., 2023; Hafeez et al., 2023).

Traditional monitoring and spraying methods often lack precision, impeding efforts to optimize agricultural outcomes. Knowledge gaps persist in areas such

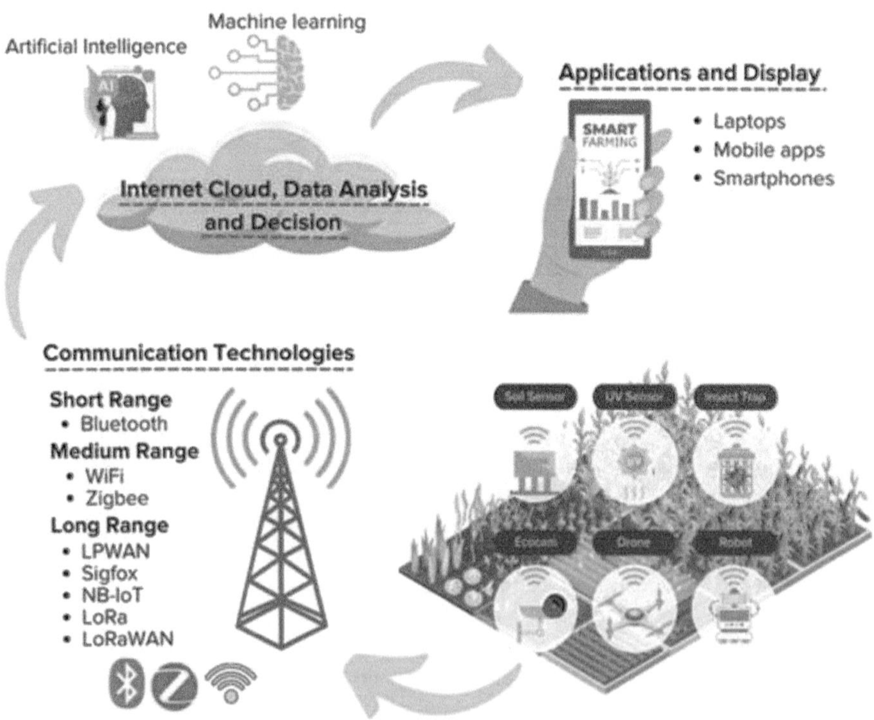

Figure 2. Modern technologies used in crop field monitoring (Ali et al., 2023).

as soil chemical profiles, early stress indicators, and accurate pest infestations (Tomaszewski and Kołakowski, 2023; Ngoc et al., 2023). Empowering farmers to adopt advanced technologies like robotic systems and drones is essential for improving crop protection and yield. Advance technologies offer invaluable insights into farm operations, providing real-time data on land use, biodiversity, and the potential threats like diseases or pests (Kaliannan and Al Saidi, 2024). This empowers farmers to proactively implement sustainable practices such as optimized irrigation, fertilization, and pesticide application. By adhering to precise chemical dosages, farmers can mitigate environmental risks and enhance operational efficiency (Hafeez et al., 2023; Meshram et al., 2022).

3. Technological Advances in Robotic Systems and Drones Used in Agriculture

Technological advancements have transformed agricultural equipment from basic machines requiring extensive manual labour into sophisticated, adaptable robotic systems and drones (Shah et al., 2023). Autonomous agricultural systems are characterized by their high efficiency and precision, and they can be fully customized to address the specific needs of each farm (Zheng et al., 2024). This results in significant savings in time, energy, and costs. Moreover, these systems reduce the need for human involvement, thereby minimizing the risk of work-related accidents such as injuries and exposure to harmful substances (Cheng et al., 2023). Agricultural robots and drones significantly enhance the sustainability, intelligence, and productivity of agriculture by providing farmers with a more comprehensive understanding of their crop needs. This improved insight enables farmers to make well-informed decisions and timely interventions, helping to prevent adverse situations from worsening (Cheng et al., 2023; Ngoc et al., 2023; Basri et al., 2021).

A variety of agricultural robots are commercially accessible, encompassing ground-based models for tasks such as plowing and weeding, aerial drones for spraying and monitoring, and aquatic robots for underwater and hydroponic crop applications (Oliveira et al., 2021). The specific design and components of these systems are tailored to their intended functions and operational needs. However, shared elements underpinning automation and enhanced efficiency are evident across these robotic platforms (Cheng et al., 2023; Oliveira et al., 2021).

4. Typical components in agricultural robotic systems

a. Mechanical systems

Agricultural robots feature robust mechanical systems designed to support various components and endure changes in external conditions. These systems typically include a sturdy chassis and articulated arms or manipulators, which allow the robots to perform a range of agricultural tasks, such as selective weeding, precise planting, and targeted spraying (Zheng et al., 2024). Hydraulic actuators are also part of the mechanical systems, converting energy into motion and operating through electricity or compressed air. Additionally, agricultural robots may be equipped with

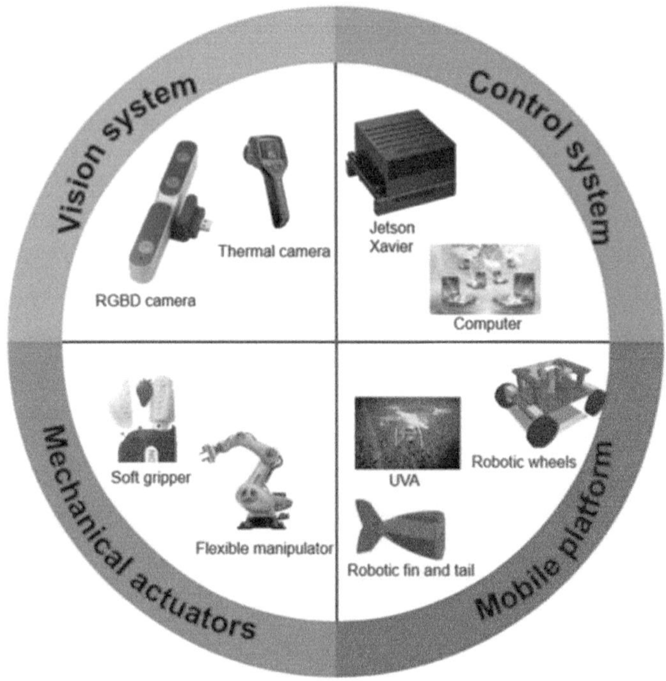

Figure 3. Typical components in agricultural robotic systems (Cheng et al., 2023).

supplementary equipment, such as water tanks, fertilizers, pesticides, and seeds (Figure 3) (Ashwini et al., 2024; Zheng et al., 2024; Ghobadpour et al., 2022).

b. Energy systems

Currently, agricultural robotic systems primarily rely on natural gas and diesel as their main energy sources. However, there is a growing shift towards using electricity and solar energy. This transition is driven by the need to improve the resistance, efficiency, and autonomy of robots, which can be achieved though advancements in battery and tank capacities. This trend has been documented in studies by Yang et al. (2023) and Ghobadpour et al. (2022).

c. Data acquisition systems

Robotic systems, like drones, offer exceptional capabilities for quickly covering large agricultural areas, delivering comprehensive, and detailed information about crops (Kumar et al., 2024). These systems can assess plant health and vitality, determine nutritional and water requirements, and identify the presence of diseases and pests. The acquisition and collection of diverse data are achieved through a variety of advanced sensors. These sensors are crucial for enabling automation and informed decision-making to optimize agricultural practices, such as irrigation, fertilization, and chemical treatments (Jin and Han, 2024). Adjustments are made based on the specific needs of each plant, while also ensuring environmental sustainability (Ashwini et al., 2024; Yang et al., 2023; Paul et al., 2022).

c.1. Remote sensing vision sensors

The sensors used in agricultural robotics are equipped with advanced cameras that capture high-resolution images and videos across various spectra, including RGB, infrared, thermal, multispectral, and hyperspectral. These images are carefully analyzed to provide farmers with precise and valuable information about issues, obstacles, or areas needing immediate attention. Additionally, laser technology, such as LiDAR, is integrated into these sensors to improve precise distance measurement and mapping of agricultural areas (Figure 3) (Jin and Han, 2024; Bai et al., 2023; Nehme et al., 2021).

c.2. Proximity sensors

Equipped with a range of sensors including inductive, ultrasonic, infrared, capacitive, and Hall effect technologies, these autonomous devises prevent collisions with obstacles. This safeguards against operational disruptions and costly repairs, ensuring uninterrupted agricultural operations (Xie et al., 2023; Yang et al., 2023).

c.3. Position and inertia sensors

GPS, IMU, magnetometers, gyroscopes, and accelerometers are essential components for precise navigation and stability of agricultural robots. These sensors enable real-time tracking of robot's position, speed, and orientation, ensuring optimal performance even in challenging field conditions (Bai et al., 2023; Ghobadpour et al., 2022).

c.4. Other important sensors (temperature, pressure, pH, stress, soil profile, etc.)

Sensors for temperature, pressure, pH, stress levels, and soil profiles provide critical data for optimizing crop management. These measurements offer farmers real-time insights into crop health, enabling timely interventions to address issues related to weather, soil conditions, and plant stress. This data-driven approach contributes to improved agricultural practices and increased yields (Ashwini et al., 2024; Xie et al., 2022).

d. Data processing systems, artificial intelligence, and machine learning

The central processing unit (CPU) of an agricultural robot accumulates and processes the raw data collected by the sensors. Modern systems are now integrating advanced artificial intelligence and machine learning algorithms, enabling the efficient analysis and interpretation of large volumes of agricultural data. This integration results in detailed reports on the state of the crops, providing valuable insights for effective decision making (Balaska et al., 2023; Fountas et al., 2022).

Artificial intelligence optimizes agricultural operations by providing real-time insights into crop conditions, such as irrigation needs, pest infestations, nutrient deficiencies, and stress levels (Fountas et al., 2022). This predictive capability minimizes idle time by ensuring timely interventions. Machine learning empowers these robotic systems to continuously adapt to changing field conditions, optimizing inspection route and task execution. This advanced technology offers intelligent, sustainable, and cost-effective solutions for precision agriculture, enabling early intervention and maximizing crop yield while preserving the environment (Fountas et al., 2022; Shaikh et al., 2022). The impact of artificial intelligence and machine learning on agriculture is significant and holds immense promise for the industry. These technologies enable the autonomous optimization of plant protection treatments, significantly reducing human errors and their potential negative effects on food, health, and the ecosystem (Rai et al., 2023; Fountas et al., 2022).

e. Communication systems

Robust and secure communication systems are essential for the effective operation of agricultural robots, facilitating seamless interaction between machine components and human operators. Farmers can access comprehensive field data through user friendly interfaces, including mobile applications, touchscreens, enabling informed decision-making and remote control. These interfaces rely on a combination of Wi-Fi, 4G/5G, and/or Bluetooth connectivity to ensure reliable communication (Rai et al., 2023; Paul et al., 2022).

5. Agricultural Drones and Their Uses: Examples of Advanced Robotic Systems

Agricultural drones, also known Unmanned Aerial Vehicles (UAVs) or Agrotech, are pilotless aircraft employed in agriculture. While initially introduced in 1980s, their transformative impact on modern farming has become increasingly evident in recent years. By surveying vast agricultural areas, drones collect critical data and enable precision farming technologies like targeted spraying and precise monitoring. This leads to substantial time and resource savings (Hafeez et al., 2023; Zhichkin et al., 2023; Subramanian et al., 2021).

Like other agricultural robotic systems:

- Drones consist of a mechanical structure consisting of a chassis and typically 4 to 8 arms.
- Drones are equipped with an energy system that typically uses lithium-ion polymer (LiPo) batteries.
- Drones are propelled by electric motors and propellers positioned at the ends of their arms, ensuring stability during flight (Gurbanov et al., 2024; Shah et al., 2023; Hafeez et al., 2023).
- Drones employed in environmental information acquisition systems equipped with a versatile sensor suite. Thise includes RGB sensors for visual crop assessment, multispectral and hyperspectral cameras for detailed plant health assessment, and thermal cameras for monitoring soil and crop temperature, as well as detecting pests and diseases. Proximity sensors are an advanced

technology used to detect objects and barriers, helping to prevent air and land mishaps. LiDAR sensors analyze topography and create 3D maps of agricultural regions. GPS and GLONASS position sensors provide accurate localization and navigation for autonomous or pre-programmed missions conducted by farmers. Additionally, the IMU (Internal Measurement Unit) is used for tracking and maintaining the drones' orientation and course (Shah et al., 2023).

- A wireless communication system employs radio frequency, Wi-Fi, and 4G/5G technology to transmit collected telemetry and other data while receiving commands and instructions.
- Leveraging artificial intelligence and machine learning, data analysis systems provide comprehensive insights into crop conditions, enabling both autonomously and farmer-controlled decision-making (Lawrence et al., 2025; Pilli et al., 2023).
- The spreading system comprises tanks and control units, enabling precise application of water, pesticides, and fertilizers.
- In addition to the aforementioned systems, drones require specialized landing structures designed to effectively absorb impacts. These systems also include autopilot software that can autonomously plan aerial missions, follow intelligent and secure paths, and perform specific agricultural tasks without the need for direct human intervention (Zhichkin et al., 2023; Hasan et al., 2020).

6. Accurate Overview of the Precise Spraying Power of Advanced Robotic Systems and Drones

The precise and efficient application of pesticides and fertilizers is crucial for sustainable agriculture. Traditional methods often result in overuse of these chemicals. This is vital for protecting the health of both plants and farmers, as well as promoting environmental sustainability. Recent scientific breakthroughs have led to the development of advanced and highly connected robots, such as drones, which offer a promising solution to meet these demands (Figure 4) (Shah et al., 2023; Basri et al., 2021; Subramanian et al., 2021).

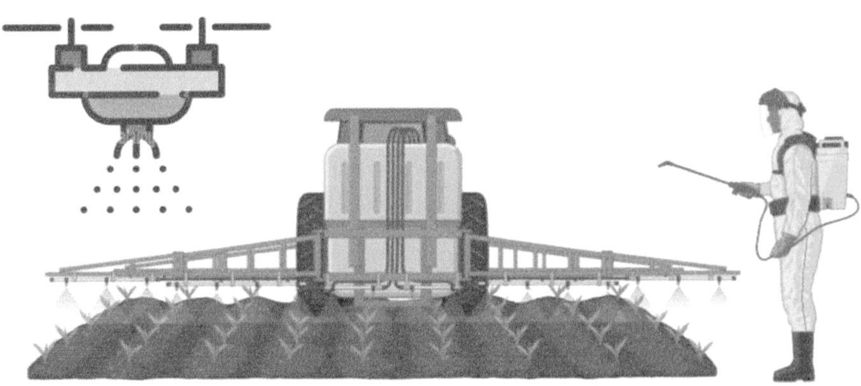

Figure 4. Development of spray technologies.

Integrated spraying in modern agricultural robots precisely deliver agrochemicals and fertilizers to target plant areas, including leaves, stems, and surrounding soil. This approach optimizes crop protection and nutrition while minimizing environmental impact (Shahrooz et al., 2020).

Agricultural robotic systems can autonomously perform spraying missions using artificial intelligence and machine learning algorithms. Alternatively, farmers can pre-plan and review these missions based on detailed summaries provided by their monitoring systems (Yu et al., 2020).

Advanced agricultural robots are equipped with specialized tanks for storing variety of inputs, including organic, mineral, and biological fertilizers, as well an array of plant protection products such as acaricides, aphicides, bactericides, corvicides, fungicides, herbicides, insecticides, molluscicides, nematicides, parasiticides, repellents, redenticides, virucides, and growth regulators. Spraying systems incorporate nozzles to control droplet size, and pumps to deliver various inputs, including fertilizers and biological control products. Application methods can vary, utilizing technologies such as electrostatic, ultrasonic, hydraulic, high-pressure, or low-pressure spraying (Shahrooz et al., 2020; Yu et al., 2020).

The optimal choice of robotic spraying system is influenced by several factors, including terrain, climate, budget, and crop type (cereals, market crops, oilseeds, fruits, vegetables, or industrial crops) (Basri et al., 2021).

Global food security is heavily reliant on effective spraying practices. Given their agility and ability to access challenging terrains, drones have emerged as valuable tool for many countries, revolutionizing large-scale farming operations. Drones offer advantage of precisely targeting high-quality agrochemicals, which helps protect ecological biodiversity (Shah et al., 2023). This contrasts with conventional heavy land machinery, which can cause plant destruction, soil compaction, reduced porosity and aeration, and disruption of biological activity. Additionally, traditional machinery often leads to issues such as over-spraying and inconsistent pesticide application, posing risks to natural resources like underground water through contamination (Mustafi et al., 2021; Shahrooz et al., 2020).

Advanced aerial robots are essential for addressing agricultural challenges due to their sophisticated sensing, control, and navigation capabilities. These robots can accurately identify areas of crops affected by malnutrition, diseases, pests, weeds, or other pathogens. By precisely targeting these areas, the robots can apply appropriate quantities of fertilizers or plant protection products thereby minimizing the risk of contamination to healthy surrounding plants (Lawrence et al., 2025; Shah et al., 2023; Meshram et al., 2022). Strategic drone-based spraying optimizes plant growth density while reducing labor, chemical inputs, and costs. Moreover, drones significantly improve worker safety in agriculture and help mitigate pollution effects. However, these autonomous machines significantly enhance the safety of the agricultural work setting (Das et al., 2025). They reduce the risk of farmers being exposed to hazardous substances used for plant protection, which scientific studies have linked to increased oxidative stress, a higher risk of cancer, and disruption of the endocrine system (Shahrooz et al., 2020; Yu et al., 2020).

7. Examination of the Meticulous Surveillance Capabilities of Sophisticated Robotic Devices and Drones

The use of sophisticated robotic systems in the agriculture industry is diverse due to their ability to operate continuously without fatigue. Currently, agricultural robots including monitoring drones, harvesting robots, welding robots, planting robots, and irrigation robots are all equipped with highly accurate monitoring capabilities (Das et al., 2025). Advanced, sensitive, and precise sensors are integral to optimizing robot performance. These devices provide real-time data, enabling timely notifications to farmers (Bale et al., 2024; Hafeez et al., 2023).

Agricultural robots utilize advanced monitoring technology to assess plant health and identify environmental hazards that could affect fruit quality. By identifying these risks, the robots enable the implementation of corrective and preventive measures to prevent the worsening and spread of adverse conditions. This initiative-taking approach positively impacts ecological balance, human well-being, and agricultural and financial outcomes (Das et al., 2025; Balaska et al., 2023).

a. Surveillance drones

Agricultural drones are capable of conducting routine aerial crop surveys, providing comprehensive data with exceptional speed and accuracy (Zhichkin et al., 2023). Drones equipped with a variety of sensors capture detailed images and videos, which are analyzed to provide farmers with precise data on topography, vegetation density, temperature, light conditions, wind speed, and soil properties including moisture, acidity, salinity, and trace elements (Lawrence et al., 2023).

The advanced monitoring systems in drones are essential for identifying areas that are either over-watered or under-watered, optimizing irrigation and conserving water. These systems are also crucial for detecting plant diseases and their early signs, enabling more effective pest control strategies (Das et al., 2025; Lawrence et al., 2023). By applying plant protection products in a smart, carefully measured manner, the negative impacts can be minimized. This approach ensures both the productivity and long-term viability of agriculture, with a focus on balancing both goals (Das et al., 2025; Cristinel et al., 2022). Agricultural surveillance drones excel at identifying water and salt stress, predicting crop maturity, and assessing damage from adverse weather events. By providing timely insights into crop conditions, they optimize harvest planning and resource allocation (Abbas et al., 2023; Ngoc et al, 2023).

Agricultural monitoring drones facilitate the creation of comprehensive data repositories, allowing farmers to get a deeper understanding of their crops' conditions and requirements. This knowledge enables them to implement timely, efficient, and environmentally friendly management and protection methods (Figure 5) (Abbas et al., 2023; Ngoc et al., 2023; Cristinel et al., 2022).

b. Weeding robots

Equipped with advanced vision sensors, agricultural robots can accurately distinguish between crops and weeds (Kumar et al., 2024). This capability allows for precise

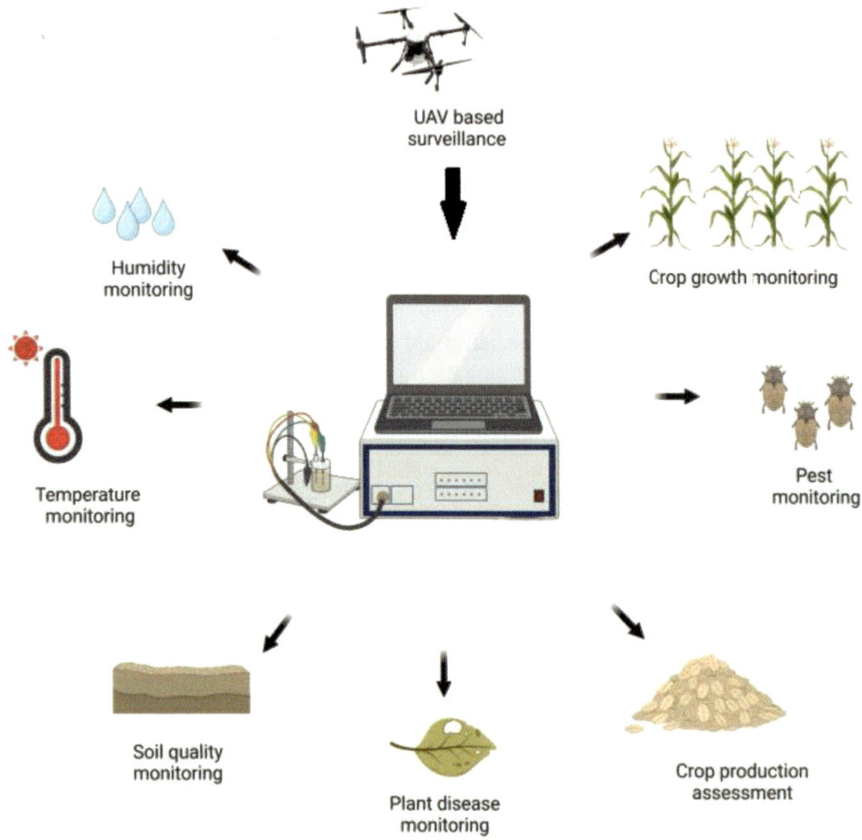

Figure 5. Information collected by drone monitoring sensors (Rajak et al., 2023).

weed removal without damaging surrounding plants (Sow et *al*., 2024). By mitigating the proliferation of weeds, these systems significantly reduce the reliance on harmful herbicides, promoting sustainable and cost-effective agricultural practices (Zhang et al., 2022).

c. Robots for planting seeds

Advanced monitoring system guarantee precise and consistent seed placement, optimizing planting operations. Despite rapid and repetitive tasks, these robots maintain millimetric accuracy in selecting optimal planting locations, enhancing efficiency, productivity, and labor savings (Liu et al., 2024).

d. Crop harvesting robots

Equipped with advanced sensors, these robots accurately determine fruit and vegetable ripeness, enabling optimal harvest timing and preserving nutritional quality (Mail et al., 2023).

e. Crop irrigation robots

Advanced monitoring sensors on these robots assess crop water status and calculate precise irrigation requirements for each plot. This data-driven approach optimizes water distribution, minimizing waste (Tomaszewski and Kołakowski, 2023 ; Nitnaware and Khandait, 2022).

8. Advantages, Disadvantages and Limitations of Robotic Systems and Drones in Advanced Spraying and Monitoring

Robotic and drone technologies have revolutionized agriculture by enhancing efficiency and sustainability. These advanced systems address numerous challenges, providing precise spraying, monitoring and data analysis capabilities (Yerebakan and Hu, 2024). Compared to traditional methods, they offer significant advantages in terms of resource optimization and environmental impact (Tomaszewski and Kołakowski, 2023; Basri et al., 2021). By automating tasks, robots and drones reduce human error, fatigue, and exposure to hazardous conditions. These technologies excel in handling large-scale, repetitive, and physically demanding agricultural operations (Hafeez et al., 2023).

Agricultural drones conduct routine aerial surveillance, capturing comprehensive data on crop health, including vegetation density, nutrient status, irrigation needs, signs of stress (Yerebakan and Hu, 2024). This information empowers farmers to make informed decisions regarding irrigation, and pest management, ultimately optimizing crop yields and minimizing resource inputs (Abbas et al., 2023; Cheng et al., 2023; Shahrooz et al., 2020).

Agricultural spraying drones significantly reduce the environmental damage caused by agrochemicals by enhancing the balance between agriculture and ecology. Their advanced technology targets stressed or diseased plants and areas infested with harmful organisms, applying precise and appropriate doses of fertilizers and plant protection treatments (Tomaszewski and Kołakowski, 2023 ; Ngoc et al., 2023). This approach significantly reduces waste from unnecessary and excessive spraying, as well as the negative impact on healthy plants, food production, wildlife, and natural flora (Lawrence et al., 2025; Meshram et al., 2022). The use of drones for spraying and advanced monitoring greatly improves crop quality and profitability, leading to a significant decrease in production costs (Yerebakan & Hu, 2024). Additionally, it enhances farmer welfare by shifting their focus to less physically demanding and more strategic tasks (Balaska et al., 2023).

Despite their potential, agricultural robots and drones face several limitations (Figure 6) (Zheng et al., 2024). Battery life, spare parts availability, limited technical support in rural areas hinder their widespread adoption. Additionally, challenges related to sensor accuracy, data transmission, and payload capacity require ongoing technological advancements. Drone operators must comply with legal regulations regarding the permitted range and altitude of flight, as well as confidentiality and ethical considerations related to surveillance recordings. This is particularly

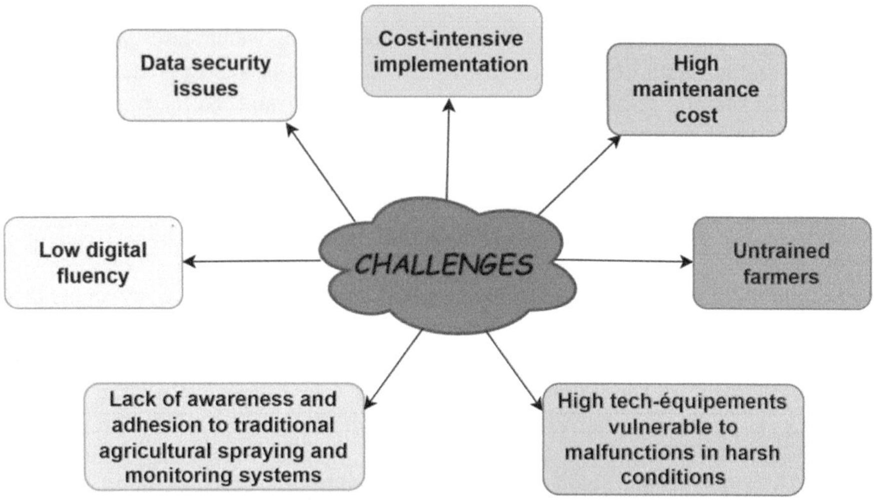

Figure 6. Challenges that hinder the use of robotic systems and drones for surveillance and spraying.

important when operating near local communities. These requirements are outlined in the works of Cheng et al. (2023), Hafeez et al. (2023), Mustafi et al. (2021), and Shahrooz et al. (2020). A major obstacle to the adoption of advanced robotic systems and agricultural drones is the resistance of some farmers to change their traditional methods (Bazargani and deemyad, 2024). This reluctance is primarily due to the high initial costs and ongoing expenses of integrating these intelligent technologies into agriculture, as well as the need to establish the necessary infrastructure for their operation (Ashwini et al., 2024; Cheng et al., 2023; Oliveira et al., 2021).

Implementing these technologies requires significant investment in training and development. Farmers need comprehensive instruction on selecting appropriate robotic systems, operating drones, and ensuring safe and effective use to maximize benefits and minimize risks (Bazargani and deemyad, 2024). Additionally, attention must be given to the maintenance and troubleshooting of software issues to ensure the longevity of the equipment, minimize work interruptions, and reduce maintenance costs in the event of mechanical and technical malfunctions. Lastly, it is crucial to consider the impact of adverse weather conditions on these emerging technologies (Fountas et al., 2022; Ghobadpour et al., 2022; Oliveira et al., 2021).

One of the key challenges hindering the adoption of advanced robotic monitoring and spraying systems is the need to raise awareness among farmers about the benefits of drones and agricultural robots. Demonstrating their concrete positive impacts on agricultural yield and quality is essential (Tomaszewski and Kołakowski, 2023; Ngoc et al., 2023). In the long term, these innovative technologies can significantly reduce production costs, particularly those related to chemicals, fuels, and labor, while also saving considerable time and effort (Cheng et al., 2023; Ghobadpour et al., 2022; Oliveira et al., 2021).

9. Conclusion

The influence of robotic systems and spraying and monitoring drones on plant protection is indisputable. These technologics enhance the quantity and quality of crop harvests in a more cost-effective and ecologically friendly manner. Due to these technical advancements, the monitoring process becomes highly efficient and precise in identifying threats to plant health. This enables rapid and focused intervention in the applications of agrochemicals, leading to a significant savings in their use.

While offering significant benefits, the integration of spraying and monitoring robots and drones into agricultural practices requires substantial investment and careful planning. Factors such as initial equipment costs, ongoing maintenance, operator training, and technological advancements must be considered for successful implementation.

Overcoming the challenges of robotic integration requires sustained farmer education on benefits of these technologies. Concurrently, research efforts must focus on developing more advanced, cost-effective, durable robots and drones equipped with sensitive sensors, high-capacity batteries, and efficient spraying systems. To maximize the effectiveness of these systems, they must incorporate advanced spraying capabilities, including the potential for biopesticide delivery. Robust wireless communication, capable of withstanding challenging weather conditions, is essential for reliable data transmission and control.

While robotic systems and drones offer significant advantages, they are complementary to human labor rather than replacements. Farmers remain essential for overseeing operations, making informed decisions based on data provided by these technologies, and ensuring the appropriate application of treatments.

References

Abbas, A., Zhang, Z., Zheng, H., Alami, M. M., Alrefaei, A. F., Abbas, Q. et al. (2023). Drones in plant disease assessment, efficient monitoring, and detection: a way forward to smart agriculture. Agronomy, 13(6): 1524.

Ali, A., Hussain, T., Tantashutikun, N., Hussain, N., and Cocetta, G. (2023). Application of smart techniques, internet of things and data mining for resource use efficient and sustainable crop production. Agriculture, 13(2): 397.

An, Z., Wang, C., Raj, B., Eswaran, S., Raffik, R., Debnath, S., and Rahin, S. A. (2022). Application of new technology of intelligent robot plant protection in ecological agriculture. Journal of Food Quality, 2022(1): 1257015.

Ashwini, T. R., Potdar, M. P., Sivarajan, S., and Odabas, M. S. (2024). Transforming Agriculture with Smart Farming: A Comprehensive Review of Agriculture Robots for Research Applications. Intelligent Robots and Drones for Precision Agriculture, 45–61.

Bai, Y., Zhang, B., Xu, N., Zhou, J., Shi, J., and Diao, Z. (2023). Vision-based navigation and guidance for agricultural autonomous vehicles and robots: A review. Computers and Electronics in Agriculture, 205: 107584.

Balaska, V., Adamidou, Z., Vryzas, Z., and Gasteratos, A. (2023). Sustainable crop protection via robotics and artificial intelligence solutions. Machines, 11(8): 774.

Bale, A. S., Varsha, S. N., Naidu, A. S., Vinay, N., and Tiwari, S. (2024). Autonomous Aerial Robots Application for Crop Survey and Mapping. In Precision Agriculture for Sustainability (pp. 123–145). Apple Academic Press.

Barua, R., Bhowmik, S., Banerjee, A., and Banerjee, D. (2024). The Emerging Advancement of Robotics in the Modern Agriculture Field. In Agriculture and Aquaculture Applications of Biosensors and Bioelectronics (pp. 253–265). IGI Global.

Basri, R., Islam, F., Shorif, S. B., and Uddin, M. S. (2021). Robots and drones in agriculture—a survey. Computer vision and machine learning in agriculture, 9–29.

Bazargani, K., and Deemyad, T. (2024). Automation's impact on agriculture: opportunities, challenges, and economic effects. Robotics, 13(2): 33.

Bilal, M., Iqbal, H. M. N., and Barceló, D. (2019). Persistence of pesticides-based contaminants in the environment and their effective degradation using laccase-assisted biocatalytic systems. Science of The Total Environment, 133896.

Cheng, C., Fu, J., Su, H., and Ren, L. (2023). Recent advancements in agriculture robots: Benefits and challenges. Machines, 11(1): 48.

Cristinel, F., Daniela, B. L., Vitali, Z., and Silvius, S. (2022). The Future of Agriculture Drone Tehnology Trends and Prospects. Economic and Social Development: Book of Proceedings, 68–77.

Daniel Lawrence, I., Pavitra, A. R. R., Karu, R., and Saravanan, M. P. (2025). An Advanced Application of UAV–Drone Technologies in Precision Agriculture for Seed Dropping, Fertilizers and Pesticides Spraying and Field Monitoring. Computer Vision in Smart Agriculture and Crop Management, 139–157.

Das, N., Kumar, A., and Kumar, R. (2025). Drone Application and Use Cases in Smart Agriculture and Crop Surveillance: Future Research Directions. Computer Vision in Smart Agriculture and Crop Management, 283–302.

Deguine, J. P., Aubertot, J. N., Bellon, S., Côte, F., Lauri, P. E., Lescourret, F. et al. (2023). Agroecological crop protection for sustainable agriculture. Advances in Agronomy, 178: 1–59.

Faizan, M., Singh, A., Eren, A., Sultan, H., Sharma, M., Djalovic, I. et al. (2024). Small molecule, big impacts: Nano-nutrients for sustainable agriculture and food security. Journal of Plant Physiology, 154305.

Fan, X., Chai, X., Zhou, J., and Sun, T. (2023). Deep learning based weed detection and target spraying robot system at seedling stage of cotton field. Computers and Electronics in Agriculture, 214: 108317.

Fountas, S., Malounas, I., Athanasakos, L., Avgoustakis, I., and Espejo-Garcia, B. (2022). AI-assisted vision for agricultural robots. AgriEngineering, 4(3): 674–694.

Fountas, S., Mylonas, N., Malounas, I., Rodias, E., Hellmann Santos, C., and Pekkeriet, E. (2020). Agricultural robotics for field operations. Sensors, 20(9): 2672.

Ghobadpour, A., Monsalve, G., Cardenas, A., and Mousazadeh, H. (2022). Off-road electric vehicles and autonomous robots in agricultural sector: trends, challenges, and opportunities. Vehicles, 4(3): 843–864.

Gurbanov, P., Shichiyakh, R., Kumar, K. V., Korrai, S., Chandra, S., and Lydia, E. L. (2024). Smart agriculture farming using drone automation technology. In Proceedings of Eighth International Conference on Information System Design and Intelligent Applications (pp. 293–304). Singapore: Springer Nature Singapore.

Hafeez, A., Husain, M. A., Singh, S. P., Chauhan, A., Khan, M. T., Kumar, N. et al. (2023). Implementation of drone technology for farm monitoring & pesticide spraying: A review. Information processing in Agriculture, 10(2): 192–203.

Hasan, K. M., Hasan, M. T., Newaz, S. S., and Ahsan, M. S. (2020, June). Design and development of an autonomous pesticides spraying agricultural drone. In 2020 IEEE Region 10 Symposium (TENSYMP) (pp. 811–814). IEEE.

He, X., Yang, F., and Qiu, B. (2024). Agricultural Environment and Intelligent Plant Protection Equipment. Agronomy, 14(5): 937.

Indu, Baghel, A. S., Bhardwaj, A., and Ibrahim, W. (2022). Optimization of pesticides spray on crops in agriculture using machine learning. Computational Intelligence and Neuroscience, 2022(1): 9408535.

Jeger, M., Beresford, R., Bock, C., Brown, N., Fox, A., Newton, A. et al. (2021). Global challenges facing plant pathology: multidisciplinary approaches to meet the food security and environmental challenges in the mid-twenty-first century. CABI Agriculture and Bioscience, 2(1): 1–18.

Jin, T., and Han, X. (2024). Robotic arms in precision agriculture: A comprehensive review of the technologies, applications, challenges, and future prospects. Computers and Electronics in Agriculture, 221: 108938.

Kaliannan, V., and Al Saidi, F. K. S. (2024). Intelligent computing with drones and robotics for precision agriculture. In Intelligent Robots and Drones for Precision Agriculture (pp. 1–17). Cham: Springer Nature Switzerland.

Kumar, S., Kumari, S., Rana, S. S., Rana, R. S., Anwar, T., Qureshi, H. et al. (2024). Weed management challenges in modern agriculture: The role of environmental factors and fertilization strategies. Crop Protection, 185: 106903.

Lawrence, I. D., Agnishwar, J., and Vijayakumar, R. (2023). Drone technology in agriculture for surveillance and inspection. European Chemical Bulletin, 12: 1253–1263.

Liu, X., Huang, H., Chen, M., Fang, Y., and Shu, Y. (2024). Optimization and performance analysis of a novel automatic planting-irrigating integrated robot. Proceedings of the Institution of Mechanical Engineers, Part C: Journal of Mechanical Engineering Science, 09544062241278191.

Lykogianni, M., Bempelou, E., Karamaouna, F., and Aliferis, K. A. (2021). Do pesticides promote or hinder sustainability in agriculture? The challenge of sustainable use of pesticides in modern agriculture. Science of the Total Environment, 795: 148625.

Meshram, A. T., Vanalkar, A. V., Kalambe, K. B., and Badar, A. M. (2022). Pesticide spraying robot for precision agriculture: A categorical literature review and future trends. Journal of Field Robotics, 39(2): 153–171.

Misra, S., and Ghosh, A. (2024). Agriculture paradigm shift: a journey from traditional to modern agriculture. In Biodiversity and Bioeconomy (pp. 113–141). Elsevier.

Mitra, D. (2021). Emerging plant diseases: research status and challenges. Emerging trends in plant pathology, 1–17.

Moshayedi, A. J., Khan, A. S., Yang, Y., Hu, J., and Kolahdooz, A. (2024). Robots in Agriculture: Revolutionizing Farming Practices. EAI Endorsed Transactions on AI and Robotics, 3.

Mustafi, S., Ghosh, P., Roy, K., Dan, S., Mukherjee, K., and Nath Mandal, S. (2021). Drones for intelligent agricultural management. In IoT-based Intelligent Modelling for Environmental and Ecological Engineering: IoT Next Generation EcoAgro Systems (pp. 81–100). Cham: Springer International Publishing.

Mail, M. F., Maja, J. M., Marshall, M., Cutulle, M., Miller, G., and Barnes, E. (2023). Agricultural harvesting robot concept design and system components: A review. AgriEngineering, 5(2): 777–800.

Nehme, H., Aubry, C., Solatges, T., Savatier, X., Rossi, R., and Boutteau, R. (2021). Lidar-based structure tracking for agricultural robots: Application to autonomous navigation in vineyards. Journal of Intelligent & Robotic Systems, 103(4): 61.

Ngoc, T. T. H., Khanh, P. T., and Pramanik, S. (2023). Smart agriculture using a soil monitoring system. In Handbook of Research on AI-Equipped IoT Applications in High-Tech Agriculture (pp. 200–220). IGI Global.

Nitnaware, P. A., and Khandait, M. A. (2022). Smart Soil Moisture Sensing & Automatic Water Irrigation Robot.

Oberti, R., and Schmilovitch, Z. E. (2021). Robotic spraying for precision crop protection. Innovation in agricultural robotics for precision agriculture: A roadmap for integrating robots in precision agriculture, 117–150.

Oliveira, L. F., Moreira, A. P., and Silva, M. F. (2021). Advances in agriculture robotics: A state-of-the-art review and challenges ahead. Robotics, 10(2): 52.

Papadopoulos, G., Arduini, S., Uyar, H., Psiroukis, V., Kasimati, A., and Fountas, S. (2024). Economic and Environmental Benefits of Digital Agricultural Technologies in Crop Production: A review. Smart Agricultural Technology, 100441.

Patel, S. K., Sharma, A., and Singh, G. S. (2020). Traditional agricultural practices in India: an approach for environmental sustainability and food security. Energy, Ecology and Environment, 5(4): 253–271.

Paul, K., Chatterjee, S. S., Pai, P., Varshney, A., Juikar, S., Prasad, V. et al. (2022). Viable smart sensors and their application in data driven agriculture. Computers and Electronics in Agriculture, 198: 107096.

Pilli, A., Pabbathi, B., Gokarakonda, Y., Pramod, T., Sami, M., Nori, V. et al. (2023, July). Structural analysis of agricultural drone. In AIP Conference Proceedings (Vol. 2788, No. 1). AIP Publishing.

Rai, A. K., Kumar, N., Katiyar, D., Singh, O., Sreekumar, G., and Verma, P. (2023). Unlocking productivity potential: the promising role of agricultural robots in enhancing farming efficiency. International Journal of Plant & Soil Science, 35(18): 624–633.

Rajak, P., Ganguly, A., Adhikary, S., and Bhattacharya, S. (2023). Internet of Things and smart sensors in agriculture: Scopes and challenges. Journal of Agriculture and Food Research, 14: 100776.

Reddy, C. A., Oraon, S., Bharti, S. D., Yadav, A. K., and Hazarika, S. (2024). Advancing disease management in agriculture: A review of plant pathology techniques. Plant Science Archives.

Shah, S. A., Lakho, G. M., Keerio, H. A., Sattar, M. N., Hussain, G., Mehdi, M. et al. (2023). Application of drone surveillance for advance agriculture monitoring by Android application using convolution neural network. Agronomy, 13(7): 1764.

Shahrooz, M., Talaeizadeh, A., and Alasty, A. (2020, November). Agricultural spraying drones: Advantages and disadvantages. In 2020 Virtual symposium in plant omics sciences (OMICAS) (pp. 1–5). IEEE.

Shaikh, T. A., Mir, W. A., Rasool, T., and Sofi, S. (2022). Machine learning for smart agriculture and precision farming: towards making the fields talk. Archives of Computational Methods in Engineering, 29(7): 4557–4597.

Song, B., Seiber, J. N., Duke, S. O., and Li, Q. X. (2020). Green plant protection innovation: challenges and perspectives. Engineering (Beijing, China), 6(5): 483.

Sow, S., Ranjan, S., Seleiman, M. F., Alkharabsheh, H. M., Kumar, M., Kumar, N. et al. (2024). Artificial Intelligence for Maximizing Agricultural Input Use Efficiency: Exploring Nutrient, Water and Weed Management Strategies. Phyton (0031-9457), 93(7).

Sparrow, R., and Howard, M. (2021). Robots in agriculture: prospects, impacts, ethics, and policy. Precision Agriculture, 22: 818–833.

Subramanian, K. S., Pazhanivelan, S., Srinivasan, G., Santhi, R., and Sathiah, N. (2021). Drones in insect pest management. Frontiers in Agronomy, 3: 640885.

Tomaszewski, L., and Kołakowski, R. (2023). Mobile services for smart agriculture and forestry, biodiversity monitoring, and water management: Challenges for 5G/6G networks. In Telecom MDPI 4(1): 67–99, MDPI.

Upadhyay, A., Zhang, Y., Koparan, C., Rai, N., Howatt, K., Bajwa, S. et al. (2024). Advances in ground robotic technologies for site-specific weed management in precision agriculture: A review. Computers and Electronics in Agriculture, 225: 109363.

Wei, W., Xiao, M., Duan, W., Wang, H., Zhu, Y., Zhai, C., and Geng, G. (2024). Research Progress on Autonomous Operation Technology for Agricultural Equipment in Large Fields. Agriculture, 14(9): 1473.

Xie, D., Chen, L., Liu, L., Chen, L., and Wang, H. (2022). Actuators and sensors for application in agricultural robots: A review. Machines, 10(10): 913.

Yang, Q., Du, X., Wang, Z., Meng, Z., Ma, Z., and Zhang, Q. (2023). A review of core agricultural robot technologies for crop productions. Computers and Electronics in Agriculture, 206: 107701.

Yerebakan, M. O., and Hu, B. (2024). Human–robot collaboration in modern agriculture: A review of the current research landscape. Advanced Intelligent Systems, 2300823.

Yu, S. H., Kim, Y. K., Jun, H. J., Choi, I. S., Woo, J. K., Kim, Y. H. et al. (2020). Evaluation of spray characteristics of pesticide injection system in agricultural drones. Journal of Biosystems Engineering, 45: 272–280.

Zhang, W., Miao, Z., Li, N., He, C., and Sun, T. (2022). Review of current robotic approaches for precision weed management. Current Robotics Reports, 3(3): 139–151.

Zheng, L. I. U., Chengqian, J. I. N., Yugang, F. E. N. G., and Tengxiang, Y. A. N. G. (2024). Research status and trends of intelligent technology in plant protection machinery. Journal of Intelligent Agricultural Mechanization, 5(1): 40–50.

Zhichkin, K., Nosov, V., Zhichkina, L., Anichkina, O., Borodina, I., and Beketov, A. (2023). Efficiency of using drones in agricultural production. In E3S Web of Conferences (Vol. 381, p. 01048). EDP Sciences.

Chapter 4

Artificial Intelligence and Sensing Techniques for the Management of Insect Pest and Diseases in Cotton

Muhammad Ashfaq,[1] *Iqra Mushtaq,*[1] *Samah Bashir Kayani,*[2] *Muhammad Mahmood Ahmed,*[3] *Muhammad Shaaban,*[4] *Nimra Zubair Khan*[1] and *Mirza Abid Mehmood*[1,*]

1. Introduction

Cotton (*Gossypium hirsutum* L.) is a crop that is of significant economic importance. Cotton is a major natural textile fiber source and an important oil crop cultivated around the world (Zhang et al., 2017). Almost, 49 species of cotton are grown in subtropical and tropical regions of the world. The cotton industry is a large and profitable sector that produces billions of dollars, including the entire process from raw fiber to end textile products (Chopda et al., 2018). As of 2017, cotton was planted in over 80 countries, covering an area of 32.4 million hectares (Chopda et al., 2018).

Diseases and insect pests can cause significant economic losses in cotton production. If they are not controlled in time, they can lead to infestations, reducing the quality and yield crop (El-Wakeil and Abdallah, 2012). In Brazil, for instance, insect pests cause annual losses in agricultural production, which is equivalent to around 17.7 billion US dollars an average of 7.7% (Oliveira et al., 2014). Multiple pathogenic and entomological issues are the reasons for economic losses worldwide (Anees and

[1] Plant Pathology, Institute of Plant Protection, Muhammad Nawaz Shareef University of Agriculture, Multan, Punjab, Pakistan.
[2] Govt. Graduate College (W) Mumtazabad, Multan, Punjab, Pakistan.
[3] Department of Bioinformatics, Institute of Biochemistry, Biotechnology and Bioinformatics, The Islamia University of Bahawalpur, Pakistan.
[4] College of Agriculture, Henan University of Science and Technology, Luoyang, China.
* Corresponding author: abid.mehmood@mnsuam.edu.pk

DOI: 10.1201/9781003546702-4

Shad, 2020; Chohan et al., 2020a; Karar et al., 2020). Different agrochemicals are used to manage insect pests and infectious diseases in cotton, but they eventually increase production costs and affect the environment significantly (Gill and Garg, 2014). Integrated pest management (IPM) is implemented to address these issues. Its aims to reduce the impact of pesticide application on the environment globally and reduces the risks on health for living organisms (Simberloff and Rejmánek, 2011). IPM has two important aspects—first one is monitoring of insect pests and second one is disease prevention which can be helped by using AI and different sensing techniques (Toscano-Miranda et al., 2022). Numerous industrial and medical sectors have undergone revolutionary changes due to technological advancements in object recognition and computer vision (Andreopoulos and Tsotsos, 2013). Additionally, the development of precision agriculture has been boosted by computer vision in identifying various diseases and insect pests (Habib et al., 2020).

Several expert systems have examined insects, pests, and diseases, which has helped farmers make better management and prevention decisions (Abu-Nasser and Abu-Naser, 2018; AlZamily, 2018; Boissard et al., 2008). The main objectives of smart agriculture include increasing productivity, enhancing adaptation and stability to climate change, reducing the emission of greenhouse gases, and improving food security. Smart Agriculture (SA) is a modern farming technique that addresses the agricultural system using computer grids, context-aware computing, and sensor networks. Farmers can make practical decisions by finding and evaluating the data from the agricultural site (García et al., 2020; Toscano Miranda, 2024; Toscano-Miranda et al., 2022). Numerous technologies are used in the deployment of SA, including data analytics, robotic automation, and remote sensing (O'Grady et al., 2019). Additional applications of artificial intelligence (AI), include the management of weeds, yield prediction, disease management, pest control, soil and irrigation management, crop management, and monitoring agricultural products and storage (Vijayakumar et al., 2022). In this chapter, specifically, the focus will be on managing diseases and pests by AI.

Patrício and Rieder (2018) conducted a literature review (SLR) on grain crops by using the application of AI and computer vision in precision agriculture, including wheat, maize, barley, soybean, and rice. Hisstudy mainly focused on disease detection, grain quality, and phenotypic characterization, and found no SLRs that specifically addressed the use of these technologies in the pest management of cotton. A scientist conducted a research on SLR for monitoring the different plant diseases and pests on the basis of remote-sensing technology, with a particular emphasis on sensing technologies and feature extraction. Kandalkar et al., 2013 conducted a brief literature review on how image processing has been utilized on different crops, except cotton, to detect and monitor insect pests. Meanwhile, in the field of agriculture, he performed a comprehensive literature survey on AI with a limited focus on cotton cultivation (Shaikh et al., 2022). Iqbal et al., 2018, performed a systematic literature review (SLR), using another image processing technique for the classification of citrus diseases, and robotic detection. It is necessary to read the articles that have been used in AI techniques which are used to manage the cotton insect pests to understand the current status of this area and identify future research directions and continuous advancement in the field.

2. Major Insect Pests of Cotton

Insects are considered pests when their damage reduces the yield of agricultural products (Dent and Binks, 2020). Following insect pests are common, cause significant damage and have been well studied(Presley, 1954; Toscano-Miranda et al., 2022).

2.1 Boll weevil

It is the most important pest of cotton that affects the production of cotton worldwide (Ben Guerrero et al., 2020; Coelho et al., 2016; Grigolli et al., 2017). Boll weevil adults can feed on different plant parts like leaf petioles, terminal growth, and fruiting.

2.2 Whitefly

Whiteflies damages the cotton crop and their alternative host plants, like rose, paprika, soybean, and tomato (Barbedo, 2014; El-Wakeil and Abdallah, 2012; Martin and Moisan, 2008; Xia et al., 2012). By feeding the sap of plants, whitefly can reduce their performance, or even kill it. Furthermore, viruses can also transmitted by whiteflies (El-Wakeil and Abdallah, 2012) in cotton crop.

2.3 Cotton Aphid

Aphids are sucking insects and phloem feeders who attack on cotton plants by both direct and indirect methods. They deplete essential nutrients required for plant development. Additionally, during the feeding process, aphids inject saliva into various parts of the cotton plant, which contains substances highly toxic to the plants (Chohan et al., 2020a).

2.4 Thrips

Host plants (cotton, tomato, avocado, broccoli, and lettuce) are infected by thrips (Shahzadi et al., 2016; Solis-Sánchez et al., 2011; Xia et al., 2015). Due to their attack, plants become stunted and this can have an impact on the performance of the crops (El-Wakeil and Abdallah, 2012). Mostly, at the seedling stage, thrips can infest the cotton crop.

2.5 Bollworms

American bollworms (*Helicoverpa armigera*) and Pink bollworms (*Pectinophora gossypiella*) are the types of bollworms that can infect cotton, tomato, and okra. Bollworms may cause severe damage to plants. At fruiting stage, larvae can feed on the cotton bolls and causes the huge loss to the cotton crop (El-Wakeil & Abdallah, 2012).

2.6 Cotton Mealybug

Mealybug has a wide host range, from woody plants to herbaceous weeds. By sucking the cell sap of phloem tissues, it can attack host plants and secrete honeydew which makes sooty molds on the leaf surface of cotton plants. By sucking the cell sap of phloem tissues, it can attack host plants and secrete honeydew secretions which causes the appearances of sooty molds on the leaf surface, affect the photosynthetic process of plant and eventually cause death of plant tissues. The main symptoms of cotton mealy bug infestations are presence of white powdery substances on curled leaves and shoots, bushy branches, less number of bolls, unopened flower, stunting, deformation and eventually death of cotton plants (Noureen et al., 2016).

3. Major Diseases of Cotton

Due to bacteria, fungi, and nematodes, many diseases occur in cotton crop (Navinkumar et al., 2021; Presley, 1954; Toscano-Miranda et al., 2022). Based on their damage, the most important diseases are discussed below.

3.1 Bacterial Blight

It is the most damaging disease that causes defoliation and rotting by blackening the veins of leaves, resulting in a "blighting" appearance. It causes infection at an early stage of cotton's growth and can spread rapidly by water during the irrigation in the field (Cox Jr et al., 2019).

3.2 Fusarium Wilt

This disease can cause necrosis, stunted plant, brown discoloration of the vascular system, wilting of the plant, and eventually death of the cotton plant. It's difficult to control the pathogen that can cause the Fusarium wilt disease. It can spread through soil, plant debris, and seeds, for long-term survival (Cox Jr et al., 2019).

3.3 Verticillium Wilt

It is the second vascular wilt disease that can infect cotton crop. Usually, fungi can develop inside the vascular tissues, but in an advanced stage of infection, they can grow outside the vascular tissues. Initially, when temperature is high, infected young plants show epinasty, yellowing, leaves defoliation and stunted growth. Plant leaves also show a mosaic pattern between veins, eventually dying and defoliating (Chohan et al., 2020a).

3.4 Alternaria Leaf Spot

The foliar disease, known as leaf blight is widespread across the world's cotton-growing regions. Initially, small, circular gray-brown to tan lesions on leaves with purple color margins appear. On the older leaves, spot size ranges from 1 to 10 mm, showing concentric rings. The damaged leaves grow brittle and withered, cracking frequently and resembling shot holes. The disease is more severe on the lower leaves, resulting in early defoliation (Chohan et al., 2020a).

3.5 Root Rot

This fungus attacks on the roots of plants and block the vascular system of cotton plant, which inhibit the water movement towards the upper plant parts. After its attack plant leaves turn brown or yellow, wilt in a few days and cause the plant death (Pammel, 1888; Uppalapati et al., 2010). The onset of symptoms during prolonged vegetative growth might persist throughout the growing season and become more noticeable during flowering and fruit development (Smith et al., 1962).

3.6 Grey Mildew

It is a foliar disease which causes the boll absorption, boll malformation, and lower fiber quality of cotton crop (Xavier et al., 2019). Firstly symptoms appear on the lower side of the leaves and then move towards the upper surface and light green to yellow-green translucent spots appears (Chohan et al., 2020a).

3.7 Cotton Leaf curl Disease

Cotton leaf curl disease shows typical symptoms such as vein swelling, downward or upward leaf curling, stunted growth, leaf enation having cup-shaped formation on the lower side of cotton leaf (Chohan et al., 2020b; Sattar et al., 2013).

4. Pest and Disease Management in Cotton & Artificial Intelligence

Artificial intelligence facilitates decision-making process in pest management, including control and monitoring. AI is being used in pest management in a few ways, including pest identification, and pest spread prediction, pest control (Xia et al., 2015; Yao Qing et al., 2014; Chen et al., 2020; Hudgins et al., 2017; Ji et al., 2020). By using AI, plant disease management and disease recognition is done (Habib et al., 2020; Velásquez et al., 2020). Furthermore, it is used in soil management, irrigation process (Navinkumar et al., 2021; Talaviya et al., 2020) and for weed management (Monteiro et al., 2021; Partel et al., 2019; Sudars et al., 2020).

4.1 Insect Pests & AI

In 2016, Sangari and Saraswady, employed an AI tool Marker-Controlled Watershed Transformation (MWT) for pest-image segmentation, and compared the results with fuzzy c-means (FCM) clustering, to identify whiteflies. The results show that MWT performs better than FCM in convergence rate. They assessed the peak signal-to-noise ratio (PSNR), normalized absolute error (NAE), structural content (SC), average difference (AD), normalized correlation coefficient (NK), and nonlinear assessment to measure the image distortion. Shahzadi et al., 2016 proposed a novel approach through a rule-based framework for identifying whiteflies and other insect pests. This system relied on a different combination of sensors like temperature, leaf-wetness duration, and humidity level. The expert system was trained using three types of inputs: field study observations, domain experts' advice, and research findings. In 2014, a scientist recognized *H. armigera* using various techniques like wavelet transform (DWT) for feature extraction, a back propagation neural network (BPNN) for pest classification, and a saliency map for image segmentation (Kandalkar et al., 2014). To identify important factors that influence *H. armigera* incidence, a scientist employed Shannon's Information Theory (SIT) (Pratheepa et al., 2016). Their findings demonstrated a negative correlation between pest population and crop stage, indicating that *H. armigera* proliferated during the early stages of boll formation and fruiting, followed by a decrease in the duration of the boll-bursting stage. It is important to understand how environmental factors (relative humidity, and the number of rainy days per week) affect insect population change due to obvious seasonal effects. Additionally, their study showed that SIT was more successful than regression analysis at identifying factors from pest surveillance data. A remote sensing technique was used to evaluate the severity of the mealybug (Singh et al., 2016). This approach showed the relationship among severity index and spectral vegetation indices (SVIs) using a multiple linear regression model used for data analysis. Notably, these indices demonstrated a significant correlation between mealybug-infested and healthy cotton crops. Ranjitha et al., 2014 used the Pearson correlation to predict the damage of thrips. They examined the relationship between SVIs and canopy reflectance to predict the impact of thrips on cotton crop. Alves et al., 2020 employed convolutional neural networks (CNNs) to classify thirteen distinct insect pests, including *Aphis gossypii, Helicoverpa armigera,* and *Aphis grandis*. Notably, a modified RestNet34*, increased the accuracy relative to other algorithms like support vector machine with local binary patterns deep residual learning model (SVM-LBP), ResNet34, ResNet50, and AlexNet. Table 1 depicts the merits and demerits of AI based insect pest detection systems.

4.2 Diseases & AI

Previously, in 2018 a scientist introduced an active-contour model (ACM) for detecting *Phyllosticta gossypina* (Zhang et al., 2018). This ACM relied on local information and global gradient to identify the disease within images. Their study revealed that ACM exhibited superior accuracy in segmentation and operated with shorter running times compared to alternative methods such as Chan-Vese, local binary fitting, and geodesic active contour. Even with varying levels of light, shadow, and fuzzy edges, ACM can separate cotton leaves from a complicated background. The results showed ACM is the most suitable approach under normal condition for segmenting of infected leaves. To differentiate different diseases, Otsu's segmentation technique was used. This technique revealed that they achieved high accuracies. For classification of cotton leaf spot, they used the back propagation neural network (BPNN) process. On the basis of shape and color of leaf spot, this research highlighted the morphology of leaf spot as these traits provided crucial visual cues for the disease diagnosis. Multiple feature of the leaf spot on diseased leaf were captured using general descriptors like as object width and length, object count and

Table 1. AI systems for the detection of pests and their merits and demerits.

Proposed system	Crop	Insects	Merits	Demerits	References
Marker-Controlled Watershed Transformation (MWT)	Cotton	White fly	Better convergence rate	Its tendency of over-segmentation and a noisy image with many local minima	(Sun and Wang, 2010)
Convolutional neural networks (CNNs)	Cotton	Boll weevil, Aphid, American bollworm	Superior performance in image processing	Lengthy training time	(Kattenborn et al., 2021)
Ontology-based Expert System.	Cotton	Various pest and diseases	Better convergence rate	Lengthy training time	(Titiya and Shah, 2018)
Recurrent Neural Networks (RNN)	Cotton	Multiple diseases and pests	It is used to predict the diseases and pests by using a long-short-term memory (LSTM) network	RNNs can suffer from the problem of exploding gradients, so there is a problem to train the network effectively	(Xiao et al., 2018)
ZigBee/FFT Waveform	Sugarcane	Borer and Whiteflies	Low cost, Strong anti-interference, Low power	Low-rate data transmissions	(Srivastav et al., 2013)
BP Neural network, web-based Expert system	Cotton	Multiple pests and diseases	The accuracy of the system is about 80–90%.	The system performance is limited by the training sample set	(Li et al , 2011)
Arduino board with ZigBee Device	Many Crops	Snail pest	Simple, Easy to use	Having limited memory and processing power	(Ferro et al., 2014)
ZigBee with Dongle	Red Palm	Palm Weevil	Low-cost, Low-consumption	To operate ZigBee-compliant devices, system expertise is required and is not as secure as wifi-based systems	(Rach et al., 2013)

overall image area. Support Vector Machine (SVM), is another unique approach used as the classification algorithm as it was used by two scientists for the classification of five cotton leaf curl diseases such as *Fusarium* wilt, *Alternaria*, *Cercospora*, bacterial blight and grey mildew (Sarangdhar and Pawar, 2017). This study showed 83.26% accuracy in the overall classification. Prashar et al., 2017 developed an automated method for identifying diseases like *Alternaria*, root rot and bacterial blight in cotton crop. SVM and multiple invariant feature descriptors were used for the classification of disease pattern. For preprocessing of any image, it is necessary to set them in uniform size. Finally, the images were converted into two-dimensional grayscale, and a Gaussian filter was used to reduce the noise. Histogram of Oriented Gradients (HOG) was used as a feature descriptor. Lastly, for classification, SVM was also used as feature descriptor, resulting in an 85% accuracy rate. For the detection of ramularia leaf blight, a scientist used multispectral classifications, employing four different types of classifiers as multinomial logistic regression software with boosting (MLRb), support vector machine (SVM), random forest (RF) and multinomial logistic regression (MLR), (Xavier et al., 2019). The MLR classifier employs a linear predictor function with minimum training data to estimate the classification parameters. Nevertheless, after being updated for nonlinear class predictors, the SVM performed admirably in the classification of multispectral remote sensing images. Moreover, the outcomes showed that MLRb and RF were more effective in overfitting with the training data or required larger datasets. MLRb's susceptibility because of its boosting mechanism, whereas RF's challenges stemmed from its ensemble approach using bootstrap aggregating (bagging). The methods utilized to map cotton root rot were: (i) unsupervised classification by using the normalized difference vegetation index (NDVI), and iterative self-organizing data analysis (ISODAT) for multispectral image classification; (ii) two supervised classification methods, SVM and BPNN were also used. All methods showed similar effectiveness and accuracy in detecting cotton root rot for site-specific disease management (Yang et al., 2015). Particularly, NDVI-based classification was a highlight for its ease of implementation without requiring complex image processing. The findings showed that the following methods were successful in detecting root rot: BPNN (95.5%), NDVI with unsupervised classification (94.5%), supervised classifiers SVM (95%), and ISODATA on multispectral imaging (94%). In order to organize disease related images of cotton divide into three groups, the k-means clustering technique was used. Many features like mean, variance, standard deviation, energy, entropy, correlation and kurtosis have been extracted from cluster. For classification these features were then input into an SVM and BPNN. These classifiers performance comparison shows the results as Alternaria leaf spot was classified with 84.3% by SVM and 77.4% accuracy by BPNN: grey mildew classified with 98.7% by SVM and 87.8% by BPNN, and rust foliar fungal diseases classified with 93.2% by SVM and 90.1% by BPNN. The overall average accuracy was 85.1% for the BPNN classifier and 92.06% for the SVM classifier, indicating that SVM can detect diseases more accurately than BPNN. Some scientists work on utilizing the k-means clustering to detect bacterial blight: as for classification combined k-means with support vector machines (SVMs), and the other is using k-means for disease identification specifically (Toscano Miranda, 2024; Toscano-Miranda et al., 2022). At an early stage of detecting various diseases,

Toscano-Miranda et al., 2022 developed an expert system with fuzzy inference. Their two-step classification technique involved feature vector extraction-based disease classification and grouping of diseases caused by similar agents (viruses, bacteria, and fungi), followed by SVM reclassification.

In 2018, Toseef and Khan suggested novel fuzzy inference technique for crop disease diagnosis. They used it to diagnose 21 cotton diseases. For using fuzzy logic in decision-making, the authors highlighted three main reasons: (i) natural language is used to express the rules, which are based on expert knowledge (ii) fuzzy logic is a reliable method for representing linguistic information, and unlike classical set theory, fuzzy logic deals with the inherent ambiguity and uncertainty in the domain of the problem. To facilitate decision-making, the system incorporates seventy-three rule (Revathi and Hemalatha, 2014) that proposed a novel method like feature extraction for the detection of different diseases such as Fusarium wilt, leaf blight, bacterial blight, verticillium wilt, root rot, and micro-nutrient deficiencies. The extracted features were subsequently classified using a Support Vector Machine (SVM) other than it by using a fuzzy classifier, and Back Propagation Neural Network (BPNN) also achieving accuracy rates of 91%, 94%, and 93%, respectively. The study found that combining both techniques like EPSO with fuzzy classifiers provided the highest accuracy (Chopda et al., 2018) by utilizing variables like soil moisture, and temperature sensors to forecast diseases like grey mildew and anthracnose. They used DT classifier because of its simplicity in classification, which consists of a series of questions about the test data's attributes. The results revealed that this system could effectively predict the occurrence and severity of these diseases using different environmental parameters (soil moisture, temperature) by collecting data from the previous year. Recently, some convolutional neural networks (CNNs) were used to detect cotton diseases. Specifically, GoogleNet and ResNet50 for achieving classification accuracies of 86.6% and 89.2%, respectively (Caldeira et al., 2021). Their results showed that these CNN-based methods performed better than traditional approaches such as SVM, neuro-fuzzy systems ANN, and KNN, in image processing tasks. Similarly, Liang (2021) utilized CNNs, including VGG, DenseNet, ResNet, and S-DenseNet, optimized with a spatial structure optimizer (SSO), which showed higher classification precision, especially with small sample sizes. Table 2 depicts the merits and demerits of AI based disease detection systems

5. Major Precision Approaches for Monitoring Pests

In recent years, a significant challenge for pest management has been assisting farmers with robotic AI sensor-based techniques and the Internet of Things (IoT). The absence of these technologies forces farmers to rely on their scouting efforts and seek assistance from crop consultants and extension specialists to mitigate the losses caused by insect pests. A high crop loss rate of approximately 40% is attributed to pest attacks. This is due to scouts′ inability to detect insect pest threshold levels in both productive fields and nearby crops (Ahmad et al., 2020; Alves et al., 2020; Gholve et al., 2012; Thenmozhi and Reddy, 2019). The global cotton ecosystem hosts approximately 1,000 pest species (Alves et al., 2020), with nearly 125 species in each country (Gholve et al., 2012). On a large area of the field pest, scouting without sensors

Table 2. AI systems for the detection of diseases and their merits and demerits.

Proposed System	Crop	Diseases	Merits	Demerits	Reference
IoT and DIP-based Framework	Multiple crops	Multiple Diseases	Within this system, sensors continuously monitor environmental parameters and digital cameras capture images of the plant leaves	Security issues, Privacy concerns, System complexity, Significant risk of system corruption	(Thorat et al., 2017)
Support Vector Machine (SVM)	Cotton	Bacterial blight	SVM classifiers gave more accurate disease detection than BPNN	Not suitable for large data sets	(Jakkula, 2006)
	Cotton	Alternaria leaf spot	More accurate disease detection process	Not perform very well when the data set has more noise	(Jakkula, 2006)
	Cotton	Grey mildew	Provide high quality results	When the data set has more noise like targeted classes are overlapped, SVM performs poorly	(Jakkula, 2006)
	Cotton	Cercospora	Provide accurate detection	Targeted classes get overlapped	(Jakkula, 2006)
	Cotton	Fusarium wilt	Highly effective in high-dimensional spaces and SVMs are relatively memory-efficient	For a large data set, SVM is not suitable	(Jakkula, 2006)
Iterative self-organizing data analysis (ISODATA)	Cotton	Cotton root rot	Effective and accurate	This reflects upon a general disadvantage of K-mean and the system is not best for remote-sensing images	(Hoque et al., 2019)
Back-propagation neural network (BPNN)	Cotton	Bacterial leaf blight	Fast, simple, and easy to program	It is quite sensitive to noisy data and irregularities	(Wang et al., 2015)
Fuzzy inference system	Cotton	Cotton leaf curl	Identify the diseases at an early stage, efficient machine control, and cost effective.	It is sensitive to irregularities	(Sabri et al., 2013)
Decision-tree (DT)	Cotton	Anthracnose and grey mildew	Based on data from the previous year, the system was able to predict the disease with parameters including soil moisture and temperature.	Decision trees are largely unstable as some minor alteration in data can cause major changes in the structure, which ultimately convey a wrong result from what users will get in a normal occasion	(Charbuty & Abdulazeez, 2021)

makes it impossible to detect some species with the human eye (Luttrell et al., 1994). The disadvantages of manual scouting, without IoT applications and AI sensor-based technologies, include pesticide resistance, health risks, increased costs for pesticide repetition, and environmental pollution (Dhananjayan and Ravichandran, 2018; Shah and Jain, 2019). Manual scouting, preservation of beneficial arthropods (parasitoids and predators) as biological control agents and use of use of selective pesticides may get neglected (Calvo-Agudo et al., 2022; Machado et al., 2019; Van Goethem et al., 2019).

Researchers should develop communication technologies and systems based on artificial intelligence (AI) sensors to detect, categorize, and monitor pests, according to farmers' requirements (Karar et al., 2021; Van Goethem et al., 2019). Although numerous computing technologies are present, in order to improve future systems, it is essential to analyze the present condition, limitations, and hazards of the existing methodologies. To better identify and categorize insect pests and recommend suitable pesticides, farmers may find it useful to employ artificial intelligence (AI) models such as K-nearest neighbors (KNN), decision trees, logistic regression and deep convolutional neural networks (CNNs) (Geetharamani and Pandian, 2019). Some pests are small and difficult to identify so these AI techniques are highly useful (Wang et al., 2018). The possibility of detecting insects in images taken with inexpensive RGB cameras further highlights the importance of AI approaches (Tenório et al., 2019). However, when it comes to developing and using AI methods for detecting, identifying, and monitoring insect pests, particularly in cotton fields, there are a few review papers, that reflect the viewpoints and opinions of current research.

5.1 Pest Recognition Features

For proper identification of pests, accurate pest features are essential. A common approach is to use image segmentation which involves grouping pixels with similar characteristics. During this process, features such as morphology, color, texture, distribution, and local characteristics can be retrieved (Boissard et al., 2008; Cho et al., 2007). The distribution of color is important for insect classification because different insect species and their body parts show different colors. A method for comparing the color of the query image with the retrieved images is color indexing (Khosla et al., 2015). Color correlograms (Vinayak and Jindal, 2017), Color moments (Zhang et al., 2012), and color histograms (Tenório et al., 2019) are techniques for extracting color features. The histogram approach offers a potential image signature because of its sensitivity and precision to pest position and orientation dynamics. However, this approach has limitations in distinguishing the spatial relationship between color patches. To address these limitations, color coherence vectors, and color distribution have been proposed, which capture spatial relationships and overall color distributions, particularly when used together. Furthermore, color moments have proven effective in content-based image extraction systems and are more accurate for describing color distributions than the histogram technique (Rasli et al., 2012). Contour length, area, border diameter, perimeter, and curvature are the morphological and local features. Characteristics of local features include roundness, aspect ratio, compactness, form factor, and extension (Gassoumi et al., 2000; Kandalkar et al., 2014). An essential characteristic must exhibit a low degree of correlation with other categories, demonstrate high significance to human observers, and display considerable variation from other insect categories. Textural qualities are examined by measuring parameters such as correlation, homogeneity, entropy, energy, and contrast, mostly using the image gray-level matrix (Jige and Ratnaparkhe, 2017). To avoid unnecessary complexity and ensure algorithmic effectiveness, it is critical to accurately identify the key criteria for distinguishing one insect pest from another.

5.2 AI Performance Indicators

Artificial Neural Networks (ANN), Recurrent Neural Networks (RNN), Deep Belief Networks (DBN), Deep Boltzmann Machines (DBM), Convolutional Neural Networks (CNN), and Support Vector Machines (SVM) rely on the clarity of sensor data. Before data collection precise sensor calibration is essential for the purpose of accurate data collection and processing by using AI models to provide informative data about diseases or pest (Lin et al., 2019; Lin et al., 2020).

6. Intelligent Sensor Systems for Monitoring and Counting Cotton Pests

The rapidly emerging transformation technology is the Internet of Things (IoT). IoT has quickly penetrated various sectors, including health, communications, manufacturing, energy, and agriculture, leading to significant advancements and improvements in product and service quality across these domains (Lin et al., 2017; Sisinni et al., 2018). In farming, the features of LORA communication modules, drones, and IT sensors are unique in potential. Consequently, the combination of IoT and agriculture is gaining notable popularity. Pest identification and prevention greatly benefit from the application of IoT technology. Nevertheless, the adoption rate among farmers remains low despite its proven advantages. As illustrated in Figure 1, various factors contribute to this reluctance. These barriers, cannot be categorized because they all are equally significant. The main concern is that traditional farmers perceive the implementation of new technology as burdensome and complex. Therefore, for further research, there is a need for IoT-based prototypes and architectures for the management of pests. Such research is crucial for improving both the quality and production of food. As technology advances, there is an urgent need for a predictive system that would assist farmers in increasing agricultural output, quality, and profitability.

In that research, we propose an IoT framework to protect cotton crops from cotton bollworms and flying insects by combining sensors and drones. Previously, we introduced a locally assembled sensor prototype to detect cotton bollworm pests. We validated its performance through real-time implementation in the field and the lab. In this study, we propose an IoT-based system for detecting and responding

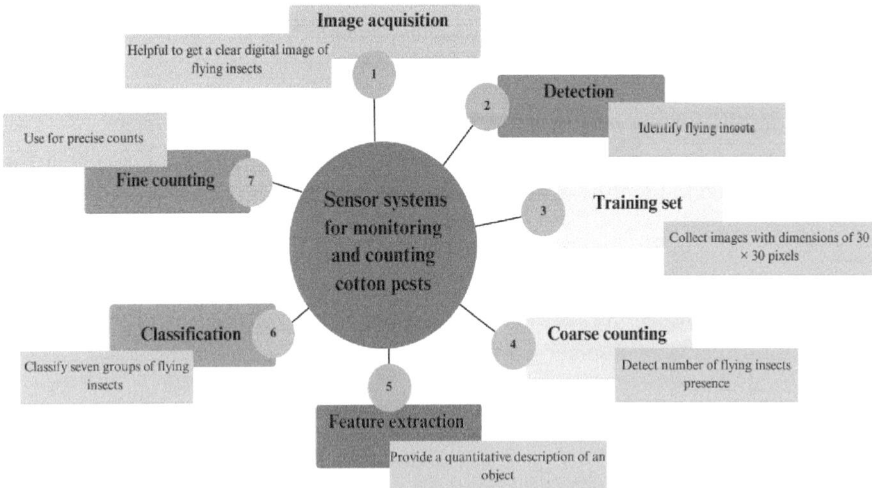

Figure 1. Sensor system for counting and monitoring pest.

to cotton plant pests that will be very useful for farmers by reducing losses and enhancing crop quality.

i. Image acquisition

For accurate counting and recognition processes clear digital image of flying insects is essential. Therefore, the trapping device and image acquisition apparatus needed to be compatible with other species of flying insects (Zhong et al., 2018).

ii. Detection

The objective of this model is to identify the different types of flying insects. It is a process sensitive to variations like impurities, camera defocus, and other factors(Ding and Taylor, 2016; Xia et al., 2012). Reliability and robust anti-interference abilities are essential for the detection algorithm. Although deep learning approaches like YOLO perform better, obtaining sufficient samples for specific insect species can be challenging. Therefore, we suggest combining all species of flying insects into a single class, using deep learning techniques for preliminary insect detection and approximate counting, and using a Support Vector Machine (SVM) for refined classification. This system facilitates the seamless addition or modification of identified flying insect categories without retraining the entire system (Zhong et al., 2018).

iii. Training set

A dataset for the YOLO (You Only Look Once) model consisted of over 10,000 manual labeling insects using rectangular annotations. To create a training set for a Support Vector Machine (SVM) classifier, images with dimensions of 30 × 30 pixels were collected. To assure accuracy, the dataset includes both positive and negative samples, all are validated by multiple experts (Zhong et al., 2018).

iv. Coarse counting

This technique can be used to count the number of flying insects in a given area, but it cannot be used to identify any specific type of flying insect (Zhong et al., 2018).

v. Feature extraction

Novel mathematical techniques provide a quantitative description of an object during the feature extraction process (Gassoumi et al., 2000). To obtain comprehensive feature information, a number of features are chosen to build the feature space (Zhong et al., 2018).

vi. Classification

Fruit flies, chafers, mosquitoes, bees, moths, and other flying insects are categorized into seven groups of detection results from YOLO, using support vector machines (SVMs) based on the extracted features (Zhong et al., 2018).

vii. Fine counting

Accurately monitoring the flying insect population dynamics and their intensity is possible only by collecting accurate counts of every flying insect species (Zhong et al., 2018).

6.1 System Components of Remote Monitoring Devices

Remote monitoring platforms have some features, including the ability to collect, store, and analyze data, and circulate the information. The data-gathering system consists of various types of sensors, infrared, and hyper spectral cameras and RGB to capture the images of different insect pests and diseases. When choosing a camera, remember that the resolution is important as low resolution may affect the quality of the images processed (Buschbacher et al., 2020). In addition, the data-gathering system consists of environmental information, like rainfall, light, surface and underground temperature sensors, and soil moisture content to detect pests instantly and predict population changes (Zhong et al., 2018). It was necessary to conduct data analysis after data collection and handling to generate useful information. The obtained data could be shared with the users like farmers to inform them to take action about managing pests. We need to combine this information with the location data, using GNSS to find which field sites should be affected by pollination or pest damage (Van Goethem et al., 2019). Now-a-days advanced tracking systems of pest, should be able to handle different types of sensors and work in two ways: giving users updates about the fields and letting them check the data at any time as shown in Figure 2 (Kiobia et al., 2023).

The "self-cleaning trap" and "remote whitefly monitor" are two systems used in cotton fields to detect flying insects (Ascolese et al., 2022; Parab et al., 2022). Sticky traps are used by the "self-cleaning trap" and "remote whitefly monitor" to fix the insects to the position of the camera sensor. In contrast to the "self-cleaning trap,"

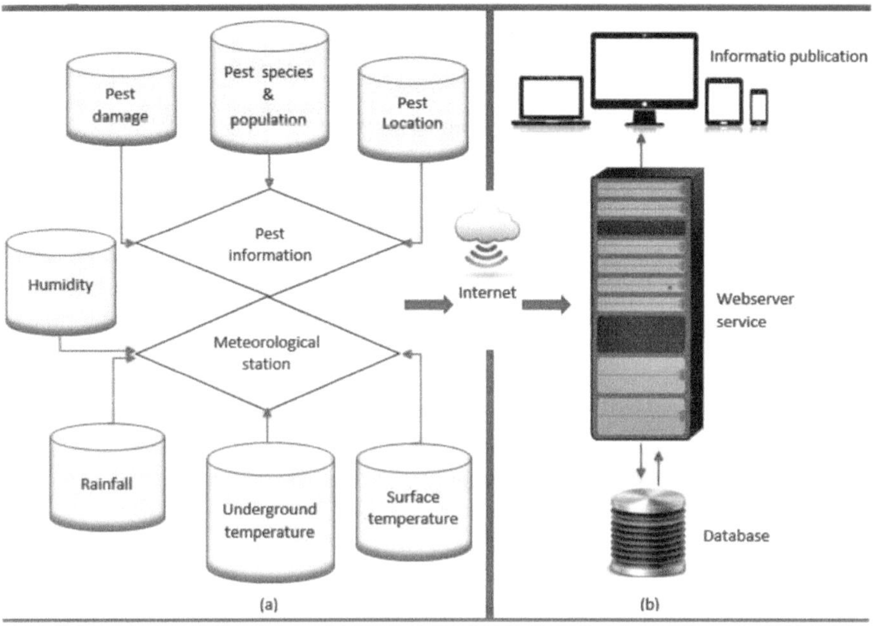

Figure 2. The components of the insect pest monitoring platform. **(a)** System for information collection, and **(b)** System for information analysis and display (Kiobia et al., 2023).

which uses pheromones to attract pests to its position, the "remote whitefly monitor" uses the sticky trap itself as an attractive mechanism for the camera (yellow, blue, or white color). One to three times a day, the whitefly monitor can count the number of pests and provide an online report. Further improvement of these devices is possible, especially in accurately classifying insects and attracting the various pests. In addition, these devices may have models that may reasonably predict the dynamics of pests, including weather patterns.

6.2 Counting of Pests on Leaves

Pest images can be recorded directly from the leaf surface, which reduces the cost of insect pest attractants like yellow sticky traps and pheromones. It was shown how to identify and estimate the population of whiteflies directly from the cotton leaf (Jige and Ratnaparkhe, 2017). In this method, by using MATLAB field-based machine-learning techniques, a bounding box and region props algorithm was used to count whiteflies. Other steps in the collection of data and analysis pipeline involved grey color thresholding, background removal, the conversion of RGB images to HSV, and image acquisition. For pest counting the proposed technique was fast and cost-effective. With a 98% success rate, the approach was also very reliable. If numerous species are detected during the feature extraction process, this method should also provide the ratios, skew, kurtosis, number of pixels standard deviation, and interior density (Alves et al., 2020). Recognizing and counting whiteflies through image processing in the field face challenges such as weather conditions like high moisture, dust, and leaf veins that can resemble whiteflies due to their light color. Further, the small sizes of whiteflies make it difficult to detect and analyze them, especially when they cluster together (Sun et al., 2017; Tian et al., 2021). In another study, agronomists count, analyze, and capture images of Silverleaf whitefly nymphs on cotton leaves by transferring the machine vision models into a mobile device (Long et al., 2019). Machine learning and segmentation were used in the process. Two different learning methods, one based on segmentation and the other on Faster R-CNN, were able to identify silverleaf whitefly nymphs with an accuracy of up to 67% and 79%, respectively. Smartphones from different companies like Sony, Samsung and Apple were tested. According to the updates, deep learning models that used only the iPhone or the iPhone along with other smartphone models had the highest F-scores (71.7% to 75.8%). In contrast, models that used Samsung and Sony, either separately or in combination, had lower F-scores (between 44.0% and 55.6%). When run on the smart phone under test, the model's accuracy fell short of the 90–95% F-score threshold. For training, by using a single image sensor detection, and performance evaluation in a real-world cotton field situation led to lower image quality, which in turn prevented the system from reaching the required F-score threshold.

For the classification, a faster region-based convolutional neural network, cloud computing approaches, and the Internet of Things (IoT) are allowed using an iOS 13.5 smartphone (Karar et al., 2021; Long et al., 2019). Nevertheless, it remained difficult due to the complex resemblance between the pests' leaf backgrounds and the plants. With their expensive price tag, high-resolution digital devices were needed to deal with some of these issues. Leaf veins are commonly removed from photos using the eccentricity and primary and minor axis lengths (Wang et al., 2018).

6.3 Counting Pests on Sticky Traps

For capturing the targeted insects and minimizing the detection difficulties in complex backgrounds, sticky traps are crucial. The insects are stuck in the sticky traps until they can be examined (Liu et al., 2016). A generalized method, using sticky traps automatically counts the number of pests on cotton crop, specifically whiteflies (Jige and Ratnaparkhe, 2017; Parab et al., 2022; Qiao et al., 2008). The process involves capturing images, subtracting backgrounds, converting color spaces, conducting the morphological tracking system and labeling procedures and performing thresh-holding operations. The camera can scan the different types of insects on the sticky trap that rotates non-automatically (Qiao et al., 2008). The proposed system worked by detecting the cotton pests present on yellow sticky tape, uploading them to a web server, then counting the pests, and updating the insect count on the web server. This study showed the real-time identification and counting of cotton pests using the Faster-RCNN detection. From multiple remote stations, the system has the potential to generate daily reports on pest monitoring. Different techniques are appropriate for field operations, particularly in underdeveloped countries, considering these remote devices are powered by low-cost batteries and solar energy.

Due to its low maintenance requirements, automating the rotation of the yellow tape can be helpful, particularly in reducing the time desired to repair fixed stick traps in the field. The system is developed on the basis of recognizing and counting whiteflies. Improving the system's ability to identify and classify various insect pests could have a major effect on the management of pests. In addition, autonomous devices in the field may consist of some additional features like micro-climates to help farmers about the prediction of pest dynamics.

6.4 Counting Pests on Paper

In this method, insects are collected onto a piece of paper for imaging. Attractive substances, such as pheromones or light, are used to capture the insects. Researchers have developed a novel system, that utilizes a smartphone camera to capture the images of insects attracted by pheromones (Dalmia et al., 2020). A supervised Neural Network approach called Single Shot Multi-Task Detector (SSD) was utilized to count the pests and process the images. The additional benefit of the method was that it showed which insecticide spraying was necessary based on the population levels of the pests. The study identified and examined the challenges posed by low-resolution cameras and diverse data which are affected by various agroclimatic zones worldwide. This technique was employed in the cotton field to identify the pink bollworms. Due to the system's reliance on human operation and mobile phone usage, variations in the system's performance could be caused by variations in the different versions of mobile phone cameras or human error. In the trap, killing the insects without crushing it before taking the image is one way to avoid problems with a smartphone camera. An experienced and dedicated farmer or consultant might be necessary for this work. Furthermore, farmers wishing to execute a pest control strategy were given no exact advice on the number of traps required to capture pests in each location.

7. Role of Precision in the Forecasting of Pests

Precision agricultural techniques facilitate the forecasting of pest activity. Automatic pest management systems claim that a combination of infrared, thermal, audio, and image-based sensors can detect and identify pests. By using sensors and software to analyze the data, agricultural productivity and sustainability can be enhanced. High-resolution crop images captured by aerial or satellite platforms were processed to get additional information, to make decisions about pest control. Figure 3 depicts the importance of pest monitoring and management for the crop's effective production. To identify different pests on a large scale, it is thought that a pest monitoring program is essential. Modern pest monitoring uses advanced tools such as GPS, artificial intelligence (AI), and the Global Navigation Satellite System (GNSS).

i. AI-based monitoring of pests:

Artificial intelligence (AI) has become an essential tool for solving numerous problems in agriculture due to its extensive set of skills as shown in Figure 3. Artificial intelligence models based on deep learning (DL) help to detect various pests (Brisco et al., 1998).

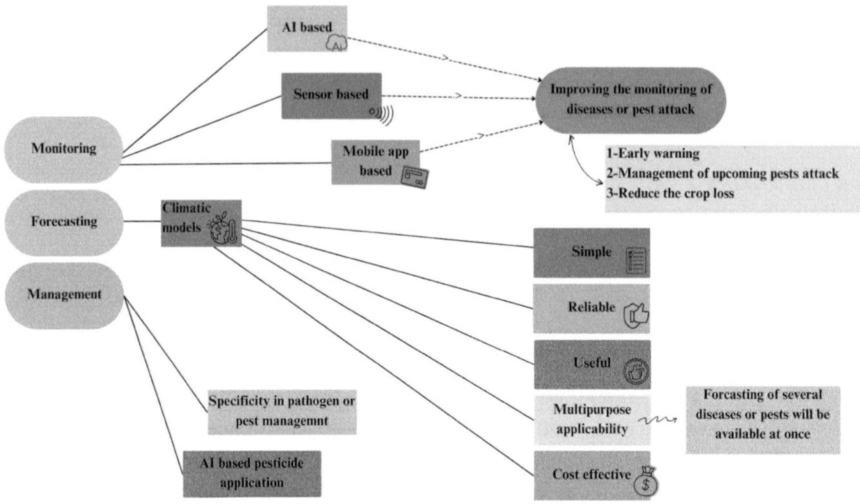

Figure 3. Role of precision in the forecasting of pests.

ii. Sensors-based monitoring of pests:

Surveys of various pests in the field have only started to use PA technology based on sensors. The use of global navigation satellite systems (GNSS) for wildlife telemetry, thermal infrared imaging, Use of video cameras to monitor flying insects, chemiluminescent tags to track movement of insect at night, habitat mapping, and radar to track pest migration are just a few examples.

iii. Mobile app for monitoring pests

For automated agricultural pest management various smartphone applications are available. Some of the most widely used applications are the Agri-App, Plantix Mobile App, Trapview System and Kheti-Badi (Toscano-Miranda et al., 2022). Pest prediction allows farmers to achieve high-quality results avoiding applying pesticide treatments, and advising on the biology of insects' occurrence (Magarey et al., 2015). Utilizing biophysical methodologies for forecasts and early warnings enables the management of upcoming pest attacks, improving pest management, reducing losses in crop, and also lower the cost of cultivation.

7.1 Climatic Model for Forecasting the Pest Attack

An essential organic component of pest population management is the impact of climate on the quantity of natural enemies. For a particular pest, by combining the environmental requirements of a particular pest species with climate models, it may be possible to predict the spectrum of possible worldwide changes. For forecasting the pest attack, it is cost-effective, simple, multi applicable, and reliable method.

7.2 Other Models for Forecasting the Attack

To protect the plants and control pests, pest forecasting primarily depends on artificial intelligent system (Gaikwad et al., 2021). Some of the most essential methods for pest forecasting are:

1. Machine learning (ML) allows users to analyze trends and manage complex data in many applications, including predictive analytics, image processing, and data mining. When the next generation of pests emerges in summer, an ML model will be used for forecasting.
2. In the context of climate change, physiologically based demographic models (PBDMs) could use. The weather, location, and pest abundance are observed using a PBDM model. Most importantly, in studying the dynamics of insect populations, global climate indicators and local weather conditions are the most important aspects. For pest forecasting, it is also necessary to predict the combined effects of both endogenous and exogenous components (Kanwal et al., 2022; Ordano et al., 2015).

8. Challenges to the Implementation of IoT-Filled Devices

Remote devices could face several problems while used for the detection of pests in the field, such as excessive network problems, insufficient security, software failure, service expiration, physical hardware flaws, power consumption, and variations in environmental conditions. The source of a specific challenge may vary over time and depend on the device's design. If necessary, the on-site installation of appropriate operational IoT devices should demonstrate high performance with small flaws.

9. Urgent Need for IoT Towards Worldwide Major Cotton Pests

The study recommended prioritizing major pests in pest management strategies using sensor-based IoT and AI technologies. The priority given to each pest can depend on how much economic damage it causes globally. For example, in regions including China, Southeast Asia, Central America, Pakistan, Central Africa, North America, South America, India, and South Africa, whiteflies have been reported (Afzal et al., 2023; Aslam et al., 2023). Similarly, from southern Texas to Argentina, boll weevils on cotton crop have been detected (Burbano-Figueroa et al., 2021). Cotton mealybugs have been also reported in over 26 countries across many ecological zones (Abbas et al., 2010). Within the cotton ecosystem, several studies have reported a significant increase in pest damage. For instance, in early 1999, the cotton flea hopper was ranked ninth to fourth and sixth by 2007 in the USA (Hamons et al., 2021). Based on local production practices, pest types, pest populations, crop stages, and pest-friendly circumstances in ecological zones in specific regions, the amount of output losses in cotton can vary.

Cotton growers may lose over 50% of their crop, which is attacked by destructive pests, which can cause significant damage. Farmers may use up to fifteen insecticide treatments every season to combat severe pests like boll weevils based on specific decisions or instructions regarding the estimated insect population threshold (Crow et al., 2021; Jaleel et al., 2014). Unfortunately, the management of significant cotton pests is complex since many farmers and consultants rely on informal information to control the dynamics of these pests (Bhosle et al., 2009; Furtado et al., 2009; Gupta, 2003; MahaLakshmi and NVVSD, 2020; Ramalho et al., 2012). Farmers have recently switched from deep plowing to no-tillage farming practices associated with increased prevalence and consistency of severe pest attacks (Kiobia et al., 2023). The severity of pest attacks has been linked to the low participation of farmers (approximately 8.33%) in using traps for mass pest trapping (Nadeem et al., 2023). Farmers have consistently struggled to manage severe pest infestations because pests can migrate from various crop hosts to cotton and the favorable temperatures enhance the reproduction of pests (Sain et al., 2021). Preventive seedling protection is a primary strategy to mitigate cotton infestations and reduce the excessive use of pesticides, particularly against unidentified pests. However, this strategy may negatively impact non-target insects, including honeybees and other pollinators (Baron et al., 2017; Krupke and Long, 2015). To minimize sub-lethal effects on non-target insects, it could facilitate the classification of multiple pests and recommend insect-specific insecticides by establishing a record of pest images and integrating it with sensor-based systems or artificial intelligence models.

10. Recommendations for Future Research

The studies reviewed here primarily addressed the detection of beneficial insects (such as parasitoids and predators) and cotton pests through imaging of stems, the undersides of leaves, bloom roots, and both external and internal parts of the bloom. However, research remains limited in these areas. This review reveals a significant gap in studies focused on detecting soil-dwelling insects that harm plants during part of their life or growth cycles. Furthermore, identifying pests from images often having complex backgrounds, pests having similar-looking species, and overlapping issues present a substantial challenge in cotton pest identification. Consequently, future research should prioritize investigating potential attractants for these pest species and using sticky boards within sensor regions to capture them for imaging. Some insects may spend some part of their life underground have limited flight range or be flightless. AI techniques for detecting and classifying cotton pests and beneficial insects have used CNN models. Additionally, there has been a lack of comprehensive studies on pest detection, classification, and quantification that account for abiotic factors. Abiotic factors can influence the behavior of some insect pests, which can affect the way artificial intelligence models work. Therefore, further research is needed to explore these abiotic aspects. While many studies have successfully identified pests, they often fail to estimate insect population dynamics.

11. Conclusion

In conclusion, adequate precision can be achieved in the identification, classification, and counting of beneficial insects or cotton insect pests using AI models and IoT sensor-based approaches. Recent developments in remote sensing technology have made it possible to detect innovations in a wider range of agricultural applications, particularly in the management of pests and plant diseases. The information technology revolution, as well as the interpretation and data analysis of these high-tech applications, offer significant benefits in terms of expanding and boosting the input knowledge that supports the advancement of numerous study fields. Remote sensing is now considered general-purpose data for a large group of users with varying needs. Remote sensing has emerged as one of the most promising technologies for integrated pest control due to its ability to investigate many environmental characteristics and available natural sources overall. To address these issues and advance AI and IoT in cotton production, some solutions include improved CNN algorithms, low-bandwidth devices, sticky boards, and more study on insect-specific attractants that attract insects to the sensor location.

References

Abbas, G., Arif, M. J., Ashfaq, M., Aslam, M., and Saeed, S. (2010). The impact of some environmental factors on the fecundity of *Phenacoccus solenopsis* Tinsley (Hemiptera: Pseudococcidae): a serious pest of cotton and other crops. *Pakistan Journal of Agricultural Sciences*, 47(4): 321–325. https://www.pakjas.com.pk/papers/55.pdf.

Abu-Nasser, B. S., and Abu-Naser, S. S. (2018). Cognitive system for helping farmers in diagnosing watermelon diseases. International Journal of Academic Information Systems Research (IJAISR), 2(7): 1–7. https://www.researchgate.net/profile/Samy-AbuNaser/publication/327200994_Cognitive_System_for_Helping_Farmers_in_Diagnosing_Watermelon_Diseases/links/5b7fbe17a6fdcc5f8b639f31/Cognitive-System-for-Helping-Farmers-in-Diagnosing-Watermelon-Diseases.pdf.

Afzal, M., Saeed, S., Riaz, H., Ishtiaq, M., and Rahman, M. H. (2023). A critical review of whitefly (*Bemisia tabaci gennadius*) cryptic species associated with the cotton leaf curl disease. J. Innov. Sci, 9, 24–43. https://researcherslinks.com/current-issues/A-Critical-Review-of-Whitefly-Bemisia-tabaci-gennadius-Cryptic/27/8/5963/html.

Ahmad, M., Muhammad, W., and Sajjad, A. (2020). Ecological Management of Cotton Insect Pests. pp. 213–238. *In*: Ahmad, S., and Hasanuzzaman, M. (Eds.). Cotton Production and Uses. Springer Singapore. https://doi.org/10.1007/978-981-15-1472-2-12.

Alves, A. N., Souza, W. S., and Borges, D. L. (2020). Cotton pests classification in field-based images using deep residual networks. Computers and Electronics in Agriculture, 174: 105488. https://www.sciencedirect.com/science/article/pii/S0168169919311342.

AlZamily, J. Y. (2018). A rational system for diagnosing banana disorders. International Journal of Academic Information Systems Research (IJAISR), 2: 1–8.

Andreopoulos, A., and Tsotsos, J. K. (2013). 50 years of object recognition: Directions forward. Computer Vision and Image Understanding, 117(8): 827–891. https://www.sciencedirect.com/science/article/pii/S107731421300091X.

Anees, M., and Shad, S. A. (2020). Insect Pests of Cotton and Their Management. In S. Ahmad & M. Hasanuzzaman (Eds.). Cotton Production and Uses (pp. 177–212). Springer Singapore. https://doi.org/10.1007/978-981-15-1472-2-11.

Ascolese, R., Gargiulo, S., Pace, R., Nappa, P., Griffo, R., Nugnes, F. et al. (2022). E-traps: A valuable monitoring tool to be improved. EPPO Bulletin, 52(1): 175–184. https://doi.org/10.1111/epp.12838.

Aslam, M. Q., Hussain, A., Akram, A., Hussain, S., Naqvi, R. Z., Amin, I. et al. (2023). Cotton Mi-1.2-like Gene: A potential source of whitefly resistance. Gene, 851: 146983. https://www.sciencedirect.com/science/article/pii/S0378111922008034.

Barbedo, J. G. A. (2014). Using digital image processing for counting whiteflies on soybean leaves. Journal of Asia-Pacific Entomology, 17(4): 685–694. https://www.sciencedirect.com/science/article/pii/S1226861514000855.

Baron, G. L., Raine, N. E., and Brown, M. J. F. (2017). General and species-specific impacts of a neonicotinoid insecticide on the ovary development and feeding of wild bumblebee queens. Proceedings of the Royal Society B: Biological Sciences, 28: 1–13. 20170123. https://doi.org/10.1098/rspb.2017.0123.

Ben Guerrero, E., Salvador, R., and Talia, P. (2020). Evaluation of hydrolytic enzyme activities from digestive fluids of *Anthonomus grandis* (Coleoptera: Curculionidae). Archives of Insect Biochemistry and Physiology, 105(2): 1–9. https://doi.org/10.1002/arch.21730.

Bhosle, B. B., More, D. G., Patange, N. R., Sharma, O. P., and Bambawale, O. M. (2009). Efficacy of different seed dressers against early season sucking pest of cotton. Pesticide Research Journal, 21(1): 75–79. https://www.indianjournals.com/ijor.aspx.

Boissard, P., Martin, V., and Moisan, S. (2008). A cognitive vision approach to early pest detection in greenhouse crops. Computers and Electronics in Agriculture, 62(2): 81–93. https://www.sciencedirect.com/science/article/pii/S0168169907002256.

Brisco, B., Brown, R. J., Hirose, T., McNairn, H., and Staenz, K. (1998). Precision Agriculture and the Role of Remote Sensing: A Review. Canadian Journal of Remote Sensing, 24(3): 315–327. https://doi.org/10.1080/07038992.1998.10855254.

Burbano-Figueroa, O., Sierra-Monroy, A., Grandett Martinez, L., Borgemeister, C., and Luedeling, E. (2021). Management of the Boll Weevil (Coleoptera: Curculionidae) in the Colombian Caribbean: A Conceptual Model. Journal of Integrated Pest Management, 12(1): 1–49. https://academic.oup.com/jipm/article abstract/12/1/15/6209865.

Buschbacher, K., Ahrens, D., Espeland, M., and Steinhage, V. (2020). Image-based species identification of wild bees using convolutional neural networks. Ecological Informatics, 55: 1–10. https://www.sciencedirect.com/science/article/pii/S1574954119303280.

Caldeira, R. F., Santiago, W. E., and Teruel, B. (2021). Identification of cotton leaf lesions using deep learning techniques. Sensors, 21(9): 1–42. https://www.mdpi.com/1424-8220/21/9/3169.

Calvo-Agudo, M., Tooker, J. F., Dicke, M., and Tena, A. (2022). Insecticide-contaminated honeydew: Risks for beneficial insects. Biological Reviews, 97(2): 664–678. https://doi.org/10.1111/brv.12817.

Castañeda-Miranda, R., García-Escalante, J. J., Torres-Pacheco, I., Guevara-González, R. G., Castañeda-Miranda, C. L., and Alaniz-Lumbreras, P. D. (2011). Scale invariant feature approach for insect monitoring. Computers and Electronics in Agriculture, 75(1): 92–99. https://www.sciencedirect.com/science/article/pii/S016816991000195X.

Charbuty, B., and Abdulazeez, A. (2021). Classification based on decision tree algorithm for machine learning. Journal of Applied Science and Technology Trends, 2(01): 20–28. https://www.jastt.org/index.php/jasttpath/article/view/65.

Chen, P., Xiao, Q., Zhang, J., Xie, C., and Wang, B. (2020). Occurrence prediction of cotton pests and diseases by bidirectional long short-term memory networks with climate and atmosphere circulation. Computers and Electronics in Agriculture, 176: 1–8. https://www.sciencedirect.com/science/article/pii/S0168169918317794.

Cho, J., Choi, J., Qiao, M., Ji, C. W., Kim, H. Y., Uhm, K. B. et al. (2007). Automatic identification of whiteflies, aphids and thrips in greenhouse based on image analysis. Red, 1(1): 46–53. https://citeseerx.ist.psu.edu/document?repid=rep1&type=pdf&doi=c5b04e3e33461e68f18ad8808eb143e9ab4b14pdf.

Chohan, S., Perveen, R., Abid, M., Tahir, M. N., and Sajid, M. (2020a). Cotton diseases and their management. pp. 239–270. *In*: Ahmad, S., and Hasanuzzaman, M. (Eds.). Cotton Production and Uses. Springer Singapore. https://doi.org/10.1007/978-981-15-1472-2_13.

Chohan, S., Perveen, R., Abid, M., Tahir, M. N., and Sajid, M. (2020b). Cotton diseases and their management. pp. 239–270. *In*: Ahmad, S., and x Hasanuzzaman, S. (Eds.). Cotton Production and Uses. Springer Singapore. https://doi.org/10.1007/978-981-15-1472-2_13.

Chopda, J., Raveshiya, H., Nakum, S., and Nakrani, V. (2018). Cotton crop disease detection using decision tree classifier. 2018 International Conference on Smart City and Emerging Technology (ICSCET), 1–5. https://ieeexplore.ieee.org/abstract/document/8537336.

Coelho, R. R., de Souza Júnior, J. D. A., Firmino, A. A., de Macedo, L. L., Fonseca, F. C., Terra, W. R. et al. (2016). Vitellogenin knockdown strongly affects cotton boll weevil egg viability but not the number of eggs laid by females. Meta Gene, 9: 173–180. https://www.sciencedirect.com/science/article/pii/S2214540016300226.

Cox, Jr., K. L., Babilonia, K., Wheeler, T., He, P., and Shan, L. (2019). Return of old foes-Recurrence of bacterial blight and Fusarium wilt of cotton. Current Opinion in Plant Biology, 50: 95–103. https://www.sciencedirect.com/science/article/pii/S1369526618301195.

Crow, W. D., Catchot, A. L., and Bao, D. (2021). Efficacy of Counter on Thrips in Cotton, 2014. Arthropod Management Tests, 46(1): tsaa115. https://academic.oup.com/amt/article-abstract/46/1/tsaa115/6103911.

Dalmia, A., White, J., Chaurasia, A., Agarwal, V., Jain, R., Vora, D. et al. (2020). Pest management in cotton farms: An ai-system case study from the global south. proceedings of the 26th ACM SIGKDD International Conference on Knowledge Discovery & Data Mining, 3119–3127. https://doi.org/10.1145/3394486.3403363.

Dent, D., and Binks, R. H. (2020). Insect pest management. Cabi. 3: 1–153. https://books.google.com.pk/books.

Dhananjayan, V., and Ravichandran, B. (2018). Occupational health risk of farmers exposed to pesticides in agricultural activities. Current Opinion in Environmental Science & Health, 4: 31–37. https://www.sciencedirect.com/science/article/pii/S2468584417300661.

Ding, W., and Taylor, G. (2016). Automatic moth detection from trap images for pest management. Computers and Electronics in Agriculture, 123: 17–28. https://www.sciencedirect.com/science/article/pii/S0168169916300266.

El-Wakeil, N., and Abdallah, A. (2012). Cotton pests and the actual strategies for their management control. Agric. Res. Updates, 7: 161–218. https://www.researchgate.net/profile/NabilWakeil/publication/265972234_Cotton_pests_and_the_actual_strategies_for_their_management_control/links/542298850cf238c6ea690ebb/Cotton-pests-and-the-actual-strategies-for-their-management-control.pdf.

Ferro, E., Brea, V. M., Cabello, D., Lopez, P., Iglesias, J., and Castillejo, J. (2014). Wireless sensor mote for snail pest detection. SENSORS, 2014 IEEE, 114–117. https://ieeexplore.ieee.org/abstract/document/6984946.

Furtado, R. F., da Silva, F. P., de Carvalho Lavôr, M. T. F., and Bleicher, E. (2009). Susceptibilidade de cultivares de *Gossypium hirsutum* L. r. Latifolium Hutch a Aphis gossypii Glover. Revista Ciência Agronômica, 40(3): 461–464. https://www.redalyc.org/pdf/1953/195317389020.pdf.

Gaikwad, S. V., Vibhute, A. D., Kale, K. V., and Mehrotra, S. C. (2021). An innovative IoT based system for precision farming. Computers and Electronics in Agriculture, 187: 1–10. https://www.sciencedirect.com/science/article/pii/S0168169921003082.

García, R., Aguilar, J., Toro, M., Pinto, A., and Rodríguez, P. (2020). A systematic literature review on the use of machine learning in precision livestock farming. Computers and Electronics in Agriculture, 179: 1–7. https://www.sciencedirect.com/science/article/pii/S0168169920317099.

Gassoumi, H., Prasad, N. R., and Ellington, J. J. (2000). Neural network-based approach for insect classification in cotton ecosystems. International Conference on Intelligent Technologies, 1–7. http://gassoumi.tripod.com/me/thai2000.pdf.

Geetharamani, G., and Pandian, A. (2019). Identification of plant leaf diseases using a nine-layer deep convolutional neural network. Computers & Electrical Engineering, 76: 323–338. https://www.sciencedirect.com/science/article/pii/S0045790619300023.

Gholve, V. M., Jogdand, S. M., Jagtap, G. P., and Dey, U. (2012). *In-vitro* evaluation of fungicides, bioagents and aqueous leaf extracts against Alternaria leaf blight of cotton. https://www.researchgate.net/profile/Gp-Jagtap/publication/332291970.

Gill, H. K., and Garg, H. (2014). Pesticide: Environmental impacts and management strategies. Pesticides-Toxic Aspects, 8(187): 10–5772. https://books.google.com.pk/books.

Grigolli, J. F. J., Souza, L. A., Fernandes, M. G., and Busoli, A. C. (2017). Spatial Distribution of Adult *Anthonomus grandis* Boheman (Coleoptera: Curculionidae) and Damage to Cotton Flower Buds

Due to Feeding and Oviposition. Neotropical Entomology, 46(4): 442–451. https://doi.org/10.1007/s13744-016-0471-1.

Gupta, S. K. (2003). Mite pests of agricultural crops in India, their management and identification. Mites, Their Identification and Management, 48–61.

Habib, M. T., Majumder, A., Jakaria, A. Z. M., Akter, M., Uddin, M. S., and Ahmed, F. (2020). Machine vision based papaya disease recognition. Journal of King Saud University-Computer and Information Sciences, 32(3): 300–309. https://www.sciencedirect.com/science/article/pii/S1319157818302404.

Hamons, K., Raszick, T., Perkin, L., Sword, G., and Suh, C. (2021). Cotton Fleahopper1 Biology and Ecology Relevant to Development of Insect Resistance Management Strategies. Southwestern Entomologist, 46(1): 1–16. https://bioone.org/journals/southwestern-entomologist/volume-46/issue-1/059.046.0101/Cotton-Fleahopper1-Biology-and-Ecology-Relevant-to-Development-of-Insect/10.3958/059.046.0101.

Hoque, M. E., Kipli, K., Zulcaffle, T. M. A., Mat, D. A. A., Joseph, A., Zamhari, N. et al. (2019). Segmentation of retinal microvasculature based on iterative self-organizing data analysis technique (ISODATA). 2019 International UNIMAS STEM 12th Engineering Conference (EnCon), 59–64. https://ieeexplore.ieee.org/abstract/document/8861259/.

Hudgins, E. J., Liebhold, A. M., and Leung, B. (2017). Predicting the spread of all invasive forest pests in the United States. Ecology Letters, 20(4): 426–435. https://onlinelibrary.wiley.com/doi/abs/10.1111/ele.12741.

Iqbal, Z., Khan, M. A., Sharif, M., Shah, J. H., ur Rehman, M. H., and Javed, K. (2018). An automated detection and classification of citrus plant diseases using image processing techniques: A review. Computers and Electronics in Agriculture, 153: 12–32. https://www.sciencedirect.com/science/article/pii/S0168169917311870.

Jakkula, V. (2006). Tutorial on support vector machine (svm). School of EECS, Washington State University, 37: 1–13. https://course.khoury.northeastern.edu/cs5100f11/resources/jakkula.pdf.

Jaleel, W., Saeed, S., Naqqash, M. N., and Zaka, S. M. (2014). Survey of Bt cotton in Punjab Pakistan related to the knowledge, perception and practices of farmers regarding insect pests. International Journal of Agriculture and Crop Sciences, 7(1): 1–10. https://www.researchgate.net/profile/Muhammad-Naqqash-2/publication/291354852.

Ji, W., Han, K., Lu, Y., and Wei, J. (2020). Predicting the potential distribution of the vine mealybug, Planococcus ficus under climate change by MaxEnt. Crop Protection, 137: 1–8. https://www.sciencedirect.com/science/article/pii/S0261219420302015.

Jige, M. N., and Ratnaparkhe, V. R. (2017). Population estimation of whitefly for cotton plant using image processing approach. 2017 2nd IEEE International Conference on Recent Trends in Electronics, Information & Communication Technology (RTEICT), 487–491. https://ieeexplore.ieee.org/abstract/document/8256644/.

Kandalkar, G., Deorankar, A. V., and Chatur, P. N. (2014). Classification of agricultural pests using dwt and back propagation neural networks. Int. J. Comput. Sci. Inf. Technol, 5(3): 4034–4037. https://citeseerx.ist.psu.edu/document.

Kandalkar, G. Y., Deorankar, A. V., and Chatur, P. N. (2013). A Review of Literature on Application of Image Processing for Identification of Agricultural Pests on Various Crops. International Journal of Engineering Science and Innovative Technology (IJESIT), 2: 6–25.

Kanwal, S., Khan, M. A., Saleem, S., Tahir, M. N., Muntaha, S. T., Samreen, T. et al. (2022). Integration of precision agriculture techniques for pest management. Environmental Sciences Proceedings, 23(1): 1–19. https://www.mdpi.com/2673-4931/23/1/19.

Karar, H., Bashir, M. A., Haider, M., Haider, N., Khan, K. A., Ghramh, H. A. et al. (2020). Pest susceptibility, yield and fiber traits of transgenic cotton cultivars in Multan, Pakistan. PloS One, 15(7): 1–10. https://journals.plos.org/plosone/article.

Karar, M. E., Alsunaydi, F., Albusaymi, S., and Alotaibi, S. (2021). A new mobile application of agricultural pests recognition using deep learning in cloud computing system. Alexandria Engineering Journal, 60(5): 4423–4432. https://www.sciencedirect.com/science/article/pii/S1110016821001642.

Kiobia, D. O., Mwitta, C. J., Fue, K. G., Schmidt, J. M., Riley, D. G., and Rains, G. C. (2023). A review of successes and impeding challenges of IoT-based insect pest detection systems for estimating agroecosystem health and productivity of cotton. Sensors, 23(8): 4127.

Kattenborn, T., Leitloff, J., Schiefer, F., and Hinz, S. (2021). Review on Convolutional Neural Networks (CNN) in vegetation remote sensing. ISPRS Journal of Photogrammetry and Remote Sensing, 173: 24–49. https://www.sciencedirect.com/science/article/pii/S0924271620303488.

Khosla, G., Rajpal, N., and Singh, J. (2015). Evaluation of Euclidean and Manhanttan metrics in content based image retrieval system. 2015 2nd International Conference on Computing for Sustainable Global Development (INDIACom), 12–18. https://ieeexplore.ieee.org/abstract/document/7100191.

Krupke, C. H., and Long, E. Y. (2015). Intersections between neonicotinoid seed treatments and honey bees. Current Opinion in Insect Science, 10: 8–13. https://www.sciencedirect.com/science/article/pii/S2214574515000553.

Li, H., Ji, R., Zhang, J., Yuan, X., Hu, K., and Qi, L. (2011). WEB-based intelligent diagnosis system for cotton diseases control. pp. 483–490. *In*: Li, D., Liu, Y., and Chen, Y. (Eds.). Computer and Computing Technologies in Agriculture IV (Vol. 346). Springer Berlin Heidelberg. https://doi.org/10.1007/978-3-642-18354-6_57.

Liang, X. (2021). Few-shot cotton leaf spots disease classification based on metric learning. Plant Methods, 17(1): 1–14.

Lin, J., Yu, W., Zhang, N., Yang, X., Zhang, H., and Zhao, W. (2017). A survey on internet of things: Architecture, enabling technologies, security and privacy, and applications. IEEE Internet of Things Journal, 4(5): 1125–1142. https://ieeexplore.ieee.org/abstract/document/7879243.

Lin, Y.-B., Lin, Y.-W., Lin, J.-Y., and Hung, H.-N. (2019). SensorTalk: An IoT device failure detection and calibration mechanism for smart farming. Sensors, 19(21): 1–37. https://www.mdpi.com/1424-8220/19/21/4788.

Lin, Y.-W., Lin, Y.-B., and Hung, H.-N. (2020). CalibrationTalk: A farming sensor failure detection and calibration technique. IEEE Internet of Things Journal, 8(8): 6893–6903. https://ieeexplore.ieee.org/abstract/document/9253648/.

Liu, T., Chen, W., Wu, W., Sun, C., Guo, W., and Zhu, X. (2016). Detection of aphids in wheat fields using a computer vision technique. Biosystems Engineering, 141: 82–93. https://www.sciencedirect.com/science/article/pii/S1537511015300866.

Long, D., Grundy, P., and McCarthy, A. (2019). Machine vision App for automated cotton insect counting: Initial development and first results. Australian Association of Cotton Scientists 2019 Australian Cotton Research Conference. 1-37. https://research.usq.edu.au/item/q5916/machine-vision-app-for-automated-cotton-insect-counting-initial-development-and-first-results.

Luttrell, R. G., Fitt, G. P., Ramalho, F. S., and Sugonyaev, E. S. (1994). Cotton Pest Management: Part 1. A Worldwide Perspective. Annual Review of Entomology, 39(1): 517–526. https://doi.org/10.1146/annurev.en.39.010194.002505.

Machado, A. V., Potin, D. M., Torres, J. B., and Torres, C. S. S. (2019). Selective insecticides secure natural enemies action in cotton pest management. Ecotoxicology and Environmental Safety, 184: 1–8. https://www.sciencedirect.com/science/article/pii/S0147651319310000.

Magarey, R. D., Borchert, D. M., Fowler, G. A., Hong, S. C., and Venette, R. (2015). The NCSU/APHIS plant pest forecasting system (NAPPFAST). Pest Risk Modelling and Mapping for Invasive Alien Species. Wallingford: CABI, 82–96. https://books.google.com.pk/books.

MahaLakshmi, M. S., and NVVSD, P. (2020). Insecticide resistance in field population of cotton leaf hopper, Amrasca devastans (Dist.) in Guntur, Andhra Pradesh, India. Int. J. Curr. Microbiol. App. Sci, 9(6), 3006–3011. https://www.researchgate.net/profile/Lakshmi-Msm/publication/343297909.

Martin, V., and Moisan, S. (2008). Early pest detection in Greenhouses. International Conference on Pattern Recognition. https://citeseerx.ist.psu.edu/document.

Monteiro, A. L., de Freitas Souza, M., Lins, H. A., da Silva Teófilo, T. M., Júnior, A. P. B., Silva, D. V. et al. (2021). A new alternative to determine weed control in agricultural systems based on artificial neural networks (ANNs). Field Crops Research, 263: 1–9. https://www.sciencedirect.com/science/article/pii/S0378429021000216.

Nadeem, A., Tahir, H. M., Khan, A. A., Hassan, Z., and Khan, A. M. (2023). Species composition and population dynamics of some arthropod pests in cotton fields of irrigated and semi-arid regions of Punjab, Pakistan. Saudi Journal of Biological Sciences, 30(2): 1–44. https://www.sciencedirect.com/science/article/pii/S1319562X22004375.

Navinkumar, T. M., Kumar, R. R., and Gokila, P. V. (2021). Application of artificial intelligence techniques in irrigation and crop health management for crop yield enhancement. Materials Today: Proceedings, 45: 2248–2253. https://www.sciencedirect.com/science/article/pii/S2214785320378330.

Noureen, N., Hussain, M., Fatima, S., and Ghazanfar, M. (2016). Cotton mealybug management: A review. Journal of Entomology and Zoology Studies, 4(4): 657–663. https://www.entomoljournal.com/archives/2016/vol4issue4/PartI/4-3-194-773.pdf.

O'Grady, M. J., Langton, D., and O'Hare, G. M. P. (2019). Edge computing: A tractable model for smart agriculture? Artificial Intelligence in Agriculture, 3: 42–51. https://www.sciencedirect.com/science/article/pii/S2589721719300339.

Oliveira, C. M., Auad, A. M., Mendes, S. M., and Frizzas, M. R. (2014). Crop losses and the economic impact of insect pests on Brazilian agriculture. Crop Protection, 56: 50–54. https://www.sciencedirect.com/science/article/pii/S026121941300269X.

Ordano, M., Engelhard, I., Rempoulakis, P., Nemny-Lavy, E., Blum, M., Yasin, S. et al. (2015). Olive fruit fly (*Bactrocera oleae*) population dynamics in the Eastern Mediterranean: Influence of exogenous uncertainty on a monophagous frugivorous Insect. PloS One, 10(5): 1–12. https://journals.plos.org/plosone/article.

Pammel, L. H. (1888). Root rot of cotton or" cotton blight". Texas FARMER Collection. https://oaktrust.library.tamu.edu/bitstream/handle/1969.1/2833/Bull0004.pdf.

Parab, C. U., Mwitta, C., Hayes, M., Schmidt, J. M., Riley, D., Fue, K. et al. (2022). Comparison of single-shot and two-shot deep neural network models for whitefly detection in IoT Web application. AgriEngineering, 4(2): 507–522. https://www.mdpi.com/2624-7402/4/2/34.

Partel, V., Kakarla, S. C., and Ampatzidis, Y. (2019). Development and evaluation of a low-cost and smart technology for precision weed management utilizing artificial intelligence. Computers and Electronics in Agriculture, 157: 339–350. https://www.sciencedirect.com/science/article/pii/S0168169918316612.

Patrício, D. I., and Rieder, R. (2018). Computer vision and artificial intelligence in precision agriculture for grain crops: A systematic review. Computers and Electronics in Agriculture, 153: 69–81. https://www.sciencedirect.com/science/article/pii/S0168169918305829.

Prashar, K., Talwar, R., and Kant, C. (2017). Robust automatic cotton crop disease recognition (ACDR) method using the hybrid feature descriptor with SVM. In 4th 2016 International Conference on Computing on sustainable Global Development, pp. 1–3.

Pratheepa, M., Verghese, A., and Bheemanna, H. (2016). Shannon information theory a useful tool for detecting significant abiotic factors influencing the population dynamics of Helicoverpa armigera (Hübner) on cotton crop. Ecological Modelling, 337: 25–28. https://www.sciencedirect.com/science/article/pii/S0304380016302022.

Presley, J. T. (1954). Cotton diseases and methods of control. US Department of Agriculture. https://books.google.com.pk/books.

Qiao, M., Lim, J., Ji, C. W., Chung, B.-K., Kim, H.-Y., Uhm, K.-B. et al. (2008). Density estimation of Bemisia tabaci (Hemiptera: Aleyrodidae) in a greenhouse using sticky traps in conjunction with an image processing system. Journal of Asia-Pacific Entomology, 11(1): 25–29. https://www.sciencedirect.com/science/article/pii/S1226861508000071.

Rach, M. M., Gomis, H. M., Granado, O. L., Malumbres, M. P., Campoy, A. M., and Martín, J. J. S. (2013). On the design of a bioacoustic sensor for the early detection of the red palm weevil. Sensors, 13(2): 1706–1729. https://www.mdpi.com/1424-8220/13/2/1706.

Ramalho, F. S., Fernandes, F. S., Nascimento, A. R. B., Júnior, J. N., Malaquias, J. B., and Silva, C. A. D. (2012). Feeding damage from cotton aphids, Aphis gossypii Glover (Hemiptera: Heteroptera: Aphididae), in cotton with colored fiber intercropped with fennel. Annals of the Entomological Society of America, 105(1): 20–27. https://academic.oup.com/aesa/article-abstract/105/1/20/110268.

Ranjitha, G., Srinivasan, M. R., and Rajesh, A. (2014). Detection and estimation of damage caused by thrips *Thrips tabaci* (Lind) of cotton using hyperspectral radiometer. Agrotechnology, 3(1): 123. https://www.researchgate.net/profile/Srinivasan-M-R-2/publication/291180706.

Rasli, R. M., Muda, T. Z. T., Yusof, Y., and Bakar, J. A. (2012). Comparative analysis of content based image retrieval techniques using color histogram: A case study of GLCM and K-means clustering. 2012 Third International Conference on Intelligent Systems Modelling and Simulation, 283–286. https://ieeexplore.ieee.org/abstract/document/6169714/.

Revathi, P., and Hemalatha, M. (2014). Cotton leaf spot diseases detection utilizing feature selection with skew divergence method. International Journal of Scientific Engineering and Technology, 3(1): 22–30. https://www.indianjournals.com/ijor.aspx.

Sabri, N., Aljunid, S. A., Salim, M. S., Badlishah, R. B., Kamaruddin, R., and Malek, M. A. (2013). Fuzzy inference system: Short review and design. Int. Rev. Autom. Control, 6(4): 441–449. https://www.researchgate.net/profile/Naseer-Sabri/publication/280739444.

Sain, S. K., Monga, D., Hiremani, N. S., Nagrale, D. T., Kranthi, S., Kumar, R. et al. (2021). Evaluation of bioefficacy potential of entomopathogenic fungi against the whitefly (*Bemisia tabaci* Genn.) on cotton under polyhouse and field conditions. Journal of Invertebrate Pathology, 183: 1–8. https://www.sciencedirect.com/science/article/pii/S0022201121000859.

Sangari, A. S., and Saraswady, D. (2016). Analyzing the optimal performance of pest image segmentation using non linear objective assessments. International Journal of Electrical and Computer Engineering (IJECE), 6(6): 2789–2796. http://download.garuda.kemdikbud.go.id/article.php.

Sarangdhar, A. A., and Pawar, V. R. (2017). Machine learning regression technique for cotton leaf disease detection and controlling using IoT. 2017 International Conference of Electronics, Communication and Aerospace Technology (ICECA), 2: 449–454. https://ieeexplore.ieee.org/abstract/document/8212855.

Sattar, M. N., Kvarnheden, A., Saeed, M., and Briddon, R. W. (2013). Cotton leaf curl disease – an emerging threat to cotton production worldwide. Journal of General Virology, 94(4): 695–710. https://doi.org/10.1099/vir.0.049627-0.

Shah, N., and Jain, S. (2019). Detection of disease in cotton leaf using artificial neural network. 2019 Amity International Conference on Artificial Intelligence (AICAI), 473–476. https://ieeexplore.ieee.org/abstract/document/8701311.

Shahzadi, R., Ferzund, J., Tausif, M., and Suryani, M. A. (2016). Internet of things based expert system for smart agriculture. International Journal of Advanced Computer Science and Applications, 7(9): 341–350. https://pdfs.semanticscholar.org/c57a/04cded2a7beba80f34c4c2cc965fea3ccb7d.pdf.

Shaikh, T. A., Rasool, T., and Lone, F. R. (2022). Towards leveraging the role of machine learning and artificial intelligence in precision agriculture and smart farming. Computers and Electronics in Agriculture, 198: 1–7. https://www.sciencedirect.com/science/article/pii/S0168169922004367.

Simberloff, D., and Rejmánek, M. (2011). Encyclopedia of biological invasions. Univ of California Press. https://books.google.com.pk/books.

Singh, S. K., Dutta, S., and Dharaiya, N. (2016). A study on geospatial technology for detecting and mapping of *Solenopsis mealybug* (Hemiptera: Pseudococcidae) in cotton crop. Journal of Applied and Natural Science, 8(4): 2175–2181. http://journals.ansfoundation.org/index.php/jans/article/view/1108.

Sisinni, E., Saifullah, A., Han, S., Jennehag, U., and Gidlund, M. (2018). Industrial internet of things: Challenges, opportunities, and directions. IEEE Transactions on Industrial Informatics, 14(11): 4724–4734. https://ieeexplore.ieee.org/abstract/document/8401919.

Smith, H. E., Elliott, F. C., and Bird, L. S. (1962). Root Rot Losses of Cotton Can Be Reduced. Miscellaneous Publication/Texas Agricultural Experiment Station; No. 361. https://oaktrust.library.tamu.edu/bitstream/handle/1969.1/160844/361/MP0361.pdf

Srivastav, N., Chopra, G., Jain, P., & Khatter, B. (2013). Pest Monitor and control system using WSN with special referance to Acoustic Device. Proc. 27th ICEEE, 92–99.

Sudars, K., Jasko, J., Namatevs, I., Ozola, L., and Badaukis, N. (2020). Dataset of annotated food crops and weed images for robotic computer vision control. Data in Brief, 31: 105833. https://www.sciencedirect.com/science/article/pii/S2352340920307277.

Sun, C., and Wang, X. (2010). Spot segmentation and verification based on improve marker controlled watershed transform. 2010 3rd International Conference on Computer Science and Information Technology, 8: 63–66. https://ieeexplore.ieee.org/abstract/document/5563982.

Sun, Y., Cheng, H., Cheng, Q., Zhou, H., Li, M., Fan, Y. et al. (2017). A smart-vision algorithm for counting whiteflies and thrips on sticky traps using two-dimensional Fourier transform spectrum. Biosystems Engineering, 153: 82–88. https://www.sciencedirect.com/science/article/pii/S1537511016303439.

Talaviya, T., Shah, D., Patel, N., Yagnik, H., and Shah, M. (2020). Implementation of artificial intelligence in agriculture for optimisation of irrigation and application of pesticides and herbicides. Artificial Intelligence in Agriculture, 4: 58–73. https://www.sciencedirect.com/science/article/pii/S258972172030012X.

Tenório, G. L., Martins, F. F., Carvalho, T. M., Leite, A. C., Figueiredo, K., Vellasco, M. et al. (2019). Comparative Study of Computer Vision Models for Insect Pest Identification in Complex Backgrounds. 2019 12th International Conference on Developments in eSystems Engineering (DeSE), 551–556. https://ieeexplore.ieee.org/abstract/document/9073250.

Thenmozhi, K., and Reddy, U. S. (2019). Crop pest classification based on deep convolutional neural network and transfer learning. Computers and Electronics in Agriculture, 164: 1–8. https://www.sciencedirect.com/science/article/pii/S0168169919310695.

Thorat, A., Kumari, S., and Valakunde, N. D. (2017). An IoT based smart solution for leaf disease detection. 2017 International Conference on Big Data, IoT and Data Science (BID), 193–198. https://ieeexplore.ieee.org/abstract/document/8336597.

Tian, M., Lu, P., Liu, X., and Lu, X. (2021). The cotton aphids counting system with super resolution and compressive sensing. Sensing and Imaging, 22(1): 1–18. https://idp.springer.com/authorize/casa.

Titiya, M. D., and Shah, V. A. (2018). Ontology based expert system for pests and disease management of cotton crop in India. International Journal of Web Portals (IJWP), 10(2): 32–49. https://www.igi-global.com/article/ontology-based-expert-system-for-pests-and-disease-management-of-cotton-crop-in-india/208168.

Toscano Miranda, R. E. (2024). Smart insect-pest management for cotton crops. https://repository.eafit.edu.co/items/3fb6f276-3a35-4bf8-8f59-3c2e5ce4d781.

Toscano-Miranda, R., Toro, M., Aguilar, J., Caro, M., Marulanda, A., and Trebilcok, A. (2022). Artificial-intelligence and sensing techniques for the management of insect pests and diseases in cotton: A systematic literature review. The Journal of Agricultural Science, 160(1-2): 16–31. https://www.cambridge.org/core/journals/journal-of-agricultural-science.

Toseef, M., and Khan, M. J. (2018). An intelligent mobile application for diagnosis of crop diseases in Pakistan using fuzzy inference system. Computers and Electronics in Agriculture, 15: 1–11.

Uppalapati, S. R., Young, C. A., Marek, S. M., and Mysore, K. S. (2010). Phymatotrichum (cotton) root rot caused by *Phymatotrichopsis omnivora*: Retrospects and prospects. Molecular Plant Pathology, 11(3): 325–334. https://doi.org/10.1111/j.1364-3703.2010.00616.x.

Van Goethem, S., Verwulgen, S., Goethijn, F., and Steckel, J. (2019). An IoT solution for measuring bee pollination efficacy. 2019 IEEE 5th World Forum on Internet of Things (WF-IoT), 837–841. https://ieeexplore.ieee.org/abstract/document/8767298.

Velásquez, D., Sánchez, A., Sarmiento, S., Toro, M., Maiza, M., and Sierra, B. (2020). A method for detecting coffee leaf rust through wireless sensor networks, remote sensing, and deep learning: Case study of the caturra variety in Colombia. Applied Sciences, 10(2): 1–51. https://www.mdpi.com/2076-3417/10/2/697.

Vijayakumar, S., Kumar, R. M., Choudhary, A. K., Deiveegan, M., Tuti, M. D., Sreedevi, B. et al. (2022). Artificial intelligence (AI) and its application in agriculture. Chronicle of Bioresource Management, 1–3. https://ojs.pphouse.org/index.php/CBM/article/view/3952.

Vinayak, V., and Jindal, S. (2017). CBIR system using color moment and color auto-Correlogram with block truncation coding. International Journal of Computer Applications, 161(9): 1–7. https://www.researchgate.net/profile/Sonika-Jindal/publication/315302511.

Wang, L., Zeng, Y., and Chen, T. (2015). Back propagation neural network with adaptive differential evolution algorithm for time series forecasting. Expert Systems with Applications, 42(2): 855–863. https://www.sciencedirect.com/science/article/pii/S0957417414004941.

Wang, Z., Wang, K., Liu, Z., Wang, X., and Pan, S. (2018). A cognitive vision method for insect pest image segmentation. IFAC-PapersOnLine, 51(17): 85–89. https://www.sciencedirect.com/science/article/pii/S2405896318311820.

Xavier, T. W., Souto, R. N., Statella, T., Galbieri, R., Santos, E. S., S. Suli, G. et al. (2019). Identification of Ramularia leaf blight cotton disease infection levels by multispectral, multiscale UAV imagery. Drones, 3(2): 1–37. https://www.mdpi.com/2504-446X/3/2/33.

Xia, C., Chon, T.-S., Ren, Z., and Lee, J.-M. (2015). Automatic identification and counting of small size pests in greenhouse conditions with low computational cost. Ecological Informatics, 29: 139–146. https://www.sciencedirect.com/science/article/pii/S1574954114001228.

Xia, C., Lee, J.-M., Li, Y., Chung, B.-K., and Chon, T.-S. (2012). In situ detection of small-size insect pests sampled on traps using multifractal analysis. Optical Engineering, 51(2): 027001–027001. https://www.spiedigitallibrary.org/journals/optical-engineering/volume-51/issue-2/027001.

Xiao, Q., Li, W., Chen, P., and Wang, B. (2018). Prediction of Crop Pests and Diseases in Cotton by Long Short Term Memory Network. pp. 11–16. *In*: Huang, D.-S., Jo, K.-H., and Zhang, X.-L. (Eds.). Intelligent Computing Theories and Application (Vol. 10955). Springer International Publishing. https://doi.org/10.1007/978-3-319-95933-7-2.

Yang, C., Odvody, G. N., Fernandez, C. J., Landivar, J. A., Minzenmayer, R. R., and Nichols, R. L. (2015). Evaluating unsupervised and supervised image classification methods for mapping cotton root rot. Precision Agriculture, 16(2): 201–215. https://doi.org/10.1007/s11119-014-9370-9.

Yao Qing, Y. Q., Xian DingXiang, X. D., Liu QingJie, L. Q., Yang BaoJun, Y. B., Diao GuangQiang, D. G., and Tang Jian, T. J. (2014). Automated counting of rice planthoppers in paddy fields based on image processing. https://www.cabidigitallibrary.org/doi/full/10.5555/20143356701.

Zhang, J., Kong, F., Wu, J., Han, S., and Zhai, Z. (2018). Automatic image segmentation method for cotton leaves with disease under natural environment. Journal of Integrative Agriculture, 17(8): 1800–1814. https://www.sciencedirect.com/science/article/pii/S209531191861915X.

Zhang, J., Qi, L., Ji, R., Yuan, X., and Li, H. (2012). Classification of cotton blind stinkbug based on Gabor wavelet and color moments. Transactions of the Chinese Society of Agricultural Engineering, 28(1): 133–138. https://www.ingentaconnect.com/content/tcsae/tcsae/2012/00000028/00000001/art00025.

Zhang, R., Ren, Z., Sun, J., Tang, W., Ning, D., and Qian, Y. (2017). Method for monitoring the cotton plant vigor based on the WSN technology. Computers and Electronics in Agriculture, 133: 68–79. https://www.sciencedirect.com/science/article/pii/S0168169916311644.

Zhong, Y., Gao, J., Lei, Q., and Zhou, Y. (2018). A vision-based counting and recognition system for flying insects in intelligent agriculture. Sensors, 18(5): 1–37. https://www.mdpi.com/1424-8220/18/5/1489.

CHAPTER 5

AI-Driven Synergies
A New Era of Pest Control Integration

Anurag Yadav[1,*] and *Kusum Yadav*[2]

1. Introduction

In the battle against agricultural pests, shifting away from sole reliance on chemical pesticides is crucial for crop protection and environmental sustainability. Integrated Pest Management (IPM) provides a framework for this shift, emphasizing a combination of biological controls, cultural practices, and judicious use of pesticides to optimize pest control. Unlike reactive approaches, IPM advocates for proactive management to deter pest problems and minimize chemical intervention (Keval et al., 2019). It involves ongoing monitoring of pest presence and potential threats and fostering a healthy agroecosystem resilient to infestations. Natural enemies of pests, including predators, parasitoids, or pathogens, become pivotal weapons (Raffel et al., 2008). IPM carefully integrates chemical pesticides as backup tools, deployed only when infestations exceed economically damaging levels and other techniques prove insufficient. This balanced approach minimizes the risks of pest resistance, pollution, and harm to beneficial organisms that often accompany the overuse of chemicals.

While IPM offers a framework for sustainable and responsible pest control, it presents several challenges that can impede its success. Unlike traditional pest control methods that rely on scheduled chemical applications, IPM requires an intricate understanding of pest biology and a willingness to react dynamically. Sudden pest outbreaks due to weather shifts or migrating pests can leave IPM practitioners needing to make rapid crop protection decisions (Agatz et al., 2020). Moreover, choosing the most appropriate control strategies based on pest types, available tools,

[1] Department of Microbiology, C.P. College of Agriculture, Sardarkrushinagar Dantiwada Agricultural University, Sardarkrushinagar, Dist. Banaskantha, Gujarat-385506, India.
[2] Department of Biochemistry, University of Lucknow, Lucknow, Uttar Pradesh-226007, India.
* Corresponding author: anuragyadav123@gmail.com

DOI: 10.1201/9781003546702-5

and the specific farming environment demands expertise. Maintaining the proactive vigilance crucial for IPM, which includes continuous monitoring of pest populations and beneficial insects, can also prove labor-intensive. Such unique challenges highlight the knowledge-intensive nature of IPM and underscore the need for robust support systems to ensure its success.

Artificial intelligence (AI) stands to revolutionize the decision-making processes at the heart of IPM. By harnessing the power of data analysis, AI promises to make IPM more proactive, precise, and effective. AI-powered models excel at pattern recognition, sifting through historical data and real-time sensor readings to forecast pest outbreaks and pinpoint areas of high infestation risk (Senoo et al., 2024). This early warning capability gives farmers crucial time to implement preventative measures before extensive crop damage occurs. Additionally, AI-driven image recognition aids in rapid and accurate pest and beneficial insect identification, crucial for ensuring treatment plans are appropriately targeted.

Perhaps more importantly, AI analyzes huge datasets, including pest behavior, weather patterns, and the outcomes of past interventions, to suggest the most effective control strategies, optimal timing, and dosage. This minimizes guesswork and helps reduce the potential overuse of pesticides. Additionally, by detecting patterns in pest populations, AI models can warn of emerging resistance to specific control methods, allowing for a rapid shift in strategies. Overall, AI empowers IPM practitioners with insights gleaned from data, enhancing the impact of their actions and facilitating a shift towards sustainable agricultural practices.

2. How AI Optimizes Biological Control?

2.1 Matching Predators and Prey

IPM emphasizes sustainable practices for pest control, often relying on introducing natural enemies like predators, parasitoids, or pathogens to suppress pest populations. This approach offers a significant advantage over heavy reliance on chemical pesticides, which can harm human health and the environment. However, selecting the most effective biological control agent (BCA) for a specific situation can be complex. AI emerges as a revolutionary tool, leveraging data analysis to provide tailored recommendations for BCA deployment within IPM programs.

AI algorithms can be trained on vast datasets encompassing intricate details about pest behavior, environmental conditions, and the efficacy of various BCAs against specific pests (Javaid et al., 2023). This empowers them to identify the most promising predator-prey matches, significantly enhancing the effectiveness of BCA implementation. Artificial neural networks (ANNs) and metaheuristic optimization algorithms, like the prey-predator algorithm (PPA), are particularly well-suited for this task (Fu et al., 2022). The effectiveness of these AI techniques has been demonstrated in diverse fields beyond pest management. For instance, ANNs have been successfully employed to predict the thermal conductivity of nanocomposite fibers (Hamadneh et al., 2019), while PPAs have been utilized to optimize microchannel heat sinks (Hamadneh et al., 2018). Furthermore, AI has shown promise in studying the spread of infectious diseases, with ANNs being used to model the spread of COVID-19 (Malik et al., 2021).

The combination of different algorithms, such as co-evolutionary algorithms and deep-reinforcement learning, has been explored to enhance the understanding of predator-prey systems (Wang et al., 2019a). These approaches allow for the simulation of complex interactions between predators and prey, enabling the organisms to adapt their behavior based on experiences gained through interactions with the environment (Wang et al., 2019b). Moreover, studying predator-prey relationships extends beyond artificial intelligence and optimization algorithms. Research has delved into areas like functional responses of predators, prey selection, and the impact of environmental factors on trophic interactions (Schmitz, 2017). Understanding the dynamics of predator-prey systems involves considering factors such as prey availability, predator size, and the architecture of food peaks (van Velzen et al., 2022). Furthermore, mathematical models have been developed to analyze the impact of seasonal variations, infections, and environmental conditions on predator-prey interactions (Sen et al., 2023). These models provide insights into the complex dynamics of ecosystems and how factors like weather variations and infections can influence the balance between predator and prey populations.

2.2 Predicting Pest Release Timing

Highlight AI's ability to model pest life cycles and population dynamics, predicting optimal timing for releasing natural enemies.

Timing is everything when it comes to maximizing the success of biological control strategies. Releasing natural enemies at the wrong point in a pest's lifecycle can significantly reduce their efficacy. AI's ability to model pest population dynamics offers a solution, providing insights for the precision deployment of biological control agents. By analyzing vast amounts of data, including historical trends, weather patterns, and sensor readings, AI models pinpoint the most vulnerable stages in a pest's development (e.g., egg or early larval stages), factoring in pest behavior and the biology of specific natural enemies (Domingues et al., 2022; Singh et al., 2022; Yadav et al., 2023). This information allows AI to predict ideal windows for introducing control agents, increasing the chances of establishment and ensuring maximum impact on the target pest.

These insights also extend to managing pest resistance; AI models can sometimes even suggest release patterns strategically designed to mitigate the risk of pests becoming resistant to particular biological controls. In essence, AI acts as a powerful decision-support tool, enhancing the precision and sustainability of IPM by equipping practitioners with crucial timing information for biological control deployment. Figure 1 focuses on data collection, modeling, and decision-making processes used to predict the optimal timing for pest release based on environmental and predictive factors.

Figure 1. Framework for predicting pest release timing.

2.3 Pest Monitoring and Evaluation

The success of biological control methods within IPM depends on robust monitoring and proactive evaluation. AI-powered sensors are game-changers in this domain. Armed with sophisticated image analysis, algorithms, and camera traps, these sensors provide real-time tracking of pest populations and beneficial insects' activity (Maraveas, 2022). The constant stream of field data empowers farmers by revealing whether biocontrol efforts successfully reduce pest numbers and if sufficient predator populations are established in the area. Furthermore, AI systems can flag subtle shifts in pest behavior or detect early warning signs of potential control failures, triggering timely interventions before pest outbreaks spiral out of control.

Beyond simply tracking insect populations, AI-powered sensors can collect environmental data such as temperature, humidity, and rainfall. Analyzing this alongside pest and predator data, AI models can identify whether conditions are favorable for the released biological control agents or if adjustments to the approach might be required. In some cases, AI-enhanced monitoring could highlight the decreasing efficacy of beneficial insect strains or pathogens, signaling the emergence of resistance in the pest population. AI creates a continuous feedback loop for biological control within IPM. This data-driven approach enables farmers to celebrate successes, fine-tune strategies based on real-time results, and proactively address issues before they escalate, optimizing overall effectiveness.

3. AI Techniques for Pest Control

3.1 Machine Learning and Predictive Modeling for Forecasting Pest Outbreaks and Optimizing Control Strategies

Machine learning and predictive modeling transform how we forecast pest outbreaks and optimize IPM strategies. These models draw on various techniques to predict different aspects of pest outbreaks. Regression models, such as linear regression (Ramazi et al., 2021), decision trees (Debeljak and Džeroski, 2011), and random forests (Simon et al., 2023), are employed to predict continuous values like pest population density. Meanwhile, classification models (including logistic regression, support vector machines, and artificial neural networks) help categorize outbreak risk levels (Yan et al., 2015). For analyzing pest trends over time, time series models such as ARIMA and LSTM networks offer valuable insights (Boopathi et al., 2015).

The performance of these models depends heavily on the quality and diversity of input data, including historical pest counts, weather conditions, crop growth stage, and records of previous control measures (Domingues et al., 2022). Python's rich scientific computing environment makes it an ideal platform for building these models. Libraries like Scikit-learn (Rajamani and Iyer, 2023), Pandas (Franklin et al., 2023), and Statsmodels (Duchesnay et al., 2021) provide a robust foundation for machine learning, data manipulation, and statistical modeling. When dealing with complex scenarios, such as time series analysis or image-based pest identification, deep learning frameworks like TensorFlow (Jang, 2023) or PyTorch (Stanescu and Dinu, 2023) can be leveraged for their advanced modeling capabilities.

3.2 Computer Vision and Image Recognition for Pest/Beneficial Insect Identification and Monitoring

Computer vision and image recognition play a transformative role in IPM by automating the identification and monitoring of pests and beneficial insects. These techniques involve a series of steps, beginning with image preprocessing, such as resizing, normalization, and background removal to isolate the insect of interest. Traditionally, features like color and texture were manually extracted and combined with classifiers like Support Vector Machines (SVMs) to differentiate insect species (Figure 2).

However, the rise of deep learning, particularly Convolutional Neural Networks (CNNs), has revolutionized this process. CNNs excel at autonomously learning features directly from vast image datasets. Pre-trained models like VGG16 or ResNet50 can often be fine-tuned for specific agricultural datasets, accelerating model development (Wu et al., 2023). With its rich libraries like TensorFlow, Keras, OpenCV, and Scikit-image, Python provides a robust framework for implementing computer vision solutions in IPM.

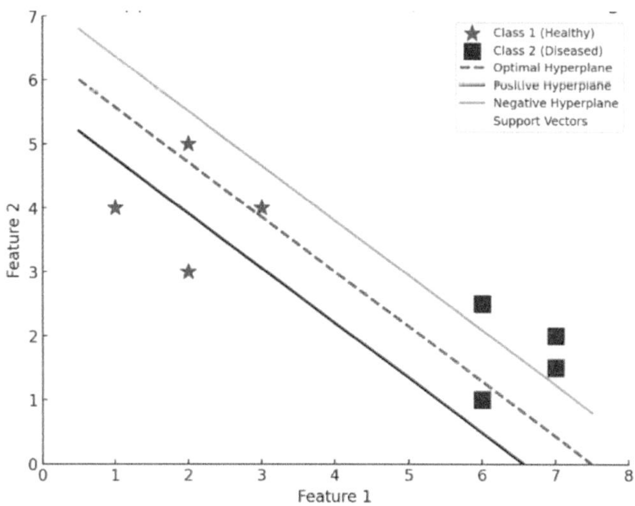

Figure 2. Support Vector Machine (SVM) classification diagram. It includes two classes represented by stars and squares, an optimal hyperplane (dashed line) that separates the two classes, positive and negative hyperplanes (solid lines), and support vectors circled near the boundary.

3.3 Optimization Algorithms to Refine Pesticide Dosages and Application Timing

Optimization algorithms can transform pesticide management in IPM by making decision-making more precise and automated. These tools consider environmental conditions, pest life cycles, and economic constraints, refining pesticide dosages, and application timings. They can address continuous challenges, such as finding the ideal dosage, and discrete ones, like determining the best schedule for pesticide applications. Table 1 and Figure 3 outlines various AI techniques used in pest control.

In simpler cases where relationships between variables are mathematically defined, gradient-based methods like linear programming work well. For example, linear programming can help optimize the cost of pesticide usage by finding the most cost-effective combination of dosage and timing. However, the complexities of real-world scenarios—such as pest population dynamics and changing weather conditions—require more advanced methods.

Evolutionary algorithms, such as genetic algorithms (GA) and differential evolution (DE), are more suitable for these complex, non-linear problems. Inspired by natural selection, these algorithms explore a wide range of solutions. GA tests multiple pesticide dosage schedules simultaneously and evolves them over time to find the best approach. DE refines dosages in dynamic environments where pest resistance and climate interact unpredictably (Bäck et al., 2023).

Nature-inspired algorithms like particle swarm optimization (PSO) and ant colony optimization (ACO) offer further benefits. PSO simulates the behavior of

Table 1. AI techniques in pest control.

AI Technique	Role	Application	References
Machine Learning	Pattern recognition, prediction	Predicting pest outbreaks, optimizing control measures	(Kamilaris and Prenafeta-Boldú, 2018)
Computer Vision	Image analysis	Identifying pest species, monitoring pest populations	(Kamilaris and Prenafeta-Boldú, 2018)
Robotics	Automated pest control	Targeted pesticide application, release of biocontrol agents	(Adetunji et al., 2023)
IoT Sensors	Data collection	Real-time monitoring of pest presence and environmental conditions	(Rustia et al., 2020)
Expert Systems	Decision support	Providing recommendations for pest control strategies	(Nieróbca and Zaliwski, 2014)

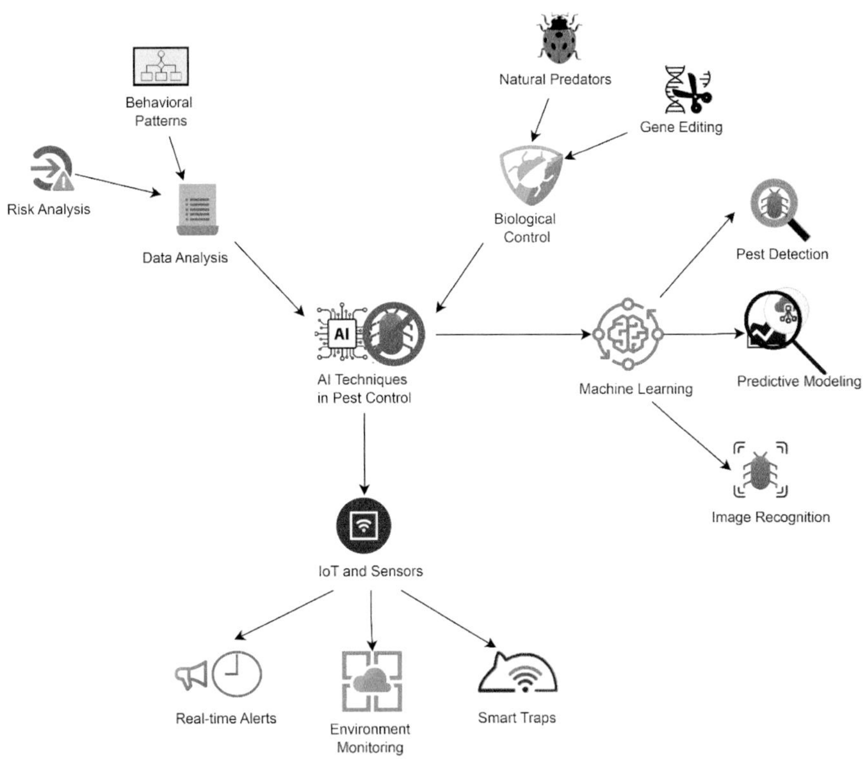

Figure 3. AI Techniques for Pest Control.

swarming organisms, which helps balance multiple objectives like pest control and environmental impact. ACO, modeled after ant foraging, is particularly effective for scheduling applications based on pest population thresholds and pesticide decay rates.

Python's scientific computing libraries, such as SciPy, facilitate the practice of these algorithms. SciPy's minimize and linprog functions help optimize pesticide dosages and schedules. More specialized libraries like DEAP and PySwarms enable evolutionary and swarm intelligence algorithms, allowing practitioners to experiment with different strategies and refine them using real-world data.

When combined with real-time inputs from pest monitoring systems and weather forecasts, these algorithms allow dynamic, adaptive pest management. Such responsiveness reduces the need for excessive pesticide use, improving sustainability and efficiency. By integrating optimization algorithms with data-driven approaches, IPM systems can become smarter and more sustainable, minimizing environmental impacts and costs while maximizing pest control effectiveness.

4. Integrating Biological and Chemical Control with AI

4.1 Intelligent Use of Pesticides

Though IPM prioritizes biological controls and preventative measures, there are instances where supplementing these with pesticides becomes necessary. AI plays a vital role in ensuring that this integration is done intelligently and cautiously to protect valuable components of the agroecosystem. AI algorithms analyze databases detailing the specific toxicity of pesticides on beneficial insects like pollinators and the natural enemies of pests (Sharma et al., 2022; Yang et al., 2022). Based on this analysis, they suggest chemicals with the lowest risk profile, reducing collateral damage. When utilizing pesticides, AI tools can guide highly localized applications or recommend strategic timing aligned with pest lifecycles but outside periods of activity for vital beneficial insects.

Moreover, AI models can incorporate knowledge of existing pest resistance within specific populations. This insight helps avoid using pesticides that are likely to fail or those that might inadvertently increase resistance among beneficial insect populations. Such data-driven insights make pesticides a calculated last resort within IPM—deployed with precision and timed to minimize disruption to the biological controls vital for long-term, sustainable pest management. Table 2 describes various AI Applications used in chemical and biological pest control.

Table 2. AI Applications in chemical vs. biological pest control.

AI Application	Chemical Pest Control	Biological Pest Control	References
Predictive Analytics	Forecasting pest resistance, optimizing chemical use	Predicting biocontrol agent effectiveness	(Kotula et al., 2021)
Monitoring Systems	Tracking pesticide levels, exposure	Monitoring biocontrol agent activity	(Adetunji et al., 2023)
Optimization Algorithms	Optimizing dosage and application timing	Enhancing deployment strategies of biocontrol agents	(Maraveas, 2022)
Decision Support Systems	Recommending pesticide types and amounts	Selecting appropriate biocontrol strategies	(Damos, 2015)
Image Recognition	Detecting pest damage	Identifying beneficial organisms and pests	(Li et al., 2021)

4.2 Understanding Synergies

Microbial dynamics in the rhizosphere are crucial for plant health, naturally deterring pests through competitive exclusion and protective metabolites. These communities support integrated pest management (IPM) by enhancing biological control agents and reducing chemical reliance, enabling synergistic, targeted pest management that fosters a resilient agroecosystem (Yadav and Yadav, 2013). The effectiveness of IPM strategies often depends on the intricate relationships between biological control agents, such as predatory insects or microbial pathogens, and the selective use of chemical pesticides. AI technologies play a pivotal role in deciphering these complex interactions, enabling a deeper understanding of potential synergies or conflicts between biological agents and pesticides.

AI-powered systems, especially those leveraging machine learning algorithms, can process large datasets to evaluate the compatibility of specific pesticides with beneficial organisms (Table 4). For instance, AI models can identify pesticides that pose minimal risk to natural predators and parasitoids, offering solutions that may even enhance the efficacy of biological control agents under certain conditions (Demirel & Kumral, 2021). By pinpointing pesticides that temporarily boost the performance of natural enemies, AI-guided decisions help reduce the reliance on broad-spectrum chemicals, which are often detrimental to non-target species and the environment (Thangamani et al., 2024). Table 3 compares chemical and biological pest control methods.

Conversely, AI models also aid in flagging pesticides that could disrupt IPM by negatively affecting natural enemies. These models analyze historical data to identify chemicals that, while controlling target pests, may also suppress beneficial insects or microbes. This disruption leads to the destabilization of the ecosystem, potentially causing pest resurgence or secondary pest outbreaks. For example, AI tools can assess pesticide-induced sub-lethal effects on predators, revealing long-term consequences for biological control programs (Skouras et al., 2021; Wang et al., 2023). Thus, AI helps avoid the application of highly toxic pesticides that may compromise the integrity of a functioning biological control system.

Beyond direct interactions, AI enhances the understanding of indirect and cascading effects of pesticide applications. Some pesticides, though effective in controlling pests, may weaken the natural enemies' population, leading to ecological imbalances. Such unintended consequences can disrupt the delicate balance of predator-prey dynamics, exacerbating pest problems in the long term (Jarvis et al., 2006). AI-driven models can predict such scenarios by analyzing ecosystem-level interactions, suggesting mitigation strategies or alternative approaches to maintain ecological harmony.

Furthermore, AI aids in optimizing the selection of pesticide formulations, tailoring them to be more compatible with biological control measures. For instance, microencapsulated pesticide formulations, designed to release active ingredients slowly, can be evaluated by AI for their reduced non-target impacts while maintaining

Table 3. Comparison of chemical and biological pest control methods.

Feature	Chemical Control	Biological Control	References
Mode of Action	Direct toxicity, disruption of physiological processes	Predation, parasitism, competition, pathogens	(Sharma et al., 2023)
Specificity	Varies (broad-spectrum to narrow-spectrum)	High (often species-specific)	(Jessie, 2017)
Environmental Impact	Potential for residue, non-target effects	Minimal residue, lower risk to non-target organisms	(Serrão et al., 2022)
Resistance Development	Higher risk	Lower risk	(Tabashnik et al., 2013)
Cost	Varies (initial application may be lower)	Potentially higher upfront costs but long-term savings	(Ehlers et al., 2020)
Speed of Action	Rapid	Gradual	
Sustainability	Lower	Higher	(Barzman et al., 2015)
Duration of Effectiveness	Short-term, requires frequent reapplication	Long-term, self-sustaining if natural enemies establish	
Impact on Soil Health	Potential negative effects on soil microbiota	Promotes biodiversity and soil health	(Ahmed et al., 2022)
Effect on Pest Population Dynamics	Can cause pest resurgence or secondary outbreaks	Helps maintain natural pest-enemy balance	(Hossain et al., 2017)
Ease of Implementation	Easy, established infrastructure for application	Requires knowledge of ecosystems and specialized tools	(Cock et al., 2009)
Human Health Impact	Potential risks from exposure to chemicals	Considered safe for humans	(Aktar et al., 2009)
Regulatory Complexity	Strictly regulated with residue and application limits	Less regulatory complexity, but species introductions may face scrutiny	(Zaller and Zaller, 2020)
Risk of Non-Target Effects	High, impacts beneficial species like pollinators	Lower risk, but species introduction must be managed carefully	(Serrão et al., 2022)
Climate Adaptability	Susceptible to environmental conditions	More resilient, especially if using local species	(Bardin et al., 2015)

effectiveness against the pest. The precision-guided approach enables the strategic integration of chemical and biological controls to preserve the integrity of the agroecosystem (Liakos et al., 2018).

Table 4. AI Applications in chemical vs. biological pest control.

AI Application	Chemical Pest Control	Biological Pest Control	References
Predictive Analytics	Forecasting pest resistance, optimizing chemical use	Predicting biocontrol agent effectiveness	(Kotula et al., 2021)
Monitoring Systems	Tracking pesticide levels, exposure	Monitoring biocontrol agent activity	(Adetunji et al., 2023)
Optimization Algorithms	Optimizing dosage and application timing	Enhancing deployment strategies of biocontrol agents	(Maraveas, 2022)
Decision Support Systems	Recommending pesticide types and amounts	Selecting appropriate biocontrol strategies	(Damos, 2015)
Image Recognition	Detecting pest damage	Identifying beneficial organisms and pests	(Li et al., 2021)

5. Benefits of AI-supported IPM

5.1 Environmental Sustainability

AI acts as a catalyst for environmental sustainability within IPM. By integrating the power of data analysis and predictive modeling, AI helps farmers prioritize biological controls, precision-target interventions and cultivate the overall health of the agroecosystem. Such a fundamental shift directly translates into significantly reducing the use of potentially harmful pesticides. AI minimizes the need for broad-spectrum chemical applications, often detrimental to biodiversity and prone to pollution (Rosner et al., 2024). Through analysis and tailored recommendations, AI guides the strategic and selective use of less damaging pesticides, protecting beneficial insects like pollinators, which uphold crucial agroecosystem services. These combined actions safeguard biodiversity and minimize pesticide runoff that risks soil, water, and air contamination.

The emphasis on preventative management and fostering ecological balance means reducing the conditions ideal for pest outbreaks. This long-term, holistic approach builds environmental sustainability and strengthens farming practices for future generations.

5.2 Economic Efficiency

AI-driven IPM can improve farm profits and economic efficiency. One of the biggest drivers of this efficiency is the reduced reliance on costly pesticides. AI systems analyze sensor data and pest populations to guide IPM tactics precisely, replacing routine chemical applications with strategic use or alternatives (Rustia et al., 2022). This targeted approach directly influences input costs, resulting in financial savings for farmers. Furthermore, healthier plants resulting from proactive, data-driven pest management foster improved yields and ultimately contribute to increased revenue. AI's precision extends to minimizing damage often caused by broad-spectrum chemicals, resulting in better-quality crops that may fetch premium prices on the market (Tannous et al., 2023).

It's not just direct savings—AI offers ways to reduce labor costs in some situations. Real-time monitoring tools like camera traps and data systems minimize the need for intensive manual crop scouting. Beyond tangible returns, AI-supported IPM protects the long-term economic viability of farming by reducing future vulnerabilities. A healthy, more robust agroecosystem lessens the need for emergency crop management interventions.

While not solely financial, the lowered environmental impact of such farming methods has implications for the overall economy. Cleaner ecosystems mean reduced public spending on cleanup, a burden often disproportionately placed on future generations. Additionally, rising consumer demand for sustainably grown produce gives those utilizing IPM an edge, enabling premium pricing for products perceived as safer and more responsibly produced. In essence, AI-supported IPM promotes both upfront and long-term economic efficiency, bolstering individual farmers' financial stability and fostering a healthy ecosystem for the future of agriculture.

5.3 Long-term Pest Resilience

IPM provides a sustainable foundation for pest control by combining diverse strategies to minimize crop damage while protecting the environment. Integration of AI is supercharging IPM practices, leading to increased precision, efficiency, and long-term crop resilience against pests.

AI drives this agricultural revolution in several ways. Firstly, using advanced image recognition tools empowers farmers and automated systems to accurately identify pests and diseases from photos, insect samples, and damage patterns. Early phytopathogen detection is vital for timely intervention, preventing devastating crop losses (Arnal Barbedo, 2019). Moreover, AI algorithms meticulously analyze data from weather stations, field sensors, and historical records. This allows them to create predictive models that forecast pest outbreaks disease spread, and assess risk levels based on temperature, humidity, and crop growth stage (Chlingaryan et al., 2018). Proactive measures can then be taken to disrupt pest lifecycles, giving farmers a crucial advantage.

Building upon this, AI systems process vast datasets encompassing pest biology, control methods, and environmental conditions. The result is tailored treatment recommendations pinpointing the most effective pesticides, biocontrol agents, or cultural practices while minimizing harm to beneficial insects and the surrounding environment (Fountas et al., 2020). AI even optimizes dosages and treatment timing to combat the development of pesticide resistance in pests. AI's precision doesn't stop there—drones and robots enable localized spraying, the strategic release of beneficial insects, and even targeted pest removal with lasers (Li et al., 2021). This drastically reduces the need for broadcast treatments, minimizing environmental impact.

These benefits are evident in real-world case studies. AI-equipped insect traps can distinguish between target and non-target insects, providing crucial, real-time population data. This early-warning system supports focused IPM strategies to combat pests like spotted lanternflies (Li et al., 2021). In vineyards, AI models detect early signs of powdery mildew in grape leaves, allowing preventative action that reduces reliance on fungicides and builds crop resilience (Sanghavi et al., 2021). Similarly, rice farmers utilize AI-driven drones for precision pesticide application, pinpointing damaged areas to reduce overall pesticide use (Talaviya et al., 2020).

The advantages of AI-supported IPM for long-term pest resilience are undeniable. Its emphasis on prevention and targeted treatments lessen dependence on chemical pesticides, fostering soil health and minimizing pest resistance. This precision often translates to economic benefits for farmers through reduced pesticide costs and greater yields due to prevented infestations. Crucially, AI-supported IPM prioritizes minimal ecological impact, protecting biodiversity, pollinator populations, and water quality, ensuring the long-term health of agroecosystems.

As the partnership between AI and IPM matures, we can expect even smarter tools, refined models, and the development of systems that autonomously monitor crops, implement tailored interventions, and adapt to evolving pest pressures. This integration will not only protect crops today but will also pave the way for a future of sustainable, long-term agricultural resilience.

6. AI-driven IPM in Greenhouse Production

Greenhouses offer environments tailored for cultivation but face challenges with pest infestations. AI-driven platforms are revolutionizing IPM within these controlled spaces, promoting a successful balance between biological and chemical control strategies (Toscano-Miranda et al., 2022). Companies like Ecoation (https://www.ecoation.com/) demonstrate AI's prowess with intelligent sensor networks and powerful algorithms. Images captured by their systems assist in real-time pest identification, alerting users to emerging problems. Predictive models anticipate potential outbreaks by analyzing pest trends alongside greenhouse environmental conditions like temperature and humidity.

This data-driven approach optimizes the timing and dosage of biological control agents such as predatory mites or parasitic wasps (Saeed Ben Youssef, 2023). It also allows for strategic integration with the controlled environment of the greenhouse; recommendations might guide maintaining ideal conditions for the success of natural pest enemies. When a well-placed chemical intervention is necessary, AI systems guide targeted applications by zone or timeframe to minimize impacts on established biocontrol programs. This process often suggests formulations that pose minimal risk to beneficial insects.

Case studies demonstrate real-world benefits. Greenhouse operators witness improved yields, healthier plant growth, and reduced reliance on harsh pesticides, thanks to AI-powered IPM. The continuous monitoring facilitated by AI-platforms translates into proactive pest management, further reducing outbreaks and fostering data-driven insights for improvement. Significantly, this controlled environment optimizes conditions for successfully deploying biological and chemical tools, giving greenhouse IPM a unique advantage when integrating strategies based on AI recommendations. Table 5 highlights some of the leading AI-enabled pest monitoring systems transforming pest management across various agricultural environments.

Integrating AI into IPM strategies has led to significant advancements in agricultural practices worldwide. These technologies reduce the dependency on

Table 5. Examples of AI-enabled pest monitoring systems.

System Name	Technology Used	Pest Targeted	Key Benefits	References
Trapview	Image recognition	Multiple pests	Real-time pest counts, alerts, decision support	(Lima et al., 2020)
FarmSense	Acoustic sensors	Crop-damaging insects	Early detection of infestations	(Bhairavi et al., 2020)
Semios	Pheromone traps + AI	Codling moth, etc.	Mating disruption, pest population monitoring	(Haviland et al., 2021)
Prospera	Computer vision	Various pests	Plant health monitoring, early disease detection	(Sree et al., 2024)
Taranis	Aerial imagery	Multiple pests	Large-scale monitoring, crop stress detection	(Kolisnyk and Lebid OV, 2023)
IPMPro	AI-driven decision-making algorithms	Multiple pest categories	Optimizes pesticide use while supporting IPM, balancing control methods	(Dwivedi et al., 2020)

chemical pesticides and improve crop yields through precise and timely interventions. Let's explore some notable examples that underscore the success of AI-driven IPM systems.

In North America and Europe, OKO platform exemplifies the transformative power of AI in agriculture. By offering Closed-Loop IPM, Yield Production Assessment, and Crop Work Quality services, ecoation has optimized pest management across several greenhouses. The platform leverages AI to enhance the efficiency of existing IPM programs, demonstrating substantial savings and improved outcomes for growers (Vänninen, 2022). The digitization of IPM activities enables rapid information dissemination, allowing for more informed decision-making and cost-effective strategies without compromising results. These initiatives have showcased how AI can optimize and reduce costs in IPM programs, providing actionable insights that lead to more profitable operations.

Another groundbreaking project, Microsoft's FarmBeats, illustrates the potential of AI, edge computing, and IoT in revolutionizing agriculture. By integrating sensor data with drone imagery, FarmBeats offers an end-to-end system that supports better farm management. Techniques developed through this project, such as TV White Spaces-based long-range sensor networks and automation of drone flights, demonstrate the feasibility of improving crop yields and reducing costs through technology. FarmBeats has been recognized for its innovative approach to increasing farm production using IoT and featured in various prestigious forums, highlighting its contribution to digital agriculture (Kharel et al., 2022).

Additionally, AI-driven models have been validated for their effectiveness in digital agriculture. Techniques involving crop images, decision trees, neural networks, and random forests have been employed to train models like DefoNet for pest and disease identification. This approach enables precise defoliation detection and offers a scalable solution for monitoring crop health, further enhancing IPM strategies through technology (Zhang et al., 2022).

These examples from different parts of the world highlight the growing importance of AI in agriculture. By integrating AI-driven tools into IPM, farmers can achieve more sustainable, efficient, and effective pest management practices. This not only leads to reduced pesticide use and environmental impact but also improves crop yields, showcasing the potential of technology in addressing some of the most pressing challenges in agriculture today.

7. Economic Analysis

Real-world applications demonstrate the transformative potential of AI in IPM. For instance, image recognition models trained on large datasets of insect images can automate pest identification and population tracking, significantly reducing the time and error associated with manual scouting (Li et al., 2021). Code plays a crucial role here, with Python libraries enabling data preprocessing, deep learning model development (using frameworks like TensorFlow or PyTorch), and deployment of these models on mobile devices or for drone image analysis.

AI also assists in disease forecasting and optimizing fungicide use. Predictive models combine real-time weather data, historical disease patterns, and crop susceptibility information to pinpoint high-risk periods (Sellam and Poovammal, 2016). Code facilitates the integration of weather APIs with farm data, building machine learning models, and creating interactive web interfaces tailored for growers.

Another exciting area for AI in IPM is optimizing the release strategies of biological control agents. Simulation models can be built to account for pest population dynamics, predator-prey interactions, and the influence of environmental conditions. Libraries in Python's scientific computing stack are essential for constructing these simulations, employing differential equation solvers, performing optimization tasks, and effectively visualizing the results for decision-making.

It's vital to remember that successful implementation hinges on the quality of data used to train these AI models and the seamless integration of code-based tools into existing agricultural practices. Additionally, focusing on user-friendly interfaces and interpretable outputs is crucial for growers to adopt them more widely.

8. Regulatory Framework

The current regulatory landscape governing AI use in agriculture remains relatively underdeveloped, posing potential hurdles for the broader adoption of AI-powered IPM systems. While existing regulations touch upon aspects like data privacy, pesticide use, and the introduction of biological control agents, a comprehensive framework specifically addressing AI integration into agriculture is lacking. This regulatory gap raises questions about liability in the case of unintended consequences from AI-guided recommendations (e.g., crop damage), the need for transparency and explainability in AI models, the complexities of cross-border data sharing within the context of varying privacy regulations, and the potential for certification standards to bolster trust in AI agricultural applications (Bronson and Knezevic, 2016).

Proactive collaboration between policymakers, researchers, technology developers, and agricultural stakeholders will be essential to navigate this evolving landscape. A balanced regulatory framework should foster innovation while addressing the ethical concerns outlined earlier and ensuring AI's safe, accountable, and transparent implementation in IPM practices.

9. Ethical and Social Considerations

AI-powered IPM presents a compelling solution for agricultural sustainability but raises critical ethical considerations requiring careful examination. The reliance on large datasets containing sensitive farm information highlights the need for transparent data ownership and privacy policies (Agatz et al., 2020). Furthermore, it is essential to be mindful of potential biases within AI models, as they risk perpetuating existing disparities if not carefully designed. To ensure equitable access to these technologies, a conscious effort must be made to address accessibility and affordability for farmers regardless of operation size or location.

While AI offers the potential for optimizing decision-making, we must be cautious about the possibility of over-reliance on automated systems, potentially overshadowing farmer knowledge and intuition. Finding the right balance between AI-guided recommendations and human expertise is critical for sustainable IPM practices. Moreover, the environmental risks of even precise, AI-guided chemical applications demand ongoing evaluation to mitigate unintended impacts on non-target organisms and biodiversity (Rose et al., 2016).

Proactive discussion and collaboration among researchers, technology developers, policymakers, and farmers are essential for ensuring AI's ethical and responsible agricultural development. Prioritizing these considerations will pave the way for a more just and sustainable future for food production.

10. Challenges and Considerations

10.1 Data Availability

One of the most significant challenges to effective AI within IPM is the need for extensive and reliable data. The most successful AI models draw upon substantial knowledge of pest lifecycles, behaviors, and responses to various management strategies (Damos, 2015; Shaikh et al., 2022). Additionally, in-depth data on specific crop characteristics, including vulnerabilities and sensitivities to pests and potential interventions, is essential for ensuring that AI-guided management protects yield and quality. For the successful integration of biocontrol agents, detailed mappings of predator-prey relationships, environmental tolerances of beneficial insects, and any sensitivities to various chemicals further empower AI systems.

Environmental variables, including weather patterns, microclimates, and soil conditions, are equally impactful, providing the AI-driven predictive ability crucial for forecasting pest outbreaks and optimizing response timing. Success depends not just on collecting data, but creating continuous feedback loops. To refine their ability to make precise recommendations, AI models need reliable information about the outcomes of specific IPM interventions, both the expected gains and less predictable failures.

Often, relevant data remains decentralized, scattered across research entities, private holdings, and individual growers. In addition, specific regions might face issues due to gaps in knowledge on minor crops or the effects of certain practices on local or emerging pests. Such gaps or inaccessible information directly restrict the quality and scope of AI solutions. Further, misrepresented data, whether a poorly labeled insect image or flawed sensor reading, can compromise AI results.

Progress hinges on concerted effort—researchers, practitioners, and tech companies in a shared quest to create robust, accurate, and accessible agricultural datasets. Standardized collection practices alongside initiatives promoting data sharing are key components for empowering AI to reach its full potential in revolutionizing IPM.

10.2 Adoption and Trust

Successful adoption of AI within IPM hinges on overcoming perceived obstacles and creating a foundation of trust between farmers and the technology. Challenges in adoption stem from concerns surrounding the complexity of AI systems, understanding their logic, cost implications, and ensuring data security. Accessibility plays a key role—intuitive interfaces, user-friendly designs, and readily available training opportunities are essential for users without exposure to complex technology. Demonstrating long-term cost savings and positive impacts on crop outcomes helps justify upfront investment into AI systems.

Perhaps more crucially, transparency in an AI system's decision-making process is critical to fostering trust. Farmers need to understand how AI analyzes data and arrives at its recommendations to integrate those insights confidently into their IPM practices (Kvakkestad et al., 2021). This means highlighting the specific data points that drove recommendations, showcasing potential alternative interventions the AI considered, and acknowledging any limitations the system may have. It is essential to reinforce the idea that AI is a powerful decision-support tool, augmenting—not replacing—the crucial role of a farmer's experience and intuition in the field. Proactive dialogue and clear policies around data privacy, ownership, and usage rights build trust and promote adoption. Figure 4, a flowchart assesses pest tolerance by predicting seasonal pest levels, estimating necessary and maximum shoot densities, and calculating their difference. A positive result suggests natural pest tolerance, while a negative result prompts IPM actions like adjusting sowing dates, using insecticide-treated seeds, or crop rotation.

Ultimately, successfully navigating these challenges lies in collaborative efforts – technologists working alongside farmers to create user-centric solutions, transparent data-sharing practices, and, importantly, consistently validating the economic gains AI can bring to IPM. Cultivating trust and accessibility empowers farmers to leverage the power of AI and transition towards a more data-driven approach to pest management within a sustainable IPM framework.

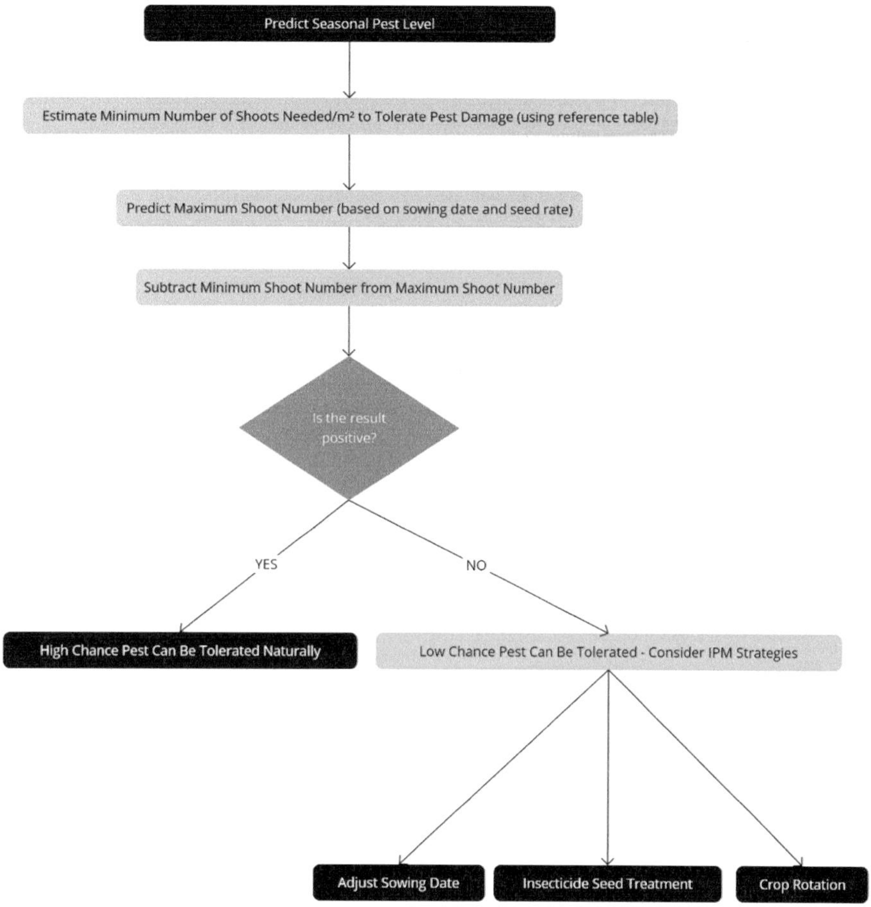

Figure 4. Enhanced decision-making framework for pest tolerance with IPM strategies.

11. Conclusion

AI fosters a shift away from reactive, chemical-heavy pest management by harnessing the power of data analysis, predictive modeling, and intelligent sensors. Instead, it empowers farmers to implement sustainable, proactive, and precise strategies informed by a complex blend of insights.

AI arms farmers with tools to predict pest outbreaks before they cause economic damage, reducing reliance on broad-spectrum chemicals. AI models allow for the strategic integration of biological control agents, optimizing their effectiveness and minimizing disruption to agricultural ecosystems. AI even allows targeted treatments tailored to the field's specific conditions and pest pressures, significantly reducing pesticide use.

Through continuous monitoring and a feedback loop with real-world results, AI systems enhance their capacity to protect healthy crops while minimizing environmental impact. AI solutions don't just benefit the current system; they promote long-term resilience by helping practitioners manage emerging pest resistance issues and prioritize ecosystem well-being.

While implementation and accessibility present challenges, AI's potential to improve crop yields, preserve biodiversity, and optimize resource usage signifies a step towards a future of sustainable agriculture. With continuous development, careful consideration of adoption issues, and a focus on transparency, AI promises to reshape how we manage pests, bolstering responsible farming practices for generations.

12. Future Potential

AI's transformative effect on IPM reaches beyond current applications—ongoing research signals vast future potential. One exciting thrust focuses on utilizing AI's analytical capabilities for proactive resistance monitoring. By combing through massive datasets of historical observations and pest population genetics, AI models could identify subtle patterns signaling the emergence of resistance to control methods. This early warning system allows farmers to adjust strategies before an issue becomes a crisis, ultimately extending the longevity of effective control methods (Toscano-Miranda et al., 2022).

Simultaneously, AI is becoming a crucial ally in accelerating the breeding of pest-resistant crop varieties. Machine learning algorithms excel at pattern recognition within extensive genetic datasets. Cross-referencing this with crop vulnerability traits could revolutionize breeding, facilitating faster identification of resilient plant lines, a long-term strategy for minimizing reliance on external pest control. Additionally, we cannot limit AI's impact on traditional field agriculture. Adapting AI solutions for resource optimization, pest monitoring, and control within controlled environments (greenhouses, vertical farming) or smaller operations can also transform their IPM practices.

These efforts underscore the dynamic transformation AI brings to agriculture. This constant stream of innovation and emerging research solidifies the promise of building intelligent, self-improving, and environmentally conscious systems to safeguard our food supply.

Predator-prey relationships are a central component of community dynamics. Classic approaches have tried to understand and predict these relationships in consumptive interactions between predator and prey species. However, characterizing the interaction this way cannot predict predator-prey relationships' complexity and context dependency. Recent approaches have begun to explore predator-prey relationships in terms of an evolutionary-ecological game in which predator and prey adapt to each other through reciprocal interactions involving context-dependent expression of functional traits that influence their biomechanics. Functional traits are any morphological, behavioral, or physiological trait of an organism associated with a biotic interaction. Such traits include predator and prey body size, predator and prey personality, predator hunting mode, prey mobility, prey anti-predator behavior, and prey physiological stress. Here, I discuss recent advances in this functional trait approach. Evidence shows that the nature and strength of many interactions depend on the relative magnitude of predator and prey functional traits.

Moreover, trait responses can be triggered by non-consumptive predator-prey interactions elicited by prey responses to predation risk. These interactions, in turn, can have dynamic feedback that can change the context of the predator-prey interaction, causing predator and prey to adapt their traits through phenotypically plastic or rapid evolutionary responses to the nature of their interaction. Research shows that examining predator-prey interactions through the lens of an adaptive evolutionary-ecological game offers a foundation to explain the variety in the nature and strength of predator-prey interactions observed in different ecological contexts.

References

Adetunji, C. O., Olaniyan, O. T., Anani, O. A., Inobeme, A., Osemwegie, O. O., Hefft, D. et al. (2023). Artificial Intelligence and Automation for Precision Pest Management. In Sensing and Artificial Intelligence Solutions for Food Manufacturing (pp. 49–70). CRC Press.

Agatz, A., Ashauer, R., Sweeney, P., and Brown, C. D. (2020). A Knowledge-Based Approach to Designing Control Strategies for Agricultural Pests. Agricultural Systems. https://doi.org/10.1016/j.agsy.2020.102865.

Ahmed, M., Javeed, A., Sajid, A. R., Ullah, S., Saleem, R. Q., Ahmad, Z. et al. (2022). Biopesticides: a healthy alternative of hazardous chemical pesticides, current development and status in China. Biomed. Lett., 8(2): 98–108.

Aktar, M. W., Sengupta, D., and Chowdhury, A. (2009). Impact of pesticides use in agriculture: their benefits and hazards. Interdiscip. Toxicol., 2(1): 1–12. https://doi.org/10.2478/v10102-009-0001-7.

Arnal Barbedo, J. G. (2019). Plant disease identification from individual lesions and spots using deep learning. Biosystems Engineering, 180: 96–107. https://doi.org/https://doi.org/10.1016/j.biosystemseng.2019.02.002.

Bäck, T. H. W., Kononova, A. V., van Stein, B., Wang, H., Antonov, K. A., Kalkreuth, R. T. et al. (2023). Evolutionary Algorithms for Parameter Optimization—Thirty Years Later. Evolutionary Computation, 31(2): 81–122. https://doi.org/10.1162/evco_a_00325.

Bardin, M., Ajouz, S., Comby, M., Lopez-Ferber, M., Graillot, B., Siegwart, M. et al. (2015). Is the efficacy of biological control against plant diseases likely to be more durable than that of chemical pesticides? Frontiers in Plant Science, 6: 566.

Barzman, M., Bàrberi, P., Birch, A. N. E., Boonekamp, P., Dachbrodt-Saaydeh, S., Graf, B. et al. (2015). Eight principles of integrated pest management. Agronomy for Sustainable Development, 35(4): 1199–1215. https://doi.org/10.1007/s13593-015-0327-9.

Bhairavi, K. S., Bhattacharyya, B., Manpoong, N. S., Das, P. P. G., Devi, E. B., and Bhagawati, S. (2020). Recent advances in exploration of acoustic pest management: A review. Journal of Entomology and Zoology Studies, 2056, 2061.

Boopathi, T., Singh, S. B., Manju, T., Ramakrishna, Y., Singh, A., Chowdhury, S. et al. (2015). Development of Temporal Modeling for Forecasting and Prediction of the Incidence of Lychee, *Tessaratoma papillosa* (Hemiptera: Tessaratomidae), Using Time-Series (ARIMA) Analysis. Journal of Insect Science, 15. https://doi.org/10.1093/jisesa/iev034.

Bronson, K., and Knezevic, I. (2016). Big Data in food and agriculture. Big Data & Society, 3(1): 2053951716648174.

Chlingaryan, A., Sukkarieh, S., and Whelan, B. (2018). Machine learning approaches for crop yield prediction and nitrogen status estimation in precision agriculture: A review. Computers and Electronics in Agriculture, 151: 61–69. https://doi.org/https://doi.org/10.1016/j.compag.2018.05.012.

Cock, M. J., van Lenteren, J. C., Brodeur, J., Barratt, I., Bigler, F., Bolckmans, K. et al. (2009). The use and exchange of biological control agents for food and agriculture.

Damos, P. (2015). Modular structure of web-based decision support systems for integrated pest management. A review. Agronomy for Sustainable Development, 35(4): 1347–1372.

Debeljak, M., and Džeroski, S. (2011). Decision Trees in Ecological Modelling. In (pp. 197–209). https://doi.org/10.1007/978-3-642-05029-9_14.

Demirel, M., and Kumral, N. A. (2021). Artificial intelligence in integrated pest management. In Artificial intelligence and IoT-based technologies for sustainable farming and smart agriculture (pp. 289–313). IGI Global.

Domingues, T., Brandão, T., and Ferreira, J. C. (2022). Machine learning for detection and prediction of crop diseases and pests: A comprehensive survey. Agriculture, 12(9): 1350.

Duchesnay, E., Lofstedt, T., and Younes, F. (2021). Statistics and machine learning in python.

Dwivedi, M., Shadab, H., and Santosh, V. (2020). Insect Pest Detection, Migration and Monitoring Using Radar and LiDAR Systems. In (pp. 61–76). https://doi.org/10.1007/978-981-15-0794-6_4.

Ehlers, G. C., Caradus, J. R., and Fowler, S. V. (2020). The regulatory process and costs to seek approval for the development and release of new biological control agents in New Zealand. BioControl, 65(1): 1–12.

Fountas, S., Mylonas, N., Malounas, I., Rodias, E., Hellmann Santos, C., and Pekkeriet, E. (2020). Agricultural robotics for field operations. Sensors, 20(9): 2672.

Franklin, M. E., Eken, S., and Osterlund, E. (2023). Current Research Updates on PANDAS and PANS. Current Developmental Disorders Reports, 10(4): 264–273.

Fu, L., He, L., and Zhang, Q. (2022). Neural Networks for Smart Agriculture. In Q. Zhang (Ed.), Encyclopedia of Smart Agriculture Technologies (pp. 1–11). Springer International Publishing. https://doi.org/10.1007/978-3-030-89123-7_164-1.

Hamadneh, N., Khan, W., and Tilahun, S. (2018). Optimization of Microchannel Heat Sinks Using Prey-Predator Algorithm and Artificial Neural Networks. Machines, 6(2): 26. https://www.mdpi.com/2075-1702/6/2/26.

Hamadneh, N. N., Khan, W. S., and Khan, W. A. (2019). Prediction of thermal conductivities of polyacrylonitrile electrospun nanocomposite fibers using artificial neural network and prey predator algorithm. Journal of King Saud University - Science, 31(4): 618–627. https://doi.org/https://doi.org/10.1016/j.jksus.2018.03.013.

Haviland, D. R., Rijal, J. P., Rill, S. M., Higbee, B. S., Burks, C. S., and Gordon, C. A. (2021). Management of navel orangeworm (Lepidoptera: Pyralidae) using four commercial mating disruption systems in California almonds. Journal of Economic Entomology, 114(1): 238–247.

Hossain, L., Rahman, R., and Khan, M. S. (2017). Alternatives of Pesticides. In (pp. 147–165). https://doi.org/10.1007/978-3-319-52683-6_9.

Jang, S.-B. (2023). A Deep Learning Performance Comparison of R and Tensorflow. The Journal of the Convergence on Culture Technology, 9(4): 487–494.

Jarvis, P., Fowler, S., Paynter, Q., and Syrett, P. (2006). Predicting the economic benefits and costs of introducing new biological control agents for Scotch broom *Cytisus scoparius* into New Zealand. Biological Control, 39(2): 135–146.

Javaid, M., Haleem, A., Khan, I. H., and Suman, R. (2023). Understanding the potential applications of Artificial Intelligence in Agriculture Sector. Advanced Agrochem, 2(1): 15–30. https://doi.org/https://doi.org/10.1016/j.aac.2022.10.001.

Jessie, W. P. (2017). Effects of interspecific competition and narrow-spectrum insecticides on the survival and development of *Diaeretiella rapae* in Oklahoma winter canola Oklahoma State University].

Kamilaris, A., and Prenafeta-Boldú, F. X. (2018). Deep learning in agriculture: A survey. Computers and Electronics in Agriculture, 147: 70–90. https://doi.org/https://doi.org/10.1016/j.compag.2018.02.016.

Keval, R., Vanajakshi, H. S., Verma, S., and Kumar, A. (2019). Analysis of Various IPM Modules Against Pea Leaf Miner (P. Horticola G.) Infesting Pea (*Pisum sativum* L.). International Journal of Bio-Resource and Stress Management. https://doi.org/10.23910/ijbsm/2019.10.1.1924a.

Kharel, T. P., Ashworth, A. J., and Owens, P. R. (2022). Linking and sharing technology: partnerships for data innovations for management of agricultural big data. Data, 7(2): 12.

Kolisnyk, O. M., Lebid, O. V., and Kovka, N. S. (2023). Application of advanced technologies in agricultural sector for increasing gross production and exporting agricultural products. Сільське господарство та лісівництво. № 4 (31). С. 75–92. DOI: 10.37128/2707-5826-2023-4-6.

Kotula, H. J., Peralta, G., Frost, C. M., Todd, J. H., and Tylianakis, J. M. (2021). Predicting direct and indirect non-target impacts of biocontrol agents using machine-learning approaches. PLoS One, 16(6): e0252448.

Kvakkestad, V., Steiro, Å. L., and Vatn, A. (2021). Pesticide Policies and Farm Behavior: The Introduction of Regulations for Integrated Pest Management. Agriculture. https://doi.org/10.3390/agriculture11090828.

Li, W., Zheng, T., Yang, Z., Li, M., Sun, C., and Yang, X. (2021). Classification and detection of insects from field images using deep learning for smart pest management: A systematic review. Ecological Informatics, 66: 101460.

Liakos, K. G., Busato, P., Moshou, D., Pearson, S., and Bochtis, D. (2018). Machine learning in agriculture: A review. Sensors, 18(8): 2674.

Lima, M. C. F., de Almeida Leandro, M. E. D., Valero, C., Coronel, L. C. P., and Bazzo, C. O. G. (2020). Automatic detection and monitoring of insect pests—A review. Agriculture, 10(5): 161.

Malik, Y. S., Sircar, S., Bhat, S., Ansari, M. I., Pande, T., Kumar, P. et al. (2021). How artificial intelligence may help the Covid-19 pandemic: Pitfalls and lessons for the future. Rev Med Virol., 31(5): 1–11. https://doi.org/10.1002/rmv.2205.

Maraveas, C. (2022). Incorporating artificial intelligence technology in smart greenhouses: Current State of the Art. Applied Sciences, 13(1): 14.

Nieróbca, A., and Zaliwski, A. S. (2014). Expert systems as a tool for decision support in integrated pest management. Agricultural Engineering, 18.

Raffel, T. R., Martin, L. B., and Rohr, J. R. (2008). Parasites as predators: unifying natural enemy ecology. Trends in ecology & evolution, 23(11): 610–618.

Rajamani, S. K., and Iyer, R. S. (2023). Machine Learning-Based Mobile Applications Using Python and Scikit-Learn. In Designing and developing innovative mobile applications (pp. 282–306). IGI Global.

Ramazi, P., Kunegel-Lion, M., Greiner, R., and Lewis, M. A. (2021). Predicting insect outbreaks using machine learning: A mountain pine beetle case study. Ecol Evol, 11(19): 13014–13028. https://doi.org/10.1002/ece3.7921.

Rose, D. C., Sutherland, W. J., Parker, C., Lobley, M., Winter, M., Morris, C. et al. (2016). Decision support tools for agriculture: Towards effective design and delivery. Agricultural Systems, 149: 165–174.

Rosner, A., Ballarin, L., Barnay-Verdier, S., Borisenko, I., Drago, L., Drobne, D. et al. (2024). A broad-taxa approach as an important concept in ecotoxicological studies and pollution monitoring. Biological Reviews, 99(1): 131–176.

Rustia, D. J. A., Chiu, L. Y., Lu, C. Y., Wu, Y. F., Chen, S. K., Chung, J. Y. et al. (2022). Towards intelligent and integrated pest management through an AIoT-based monitoring system. Pest Management Science, 78(10): 4288–4302.

Rustia, D. J. A., Lin, C. E., Chung, J.-Y., Zhuang, Y.-J., Hsu, J.-C., and Lin, T.-T. (2020). Application of an image and environmental sensor network for automated greenhouse insect pest monitoring. Journal of Asia-Pacific Entomology, 23(1): 17–28.

Saeed Ben Youssef, A. (2023). Modern trends for the application of biological control and modern technologies in agricultural projects. International Journal of Modern Agriculture and Environment.

Sanghavi, K., Sanghavi, M., and Rajurkar, A. M. (2021). Early stage detection of Downey and Powdery Mildew grape disease using atmospheric parameters through sensor nodes. Artificial Intelligence in Agriculture, 5: 223–232. https://doi.org/https://doi.org/10.1016/j.aiia.2021.10.001.

Schmitz, O. (2017). Predator and prey functional traits: understanding the adaptive machinery driving predator-prey interactions. F1000Res, 6: 1767. https://doi.org/10.12688/f1000research.11813.1.

Sellam, V., and Poovammal, E. (2016). Prediction of crop yield using regression analysis. Indian Journal of Science and Technology, 9(38): 1–5.

Sen, P., Samanta, S., Khan, M. Y., Mandal, S., and Tiwari, P. K. (2023). A Seasonally Forced Eco-Epidemic Model with Disease in Predator and Incubation Delay. Journal of Biological Systems.

Senoo, E. E. K., Anggraini, L., Kumi, J. A., Luna, B. K., Akansah, E., Sulyman, H. A. et al. (2024). IoT Solutions with Artificial Intelligence Technologies for Precision Agriculture: Definitions, Applications, Challenges, and Opportunities. Electronics, 13(10): 1894.

Serrão, J. E., Plata-Rueda, A., Martínez, L. C., and Zanuncio, J. C. (2022). Side-effects of pesticides on non-target insects in agriculture: A mini-review. The Science of Nature, 109(2): 17.

Shaikh, T. A., Rasool, T., and Lone, F. R. (2022). Towards leveraging the role of machine learning and artificial intelligence in precision agriculture and smart farming. Computers and Electronics in Agriculture, 198: 107119.

Sharma, A., Kumar, V., and Zheng, B. (2023). PESTICIDES in the ENVIRONMENT Impact, Assessment, and Remediation. Elsevier.

Sharma, R., Kumar, N., and Sharma, B. B. (2022). Applications of artificial intelligence in smart agriculture: a review. Recent Innovations in Computing: Proceedings of ICRIC 2021, 1: 135–142.

Simon, S. M., Glaum, P., and Valdovinos, F. S. (2023). Interpreting random forest analysis of ecological models to move from prediction to explanation. Scientific Reports, 13(1): 3881. https://doi.org/10.1038/s41598-023-30313-8.

Singh, A., Mehrotra, R., Rajput, V. D., Dmitriev, P., Singh, A. K., Kumar, P. et al. (2022). Geoinformatics, artificial intelligence, sensor technology, big data: emerging modern tools for sustainable agriculture. Sustainable Agriculture Systems and Technologies, 295–313.

Skouras, P. J., Darras, A. I., Mprokaki, M., Demopoulos, V., Margaritopoulos, J. T., Delis, C. et al. (2021). Toxicity, sublethal and low dose effects of imidacloprid and deltamethrin on the aphidophagous predator *Ceratomegilla undecimnotata* (Coleoptera: Coccinellidae). Insects, 12(8): 696.

Sree, K. B., Rajesh, K., Arunbabu, T., and Kumar, M. S. (2024). Artificial Intelligence in Agriculture–Paving Way Towards Future Farming.

Stanescu, L., and Dinu, G. (2023). TensorFlow vs. PyTorch in Classifying Medical Images–preliminary results. 2023 27th International Conference on System Theory, Control and Computing (ICSTCC).

Tabashnik, B. E., Brévault, T., and Carrière, Y. (2013). Insect resistance to Bt crops: lessons from the first billion acres. Nature Biotechnology, 31(6): 510–521. https://doi.org/10.1038/nbt.2597.

Talaviya, T., Shah, D., Patel, N., Yagnik, H., and Shah, M. (2020). Implementation of artificial intelligence in agriculture for optimisation of irrigation and application of pesticides and herbicides. Artificial Intelligence in Agriculture, 4: 58–73. https://doi.org/https://doi.org/10.1016/j.aiia.2020.04.002.

Tannous, M., Stefanini, C., and Romano, D. (2023). A Deep-Learning-Based Detection Approach for the Identification of Insect Species of Economic Importance. Insects. https://doi.org/10.3390/insects14020148.

Thangamani, R., Sathya, D., Kamalam, G., and Lyer, G. N. (2024). AI Green Revolution: Reshaping Agriculture's Future. In Intelligent Robots and Drones for Precision Agriculture (pp. 421–461). Springer.

Toscano-Miranda, R., Toro, M., Aguilar, J., Caro, M., Marulanda, A., and Trebilcok, A. (2022). Artificial-Intelligence and Sensing Techniques for the Management of Insect Pests and Diseases in Cotton: A Systematic Literature Review. The Journal of Agricultural Science. https://doi.org/10.1017/s002185962200017x.

van Velzen, E., Gaedke, U., and Klauschies, T. (2022). Quantifying the capacity for contemporary trait changes to drive intermittent predator–prey cycles. Ecological Monographs, 92(2): e1505.

Vänninen, I. (2022). Advances in insect pest and disease monitoring and forecasting in horticulture.

Wang, L., Zhai, Y., Zhu, J., Wang, Q., Ji, X., Wang, W. et al. (2023). Sulfoxaflor adversely influences the biological characteristics of *Coccinella septempunctata* by suppressing vitellogenin expression and predation activity. Journal of Hazardous Materials, 447: 130787.

Wang, X., Cheng, J., and Wang, L. (2019a). Deep-reinforcement learning-based co-evolution in a predator–prey system. Entropy, 21(8): 773.

Wang, X., Cheng, J., and Wang, L. (2019b). Deep-Reinforcement Learning-Based Co-Evolution in a Predator–Prey System. Entropy. https://doi.org/10.3390/e21080773.

Wu, Y., He, Y., and Wang, Y. (2023). Multi-class weed recognition using hybrid CNN-SVM classifier. Sensors, 23(16): 7153.

Yadav, A., and Yadav, K. (2013). Seasonal Population Dynamics of Rhizosphere and Non-rhizosphere Soil Microorganisms of Chir Pine Seedlings (*Pinus roxburghii* Sarg.). Microbiology Research Journal International, 3(4): 664–677. https://doi.org/10.9734/BMRJ/2013/4206.

Yadav, A., Yadav, K., Ahmad, R., and Abd-Elsalam, K. A. (2023). Emerging Frontiers in Nanotechnology for Precision Agriculture: Advancements, Hurdles and Prospects. Agrochemicals, 2(2): 220–256. https://www.mdpi.com/2813-3145/2/2/16.

Yan, Y., Feng, C.-C., Wan, M., and Chang, K. (2015). Multiple Regression and Artificial Neural Network for the Prediction of Crop Pest Risks. In (pp. 73–84). https://doi.org/10.1007/978-3-319-24399-3_7.

Yang, P., Henle, E. A., Fern, X. Z., and Simon, C. M. (2022). Classifying the toxicity of pesticides to honey bees via support vector machines with random walk graph kernels. The Journal of Chemical Physics, 157(3).

Zaller, J. G., and Zaller, J. G. (2020). Where Are the Solutions to the Pesticide Problem? Daily Poison: Pesticides-an Underestimated Danger, 223–295.

Zhang, Z., Khanal, S., Raudenbush, A., Tilmon, K., and Stewart, C. (2022). Assessing the efficacy of machine learning techniques to characterize soybean defoliation from unmanned aerial vehicles. Computers and Electronics in Agriculture, 193: 106682.

CHAPTER 6

Early Disease Prediction and Prevention Strategies

Abdelatti I. Nowwar,[1] *Samah H. Abu-Hussien*[2] and *Salem S. Salem*[1,*]

1. Introduction

Due to the abundance of pathogens in their surroundings, agricultural crops are more vulnerable to diseases (Vishnoi et al., 2023). Any anatomical or physiological abnormality brought on by a living organism is classified as a plant disease (Masood et al., 2023). Plant illnesses are brought on by environmental conditions or plant infections (Barburiceanu et al., 2021). Microbiological assault, insufficient nourishment, unfavorable environmental conditions, and insect attack are the main causes of plant ailments (Zeng et al., 2020). Plant pathogen infection is one of the primary reasons for decreased agricultural productivity globally (Nowwar et al., 2023). The disease is worse when different pathogen groups attack plants one at a time or in combination (Amin et al., 2022). Food security may be jeopardized by plant diseases that damage crops, reducing the quantity of food available and raising food prices (Shafi et al., 2023).

For the expanding global populace to have access to food and revenue sources, protecting plants from disease is more crucial (Tripathi et al., 2022). A visual inspection of the plant is among the traditional techniques for identifying plant diseases since pathogen effects aren't always evident until the plant sustains serious damage. Automated plant illness identification is a very promising field (Chelloug et al., 2023). Because early diagnosis and treatment of these disorders reduces the need for pesticides and increases crop quality and quantity (Zhu et al., 2022). Finding these disorders through manual assessment can be difficult, time-consuming, and subjective (Shovon et al., 2023).

[1] Botany and Microbiology Department, Faculty of Science, Al-Azhar University, Cairo, 11884, Egypt.
[2] Agriculture Microbiology Department, Faculty of Agriculture, Ain Shams University, P.O. Box 68, Cairo, 11241, Egypt.
* Corresponding author: salemsalahsalem@azhar.edu.eg

DOI: 10.1201/9781003546702-6

Therefore, automated equipment that help farmers monitor crops at all stages of growth are essential to meet the demands of an increasingly populous world. One of agricultural precision study's most significant uses is the identification of plant illnesses in leaves through visual inspection (Chen et al., 2022). To determine the degree of severity of plant diseases, skilled professionals visually inspect plant tissues using a traditional method (Dwivedi et al., 2021). Plant production has increased dramatically due to the extensive usage of digital-cameras, and the advancement of agricultural technology, which has resulted in the deployment of expert systems in development and control. Expert systems, however, are not very efficient because they heavily rely on the knowledge of experts to extract and describe the features of pests and diseases (Li et al., 2021).

Plant diseases may now be detected and categorised using a variety of artificial intelligence techniques. Support convolutional neural networks (CNNs), K-nearest neighbors (K-NN), logistic-regression, vector machines(SVMs), and decision trees are among the most widely used methods (Schor et al., 2016). Various image pre-processing techniques are coupled with these methods to promote feature extraction. The K-NN algorithm is one for supervised learning. It classifies the data using similarity measures. In K-NN, unlabeled things are categorized using nearby labeled objects. The decision tree is a method for learning from flow charts. Each node represents a decision attribute, whereas branches display possible node outcomes and leaves denote classifications. Decision trees do, however, have several shortcomings, such as overlapping nodes and overfitting of the data. Vector machines is a well-liked mechanism learning framework that is associated with learning algorithms in order to classify and perform regression modeling that are based on statistical learning concepts. In the previous ten years, text and image categorisation has made significant use of SVMs. Previous machine detection techniques often use traditional techniques of image processing, e.g. morphological operations, picture improvement, and noise reduction to pre-process images of diseased plant leaves (Ahmad et al., 2021). Next, among other low-level features, the color, shape, and texture of the leaves are extracted using manually constructed feature extraction techniques (Nie et al., 2019).

In recent times, object segmentation, classification, and identification have showed promise regarding the structures of deep learning (Singh et al., 2019). CNN techniques are the mostly used for deep learning problems. Despite being widely utilized in the classification of plant diseases, the basic CNN architectures AlexNet, VGGNet, GoogLeNet, DenseNet, and ResNet possess a variety of drawbacks, such is the need for a sluggish pace of estimation and a large number of characteristics. Although deep learning methods have proven to be remarkably effective in displaying both high-level and low-level properties, there is a lack of consistency in how well they reflect local spatial aspects (Xiao et al., 2023). By highlighting the advantages and disadvantages of these approaches, providing a thorough understanding of the current state of research in this field, and suggesting potential solutions to overcome the difficulties in implementing them, this chapter offers insightful information for early plant disease detection researchers, practitioners, and industry professionals.

2. The Process of Plant Disease Detection

As illustrated in Figure 1, the plant disease detection system process consists of four main stages. Getting pictures is the first step, which can be done online, with a digital camera, or with a smart-phone. The image is split up into a number of groups in the second stage, every one of them can be managed differently depending on the approach used. The subsequent phase covers feature extraction approaches, while the final part deals with sickness classification.

(i) Image Acquisition

Employing digital media such as cameras, mobile phones, etc., photographs of plant leaves are collected in this phase with the necessary size and resolution. It is also possible to take pictures from the internet. The production of an image database is entirely up to the application system developer. Improved classifier performance in the detection system's final step is the result of the picture database (Ramedani et al., 2014).

(ii) Segmentation of the Image

The goal of this stage is to simplify the portrayal of the image, in order for it to be analyzed more easily and has greater meaning (Gharge and Singh, 2016). This phase forms the basis for the extraction of features and is the core method of image processing. A range of methods can be used to segment images. Using a variety of characteristics, K number of classes are created from an object or pixel applying the k -means sorting. The process of classification involves the reduction of the observed distances total squares between the items and the related groups (Singh and Kaur, 2018).

(iii) Feature Extraction

Following division, the region of interest represents the result attained thus far. Thus, it's essential to excerpt the characteristics from this area of interest at this point. To understand the meaning of an example image, certain characteristics are required. Features can be defined using color, texture, and shape (Dey et al., 2016). Numerous feature extraction methods, such as the following, can be used to construct the system:

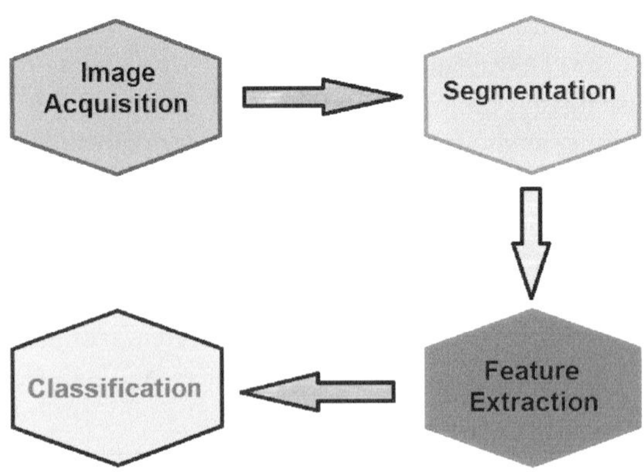

Figure 1. System stages for detecting plant diseases.

histogram-based obtaining of features, the spatial grey -level dependence matrix, gray -level co -occurrence matrix (GLCM), and color co-occurrence approach. For classifying textures, a statistical approach is the GLCM method.

(iv) Classification

The categorisation stage involves identifying whether the input imagined is healthy or not. In the event that the image is determined to be unhealthy, it has been further classified into several disorders by some previous works. Writing a classifier is a MATLAB-software function that is required to carry out classification. Various classifiers have been used by researchers in recent years, such as:

1. K -nearest neighbor (KNN),
2. Decision tree,
3. Support vector machines (SVM),
4. Back propagation neural network (BPNN),
5. Artificial neural-network (ANN),
6. Naïve Bayes.

SVM is the classifier that is most frequently employed. Even though each classifier has pros and cons, SVM is a reliable and easy-to-use method (Gavhale and Gawande, 2014).

3. The Traditional Methods for Plant Disease Detection

Diagnosing plant diseases is becoming more accurate every day. Innovative tools and methods are being created and implemented to guarantee the speed and accuracy of diagnosis. Though they have lost some of their value, the established diagnostic techniques nevertheless allow early identification of plant diseases, accurately, and with the use of newly developed instruments and technology. This facilitates the making of appropriate and prompt decisions about the management of crop diseases.

Research on plant pathology employs a variety of diagnosis techniques. These could include molecular genetic identification, mass spectrophotometry, immunological diagnostics, visual observation, markers plant testing, microscopy, mycological diagnosis, etc. (Kamilov et al., 2020; Zuparov et al., 2020).

(a) Visual observation

First, visual inspection of the cell plants is done at the site of cultivation, where a healthy plant and a diseased plant are contrasted. In this instance, observations include changes in the root system, color, height of the plant, leaf form, and branches' leaf density. Using a magnifying lens, check for the existence of sporangiophora, sclerotia, or mycelium if there are disease foci (leaves, stems spots or necrosis, etc.). The existence of a harmful fungus is indicated if any are discovered. Through visual examination, the primary pathogenesis of bacteria, worms, fungi, or viruses may frequently be easily identified based on the symptoms present in the plant sample. A preliminary diagnosis can be made if the sample exhibits any distinctive symptoms. For instance, the bacteria *Agrobacterium tumefaciens* promotes the development of host tissue from plants and results in the easily observable formation of tumors (Putnam, 1995).

When viewing, it's crucial to consider the plant's overall health (standing, turgidity); if it's withering, there may be an issue with the roots or conducting tissues. Both bacteria and phytopathogenic fungi, such as tracheomycoses and tracheobacteriosis, can cause these illnesses. In many circumstances, visual observation is insufficient to make a definitive diagnosis. That is, the diagnosis's accuracy is lowered when inferences are made solely on the basis of a single morphological trait. Microscopy and other research should be carried out in order to improve the diagnosis' dependability.

(b) Microscopy

For accurate diagnosis, high -precision microscopes are an essential and frequently used tool. The diseased plant pieces are taken to the lab for microscopic examination following visual inspection. Plant materials are subjected to initial observation of disease symptoms using a binocular stereomicroscope. Looking into the micro-sclerosis, hyphae, conidia, conidiophores, and bacterial cell clusters of pathogenic fungus under a microscope can frequently lead to a diagnosis of the affected plant sections. On the other hand, fungi may not generate spores in unfavorable weather, and fungal mycelium might not even be visible on the impacted surface (Dyakov and Elansky, 2019).

Furthermore, secondary parasites or saprophytes may establish colonies in the dead tissue even if there may not be a true infection there (Putnam, 1995). In these situations, a visual examination and preliminary microscopy are insufficient to identify the pathogen. Further mycological analysis ("moist chamber" technique, pure culture separation, etc.) is applied in this instance.

(c) Mycological diagnosis

Infected plant tissues are incubated in a room with a lot of humidity using the "moist chamber" approach. In this instance, favorable conditions allow the fungus in the diseased tissues to grow and show symptoms. Using the "moist chamber" method, sterile Petri-dishes are lined with blotting filter paper, hygienic glassware is positioned on top, and plant portions that are diseased are set on the glass. The Petri dish is then covered and kept at 24–28°C (or 17–20°C for Phytophtora species). Through microscopic examination of the hyphae, micro, and macroconidia collected using this technique, a fungus, up to the genus and species level, can be identified as a harmful fungus.

Microscopy, visual observations, and "moist chamber" techniques, however, might not always be sufficient to accurately identify all phytopathogens. Numerous phytopathogens, such as *Ustilago* spp., are dormant or latent and do not show symptoms for a while. However, a lot of infections have symptoms that are comparable to one another. Another issue is that, even in a moist chamber, some harmful fungi do not produce spores (Dyakov and Elansky, 2019).

The mycelium or spores of a fungal pathogenic species may not always be sufficient to identify it. The colonies' colour, shape, and other physical characteristics in the nutrition media were utilized as indicators in several detection procedures to help identify the species. As a result, isolating the fungus's pure culture is frequently required (Bilay, 1982). It is possible to isolate roots of plants, soil, and diseased plant surface areas from pure culture. To isolate pure culture from plant tissue that has been infected, the following tools are needed: a disinfected standard, selective, or semi -selective agar-medium; an alcohol lamp; a sterile- needle; and a binocular microscope.

The majority of ascomycetes fungus may be isolated using wort agar and PGA nutritional media, whereas oomycetes are often isolated using an agar medium made of maize, rye, or pea flour. Additionally, ideal circumstances for the formation of spores and the pathogen development to be separated must be generated when incubation is in a moist chamber. For instance, Phytophthora infestans, the bacteria that causes phytophthora in potatoes, should be isolated using moist chamber incubation at a temperature of 16–18°C, while members of the Alternaria family should be isolated using a temperature of 24–25°C (Dyakov and Elansky, 2019).

Following the isolation of a pure-culture, the pathogen's macro-morphological and cultural traits in the nutritional medium are ascertained based on the identification of micro-morphological features using molecular genetic identification or microscopy (using identifiers). Koch's postulates must be adhered to in order to use this approach to determine the kind of pathogen. In other words, Koch's postulates require that the pathogen be isolated first, then re-isolated from the artificially infected plant, and finally, that the symptoms of the pathogen that were isolated from the affected plant in a natural environment are consistent (Khasanov, 2011).

(d) Biological assays or indicator plant tests

This technique is frequently applied in conjunction with other diagnostic techniques to identify phytoplasmas and phytoviruses. Herbaceous plants and indicator trees are used for biological diagnosis. In order to infect indicator plants, mechanical inoculation or injection is employed. In this instance, a healthy indicator plant's leaves exhibiting disease symptoms are applied, or an injection of a typical buffer suspension is made. On indicator trees and shrubs, the technique of fake infection through micro-grafting is also employed. Such testing is done in highly isolated environments on indicator plants. It is necessary to prepare healthy seedlings from virus-free mother material before performing diagnostic work on marker trees (Khakimov et al., 2022).

Plants that are young and grow rapidly are chosen in order to cause viral infection. When dicotyledon indicator plants grow two or three genuine leaves, they are mechanically infected for testing purposes. In certain instances, it can be performed on primordial-leaves and cotyledons, or on the initial leaves of seeds. Plants known as monocotylidon, or grain crops, become susceptible to viruses when they develop one or two leaves and grow in dark conditions for three days (Bhat and Rao, 2020).

While there are several benefits to this method (such as ease of use and no need for expensive diagnostic equipment), the main drawbacks are the lengthy waiting period and the requirement for specialized, separate culture chambers. For herbaceous indicator plants, the waiting period may extend to 30–60 days, while for trees, it may extend to 45–2 years (Sorokopudov et al., 2018).

As indicator plants, other disease-prone cultivars or native relatives of the target plant could be employed. There are currently around 600 indicator-plants from 173 genera and 43 families. For instance, *Gomphrena globosa* and *Datura stramonium* are plant indicators of the potato X -virus, *Nicotiana glutinosa* and *D. stramonium* are plant-indicators of the tobacco mosaic virus, and *N. physaloides* and *C. foetidum* species are plant indicators of the plum orca virus (Upadyshev and Donetskikh, 2008). Indicator plants are noticed two to three days following infestation; for herbaceous indicator plants, the monitoring period is four weeks. On the plant being tested, viral illness signs start to appear during this period.

4. Plant Disease Detection Techniques

An essential component of economic growth is agriculture. Consequently, damage to the crops would lead to a large loss of output, which would ultimately affect the economy. Since leaves are the most delicate portion of plants, illness signs appear first on leaves (Van Der Plank, 2012). The crops must be monitored for illnesses from the start of the growing process until they're scheduled to harvest. Originally, the traditional naked eye monitoring method was a laborious process requiring experts to manually keep an eye on the fields of crops to keep an eye out for plant diseases (Yadav and Singh, 2022). Recently, a variety of techniques have been employed to develop automatically and semi-automatically plant disease detection systems. These techniques have so far shown to be faster, less costly, and more accurate than producers' manual perception, which is the traditional way (Table 1).

5. The Innovative Methods for Disease Detection

Since various pathogens, including nematodes, fungus, bacteria, viruses, phytoplasmas, and viroids, can impact farmed plants simultaneously and traditional approaches only identify these pathogens at a late, symptomatic stage, the ability to detect several infections in a single plant is important. At the presymptomatic to early dissemination stages, relatively new methods may provide pathogen detection that is simple, quick, frequently affordable, and trustworthy (Figure 2). On the other hand, when symptoms are elusive and limited to a small number of plants, new techniques enable the identification of pathogen primary infections.

(a) Lateral flow microarrays

Using a readily observable colorimetric signal, lateral-flow microarrays (LFM) provide quick, hybridization-based nucleic-acid identification (Carter and Cary, 2007). These arrays potentially eliminate the need for pricey laboratory equipment because they are based on nitrocellulose membrane for quick hybridization , lateral flow chromatography in miniature, and possess detection limitations similar to those of microarrays. Strong and trustworthy host and pathogen biomarkers identified by transcriptome techniques are essential for the technology's operation (Martinelli et al., 2013; Nowwar et al., 2022). Important plant metabolites of secondary and primary metabolism that can be utilized as biomarkers for various pathogen infections or environmental stresses are identified by the widely used technique of metabolomics

Table 1. Plant disease detection techniques.

References	Description	Outcomes
(Al Bashish et al., 2010)	The author divided the image of leaf into 4 clusters using k-means clustering, presuming that at least one of the clusters would contain sick pixels. SGDM were used to build a CCM approach for feature extraction. We get the texture oriented features with this strategy. Using the feed forward back propagation neural network (BPNN), classification was carried out based on neural networks.	It was discovered that the total system's accuracy for disease identification and categorization was close to 93%. On the other hand, BPNN's color characteristics produced the best overall average accuracy of 99%. The mean intensity level, angular moment, variance, product moment, entropy, contrast, and correlation were discovered to be texture-oriented properties.
(Bhange and Hingoliwala, 2015)	This work uses k-means clustering for segmentation since it is more efficient when used for larger datasets. Sum of squared distances has been used to compare the color histograms of two photographs. For the shape vector or boundary extraction, morphology is helpful. When comparing photos with spatial information, color coherence vector (CCV) is utilized. SVM is used for classification in this paper.	The authors were capable of attaining an overall system precision of 82% by using SVM. SVM have a slow training period but are quite accurate. Three feature vectors one each for the color histogram, morphology, and CCV were thus produced.
(Singh and Misra, 2017)	The evolutionary algorithm's search feature has been utilized to divide unlabeled N-dimensional points into K-clusters in order to accomplish clustering. The texture and color of an image have both been considered in the CCM approach. Two different methods one using K means-clustering and the other using the Genetic-algorithm used the minimal distance criterion.	The texture features that are computed for local homogeneity, contrast, energy, cluster prominence, and cluster shadow in the H image. The accuracy of the minimal distance criterion was 86.54% when using k-mean clustering, and 95.71% when using SVM. The accuracy was increased to 93.63% by using the Genetic-Algorithm with the shortest distance criteria.
(Kiani and Mamedov, 2017)	There are two outputs and five inputs in the algorithm. A fungal infection is mentioned in two of the inputs, and an iron deficit is mentioned in the other. If the leaf is diseased, the outputs speak of the two illnesses.	By employing the suggested algorithm, the researchers in charge of this work achieved a 96% system accuracy overall.
(Ali et al., 2017)	The distance between two colors is computed using a technique based on the delta E color difference. Features from the RGB, HSV, and local-binary patterns (LBP) histograms are used for obtaining features.	We obtain the rotational constancy in the RGB histogram. They obtain the lighting constant brought about by various lightning situations as a feature in the HSV histogram.
(Sheelavantamath, 2020)	The Otsu technique creates a bimodal histogram with the backdrop and foreground pixels assuming that the image consists of two classes of pixels. Texture-and shape-oriented characteristics were used for feature extraction.	The texture-oriented parameters that were employed were contrary, interaction, energy, unity, mean, and area, while the shape-oriented characteristics were color, axis size, deviation, rigidity, and area.

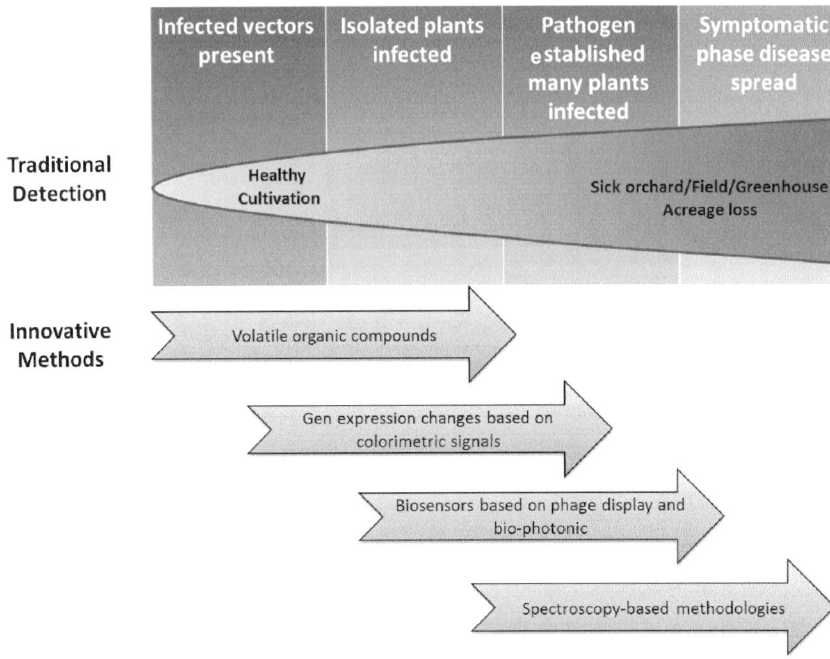

Figure 2. Traditional and innovative methods.

(Salam et al., 2023). Early attacks by pathogens, as Citrus Huanglongbing illness, can be detected using an integrated omic method (Li et al., 2024). According to Sharma and Dubey (2019), highly interactive proteins that are elevated by many environmental conditions, like heat shock proteins or dehydrins, may serve as indicators of the health status of plants.

(b) Volatile compounds as biomarkers

Plants release a large number of volatile compounds (VCs) into their immediate environment that are necessary for growth, communication, defense, and survival (Ninkovic et al., 2021). The final metabolites of the host plant are VCs that are expelled from leaf surfaces and can reveal the plant physiological health. VCs are biomolecules with a high vapor pressure and low boiling point that have a low molecular weight (Ogbodo et al., 2022). Under regular temperature and pressure, they are easily existent in the gaseous stage, and are typically existing in extremely tiny amounts under the threshold of the human sense of smell. Within the plant sciences, VCs profiling is a new field that has the potential to be applied immediately. Growers would greatly benefit from the capacity to quickly, regularly, and noninvasively check the health condition of high-value commodity crops. By identifying potential pathways for "plant-to-insect" and "plant-to-plant" dialogue, VCs profiling opens up new study directions and provides fresh perspectives on how infections and abiotic stresses affect hosts. To create and analyze these data sets, new analytical techniques, instrumentation, and multivariate data analysis (MVA) techniques are needed; the creation of this kind software and hardware tools is essential to implementing these ideas in practice.

White mold infections differed markedly from healthy controls in terms of their released volatile compounds (VCs) profiles; the main variations were in the amounts of 3-octanone and methyl salicylate (Tiwari et al., 2020). The detection of volatile compound markers shows that VCs profiling has the ability to quickly distinguish between fungal-infection and insect-vector feeding. The eating customs of beet army-worms (BAW) on both healthy and diseased leaves with white mold were also examined in the study; the BAW ate more infected leaves on preference. Additionally, the BAW spontaneously released trace amounts of methyl salicylate, linalool, and hexenyl acetate, all of which inhibited *S. rolfsii's* development. The production of induced volatiles, including 3-octanone and methyl salicylate, by diseased plants is a direct host response to fungus-related illnesses and attracts insect vectors (Razo-Belman and Ozuna, 2023). After harvesting potato tubers injected with *Fusarium coeruleum* (dry rot) and *phytophthora infestans* (late blight), 52 volatiles were found by VCs profiling. The observed variations in the volatile compound (VCs) profile of potato tubers injected with *F. coeruleum* and *P. infestans* could potentially serve as a precursor to postharvest disease in potatoes (Agho, 2024) hence facilitating the management of post-harvest cash crops.

Using volatile compounds profiling, the cucumber mosaic-virus (CMV) has also been studied by (Mauck et al., 2010). *Cucurbita pepo* cv. Dixie, a cultivar of squash, was employed as a model system to study the virus's method of action, the ensuing host response, and the consequent impact on insect vectors. Comparable to BAW's attraction to mouldy peanut plants, *Myzus persicae* and *Aphis gossypii* were attracted to CMV–infected plants despite their reduced size and worse quality. Healthy plants may closely resemble the modified volatile substances profile released by CMV–infected plants. This shows that as the result of viral infection, the plant is causing changed VCs profile (Martinelli et al., 2015).

The host response of plant pathogen infections, both induced and released, is the subject of a substantial body of research. No single VCs marker has been linked to a single plant disease, despite the fact that this list is not comprehensive nor full (Cellini et al., 2021).

6. Development of Instruments for the Identification of Plant Volatile Compound Biomarkers

Among the many chemical activities that plants regulate is ethylene, from organ senescence to seed germination, is one plant volatile compounds (VCs) marker that is frequently studied. Ethylene's ability to initiate fruit ripening is arguably its most important characteristic (Light et al., 2016). Post-harvest monitoring of agricultural commodities could be considerably aided by further development of ethylene sensors.

Single-molecule specialised analysis is possible, but it is not widely employed since it provides only a limited amount of the accessible chemical data. When multiple classes of volatile compounds (VCs) need to be concurrently identified and quantified, untargeted volatile compounds (VC) profiling can be quite useful in providing a more comprehensive view of the plant system under study.

According to Xu et al. (2016), GC-MS is the preferred analytical tool for VC profiling since it employs a gas-phase separation and detection technique to offer detailed structural and chemical data on each component of VC mixtures. Proton-transferreaction mass spectrometry (PTR-MS) (Blake et al., 2009) and selected ion flow-tube mass-spectrometry (SIFT-MS) (Smith and Spanel, 2005) are two other spectrometric-based methods for VC profiling that offer near-real-time VC sample measurement and profiling.

Present MS-based systems are too large and fragile for direct, on-site examination; however, they would be useful if they were greatly reduced in size and placed in safe enclosures with enough infrastructure for maintenance. Right now, a variety of techniques are used to gather VCs, and the specimens are shipped to a lab for chemical analysis and characterisation. Samples of VCs can be taken either actively or passively. Small, functionalized polymer traps such as stir bar sorptive-extraction (SBSE) (Yilmaz and Soylak, 2020) or solid-phase microextraction (SPME) (Maruya et al., 2015) are used in passive sampling.

Another tool for profiling volatile organic compounds is the electronic nose (e-nose). Numerous specialized metal-oxide sensors, each with selectivity towards specific types of volatile organic compounds, are used in these systems. When volatiles are introduced, individual sensors within the e-nose generate an impedance response, which is simultaneously recorded and displayed (Piri-Moghadam et al., 2016). When a VC combination is employed, E-nose can quickly provide an overlay-spectral response in almost real -time, using the entire VC pattern without distinguishing any specific elements. According to Wilson (2018), e-noses have proven effective in correlating particular physiological changes in the state of health of plants. Since e-nose is a developing technology, problems with robustness, resolution, and repeatability still need to be resolved. Because of their high complexity, the data produced by e-nose must be processed using multivariate statistical methods in order to be effectively and appropriately analyzed. If these problems are fixed, e-nose devices may be used for in-field analysis in the future. Significant advancements in sensor technology, including DMS, SIFT-MS, PTR-MS, and e-nose, emphasize the necessity of sensitive and specialized instruments for the analysis of VCs from biological systems. It goes without saying that further progress in this area will aid and facilitate advancements in the discipline.

7. Plant Disease Prevention and Management Strategies

Using "silver bullet" products to contain plant disease outbreaks has ruled plant disease management strategies. Regretfully, the majority of these treatment approaches only work temporarily. Although there has been much research and exploration into long-term plant disease prevention measures based on the ecological understanding of pathogens, large-scale farming's financial concerns have limited their application.

Globally, the most common pesticide used in the agriculture sector is herbicide, that is followed by the use of fungicides and insecticide (Mazed et al., 2022). In the United States, the usage of fumigants is really second only to the use of herbicides in the agricultural industry (Desaeger et al., 2020), higher than insecticide and fungicide use, based on the most recent data on pesticide use from the US Environmental-Protection Agency. An encouraging aspect of this data is that, according to Ortiz-Hernández et al. (2011), the quantity of organophosphate pesticides used in the

United States has decreased by more than 60% since 1990, with a calculated 38.6 million-kg in 1990 and 15 million-kg in 2007. This decline is more than positive, as organophosphates are one of the most severely dangerous pesticides currently in use. When developing a long-term program to manage plant diseases, using insecticides or even agents for bio-control as "therapeutic" techniques should be the last option because they are unsustainable.

8. Management of Plant Diseases using Nanotechnology

Nanotechnology applications in agriculture have a bright future ahead of them (Salem and Husen, 2022; Salem et al., 2023; Salem and Fouda, 2021). They can be a major source of innovation for precision agriculture farming methods and yield improvement (Abdelfattah et al., 2023; Salem and Husen, 2023; Hashem et al. 2022; Badawy et al., 2021). According to Shamsiddinovich (2024), nanotechnology holds great promise for enhancing environmental health. Using nanostructured materials, various instruments are developed for the prevention of environmental contamination and the development of treatment devices (Figure 3) (Asghar et al., 2024; Badawy et al., 2024).

Furthermore, according to Singh et al. (2024), these materials are utilized in the production of a variety of goods that are sold in the market, such as antibacterial dressings and environmentally friendly insecticides based on nanoparticles. Nanoparticles (NPs) are becoming increasingly popular as a viable substitute for agrochemicals. Their ability to eradicate specific microorganisms from soil, plants,

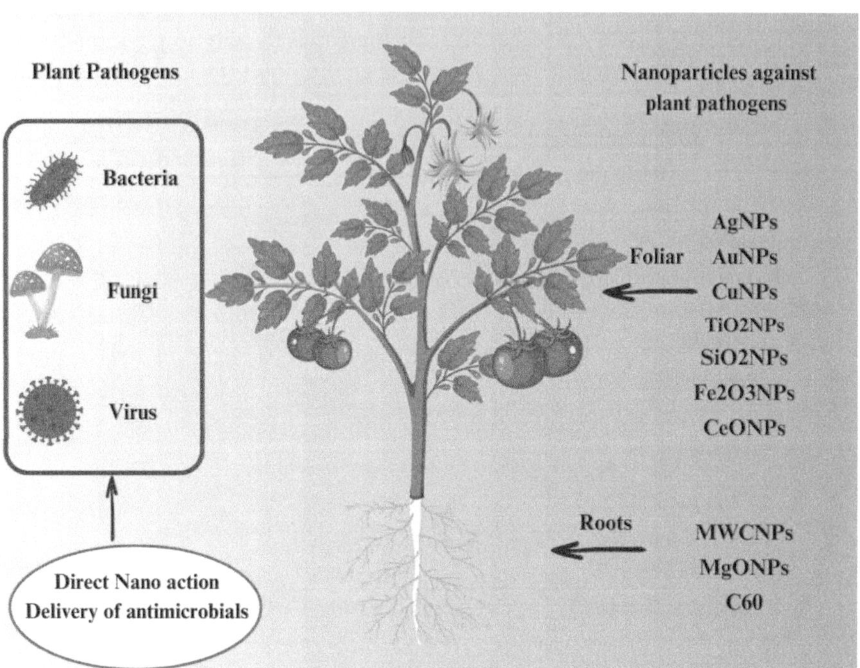

Figure 3. Control of plant pathogens by nanotechnology.

and hydroponic systems is exceptional (Handa and Kalia, 2024). There are two applications for NPs: on phyto-pathogens directly by applying them, or they can be incorporated into fungicide formulations (Khan et al., 2024).

Nanoparticles can be sprayed on leaves to eradicate infections. Additionally, they promote the growth of plants (Chen et al., 2024). Pathogens may be successfully controlled at low concentrations (Abdel-Aal Amin et al., 2024; Chen et al., 2024). Fungal cell walls are harmed by metallic NPs, which results in hyphal plasmolysis (Mosa et al., 2024). Nonetheless, other scientists have suggested that their modes of action could differ and depend on distinct processes (Figure 4) (Salas-Orozco et al., 2024). Among them are:

a- The attachment of protein sulfur groups in NPs causes a disruption in the permeability of the plasma membrane, hence impeding its correct functioning.

b- Damaging of DNA.

c- Oxidation of proteins and disturbance of the electron delivery chain.

d- It's possible for reactive oxygen-species (ROS), which harm cells, to be produced.

e- Impede the absorption of nutrients.

The aforementioned processes are interconnected and work in concert to combat phytopathogens (Mokrini and Bouharroud, 2019). For the manufacture of metallic nanoparticles, the bio-reduction reaction of several metals has been assessed (Salem, 2023a,b). Numerous *in vivo* and *in vitro* investigations have been also conducted to examine the metallic nanoparticles' potential as antibacterial agents (Attia et al., 2024; Salem et al., 2024). Because of their increased toxicity to phyto-pathogens, silver nanoparticles are regarded as successful control tools (Qamar et al., 2024).

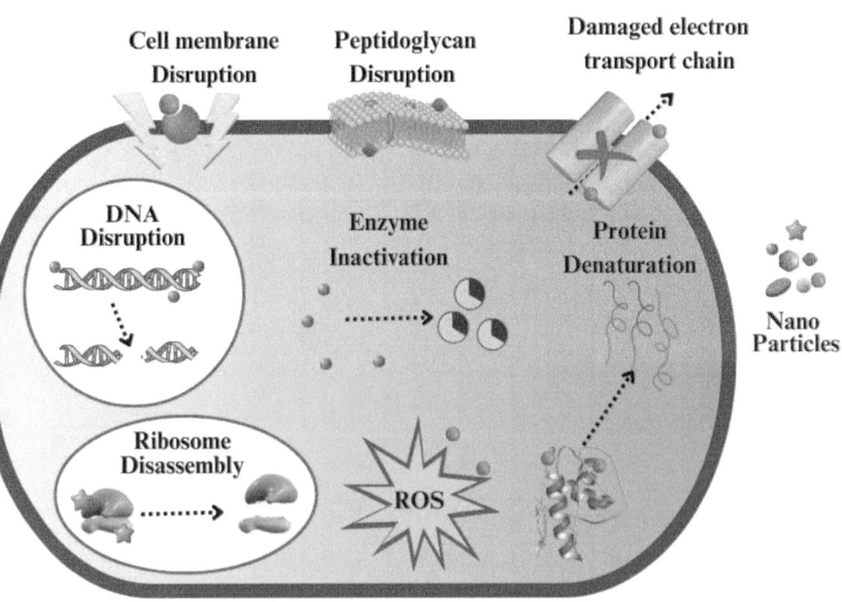

Figure 4. NPs' mode of action against plant diseases.

Using minuscule doses of nano-fungicides, researchers from all over the world have examined the NPs toxicity to the pathogens and their safety for beneficial species. Thus, safer management techniques rather than agrochemicals are advised (Zeng et al., 2008). Whereas, Rajeshkumar (2019) assessed the fungicidal effectiveness of NPs against a fungal pathogen that causes wilt disease. They observed that fungal growth was inhibited. Similarly, tests of cereals' antifungal activity were conducted against *Magnaporthe grisea & Bipolaris sorokiniana* (Jo et al., 2009), with records of the severity and progression of the disease. Because they disrupt the ion efflux transport system and normal cellular activity in fungus, silver nanoparticles are widely used in bio-systems (Shnawa et al., 2023). Also, Wang et al. (2011) assessed the toxicity of gold nanoparticles against *Salmonella* spp. and found that the effects were more harmful than those of the macroform. While silica does not directly combat microbes, it can strengthen host plants' defenses against illnesses and other environmental challenges (Deng et al., 2024).

9. Integrated Disease Management

Using a variety of environmentally friendly management techniques, integrated disease-management (IDM) aims to maintain the illness below its level of the economic borderline. Stated differently, the goal of crop management practices should be to maximize profit while minimizing loss. The population of any creature in a biological system ultimately needs to be in balance with its resources. If there is an imbalance in any way, it should be corrected by integrating several activities in a way that results in a level of detrimental factors that is acceptable. Practices for IDM might be broad or focused. General methods focus on all the diseases that might affect a particular crop, whereas specialised management aims to control just one pathogen in a crop (Oliver 2024).

10. Principles of Integrated Disease Management

10.1 Exclusion

Exclusion refers to the idea of keeping the disease away from the plant. This protocol is typically observed in areas devoid of pathogens. There are no diseases affecting the crops that are growing there. They are not investigated genetically to find a gene or group of genes that work well against the infection. The latter may start an epidemic if it manages to get into that specific area. Laws are put in place at the federal and state levels to limit the introduction and spread of specific plant infections in areas that are free of pathogens. Quarantine, inspection, and crop certification are some other measures that involve exclusion (Shakeel et al., 2020).

10.1.1 Quarantine and Inspection

In the event that a novel plant disease enters an area where the plant has been cultivated without the presence of that specific pathogen (Green Fields, 2024). When a disease is introduced into a previously uncultivated area, it can cause far greater losses than when the infection is already there. This is since:

i- the crop has never been exposed to that particular pathogen, no resistance aspect action towards the infection takes place.

ii- no microorganisms are acting antagonistically or in conflict with the pathogen.

iii- due to the large number of vulnerable hosts, the disease may proliferate and reproduce unchecked.

10.1.2 Crop Certification

Crop certification is another type of quarantine or exclusion that can be either mandatory or optional. Seed certification is required for seed import or export. The lot may be refused or prohibited from entering the nation if the seed is not thoroughly examined and does not even have an inspection certificate (Hassena et al., 2023). Farmers who would like to import or export plants are welcome to voluntarily send in their item for certification and inspection under field or storage circumstances. Following submittal, the plants are examined using suggested protocols. A certificate for the illnesses or pathogens will be provided by the inspection authority in question once the plant or planting material has been determined to be free of all known pathogens and illnesses. Following the certificate's issuing, producers are free to market or sell their goods. It is important to note that the certificate will only address the particular issue or illness that the plants were examined for (Kumar et al., 2023).

10.2 Avoidance

Avoiding the illness is the next course of action after exclusion. The goal of measures like changing the planting period, choosing a different field, and switching up the cultivating zones is to prevent any contact between the pathogen and susceptible hosts. It could be possible to stop illness from developing by altering the surrounding environment.

10.2.1 Choice of Geographical Area

Crop environmental parameters can also be targeted for managing crop diseases (Pugliese et al., 2024). A problem with a crop's cultivation site may be resolved by shifting its geographic location. The infection could adjust to local circumstances in a particular area and possibly develop virulence. Both damp and dry environments can influence the development of fungi and bacteria illnesses. That particular crop may be grown in an irrigated, dry location if a disease develops severely in a wet area, which will lessen or even prevent the disease (Ali et al., 2023; Bwire et al., 2024). One can profit from avoiding smut and ergot can be increased by simply switching the bajra crop from a wet to a dry place. Comparably, wet environments are preferred by *Xanthomonas phaseoli*, *Colletotrichum lindemuthianum*, and *Pseudomonas phaseolicola*. Growing beans in arid regions can significantly lower the prevalence of these pathogens (Ema et al., 2022).

10.2.2 Selection of Field

Choosing disease-free soil is essential to a successful crop. By choosing a site carefully, many soil-borne illnesses can be prevented. It is usually advisable to forgo crops for a few years if the soil-borne disease's causative organism is found there. The red-rot fungus (*C. falcatum*) can live for several months in soil. Consequently, there is a high likelihood of disease exacerbation and crop failure planting sugarcane in the same spot area right after the previous crop is harvested (Viswanathan et al., 2020). Similar to this, the causative organisms for many illnesses such as bacterial wilt of potatoes, bajra smut and ergot, wheat ear cockle of wheat, root-knot or other nematodes, etc., remain in soil for varied lengths of time. It is not advisable to choose vulnerable crops for such fields.

10.2.3 Choice of Sowing Time

The timing of cultivation is the next important factor in preventing a disease, assuming the crop's location is undisputed. The choice will be between early and late sowing. The rationale for this decision is to prevent a disease's favorable climatic conditions from being matched with a crop's susceptible stage. Each crop's sowing period should be modified based on the current environmental circumstances to minimize or eliminate the amount of active inoculum in the field. When planting, it's important to take into account the pathogen's peak activity period, ideal epidemiological circumstances, soil composition, and, last but not least, crop susceptibility. There shouldn't be any overlap between these requirements.

The majority of infections' advantageous conditions are avoided by sowing crops either early or late, enabling robust crop growth. For instance, it is impractical to cultivate *Pisum sativum & Cicer arietinum* in soil that has been exposed to high temperatures and wetness from rain since these conditions inspire the growth of root rot and blight diseases in both crops (Upadhyay et al., 2019).

10.2.4 Disease Escaping Varieties

When creating strategies for managing diseases, varieties with development characteristics that aid in disease avoidance or escape should be taken into consideration. In this instance, the plants' capacity to evade disease is not a result of their genetic resistance, but rather of their growth and maturity schedule. For instance, early maturing cultivars of peas may be resistant to powdery mildew and rust. Usually, the illnesses worsen by January or later. Losses are significantly decreased if pods have formed before the occurrence of major illness (Alford, 2008).

10.2.5 Seed and Planting Stock

Only seed or other propagating material (grafts, tubers, cuttings, bulbs, etc.) can transmit numerous diseases into the field. The majority of vascular pathogens, including as bacteria, fungi, viroids, phytoplasma, and viruses, are spread via contaminated propagation materials such rhizomes, rootstocks, buds, bulbs, grafting, etc. The surface of these spreading materials may potentially harbor pathogens, such as nematodes, fungal spores, and bacterial oozes. Spreading disease—free seeds in a disease—free field is frequently the best way to control outbreaks of certain diseases like smuts, potato black scurf, red rot, etc. The use of disease—free seed, early or late planting in disease—free areas, maintaining plant-plant and row-row distances, cultivating trap-crops, keeping soil-drainage, preserving protection for plants and breaks in the wind are only a few of the actions that must be included in IDM plans to avoid diseases. As a result of these efforts, infections can be avoided and rapid growth can be maintained, preventing harm from reaching economic thresholds (Hamelin et al., 2021).

10.2.6 Epidermal Coating to Avoid Pathogens

The key to managing a disease is preventing plant-pathogen interaction. Many substances (ex. dodecyl alcohol and Kaolin-based films) that can be sprayed on plants can create a membrane on their surface, creating a barrier that prevents the pathogen from attaching itself to the host. This prevents the pathogen from penetrating and, eventually, from causing infection (Qadri et al., 2020).

10.3 Eradication

Despite the measures listed, pathogens can still penetrate crops or fields because there's a chance that an inoculum already exists in the field's surrounding area. In these circumstances, the pathogen must be eliminated from the crop or field. The objectives of this approach are to reduce, deactivate, eliminate, or destroy the source of the inoculum at the level of the plant, field, or entire geographic area. Eradication can be accomplished by physical (e.g., soil sterilization by heat, mulching, hot-water treatment of propagative organs, and disease control by refrigeration) or chemical (ex. soil treatment with chemicals, fumigation, disinfestations of warehouses, and control of insect vectors) treatments, crop rotation, removal of infected plants or biological approaches (Narayanasamy 2013).

11. The Role of Artificial Intelligence (AI) in Early Plant Disease Detection

The ability of computer systems to do activities like learning, reasoning, observation, prediction, and decision-making that typically require human cognition is known as artificial intelligence (AI) (Khaleel et al., 2024). Fundamentally, AI is the development of algorithms that enable robots to mimic human intelligence in areas such as information processing and long-term adaptation. AI includes a number of subfields, such as computer vision, which gives computers sight (González-Rodríguez et al., 2024), machine learning, which finds patterns in data without direct programming, and natural language processing, which allows machines to comprehend and generate human language (Cichos et al., 2020). Expert systems serve as organized representations of human knowledge, and robotics integrates these AI skills for interaction with the environment (Khaleel et al., 2024). Several studies show that AI is effective at correctly detecting complicated illnesses. In a pioneering instance (Omaye et al., 2024) used deep learning methods to use image analysis to identify and diagnose cassava infections. They demonstrated that deep learning is useful in diagnosing a variety of cassava illnesses by achieving diagnostic accuracy above 90%. This technology not only outperformed conventional techniques in terms of speed and accuracy, but it also made it possible to use these models on mobile devices, which made field diagnosis easier. Similarly, to identify and diagnose pests and diseases in tomato plants, Nyakuri et al. (2024) used three artificial intelligence architectures: Faster SSD, R-CNN, and R-FCN. Convolutional neural networks (CNNs) are particularly well-suited for image identification applications because they can learn the spatial patterns of characteristics from input images. These architectures are part of the larger category of CNNs. According to the growth of new AI capabilities, recent research have continued to show the potential of AI in plant disease identification and diagnosis utilizing increasingly advanced, accurate, and potent models. Recently, Sharma (2024) investigated the use of visual and automated recognition techniques in AI technologies for the diagnosis of fungal diseases in plants. According to this investigation, disease patterns including brown spots, yellow-spots, and brown & yellow rust can be detected and classified by artificial neural networks. Arinichev (2022) investigated four well-known and reasonably light CNN architectures Google-Net, SqueezeNet–1.0, ResNet–18, and DenseNet–121-to further this field of study. The DenseNet –121 model stood out for its ideal balance of high operating efficiency and accuracy. With a file size appropriate for mobile devices and a comparatively small number of parameters, this model outperformed the other models under evaluation in terms of classification accuracy. Siamese Convolutional Neural Network (SNN) is a novel CNN evolution. The distinctive structure of Siamese neural networks, which is intended to learn to distinguish among pairs of inputs, sets them apart from traditional CNN models and makes them ideal for tasks involving comparison and distinction. SNNs can provide significant illness classification accuracy by analyzing differences and similarities between picture pairs, frequently resolving issues with interclass variation and the lack of labeled data that standard CNNs confront (Figure 5).

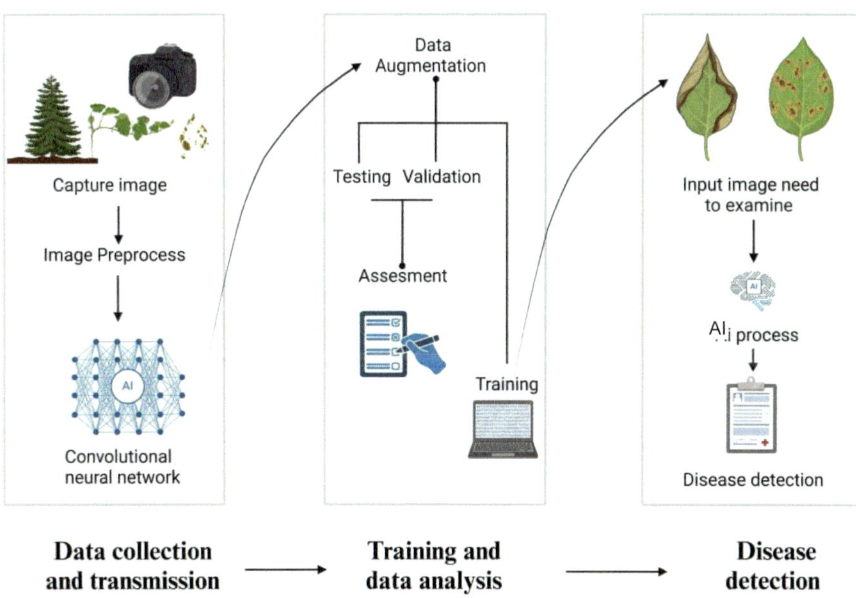

Figure 5. Graph diagram showing how SNNs work to identify plant diseases in a dataset of images.

By offering a more reliable and flexible mechanism for identifying intricate patterns linked to different plant diseases, Siamese networks represent an improvement in the precision and efficiency of identifying diseases in crops, enabling quicker and more accurate application of suitable management measures (Narain et al., 2022).

12. Obstacles in AI for Identifying Plant Diseases

- Limited Supply of Real-time Datasets: The lack of datasets that offer current and pertinent information makes it difficult to construct AI models for real-time plant disease diagnosis (Thakur et al., 2022).
- Absence of Defined Datasets: It is difficult to precisely evaluate and measure the performance of various AI systems in the absence of defined datasets (Shafik et al., 2023).
- Complex and Diverse backdrops: One of the biggest challenges for AI systems is detecting plant diseases, which frequently entails differentiating symptoms on crops from complex and varied natural backdrops (Kamilaris and Prenafeta-Boldú, 2018).
- Close-distance Data Capture: Gathering high-quality data at close ranges to detect minute disease symptoms (Thakur et al., 2022) necessitates certain tools and methods, which makes the procedure more difficult.
- Identifying Small-sized Symptoms: AI systems may find it difficult to accurately identify small-sized disease symptoms because they are easily missed.
- Changes in Lighting and Image Brightness: Robustness is a crucial concern because AI models' performance can be impacted by changes in lighting and image quality (Thakur et al., 2022).
- Disease Development and Inter-class Similarity: More advanced AI algorithms are needed to distinguish between the stages of disease development and to handle inter-class similarity problems.
- Computational Difficulties with Big Datasets: In order to effectively diagnose diseases, it is necessary to overcome the computational difficulties associated with processing huge datasets for AI model training (Shafik et al., 2023).
- Impact of Environmental Elements: Weather and soil quality are two examples of environmental elements that might affect how accurately diseases are detected, hence AI models must take these things into consideration.
- Integration of Multi-modal Data: To develop a comprehensive approach to disease detection, sophisticated approaches are needed to integrate data from multiple sources, including sensors, pictures, and climate data (Shafik et al., 2023).

13. Conclusion

In this chapter, several key methods for early disease prediction and prevention were covered. The first and most crucial step in disease control is the prompt and accurate identification of the pathogen causing a plant disease. Early diagnosis of the disease's etiology enables prompt selection of the best defense strategy and guarantees that crop losses are avoided. There are several established techniques for diagnosing plant diseases; however, new technologies and methods for pathogen identification have been developed recently in order to guarantee the promptness and accuracy of diagnostics as well as to remove the drawbacks associated with traditional diagnostics. Additionally, this chapter also offers details on these cutting-edge, commonly applied techniques for identifying infections and diagnosing disorders.

References

Abdel-Aal Amin, Mohamed, Abdelghafar Mohamed Abu-Elsaoud, Abdelatti Ibrahim Nowwar, Abdelrahman Talha Abdelwahab, Marwa Abdelfattah Awad et al. (2024). Green Synthesis of Magnesium Oxide Nanoparticles Using Endophytic Fungal Strain to Improve the Growth, Metabolic Activities, Yield Traits, and Phenolic Compounds Content of *Nigella Sativa* L. Green Processing and Synthesis 13(1). doi: 10.1515/gps-2023-0215.

Abdelfattah, N. A. H., Yousef, M. A., Badawy, A. A., and Salem, S. S. (2023). Influence of biosynthesized magnesium oxide nanoparticles on growth andphysiological aspects of cowpea (*Vigna unguiculata* L.) plant cowpea beetle and cytotoxicity. Biotechnology Journal. https://doi.org/10.1002/biot.202300301.

Agho, Collins Aimuaenvbosa. (2024). Transcriptomic and Volatile Organic Compounds Analysis of Resistance Mechanisms and Diversity of Late Blight Pathogen Populations in the Baltic Sea Region.

Ahmad, Mobeen, Muhammad Abdullah, Hyeonjoon Moon, and Dongil Han. (2021). Plant disease detection in imbalanced datasets using efficient convolutional neural networks with stepwise transfer learning. IEEE Access, 9: 140565–80.

Alford, David, V. (2008). Pest and Disease Management Handbook. John Wiley & Sons.

Ali, Awais, Tajamul Hussain, Noramon Tantashutikun, Nurda Hussain, and Giacomo Cocetta. (2023). Application of smart techniques, internet of things and data mining for resource use efficient and sustainable crop production. Agriculture, 13(2): 397.

Ali, Hussam, M. I. Lali, Muhammad Zohaib Nawaz, Muhammad Sharif, and Saleem, B. A. (2017). Symptom based automated detection of citrus diseases using color histogram and textural descriptors. Computers and Electronics in Agriculture, 138: 92–104.

Amin, Hassan, Ashraf Darwish, Aboul Ella Hassanien, and Mona Soliman. (2022). End-to-end deep learning model for corn leaf disease classification. IEEE Access, 10: 31103–15.

Arinichev, I. V. (2022). Using Digital Intelligent Technologies for the Diagnosis of Cereals Diseases in the Kuban. Agrar. Sci. J., 5: 70–73.

Asghar, Nosheen, Alamdar Hussain, Duc Anh Nguyen, Salar Ali, Ishtiaque Hussain, Aurangzeb Junejo, and Attarad Ali. (2024). Advancement in nanomaterials for environmental pollutants remediation: a systematic review on bibliometrics analysis, material types, synthesis pathways, and related mechanisms. Journal of Nanobiotechnology, 22(1): 26.

Attia, M. S., Salem, S. S., Elakraa, A. A., Abdel-Maksoud, M. A., Malik, A., Kiani, B. H. et al. (2024). Promising antagonistic effect of bimetallic silver-selenium nanoparticles against Ralstonia solanacearum-causing wilt disease in eggplant (*Solanum melongena* L.). Physiological and Molecular Plant Pathology, p. 102369.

Barburiceanu, Stefania, Serban Meza, Bogdan Orza, Raul Malutan, and Romulus Terebes. (2021). Convolutional neural networks for texture feature extraction. applications to leaf disease classification in precision agriculture. IEEE Access, 9: 160085–103.

Al Bashish, Dheeb, Malik Braik, and Sulieman Bani-Ahmad. (2010). A framework for detection and classification of plant leaf and stem diseases. pp. 113–18 in 2010 international conference on signal and image processing. IEEE.

Badawy, A. A., Husen, A., and Salem, S. S. (2024). Use of nanobiotechnology in augmenting soil–plant system interaction for higher plant growth and production. In Essential Minerals in Plant-Soil Systems (pp. 423–443). Elsevier.

Badawy, A. A., Abdelfattah, N. A. H., Salem, S. S., Awad, M. F., and Fouda, A. (2021). Efficacy assessment of biosynthesized copper oxide nanoparticles(CuO-NPs) on stored grain insects and their impacts on morphological and physiological traits of wheat (*Triticum aestivum* L.) plant. Biology, 10(3). https://doi.org/10.3390/biology10030233.

Bhange, Manisha, and Hingoliwala, H. A. (2015). Smart farming: pomegranate disease detection using image processing. Procedia Computer Science, 58: 280–88.

Bhat, Alangar Ishwara, and Govind Pratap Rao. (2020). Characterization of plant viruses. Springer Protocols Handbooks 23.

Bilay, V. I. 1982. Experimental Mycology Methods.

Blake, Robert S., Paul S. Monks, and Andrew M. Ellis. (2009). Proton-transfer reaction mass spectrometry. Chemical Reviews, 109(3): 861–96.

Bwire, Denis, Hirotaka Saito, Roy C. Sidle, and Junko Nishiwaki. (2024). Water management and hydrological characteristics of paddy-rice fields under alternate wetting and drying irrigation practice as climate smart practice: A Review. Agronomy, 14(7): 1421.

Carter, Darren J., and R. Bruce Cary. (2007). Lateral flow microarrays: a novel platform for rapid nucleic acid detection based on miniaturized lateral flow chromatography. Nucleic Acids Research, 35(10): e74.

Cellini, Antonio, Francesco Spinelli, Irene Donati, Choong-Min Ryu, and Joseph W. Kloepper. (2021). Bacterial Volatile Compound-Based Tools for Crop Management and Quality. Trends in Plant Science, 26(9): 968–83.

Chelloug, Samia Allaoua, Reem Alkanhel, Mohammed Saleh Ali Muthanna, Ahmed Aziz, and Ammar Muthanna. (2023). MULTINET: A Multi-Agent DRL and EfficientNet Assisted Framework for 3D Plant Leaf Disease Identification and Severity Quantification. IEEE Access.

Chen, Hao, Yujie Song, Yu Wang, Huan Wang, Zhaotang Ding, and Kai Fan. (2024). Zno Nanoparticles: Improving Photosynthesis, Shoot Development, and Phyllosphere Microbiome Composition in Tea Plants. Journal of Nanobiotechnology, 22(1): 389.

Chen, Junde, Weirong Chen, Adan Zeb, Shuangyuan Yang, and Defu Zhang. (2022). Lightweight inception networks for the recognition and detection of rice plant diseases. IEEE Sensors Journal, 22(14): 14628–38.

Cichos, F., Gustavsson, K., Mehlig, B., and Volpe, G. (2020). Machine learning for active matter. Nature Machine Intelligence, 2(2): 94–103.

Deng, Quanqing, Suihua Huang, Hao Liu, Qing Lu, Puxuan Du, Haifen Li et al. (2024). Silica Nanoparticles Conferring Resistance to Bacterial Wilt in Peanut (*Arachis Hypogaea* L.). Science of The Total Environment, 915: 170112.

Desaeger, Johan, Catherine Wram, and Inga Zasada. (2020). New reduced-risk agricultural nematicides-rationale and review. Journal of Nematology, 52(1): 1–16.

Dey, Amar Kumar, Manisha Sharma, and M. R. Meshram. (2016). Image processing based leaf rot disease, detection of betel vine (*Piper Betle* L.). Procedia Computer Science, 85: 748–54.

Dwivedi, Rudresh, Tanima Dutta, and Yu-Chen Hu. (2021). A leaf disease detection mechanism based on L1-norm minimization extreme learning machine. IEEE Geoscience and Remote Sensing Letters, 19: 1–5.

Dyakov, Y. T., and S. N. Elansky. (2019). General Phytopathology.

Ema, Israt Jahan, Marjia Akhter Monika, Ahasan Ullah Khan, Mohammad Monirul Hasan Tipu, Md Ruman Faruk, Shofiul Azam Tarapder et al. (2022). A review on the management of country bean (*Lablab Purpureus* L.) Diseases in Bangladesh. Journal of Agricultural Sciences (Sri Lanka) 17(3).

Gavhale, Kiran R., and Ujwalla Gawande. (2014). An overview of the research on plant leaves disease detection using image processing techniques. Iosr Journal of Computer Engineering (Iosr-Jce), 16(1): 10–16.

Gharge, Saylee, and Priyanka Singh. (2016). Image processing for soybean disease classification and severity estimation. pp. 493–500. In Emerging Research in Computing, Information, Communication and Applications: ERCICA 2015, Volume 2. Springer.

González-Rodríguez, V. E., Izquierdo-Bueno, I., Cantoral, J. M., Carbú, M., and Garrido, C. (2024). Artificial Intelligence: A Promising Tool for Application in Phytopathology. Horticulturae, 10: 197.

Green Fields, Guntur. (2024). Host Recognition for Survival of Plant Pathogens.

Hamelin, Frédéric Marie, Brady Bowen, Pierre Bernhard, and Vushrali A. Bokil. (2021). Optimal control of plant disease epidemics with clean seed usage. Bulletin of Mathematical Biology, 83: 1–24.

Handa, Megha, and Anu Kalia. (2024). Exploring the dynamics of nanoparticle-plant-microbe interactions to realize the goal of improved crop productivity and food security. Rhizosphere 100884.

Hashem, A. H., Abdelaziz, A. M., Attia, M. S., and Salem, S. S. (2022). Selenium and nano-selenium-mediated biotic stress tolerance in plants. *In*: Hossain, M. A., Ahammed, G. J., Kolbert, Z., El-Ramady, H., Islam, T., and Schiavon, M. (Eds.). Selenium and nano-selenium in environmental stress management and crop quality improvement. Springer International Publishing, Cham., pp. 209–226. https://doi.org/10.1007/978-3-031-07063-1_11.

Hassena, M., Alemu, D., Dey, B., Mohammed Hassena, Dawit Alemu, and Bhramar Dey. (2023). Seed policy provisions and operational challenges in Ethiopia. A Feed the Future Global Supporting Seed Systems for Development Activity (S34D) Report.

Jo, Young-Ki, Byung H. Kim, and Geunhwa Jung. (2009). Antifungal Activity of Silver Ions and Nanoparticles on Phytopathogenic Fungi. Plant Disease, 93(10): 1037–43.

Kamilaris, A., and Prenafeta-Boldú, F. X. (2018). Deep learning in agriculture: A survey. Computers and Electronics in Agriculture, 147: 70–90.

Kamilov, Sh G., Kh Kh Nuraliev, Sattarova, R. K., and Khakimov, A. A. (2020). The History of Mycology and Phytopathology Development in Uzbekistan.

Khakimov, Albert, Ilias Salakhutdinov, Almas Omolikov, and Samad Utaganov. (2022). Traditional and Current-Prospective Methods of Agricultural Plant Diseases Detection: A Review. P. 12002 in IOP Conference series: earth and environmental science. Vol. 951. IOP Publishing.

Khaleel, M., Jebrel, A., and Shwehdy, D. M. (2024). Artificial Intelligence in Computer Science: https:// doi. org/10.5281/zenodo. 10937515. Int. J. Electr. Eng. and Sustain, 1–21.

Khan, Masudulla, Jelena Popović-Djordjević, and Jelena Katanić Stanković. (2024). Nanomaterials as Effective Fertilizers and Fungicides in Plant Protection for Improving Crop Growth and Inducing Tolerance Against Phytopathogens. pp. 433–49 in Nanoparticles in Plant Biotic Stress Management. Springer.

Khasanov, B. A. (2011). Diseases of Agricultural Crops and Measures to Combat Them. Textbook for Graduate Students.

Kiani, Ehsan, and Tofik Mamedov. (2017). Identification of plant disease infection using soft-computing: application to modern botany. Procedia Computer Science, 120: 893–900.

Kumar, Sanjay, K., Sripathy, V., Udaya Bhaskar, K., and Banoth Vinesh. (2023). Principles of Quality Seed Production. Malavika Dadlani 109.

Li, Fengdi, Zhenyu Liu, Weixing Shen, Yan Wang, Yunlu Wang, Chengkai Ge et al. (2021). A remote sensing and airborne edge-computing based detection system for pine wilt disease. IEEE Access, 9: 66346–60.

Li, Ruimin, Xinyou Wang, Yanan Hu, and Guiyan Huang. (2024). Analysis of huanglongbing-associated RNA-Seq data reveals disturbances in biological processes within *Citrus* Spp. Triggered by Candidatus Liberibacter Asiaticus Infection. Frontiers in Plant Science, 15: 1388163.

Light, Kenneth M., John A. Wisniewski, Andrew W. Vinyard, and Matthew T. Kieber-Emmons. (2016). Perception of the plant hormone ethylene: known-knowns and known-unknowns. JBIC Journal of Biological Inorganic Chemistry, 21: 715–28.

Martinelli, Federico, Russell L. Reagan, Sandra L. Uratsu, My L. Phu, Ute Albrecht, Weixiang Zhao et al. (2013). Gene regulatory networks elucidating huanglongbing disease mechanisms. PloS One 8(9): e74256.

Martinelli, Federico, Riccardo Scalenghe, Salvatore Davino, Stefano Panno, Giuseppe Scuderi, Paolo Ruisi et al. (2015). Advanced methods of plant disease detection. A review. Agronomy for Sustainable Development, 35: 1–25.

Maruya, Keith A., Wenjian Lao, David Tsukada, and Dario W. Diehl. (2015). A passive sampler based on solid phase Microextraction (SPME) for sediment-associated organic pollutants: comparing freely-dissolved concentration with bioaccumulation. Chemosphere, 137: 192–97.

Masood, Momina, Marriam Nawaz, Tahira Nazir, Ali Javed, Reem Alkanhel, Hela Elmannai et al. (2023). MaizeNet: A deep learning approach for effective recognition of maize plant leaf diseases. IEEE Access, 11: 52862–76. doi: 10.1109/ACCESS.2023.3280260.

Mauck, Kerry E., Consuelo M. De Moraes, and Mark C. Mescher. (2010). Deceptive chemical signals induced by a plant virus attract insect vectors to inferior hosts. Proceedings of the National Academy of Sciences, 107(8): 3600–3605.

Mazed, Md. Khairul, Mansura Afroz, Md. Shamim Hossain, and Md. Ruhul Amin. (2022). Trends of crop production, land use and pesticide consumption in Bangladesh. Ecol. J., 4: 91–99.

Mokrini, Fouad, and Rachid Bouharroud. (2019). Application of nanotechnology in plant protection by phytopathogens: present and future prospects. Microbial Nanobionics: Volume 2, Basic Research and Applications 261–79.

Mosa, Mohamed A., Dina S. S. Ibrahim, Mohamed A. M. El-Tabakh, and Sozan E. El-Abeid. (2024). Recent advances in using nanotechnology in the management of soilborne plant pathogens. Nanotechnology in Plant Health 311–34.

Narain, T., Sahu, P., and Singh, A. P. (2022). Plant disease classification using siamese convolutional neural network. International Conference on Innovative Computing and Communications: Proceedings of ICICC 2022, 2: 57–65.

Narayanasamy, P. (2013). Abiotic Biological Control Agents for Crop Disease Management. Vol. 1.

Nie, Xuan, Luyao Wang, Haoxuan Ding, and Min Xu. (2019). Strawberry verticillium wilt detection network based on multi-task learning and attention. IEEE Access, 7: 170003–11.

Ninkovic, Velemir, Dimitrije Markovic, and Merlin Rensing. (2021). Plant volatiles as cues and signals in plant communication. Plant, Cell & Environment, 44(4): 1030–43.

Nowwar, Abdelatti I., Ibrahim I. Farghal, Mohamed A. Ismail, and Mohamed A. Amin. (2023). Impact of irrigation with wastewater on accumulation of heavy metals in *Phaseolus vulgaris* L. and its remediation. Journal of Soil Science and Plant Nutrition, 23(1): 761–77.

Nowwar, Abdelatti Ibrahim, Ibrahim I. Farghal, Mohamed A. Ismail, and Mohamed A. Amin. (2022). Biochemical changes on jute mallow plant irrigated with wastewater and its remediation. Egyptian Journal of Chemistry, 65(8): 271–83.

Nyakuri, J. P., Nkundineza, C., Gatera, O., and Nkurikiyeyezu, K. (2024). State-of-the-art deep learning algorithms for internet of things-based detection of crop pests and diseases: A comprehensive review. IEEE Access.

Ogbodo, John Onyebuchi, Amarachukwu Vivan Arazu, Tochukwu Chisom Iguh, Ngozichukwuka Julie Onwodi, and Tobechukwu Christian Ezike. (2022). Volatile organic compounds: A proinflammatory activator in autoimmune diseases. Frontiers in Immunology, 13: 928379.

Oliver, Richard P. (2024). Control of plant diseases. pp. 279–316 in Agrios' Plant Pathology. Elsevier.

Omaye, J. D., Ogbuju, E., Ataguba, G., Jaiyeoba, O., Aneke, J., and Oladipo, F. (2024). Cross-comparative review of Machine learning for plant disease detection: Apple, cassava, cotton and potato plants. Artificial Intelligence in Agriculture.

Ortiz-Hernández, Ma Laura, Enrique Sánchez-Salinas, Angeluz Olvera-Velona, and Jorge Luis Folch-Mallol. (2011). Pesticides in the Environment: Impacts and Their Biodegradation as a Strategy for Residues Treatment.

Piri-Moghadam, Hamed, Fardin Ahmadi, and Janusz Pawliszyn. (2016). A critical review of solid phase microextraction for analysis of water samples. TrAC Trends in Analytical Chemistry, 85: 133–43.

Van Der Plank, J. (2012). Principles of Plant Infection. Elsevier.

Pugliese, Massimo, Giovanna Gilardi, Angelo Garibaldi, and Maria Lodovica Gullino. (2024). The Impact of climate change on vegetable crop diseases and their management: the value of phytotron studies for the agricultural industry and associated stakeholders. Phytopathology® 114(5): 843–54.

Putnam, M. L. (1995). Evaluation of selected methods of plant disease diagnosis. Crop Protection, 14(6): 517–25.

Qadri, Rashad, Muhammad Azam, Imran Khan, Yaodong Yang, Shaghef Ejaz, Muhammad Tahir Akram et al. (2020). Conventional and Modern Technologies for the Management of Post-Harvest Diseases.

Qamar, Safi Ur. Rehman, Samina Tanwir, Waqar Ali Khan, Javaria Altaf, and Jam Nazeer Ahmad. (2024). Biosynthesis of silver nanoparticles using *Ocimum tenuiflorum* extract and its efficacy assessment against *Helicoverpa armigera*. International Journal of Pest Management, 70(3): 375–83.

Rajeshkumar, Shanmugam. (2019). Antifungal impact of nanoparticles against different plant pathogenic fungi. pp. 197–217 in *Nanomaterials in plants, algae and microorganisms*. Elsevier.

Ramedani, Zeynab, Mahmoud Omid, Alireza Keyhani, Shahaboddin Shamshirband, and Benyamin Khoshnevisan. (2014). Potential of radial basis function based support vector regression for global solar radiation prediction. Renewable and Sustainable Energy Reviews, 39: 1005–11.

Razo-Belman, Rosario, and César Ozuna. (2023). Volatile organic compounds: a review of their current applications as pest biocontrol and disease management. Horticulturae, 9(4): 441.

Salam, Uzma, Shakir Ullah, Zhong-Hua Tang, Ahmed A. Elateeq, Yaseen Khan, Jafar Khan et al. (2023). Plant metabolomics: An overview of the role of primary and secondary metabolites against different environmental stress factors. Life, 13(3): 706.

Salas-Orozco, Marco Felipe, Ana Cecilia Lorenzo-Leal, Idania de Alba Montero, Nuria Patiño Marín, Miguel Angel Casillas Santana, and Horacio Bach. (2024). Mechanism of escape from the antibacterial activity of metal-based nanoparticles in clinically relevant bacteria: a systematic review." Nanomedicine: Nanotechnology, Biology and Medicine, 55: 102715.

Salem, S. S. (2023a). A mini review on green nanotechnology and its development in biological effects. Arch. Microbiol. 205: 128. https://doi.org/10.1007/s00203-023-03467-2.

Salem S. S. (2023b). Application of Nano-materials. *In*: Raja, R., Hemaiswarya, S., Narayanan, M., Kandasamy, S., and Jayappriyan, K. (Eds.). Haematococcus. Springer, Singapore. https://doi.org/10.1007/978-981-99-2901-6_10.

Salem, S. S., and Mekky, A. E. (2024). Biogenic nanomaterials: Synthesis, characterization, and applications. In Biogenic Nanomaterials for Environmental Sustainability: Principles, Practices, and Opportunities (pp. 13–43). Cham: Springer International Publishing.

Salem, S. S., Saied, E., and Shah, M. P. (2024). Advanced (nano) materials. In Development in wastewater treatment research and processes (pp. 93–115). Elsevier.

Salem, S. S., and Fouda, A. (2021). Green synthesis of metallic nanoparticles and their prospective biotechnological applications: an overview. Biol. Trace Elem. Res., 199(1): 344–370. https://doi.org/10.1007/s12011-020-02138-3.

Salem, S. S., Hammad, E. N., Mohamed, A. A., and El-Dougdoug, W. (2023). A comprehensive review of nanomaterials: types, synthesis, characterization, and applications. Biointerface Res. Appl. Chem., 13(1). https://doi.org/10.33263/BRIAC131.041.

Salem, S. S., and Husen, A. (2022). Effect of engineered nanomaterials on soil microbiomes and their association with crop growth and production. *In*: Engineered Nanomaterials for Sustainable Agricultural Production, Soil Improvement and Stress Management. pp. 311–336. https://doi.org/10.1016/B978-0-323-91933-3.00010-6.

Salem, S. S., and Husen, A. (2023). Zinc Oxide Nanoparticles: Plant Response, Interaction, Phytotoxicity, and Defence. *In*: Husen, A. (Eds.). Nanomaterials and Nanocomposites Exposures to Plants. Smart Nanomaterials Technology. Springer, Singapore. https://doi.org/10.1007/978-981-99-2419-6_12.

Schor, Noa, Avital Bechar, Timea Ignat, Aviv Dombrovsky, Yigal Elad, and Sigal Berman. (2016). Robotic disease detection in greenhouses: combined detection of powdery mildew and tomato spotted wilt virus. IEEE Robotics and Automation Letters. 1(1): 354–60.

Shafi, Uferah, Rafia Mumtaz, Muhammad Deedahwar Mazhar Qureshi, Zahid Mahmood, Sikander Khan Tanveer, Ihsan Ul Haq et al. (2023). Embedded AI for wheat yellow rust infection type classification. IEEE Access, 11: 23726–38.

Shafik, W., Tufail, A., Namoun, A., De Silva, L. C., and Apong, R. A. A. H. M. (2023). A systematic literature review on plant disease detection: Motivations, classification techniques, datasets, challenges, and future trends. Ieee Access, 11: 59174–59203.

Shakeel, Qaiser, Guoqing Li, Yang Long, and Hafiz Abdul Samad Tahir. (2020). Development and implementation of IDM program for annual and perennial crops. Plant Disease Management Strategies for Sustainable Agriculture through Traditional and Modern Approaches 295–327.

Shamsiddinovich, Xusniddinov Faxriddin. (2024). Nanotechnologies and their significance in environmental protection. Web of Teachers: Inderscience Research, 2(3): 16–21.

Sharma, R. (2024). Revolutionizing Rice Agriculture: A Machine Learning Approach to Fungal Disease Management for Economic Sustainability. 2024 International Conference on Communication, Computer Sciences and Engineering (IC3SE), 798–805.

Sharma, Pallavi, and R. S. Dubey. (2019). Protein synthesis by plants under stressful conditions. pp. 405–49 in Handbook of Plant and Crop Stress, Fourth Edition. CRC Press.

Sheelavantamath, Bindu. (2020). Plant disease detection and its solution using image classification. International Journal of Futures Research and Development, 1(1): 73–79.

Shnawa, B. H., S. M. Al-Bader, and P. J. Jalil. (2023). Antifungal potency of green-fabricated nanoparticles-an overview of recent advances. One Health Triad, Unique Scientific Publishers, Faisalabad, Pakistan, 2: 251–58.

Shovon, Md Sakib Hossain, Shakrin Jahan Mozumder, Osim Kumar Pal, Muhammad F. Mridha, Nobuyoshi Asai, and Jungpil Shin. (2023). PlantDet: A robust multi-model ensemble method based on deep learning for plant disease detection. IEEE Access.

Singh, Jaskaran, and Harpreet Kaur. (2018). A review on: various techniques of plant leaf disease detection. pp. 232–38 in 2018 2nd International Conference on Inventive Systems and Control (ICISC). IEEE.

Singh, Swati, Harshita Tiwari, Ashish Verma, Priyamvada Gupta, Amrit Chattopadhaya, Ananya Singh et al. (2024). Sustainable synthesis of novel green-based nanoparticles for therapeutic interventions and environmental remediation. ACS Synthetic Biology.

Singh, Uday Pratap, Siddharth Singh Chouhan, Sukirty Jain, and Sanjeev Jain. (2019). Multilayer convolution neural network for the classification of mango leaves infected by anthracnose disease. IEEE Access, 7: 43721–29.

Singh, Vijai, and Ak K. Misra. (2017). Detection of plant leaf diseases using image segmentation and soft computing techniques. Information Processing in Agriculture, 4(1): 41–49.

Smith, D., and P. Španěl. (2005). Selected ion flow tube mass spectrometry, SIFT-MS, for on-Line trace gas analysis of breath. Breath Analysis For Clinical Diagnosis And Therapeutic Monitoring: (With CD-ROM) 3–34.

Sorokopudov, V. N., I. V Knyazeva, O. A. Sorokopudova, J. V Burmenko, and T. V Baranova. (2018). Biotechnological methods of maintaining collections of the *Genus ribes* L. pp. 123–30 in VIII International Scientific and Practical Conference on Biotechnology as an Instrument for Plant Biodiversity Conservation 1324.

Thakur, P. S., Khanna, P., Sheorey, T., and Ojha, A. (2022). Trends in vision-based machine learning techniques for plant disease identification: A systematic review. Expert Systems with Applications, 208: 118117.

Tiwari, Shikha, Adinath Kate, Debabandya Mohapatra, Manoj Kumar Tripathi, Hena Ray, Amitava Akuli et al. (2020). Volatile Organic Compounds (VOCs): Biomarkers for quality management of horticultural commodities during storage through e-sensing. Trends in Food Science & Technology 106:417–33.

Tripathi, Ruchi, Rashmi Tewari, K. P. Singh, Chetan Keswani, Tatiana Minkina, Anoop Kumar Srivastava et al. (2022). Plant mineral nutrition and disease resistance: A significant linkage for sustainable crop protection. Frontiers in Plant Science, 13: 883970.

Upadhyay, Vinod, Kuldip Medhi, Puja Pandey, Palash Thengal, Sunil Kumar Paul, and K. Kushwah. (2019). Rust disease of pea: A review. International Journal of Current Microbiology and Applied Sciences, 8(4): 416–34.

Upadyshev, M. T., and V. I. Donetskikh. (2008). New method of sanitizing berry and fruit crops from viruses by the magnetotherapy method. Russian Agricultural Sciences, 34(4): 223–26.

Vishnoi, Vibhor Kumar, Krishan Kumar, Brajesh Kumar, Shashank Mohan, and Arfat Ahmad Khan. (2023). Detection of apple plant diseases using leaf images through convolutional neural network. IEEE Access, 11: 6594–6609. doi: 10.1109/ACCESS.2022.3232917.

Viswanathan, R., R. Selvakumar, K. Manivannan, R. Nithyanantham, and K. Kaverinathan. (2020). Behaviour of soil borne inoculum of colletotrichum falcatum in causing red rot in sugarcane varieties with varying disease resistance. Sugar Tech, 22(3): 485–97.

Wang, Shuguang, Rasheeda Lawson, Paresh C. Ray, and Hongtao Yu. (2011). Toxic effects of gold nanoparticles on Salmonella Typhimurium bacteria. Toxicology and Industrial Health, 27(6): 547–54.

Wilson, Alphus Dan. (2018). Applications of electronic-nose technologies for noninvasive early detection of plant, animal and human diseases. Chemosensors, 6(4): 45.

Xiao, Zhiyong, Yongge Shi, Gailin Zhu, Jianping Xiong, and Jianhua Wu. (2023). Leaf disease detection based on lightweight deep residual network and attention mechanism. IEEE Access.

Xu, Mingjun, Zhentao Tang, Yixiang Duan, and Yong Liu. (2016). GC-based techniques for breath analysis: current status, challenges, and prospects. Critical Reviews in Analytical Chemistry, 46(4): 291–304.

Yadav, Preeti, and Parvinder Singh. (2022). Disease detection techniques in plants: transition from manual to automation. pp. 93–109 in International Workshop on New Approaches for Multidimensional Signal Processing. Springer.

Yilmaz, Erkan, and Mustafa Soylak. (2020). Type of new generation separation and preconcentration methods. pp. 75–148 in New Generation Green Solvents for Separation and Preconcentration of Organic and Inorganic Species. Elsevier.

Zeng, Hui, Xuefeng Li, Gaoyong Zhang, and Jinfeng Dong. (2008). Preparation and characterization of beta *Cypermethrin nanosuspensions* by diluting O/W Microemulsions. Journal of Dispersion Science and Technology, 29(3): 358–61.

Zeng, Qingmao, Xinhui Ma, Baoping Cheng, Erxun Zhou, and Wei Pang. (2020). Gans-based data augmentation for citrus disease severity detection using deep learning. IEEE Access, 8: 172882–91.

Zhu, Xianyu, Jinjiang Li, Runchang Jia, Bin Liu, Zhuohan Yao, Aihong Yuan et al. (2022). Lad-net: A novel light weight model for early apple leaf pests and diseases classification. IEEE/ACM Transactions on Computational Biology and Bioinformatics, 20(2):1156–69.

Zuparov, Mirakbar A., Albert A. Khakimov, Mukhiddin S. Mamiev, and Abdurakhmon N. Allayarov. (2020). *In vitro* efficacy testing of fungicides on *Botrytis cinerea* causing gray mold of tomato. International Journal on Emerging Technologies, 11(5): 50–55.

CHAPTER 7

Detection of Tomato, Potato, and Wheat Leaf Diseases via Artificial Intelligence

Faizan Ali,[1,*] *Akhtar Hameed,*[2] *Abdul Rahman,*[3]
Hafiz Muhammad Usman Aslam,[4] *Sohaib Sarfraz*[3] and
Muhammad Usman Afzal[5]

1. Introduction

Tomato, potato, and wheat are among the most important crops cultivated globally, making an integral part of the global food chain and contributing to the nutritional and economic wellbeing of billions of people (Birch et al., 2012). Each crop plays its unique importance in human diet and agricultural economies. For instance, tomatoes, rich in vitamins A and C, contribute to an annual production of about 185 million metric tons, providing essential nutrients and generating substantial income for farmers, especially in developing regions (Borguini and Ferraz Da Silva Torres, 2009). Potatoes, with a global production of over 370 million metric tons, are another critical component for global food chain as it is a rich source of carbohydrates and dietary fiber, ranking as the fourth most important food crop after rice, wheat, and maize (the 3 major cereals) (Singh et al., 2022). Wheat, a staple crop, holds even greater significance, with global production surpassing 780 million metric tons

[1] Department of Agricultural Science & Engineering, College of Agriculture, Tennessee State University, Nashville, TN 37209, USA.
[2] Institute of Plant Protection, MNS University of Agriculture Multan 61000, Pakistan.
[3] Department of Plant Pathology, University of Agriculture Faisalabad 38000, Pakistan.
[4] Department of Plant Pathology, San Luis Valley Research Centre, Colorado State University, Fort Collins, CO 80523, USA.
[5] ViviGro Sustainable Solutions, Saskatoon, SK S7J 3L8, Canada.
* Corresponding author: faizanali_95@outlook.com

DOI: 10.1201/9781003546702-7

annually (Španić et al., 2023). It provides 20% of the world's caloric intake and serves as a major source of protein, particularly in many developing nations where wheat is consumed as a staple crop. Together, these crops form the cornerstone of global agricultural production, contributing significantly to food security and economic growth.

However, the global production of these crops is persistently threatened by various leaf diseases, which have a profound impact on crop health, production margins and overall yield. Leaf diseases, caused by wide range of bacterial, fungal, and viral pathogens, primarily attack the photosynthetic tissues of plants, compromising their ability to perform bio-chemical reactions that are responsible for food and energy production for the plants. For example, late blight, caused by the pathogen *Phytophthora infestans*, affects both tomato and potato crop, leading to severe leaf necrosis, stem lesions, and, in many cases, the complete loss of crops. In wheat, diseases like leaf rust, caused by the fungus *Puccinia triticina*, the pathogen is capable of reducing yield by up to 40% under severe infestations. Such widespread losses directly correlate with significant reductions in both the quantity and quality of harvests, posing a threat to food security on a global scale. The economic impact of leaf diseases is staggering. Each year, millions of tons of crops are lost to diseases that spread rapidly under favorable conditions. For instance, the late blight of potato is responsible for billions of dollars in annual losses globally due to direct crop damage, the increased cost of fungicide applications, and post-harvest losses. Leaf rust in wheat has led to extensive yield reductions in countries like India, the United States, and Mexico, where wheat serves as a dietary staple (Huerta-Espino et al., 2011). The financial strain extends beyond individual farmers to entire economies, particularly in regions where agricultural output constitutes a significant portion of gross domestic product (GDP) and employees a large number of people in job market. In these areas, unchecked disease outbreaks can destabilize food markets, raise commodity prices, and lead to increased reliance on imports, further deteriorating food security issue.

Given the wide spread impact of these diseases, early and accurate detection is crucial for minimizing their impact. Traditionally, disease detection has relied on visual inspection of the crop fields by farmers or agricultural extension workers, who manually assess fields for signs of infection. While this method can be effective on a small scale, it is highly labor-intensive, and often ineffective in large scale disease identification or identifying diseases before they reach an advanced stage. Another major challenge is the varying skillset for each of the field workers which can lead to overlooking or misinterpreting some of the symptoms leading to inaccurate detection and management of disease. Furthermore, traditional methods are impractical for large-scale farming operations, where inspecting thousands of plants is logistically challenging. In many cases, by the time visible symptoms of disease appear, the infection has already spread, making it difficult to control and leading to greater crop losses.

In recent years, advances in modern technology have opened new avenues for disease detection. Remote sensing, drone imaging, and spectral analysis have emerged as tools to detect stress in plants before symptoms are visible to the naked eye (Barbedo, 2019). However, these technologies are often expensive and require specialized knowledge for implementation. Moreover, they may not provide real-time feedback, which is critical for timely intervention in disease management. This is where artificial intelligence (AI) has proven to be a game-changer in modern agriculture. AI, with its capacity to process massive datasets rapidly and accurately, offers a promising solution to the limitations of traditional disease detection methods (Xu et al., 2021). Through machine learning (ML) and deep learning (DL) algorithms, AI systems can analyze images of plants to detect early signs of disease, even in cases where human observation would fail to detect any presence of the disease at any given stage (Jackulin and Murugavalli, 2022). AI models are trained on vast datasets of plant images, allowing them to recognize subtle patterns and features associated with specific diseases. For example, convolutional neural networks (CNNs) approach, a type of deep learning model, are capable of identifying leaf diseases from images by learning to distinguish between healthy and diseased leaves based on their color, texture, and shape (Barburiceanu et al., 2021). The integration of AI into disease detection offers a number of key advantages (Badidi, 2023); First, AI systems can work in real time, providing immediate feedback to farmers about the health of their crops. This enables farmers to take swift action, applying targeted treatments only where needed, thus reducing the overuse of chemical pesticides and fungicides, which can have harmful environmental effects. Secondly, the AI models can be deployed at scale, making it feasible to monitor large fields or even entire regions for disease outbreaks. This scalability is especially important in the context of global food production, where rapid disease detection and intervention can prevent widespread crop losses. Additionally, AI systems can be integrated into mobile devices, drones, or field sensors, making them accessible to farmers in both developed and developing countries (Mohamed et al., 2021). One of the most key aspect of AI-driven disease detection is its potential to contribute to more sustainable agricultural practices. By providing precise, real-time information about disease presence, AI allows for more efficient use of agronomic resources such as water, fertilizers, and pesticides. This targeted approach not only reduces costs for farmers but also minimizes the environmental impact of farming practices. In the long term, widespread adoption of AI technologies in agriculture could lead to more resilient and sustainable food production systems, capable of withstanding the challenges posed by climate change and a growing global population (Pandey and Mishra, 2024). Although the AI technology is quite effective and seems to have a strong future. Despite the clear benefits, several challenges remain in the implementation of AI for disease detection. The accuracy of AI models depends heavily on the quality and diversity of the data used to train them. In many cases, publicly available datasets are limited in scope, particularly for certain crops or regions. Additionally, environmental factors such as lighting conditions, weather, and plant growth stages can introduce variability into the data, affecting the model's ability to generalize across different contexts. Nevertheless, ongoing research and development in AI and machine learning (ML) are rapidly addressing these challenges, paving the way for more robust and reliable disease detection systems.

In this chapter, we will explore the current state of AI applications in detecting leaf diseases in tomato, potato, and wheat, focusing on the methodologies, tools, and technologies that are driving innovation in this field. We will also be understanding

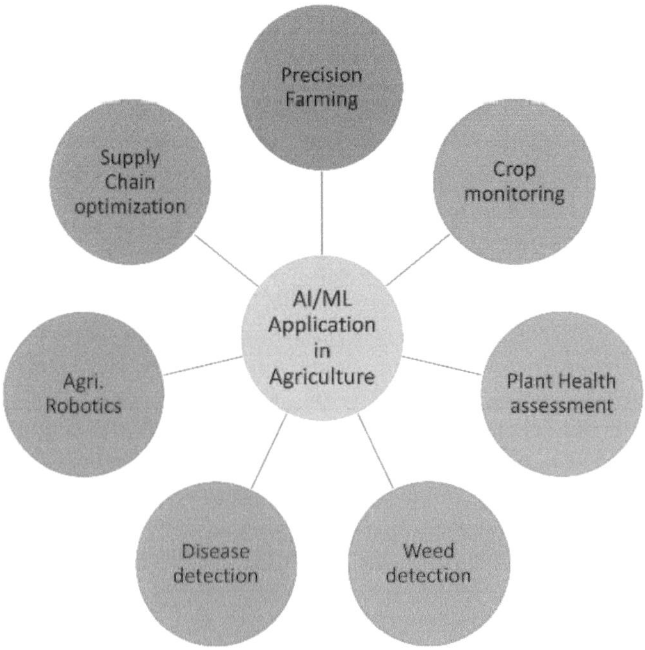

Figure 1. Diverse application for AI in agriculture and food sector.

the potential and limitation of AI in leaf disease detection, and how these technologies can be harnessed to improve crop health, increase yield, and ensure global food security.

2. Common Leaf Diseases in Tomato, Potato, and Wheat

Leaf diseases are significant threats to the productivity of tomato, potato, and wheat crops, impacting both yield and quality. Understanding the specific diseases affecting these crops, their symptoms, and management practices is crucial for developing effective detection and control strategies. Following are some of the common leaf diseases associated with each crop, including their causal agents, symptoms and economic implications.

2.1 Tomato Leaf Diseases

Early blight of tomato is a common fungal disease caused by *Alternaria solani* that affects tomato plants worldwide. The disease typically begins on older, lower leaves with small dark spots that develop concentric rings, giving a "bull's-eye" appearance (Riley et al., 2016). As the infection progresses, affected leaves turn yellow, then brown, and eventually die and drop off. The fungus can also cause stem lesions and fruit rot, particularly at the stem end. Early blight thrives in warm, moist conditions and can survive in soil and plant debris for extended periods. The disease spreads through wind, rain splash, and occasionally insects (Pankaj et al., 2020). While generally less severe than late blight, early blight can still significantly reduce plant vigor and tomato yields, especially in plants already under stress. The economic impact of this disease is considerable, with losses often exceeding 50% if left unmanaged.

Late Blight (*Phytophthora infestans*): Late blight is one of the most destructive diseases of tomato, caused by the oomycete *Phytophthora infestans* (Mazumdar et al., 2021). Characterized by water-soaked lesions that quickly expand, late blight can devastate entire fields within days, especially under cool, wet conditions. The rapid spread of the disease is facilitated by spores carried by wind and rain, making it a significant threat to tomato production. Farmers face economic losses in millions due to late blight outbreaks, *Phytophthora infestans* is notorious for its highly efficient dispersal mechanisms, primarily through spores that can travel considerable distances on wind currents, infecting new fields with alarming speed (Kasteel et al., 2023). The pathogen overwinters on infected plant debris and can persist in crop residues, which serve as inoculum sources for new infections in the following season. Infected transplants or nearby fields can also introduce the pathogen to tomato crops. Once introduced, the disease can advance rapidly, especially under favorable conditions, sometimes devastating entire fields within days. Consequently, late blight remains one of the most challenging diseases to manage in tomato cultivation, necessitating vigilant monitoring and rapid response strategies to curb its spread and protect tomato yields.

Septoria Leaf Spot (*Septoria lycopersici*): Septoria leaf spot, is characterized by small, round, grayish lesions with dark borders on the leaves (Nayyer et al., 2021). Infected leaves may eventually yellow and drop, leading to reduced photosynthetic capacity and overall plant vigor. This disease is particularly problematic in humid conditions where moisture can persist on leaf surfaces. Effective control measures include crop rotation, resistant varieties, and timely fungicide applications. Monitoring weather conditions can also aid in predicting disease outbreaks, allowing for proactive management strategies. The pathogen flourishes in warm, humid conditions, particularly when temperatures range between 60–80°F. Under such favorable circumstances, the disease can spread quickly, advancing from the lower, older leaves to newer growth as it progresses up the plant. Without intervention, Septoria leaf spot can cause severe defoliation, stripping the plant of its leaves and exposing fruits to the risk of sunscald, while drastically impacting the plant's ability to mature its fruit (Vallad et al., 2018). Although the fungus rarely infects the fruits directly, the loss of foliage not only reduces yield but also leads to uneven ripening and compromised fruit quality, presenting a significant challenge for growers aiming to protect their crops.

2.2 Potato Leaf Diseases

Late Blight (*Phytophthora infestans*) like in tomatoes, late blight poses a catastrophic threat to potato crops. Late blight of potato, caused by the notorious pathogen *Phytophthora infestans*, is a catastrophic disease with a historical legacy, having famously triggered the Irish Potato Famine of the 1840s (Majeed et al., 2022). This pathogen can decimate entire potato fields under favorable conditions, particularly during cool and moist late summer weather. The disease initially manifests as dark brown blotches with green-gray edges on the potato plant's leaves, typically bypassing major leaf veins. As the infection spreads, especially in persistently wet and cool conditions, the entire plant may turn brown and wilted, appearing as though it has been struck by frost. This rapid decline makes late blight one of the most dreaded diseases for potato growers worldwide. The pathogen's effectiveness stems from its efficient spore production. *Phytophthora infestans* generates sporangia that can be spread long distances by the wind, enabling swift transmission to new plants (Rhouma et al., 2024). Under ideal conditions, with temperatures between 60–70°F and high humidity, countless spores can develop, spreading the infection throughout a field within days. The pathogen also has the ability to overwinter in infected potato tubers, creating a reservoir for the next growing season. When rain carries zoospores into the soil, tubers become infected, leading to a range of discolorations, from brown to reddish-purple. Proper disposal of infected plant material is essential to breaking this cycle and minimizing the pathogen's impact on subsequent potato crops, underscoring the importance of comprehensive disease management for sustainable potato production.

Early Blight (*Alternaria solani*) also affects potatoes, manifesting through dark spots on leaves and affecting tubers during storage. The lesions can coalesce, leading to significant defoliation and reduced tuber quality (Chaudhary et al., 2021). Similar to tomatoes, early blight is best managed through a combination of cultural practices, such as crop rotation and sanitation, along with fungicide applications when necessary. The disease typically begins on older foliage, where small, dark, circular to oval lesions appear. As these spots enlarge, they develop distinct concentric rings, creating a target-like pattern that is characteristic of early blight. These lesions are often surrounded by a yellowish or greenish halo, which can lead to the yellowing and eventual death of affected leaves, particularly on the lower parts of the plant. The disease flourishes in warm, humid conditions, with optimal growth occurring at temperatures between 75–84°F and peak infection rates at 82–86°F and relative humidity above 80%. *A. solani* can overwinter in infected plant debris, allowing it to survive between growing seasons and re-emerge in new crops. The pathogen spreads efficiently through spores, which are easily transported by wind, rain splash, insects, animals, and agricultural equipment, facilitating rapid infection across fields (Tsedaley, 2014). Under favorable conditions, lesions can form in as little as 2–3 days, allowing early blight to progress swiftly. Beyond its impact on leaves, early blight also affects potato stems and tubers. Stem infections appear as sunken, lens-shaped lesions, while tubers develop irregularly shaped, sunken lesions with a distinct purple border. Severe early blight infections can lead to significant defoliation, which reduces photosynthetic capacity, lowers yield potential, and negatively impacts tuber quality. Effective management strategies are essential to minimizing the disease impact on potato crops, as early blight poses a persistent challenge to growers worldwide.

Blackleg of potato is a serious bacterial disease, primarily caused by species within the genera Pectobacterium and Dickeya (van der Wolf et al., 2021). This disease is commonly introduced via infected seed tubers and can inflict extensive damage on potato crops under suitable conditions. Infected plants may fail to emerge

Figure 2. Comparison of (A) Healthy potato leaf with (B) Late blight infected potato leaf; (C) healthy tomato leaf (D) with late blight infected leaf—Images retrieved from PlantVillage dataset.

or produce stunted shoots with yellow, curled leaves. One of the defining signs of blackleg is a blackened, soft decay at the stem base near the soil line, which gives the disease its characteristic name. As the infection progresses in more mature plants, healthy stems darken, and affected plants experience rapid wilting along with yellowing foliage. Blackleg thrives in wet, moist conditions and can quickly spread throughout a crop. Bacteria released from decaying seed tubers invade emerging shoots, either preventing their emergence or causing a slimy rot on newly emerged stems (Skelsey et al., 2018). The pathogen's reach extends to daughter tubers, where it causes soft rot, further impacting yield and quality. Long-distance spread occurs through the movement of infected seed potatoes, while within-field spread can happen via contaminated tools, irrigation water, rain splash, and insects. This disease presents a persistent challenge, making strict sanitation and careful seed selection essential to minimize its impact on potato crops.

2.3 Wheat Leaf Diseases

Wheat rust is a serious fungal disease impacting wheat production worldwide, with three primary types: brown rust, yellow rust, and black rust, each caused by a different species of the Puccinia fungus. Each rust type has unique symptoms, favored conditions, and implications for wheat crops, requiring targeted management practices to reduce yield losses and protect wheat quality. Brown Rust (Leaf Rust); caused by *Puccinia triticina*, primarily infects the leaves of wheat plants, forming small, reddish-brown pustules on the leaf surface (Huerta-Espino et al., 2014). This type of rust is particularly prevalent in regions with mild temperatures and frequent morning dew, which create ideal conditions for the pathogen's spore development and dispersal. While it tends to affect the foliage rather than the stems or heads, brown rust can lead to significant yield losses if left unmanaged, as infected leaves are unable to photosynthesize effectively, weakening the overall plant health and productivity. Yellow Rust (Stripe Rust); Yellow rust, or stripe rust, is caused by *Puccinia striiformis* f. sp. *tritici* and is recognizable by the long, yellow to orange stripes it produces on the leaf surfaces (Chen et al., 2014). These stripes are formed by pustules arranged in linear patterns parallel to the leaf veins. Yellow rust thrives in cool, moist environments, making it especially problematic in temperate regions or during wet, cool seasons. The disease progresses rapidly under these conditions, potentially leading to substantial crop losses as it disrupts the plant's ability to produce and store energy, ultimately impacting grain formation and reducing yield quality. Black Rust (Stem Rust); also known as stem rust, is caused by *Puccinia graminis* f. sp. *tritici* and is the most destructive form of wheat rust (Eshete). It affects not only the leaves but also the stems and heads of the wheat plant, producing elongated pustules that range in color from reddish-brown to black as they mature. Black rust flourishes in warm, humid conditions and can spread quickly, causing entire fields to exhibit significant wilting and lodging (falling over) due to weakened stems (Rehman et al., 2024). This disease is known to cause drastic reductions in both yield and grain quality, making it a priority for control in susceptible wheat-growing regions.

Powdery Mildew (*Blumeria graminis*) is a widespread disease that poses a considerable threat to wheat crops around the world (Basandrai et al., 2023). The disease is easily identifiable by its characteristic white to gray, powdery fungal growth, which typically appears on the upper surfaces of leaves, stems, and occasionally the heads of the wheat plant. Initially, powdery mildew tends to emerge on lower leaves, gradually spreading upward to the flag leaf and even the ear when conditions are favorable. Infected leaves often display yellowish spots on their undersides, directly beneath the fungal colonies, and in severe cases, these leaves may turn yellow and wither prematurely. Powdery mildew thrives in cool, humid environments, with optimal temperatures ranging from 59–71°F, and is particularly prevalent in dense crop stands where air circulation is limited. The disease affects the plant physiology in multiple ways, reducing photosynthesis, increasing respiration and transpiration, and ultimately impairing heading and grain filling. These physiological disruptions can lead to yield losses of 10–20% in typical cases, though severe infections—especially those impacting the flag leaf by heading—may result in yield reductions of up to 45%.

3. Challenges in Traditional Disease Detection Methods

The detection of plant diseases is a crucial aspect of modern agriculture, impacting crop yield, quality, and sustainability. Traditional plant disease detection methods, though historically reliable to some extent, face numerous challenges that limit their effectiveness in the context of contemporary agricultural needs. These challenges include reliance on subjective visual observation, time-consuming procedures, limited scalability, and a lack of precision in disease quantification. Additionally, traditional methods often rely on expensive laboratory tests, making them impractical for routine field use, and they struggle with early-stage disease detection. Some major challenges in traditional disease detection are:

3.1 Reliance on Visual Symptoms

One of the primary challenges associated with traditional plant disease detection is the reliance on visual symptoms. Disease identification based on visual observation is highly subjective and dependent on the expertise of the observer (John et al., 2023). The accuracy of this method can vary significantly depending on the experience of the individual conducting the assessment, which introduces a level of uncertainty. In many cases, early-stage symptoms of plant diseases are subtle and not immediately apparent to the untrained eye, making it difficult to detect diseases in their incipient stages. As a result, by the time visible symptoms are noticeable, the disease may have already spread, leading to significant crop damage that could have been mitigated through earlier detection. Moreover, some plant diseases exhibit symptoms that closely resemble those of other conditions, such as nutrient deficiencies, environmental stress, or abiotic factors, further complicating visual diagnosis (Martinelli et al., 2015). This overlap can lead to misidentification, which in turn affects the effectiveness of disease management strategies. As a consequence, farmers and agronomists are often forced to rely on trial-and-error methods for disease management, increasing the risk of crop loss and the unnecessary application of pesticides or other interventions.

3.2 Time-Consuming and Labor-Intensive

Another significant challenge of traditional plant disease detection method is their time-consuming and labor-intensive nature. Field inspections, which often require walking through vast agricultural areas to observe plant conditions, can take considerable time, especially in large-scale operations. This extensive fieldwork is not only tiring but also often results in delayed disease detection, as many diseases do not manifest their full impact until after significant damage has been done. By the time diseases are identified, the window for effective intervention may have already closed, rendering disease control efforts less effective or costlier. Additionally, traditional methods may require frequent visits to the field to track disease progression and re-assess plant health. This repetition of tasks further compounds the labor and resource requirements. For instance, in the case of some diseases that progress slowly, repeated visits to monitor slight changes in plant health may be necessary, leading to inefficiency in disease management practices.

3.3 Limited Scalability

Expert knowledge-driven disease detection methods are also hindered by their limited scalability. Traditional methods heavily rely on specialized knowledge, typically provided by agronomists or plant pathologists, who may not be available in all regions (Miller et al., 2009). This limitation makes it difficult to apply these methods on a large scale, especially in regions where access to trained professionals is scarce or economically unfeasible. As agricultural operations expand in size and geographic reach, the task of providing expert-level disease detection becomes increasingly difficult. Furthermore, scaling these traditional methods to cover a wider range of crops and diseases introduces another challenge. The complexity of plant-pathogen interactions means that a single expert might not possess sufficient knowledge across all crop species, particularly in large farming systems that grow a diverse range of plants (Garrett et al., 2018). This lack of scalability results in inefficiencies and may cause gaps in disease detection coverage, leaving certain crops or regions vulnerable to unchecked disease outbreaks.

3.4 Resource-Intensive Laboratory Tests

While laboratory-based diagnostic techniques, such as molecular assays and pathogen isolation, can offer high accuracy, they come with significant drawbacks (Verma et al., 2022). These methods are resource-intensive, requiring specialized equipment, reagents, and trained personnel. In addition, laboratory tests can be expensive and time-consuming, making them impractical for routine use in the field. The delay in obtaining results from these tests may hinder timely decision-making, especially in fast-paced agricultural environments where early intervention is critical. The reliance on laboratory-based tests also introduces logistical challenges. Samples must often be collected and transported to laboratories, which may not always be easily accessible, particularly in remote areas. This further delays the diagnosis process, making it difficult to address disease threats promptly. While laboratory tests are invaluable for confirming diagnoses and providing insights into pathogen characteristics, their role in everyday field diagnostics is limited by these logistical and financial barriers (Debnath et al., 2010).

3.5 Subjectivity and Bias

Traditional disease detection methods that depend on expert knowledge are often influenced by the subjectivity and biases of the individuals involved. Different pathologists may interpret symptoms in varying ways, leading to inconsistencies in disease identification. Furthermore, expert assessments can be influenced by experience, preconceived notions, and even cognitive biases, which may cloud objective analysis and decision-making. For example, an experienced pathologist might be more likely to diagnose a disease they are familiar with, overlooking less common pathogens that could cause similar symptoms. This subjectivity not only affects the accuracy of disease detection but also the subsequent management recommendations. Disease management strategies are often tailored to specific diseases or pathogens, and any misidentification or misinterpretation of symptoms can lead to inappropriate treatments. This increases the risk of applying ineffective or even harmful interventions, thereby exacerbating the issue rather than resolving it.

3.6 Complexity of Plant-Pathogen Interactions

Finally, traditional methods may struggle to capture the complexity of plant-pathogen interactions, especially when multiple environmental factors are involved (Manoharachary and Kunwar, 2014). Plant diseases are often influenced by a wide range of abiotic and biotic factors, including soil conditions, climate, and the presence of other pests. Traditional diagnostic methods may fail to account for these interactions, leading to oversimplified diagnoses that do not fully explain the dynamics of the disease. For instance, the same pathogen may behave differently under varying environmental conditions, making it difficult for experts to predict disease outcomes with certainty (Woolhouse, 2011). The inability to account for these complex interactions further underscores the limitations of traditional plant disease detection methods. As the agricultural landscape continues to evolve, with climate change, global trade, and new pathogens introducing additional complexity, the need for more comprehensive and accurate diagnostic methods becomes increasingly urgent. Traditional plant disease detection methods, while foundational to plant pathology, face numerous challenges that hinder their effectiveness and efficiency (Miller et al., 2009). From reliance on subjective visual symptoms and time-consuming field inspections to issues with scalability, precision, and early detection, these methods often fall short in the context of modern, large-scale agriculture. The growing complexity of plant diseases, coupled with the limitations of expert-driven approaches, has spurred the search for innovative diagnostic techniques, including molecular methods, remote sensing, and artificial intelligence-based systems. These advanced methods offer the promise of overcoming the constraints of traditional disease detection, providing more accurate, efficient, and scalable solutions for plant disease management.

4. Artificial Intelligence in Disease Detection

Artificial Intelligence (AI) has become a cornerstone in advancing disease detection technologies, revolutionizing the way diseases are identified, monitored, and managed (Sahoo et al., 2024). Through its ability to process massive datasets, identify subtle patterns, and deliver accurate, real-time results, AI has proven invaluable in diverse fields, from medicine to agriculture. Among AI's subfields, Machine Learning (ML) and Deep Learning (DL) have shown exceptional potential in transforming disease detection and diagnostics. Artificial Intelligence refers to the simulation of human intelligence in machines designed to perform tasks that typically require human cognitive abilities, such as reasoning, learning, problem-solving, and decision-making. AI systems rely on algorithms and computational frameworks that allow machines to not only perform tasks but also improve over time through data exposure (Liu et al., 2022). Within AI, key subfields play significant roles in disease detection:

Machine Learning is a subset of AI that involves designing algorithms capable of learning from data and making predictions or decisions without explicit programming for specific tasks (Helm et al., 2020). ML is particularly impactful in disease detection for; Time Series Forecasting, using historical and real-time data to predict disease outbreaks or assess their progression over time. This is critical for managing both human diseases and agricultural pathogens. Classification tasks, employing classification algorithms to categorize diseases based on symptomatology, genetic markers, or diagnostic imaging. This is especially relevant for detecting multiple diseases simultaneously in complex datasets.

Deep Learning, a more advanced subset of ML, employs artificial neural networks modeled after the human brain to analyze large and complex datasets (Sarker, 2021). DL algorithms are highly effective in disease detection due to their capacity for image analysis, identifying diseases through the analysis of medical or agricultural images, such as X-rays, CT scans, or high-resolution plant images, pattern recognition, unveiling intricate and previously undetectable patterns in disease progression, pathogen behavior, or host-pathogen interactions. Convolutional Neural Networks (CNNs), a class of DL algorithms, have become a gold standard for image-based disease detection due to their ability to handle multi-dimensional data and extract features automatically without manual intervention (Alzubaidi et al., 2021).

4.1 Integration of AI with Image Processing for Disease Detection

The integration of AI with image processing has revolutionized disease detection, allowing for unprecedented accuracy, speed, and scalability. Image processing techniques enhanced by AI enable precise analysis of visual data, such as medical imaging or plant disease symptoms captured through cameras, drones, or satellites. AI's contributions to image processing include:

4.1.1 Feature Extraction

Traditional image processing relies on manual feature extraction, which can be labor-intensive and error-prone. AI automates this process, identifying the most relevant features directly from the data (Konnik et al., 2021).

4.1.2 Subtle Pattern Recognition

Advanced DL models, particularly CNNs, can detect minute changes and subtle abnormalities in images that may be imperceptible to human experts. This capability is critical for early disease detection (Ren and Yi, 2022).

4.1.3 Automated Analysis at Scale

AI-powered systems can analyze thousands of images or datasets simultaneously, making them ideal for large-scale screening programs. In agriculture, this means monitoring vast fields for early signs of disease (Elbasi et al., 2024).

4.1.4 Multi-modal Integration

AI systems integrate image data with other datasets, such as environmental factors, genetic information, or soil quality metrics, to provide a holistic understanding of disease risks and dynamics. By automating and refining image analysis, AI has enabled significant advances in disease detection, supporting applications in human health, veterinary sciences, and agriculture.

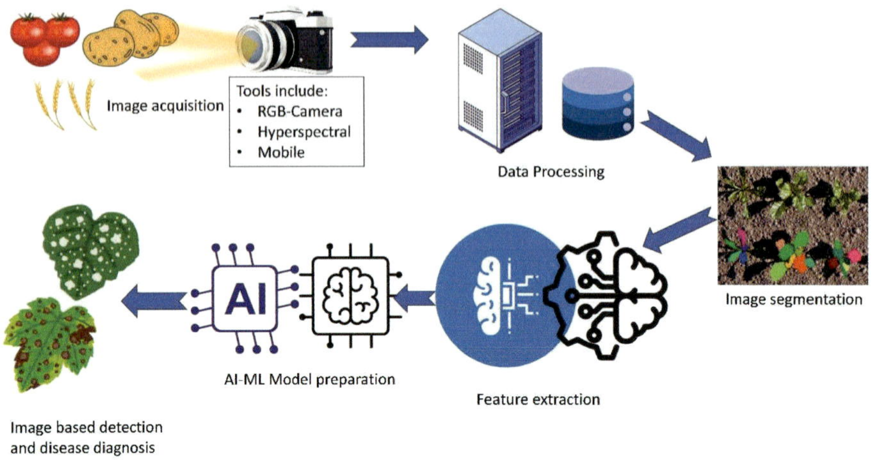

Figure 3. Workflow for building AI based disease detection model starting from collection of raw data to model testing and on-field application.

5. AI Models and Techniques Used for Leaf Disease Detection

The detection and identification of plant leaf diseases have gained substantial momentum with the integration of Artificial Intelligence (AI) into agriculture. AI-driven solutions offer robust, accurate, and automated methods to identify plant diseases at early stages, reducing crop losses and enhancing agricultural productivity. Recent advancements in machine learning (ML) and deep learning (DL) techniques have revolutionized leaf disease detection, enabling researchers and practitioners to handle diverse datasets, analyze complex disease patterns, and deliver actionable insights with remarkable precision. This section elaborates on the prominent AI models and techniques utilized for leaf disease detection, focusing on their strengths, applications, and comparative performance.

5.1 Machine Learning Approaches

Machine learning techniques have been extensively applied to leaf disease detection tasks due to their ability to classify diseases based on extracted features. These models rely on feature engineering and traditional statistical methods to establish patterns within the data.

5.1.1 Support Vector Machines (SVMs)

Support Vector Machines are widely recognized for their effectiveness in classifying plant diseases, particularly when paired with well-structured feature extraction techniques. SVMs work by creating hyperplanes that separate different disease classes in feature space, ensuring high accuracy in classification tasks. For instance, in studies involving feature descriptors such as Histogram of Oriented Gradients (HOG) and color histograms, SVMs have shown promising results in distinguishing healthy leaves from diseased ones (Sahu et al., 2023). Their ability to perform well on moderately sized datasets makes them an attractive option for small-scale agricultural studies.

5.1.2 Decision Trees and Ensemble Methods

Decision trees, often implemented in ensemble models like Random Forests, have demonstrated significant success in disease detection (Fratello et al., 2018). These models operate by splitting datasets into branches based on feature values, making decisions at each node until a final classification is achieved. Random Forests, which aggregate predictions from multiple decision trees, offer higher accuracy and robustness compared to standalone trees. Additionally, these models are valued for their interpretability, as they allow researchers to trace decisions back to individual features, offering insights into disease progression and contributing factors.

5.2 Deep Learning Approaches

Deep learning has emerged as the dominant paradigm for leaf disease detection due to its capacity to learn directly from raw image data. Unlike traditional ML approaches, deep learning models eliminate the need for manual feature engineering, enabling end-to-end learning from input data to disease classification.

5.2.1 Convolutional Neural Networks (CNNs)

CNNs have become the cornerstone of deep learning-based disease detection, excelling at processing image data (Khalid et al., 2024). By leveraging convolutional layers to extract spatial hierarchies of features, CNNs are capable of detecting intricate patterns associated with leaf diseases, such as spots, discoloration, and texture anomalies. Advanced architectures such as ResNet50, AlexNet, GoogLeNet, VGG19, and EfficientNet have been widely adopted in agricultural research (Shaheed et al., 2023). These models, pre-trained on large image datasets like ImageNet, are fine-tuned for specific leaf disease detection tasks. Studies have reported classification accuracies exceeding 97% for detecting diseases in crops such as tomatoes, potatoes, and wheat. Fine-tuning pre-trained CNN models significantly reduces the computational burden and time required to train models from scratch, making them accessible for resource-constrained environments.

Transfer learning has revolutionized disease detection, particularly in scenarios with limited labeled datasets (Chen et al., 2020). This technique involves using pre-trained models, such as ResNet or VGG, as a starting point and fine-tuning their weights on domain-specific datasets. Transfer learning enhances model performance by leveraging knowledge from large, diverse datasets, enabling robust detection of diseases even in underrepresented crops. Beyond classification, neural networks have been adapted for segmenting diseased areas on leaves. For example, You Only Look Once (YOLO)-based architectures have been applied for detecting multiple infection regions in a single image (Mzoughi and Yahiaoui, 2023). These models provide detailed spatial information about disease spread, which is invaluable for precision agriculture practices.

5.3 Comparative Analysis of Techniques

The choice between machine learning and deep learning techniques for leaf disease detection depends on various factors, including dataset size, computational resources, and specific application requirements (Wani et al., 2022). Deep learning models, particularly CNNs, consistently outperform traditional machine learning techniques in terms of accuracy. For instance, CNN-based models have achieved classification accuracies above 97% in studies involving tomato and wheat leaf diseases, whereas SVMs and Random Forests typically achieve lower but still respectable accuracy levels. Deep learning models are highly scalable and can generalize well across diverse datasets. Their ability to handle large-scale, heterogeneous data makes them suitable for detecting multiple diseases across various crop species. In contrast, traditional machine learning models may struggle with generalization, especially when dealing with complex or high-dimensional data. CNNs eliminate the need for manual feature extraction by learning features directly from raw images, streamlining the detection process. Machine learning models, on the other hand, require significant effort in feature engineering, which can be a limitation in time-sensitive scenarios. While deep learning models offer superior performance, their black-box nature often limits interpretability. Machine learning models like decision trees provide better understanding, which is crucial for detection of disease patterns and communicating findings to stakeholders. Additionally, deep learning models demand higher computational resources, making them less accessible for small-scale farmers or regions with limited infrastructure. Deep learning models thrive on large datasets, while traditional machine learning techniques perform well with smaller datasets. In resource-limited settings, this distinction can influence the choice of technique.

6. AI-Based Tools for Disease Detection

Mobile applications such as Plantix and Agrio are transforming how farmers and agricultural professionals detect and manage plant diseases (Siddiqua et al., 2022). These innovative apps harness the power of artificial intelligence (AI) and machine learning (ML) algorithms to deliver quick, accurate diagnoses by analyzing images captured with smartphone cameras. Their ability to provide reliable and actionable insights in real time is reshaping modern agricultural practices. One of the key advantages of these AI-driven apps is their accessibility (Altayeb et al., 2024). With just a smartphone, farmers can diagnose plant health issues directly in the field, eliminating delays caused by traditional diagnostic methods. These tools allow for immediate detection of plant diseases, empowering farmers to act swiftly to prevent the spread of infections. By offering diagnoses on the spot, these apps bridge the gap between cutting-edge technology and on-the-ground agricultural needs.

In addition to their convenience, these applications significantly reduce costs associated with disease management. Traditional methods often require specialized equipment, laboratory analysis, or frequent consultations with agricultural experts, all of which can be costly and time-consuming. Mobile apps offer a more affordable alternative, putting advanced diagnostic capabilities directly in the hands of farmers, without the need for expensive infrastructure.

The speed and efficiency of AI-powered apps are particularly noteworthy. Once a farmer captures an image of a diseased leaf, the app can analyze it within seconds, identifying the specific disease and recommending actionable solutions. This real-time capability ensures that farmers can implement the necessary interventions immediately, minimizing crop losses and enhancing overall productivity. Another critical factor driving the widespread adoption of these apps is their cross-platform compatibility. By targeting both iOS and Android platforms, these tools cater to nearly 98% of the global smartphone market. This broad compatibility ensures that farmers across diverse regions and economic backgrounds can benefit from this technology, fostering inclusivity and global agricultural development (Fatumo, 2023).

Beyond disease detection, many of these apps also offer additional features such as pest identification, nutrient deficiency analysis, and tailored agricultural advice. Such comprehensive functionality enhances their value proposition, making them indispensable tools for modern farming. The integration of drones equipped with advanced cameras and AI algorithms has revolutionized the way agricultural diseases are detected and monitored. By providing real-time aerial surveillance and leveraging satellite data, this technology empowers farmers with powerful tools to enhance crop health management and ensure sustainable agricultural practices.

6.1 Drone-Based Disease Detection

Modern drones equipped with high-resolution cameras and multispectral sensors have become essential in agricultural disease detection (Abbas et al., 2023). These drones can capture detailed, high-quality images of vast crop fields, enabling AI algorithms to analyze plant health and detect disease symptoms efficiently. Multispectral imaging, a standout feature of this technology, allows drones to capture data beyond the visible light spectrum, revealing stress markers, nutrient deficiencies, and disease indicators that are often invisible to the naked eye. AI-powered algorithms analyze these images in real time, identifying patterns, anomalies, or discolorations associated with specific diseases. By processing this data, the system can pinpoint problem areas, enabling farmers to implement targeted interventions. Furthermore, the speed and efficiency of drone-based surveys allow for the rapid mapping of large agricultural areas, providing a comprehensive overview of crop health across entire fields or regions (Guebsi et al., 2024). This capability significantly reduces the labor and time traditionally required for manual inspections.

Drones provide significant advantages for real-time disease mapping and surveillance, particularly in early detection and monitoring. Their rapid deployment capability ensures that fields can be surveyed frequently, allowing for the identification of potential disease outbreaks before they spread uncontrollably. Real-time data transmission facilitates immediate analysis, empowering farmers and agricultural professionals to respond swiftly to emerging issues. Additionally, drones can be programmed with automated flight paths, which, combined with AI-driven image analysis, eliminate the need for extensive manual labor. This automation not only reduces costs but also ensures consistent and accurate monitoring. With drones handling both data collection and analysis, farmers can focus on implementing corrective measures rather than spending time on data acquisition.

6.2 Satellite Data Integration

While drones excel at providing detailed, localized data, satellite imagery offers a complementary macro-level perspective. Satellite data provides broader geographic coverage and enables the detection of large-scale patterns of disease spread across regions or even entire countries (Zhang et al., 2020). By integrating AI algorithms, satellite imagery can be analyzed to identify significant trends in disease progression, assess environmental stress factors, and monitor crop health over time. Temporal analysis of satellite data is particularly valuable for tracking how diseases evolve and spread across different seasons. When combined with drone-based surveillance, this creates a multi-scale approach to agricultural monitoring, encompassing both the micro-level details of individual fields and the macro-level patterns across vast agricultural regions (Nahiyoon et al., 2024).

6.3 Examples of Successful Implementations

Tomato (*Solanum lycopersicum*) is a vital agricultural crop and is highly susceptible to various diseases caused by bacterial pathogens, microbes, and pest infestations. Machine learning (ML) and deep learning (DL) have significantly advanced the detection and classification of plant diseases, particularly in tomatoes. Francis and Deisy (2019) proposed a convolutional neural network (CNN) model to differentiate between healthy and diseased leaves of tomato and apple plants. The model consisted of four convolutional layers paired with pooling layers, a sigmoid activation function, and two fully connected dense layers. Trained on 3663 image samples from the PlantVillage dataset, the model achieved an accuracy of 87%. Basavaiah and Arlene Anthony, 2020) explored various ML algorithms to identify diseases in tomato plants. They utilized a dataset of 200 images across five classes, leveraging global feature descriptors such as texture, color, and form. Among the algorithms tested, including k-nearest neighbor (KNN), logistic regression (LR), decision trees (DT), random forests (RF), and support vector machines (SVM), the RF model delivered the highest accuracy at 94%. Furthermore, studies have involved utilizing KNN and probabilistic neural networks (PNN) to classify diseases affecting tomato leaves, including conditions such as Verticillium wilt, powdery mildew, Septoria leaf spot, leaf miners, and spider mites, the PNN model outperformed KNN, achieving a classification accuracy of 91.88%.

Potatoes, the fourth-largest crop globally, face significant challenges due to disease susceptibility. Common diseases affecting potatoes include black scurf, common scab, black leg, and pink rot (Wani et al., 2022). Patil et al. (2017) compared RF, SVM, and artificial neural networks (ANN) for identifying blight diseases in potato leaves. Using an 892-image dataset from the PlantVillage database and the University of Agricultural Sciences, India, ANN demonstrated superior performance with an accuracy of 92%, followed by SVM (84%) and RF (79%). Studies have been carried out on implementing a CNN to classify potato diseases using a dataset of

5000 images. The model demonstrated superior performance, achieving 100% and 99% accuracy across different disease categories. CNN and AlexNet architectures have been used to detect diseases in potato and mango leaves, using a dataset of 4004 potato images from the PlantVillage database, AlexNet achieved an accuracy of 98.33% (Arya and Singh, 2019).

Recent advancements in AI-based wheat leaf disease detection have provided farmers and researchers with cutting-edge tools for early diagnosis and effective crop health management. Convolutional Neural Networks (CNNs) have become a dominant technology in this domain, achieving remarkable accuracy levels exceeding 90% in identifying common wheat diseases, including powdery mildew, leaf rust, and stripe rust. One noteworthy example is the CerealConv CNN model, which attained a classification accuracy of 97.05% using a dataset comprising wheat images captured in both real field and glasshouse conditions (Long et al., 2023). In fact, this model surpassed expert pathologists in accuracy by a margin of 2% when evaluated on a subset of images. Additionally, another study utilizing the YOLOv7 algorithm reported exceptional performance, achieving 97.3% accuracy, 96.7% precision, and 96.4% recall in wheat disease detection.

These AI systems leverage images obtained through devices such as smartphones, drones, and satellites, facilitating rapid and large-scale disease surveillance. The integration of multispectral imaging technologies and advanced feature extraction methods, like the Gray Level Co-occurrence Matrix (GLCM), has significantly improved the detection of diseases at early stages—often before symptoms become visually apparent. These innovations highlight the potential of AI in transforming wheat disease management and safeguarding agricultural productivity.

7. Conclusions

Artificial Intelligence (AI) has emerged as a transformative technology in agriculture, particularly in diagnosing and managing leaf diseases in key crops such as tomatoes, potatoes, and wheat. Leveraging AI-based tools offers unprecedented precision, speed, and scalability in addressing challenges faced by modern farming practices. AI-powered image recognition systems, combined with machine learning (ML) and deep learning (DL) algorithms, have significantly enhanced the ability to detect and classify plant leaf diseases with remarkable accuracy. These systems utilize high-resolution image data captured from diverse sources such as smartphones, drones, and satellites to analyze leaf conditions. Advanced models like Convolutional Neural Networks (CNNs) can distinguish between healthy and diseased plants, identifying specific diseases with accuracy rates often exceeding 90%. These systems can process and analyze vast amounts of data quickly, making them suitable for real-time monitoring. Deep learning frameworks, such as YOLO (You Only Look Once) and ResNet, have demonstrated remarkable success in detecting and classifying diseases such as early and late blight in potatoes, bacterial leaf spots in tomatoes, and rust in wheat. These models excel in recognizing subtle differences that may not be discernible to human experts. AI systems can identify diseases in their nascent stages, even before visible symptoms manifest, enabling timely interventions and minimizing damage (Table 1). The adoption of AI in leaf disease detection has far-reaching economic

Table 1. AI systems leverage image analysis to detect diseases in tomato, potato, and wheat crops.

Crop	Disease	AI model	Accuracy	References
Tomato	Bacterial spot	Random forest (RF) and Decision Tree (DT)	94%	Basavaiah and Arlene Anthony, 2020
	Leaf spot	Convolution neural networks (CNN)	87%	Francis and Deisy, 2019
	Verticillium wilt	K-Nearest Neighbor (kNN)	92%	Balakrishna et al., 2019
	Yellow leaf curl	Support vector machine (SVM) and CNN	99%	Vadivel et al., 2022
	Early blight	RestNet	99%	Chakravarthy and Raman, 2020
	Leaf mold disease	CNN	99%	Kumar and Vani, 2019
	Septoria leaf spot	CNN	100%	Kumari et al., 2019
	Early blight and Late blight	RF	95%	Govardhan and Veena, 2019
	Stalk scar	SVM	98%	Ireri et al., 2019
Potato	Late blight and early blight	SVM and artificial neural network (ANN)	92%	Patil et al., 2017
	Black leg and common scab	CNN	100%	Arshaghi et al., 2023
	Early blight	CNN and AlexNet	98%	Arya and Singh, 2019
	Leaf roll and verticillium wilt	DenseNet	97%	Mahum et al., 2023
	Late blight, early blight	Deep Convolutional Neural Network (DCNN)	98%	Al-Amin et al., 2019
Wheat	Yellow rust, brown rust, powdery mildew	CerealConv	97%	Long et al., 2023
	Fusarium head blight, rust disease	CNN	98%	Hossen et al., 2022
	Septoria leaf blotch, mildew and Rust disease	RFE-CNN	99%	Xu et al., 2023
	Leaf rust	ResNet-50	96%	Shafi et al., 2023
	Fusarium ear blight	FCN	97%	Kumar et al., 2024
	Stripe rust	PSPNet	89%	Kumar et al., 2024

and environmental implications, making it a cornerstone of sustainable agriculture. By enabling early detection, AI mitigates the economic impact of plant diseases, saving farmers millions in potential revenue losses annually. It also minimizes unnecessary pesticide applications, reducing operational costs while maintaining crop health. AI-driven disease detection reduces excessive pesticide use, leading to environmentally friendly farming practices that promote biodiversity. Improved disease management ensures higher and more consistent crop yields, contributing to global food security. Beyond direct economic and environmental benefits, AI-based early detection systems transform agricultural operations by fostering, precision agriculture, driven by AI, enables site-specific disease treatment, optimizing the use of water, fertilizers, and pesticides. Early interventions reduce crop losses and environmental degradation, ensuring long-term agricultural productivity. The technology also enables predictive analytics, allowing farmers to anticipate disease outbreaks based on historical patterns and environmental data.

The future of AI in agriculture lies in its integration with cutting-edge technologies and data-driven frameworks. AI facilitates targeted treatments for diseased plants, enhancing efficiency while minimizing environmental impact. Advanced models can forecast disease outbreaks by analyzing weather data, soil conditions, and historical trends, enabling proactive measures. AI-powered drones and robots can conduct routine inspections, detect anomalies, and apply treatments autonomously, reducing the need for manual labor. Combining AI with large-scale agricultural datasets allows for more robust and adaptable disease detection systems. To maximize the potential of AI-based systems for leaf disease detection, researchers, technologists, farmers, and policymakers must collaborate effectively. Expansion in training datasets to include diverse plant species, growth stages, and environmental conditions to improve the generalizability of models. Development of lightweight models capable of operating on low-power devices for on-field applications. Furthermore, incorporation of multi-spectral and hyperspectral imaging techniques to identify diseases before visible symptoms appear. Researchers and technologists should work closely with agricultural experts to ensure solutions address practical farming challenges. Farmers should actively participate in system design and validation, providing insights for improved usability and adoption. A major support in form of policymakers must establish supportive frameworks for the ethical and responsible implementation of AI in agriculture, ensuring equitable access and sustainability.

Conclusively, AI has the potential to revolutionize disease detection in agriculture, driving transformative changes in how farmers diagnose, manage, and mitigate crop diseases. By fostering innovation, collaboration, and sustainability, AI-based solutions can ensure more resilient, productive, and eco-friendly farming practices, addressing critical global challenges such as food security and environmental conservation.

References

Abbas, A., Zhang, Z., Zheng, H., Alami, M. M., Alrefaei, A. F., Abbas, Q. et al. (2023). Drones in Plant Disease Assessment, Efficient Monitoring, and Detection: A Way Forward to Smart Agriculture. 13: 1524.

Al-Amin, M., Karim, D. Z., and Bushra, T. A. (2019). Prediction of rice disease from leaves using deep convolution neural network towards a digital agricultural system. 2019 22nd international conference on computer and information technology (ICCIT), IEEE, 1–5.

Altayeb, J. M., Eleyan, H., Wishah, N. D., Elmahmoum, A. E., Khalil, A. J., Abu-Nasser et al. (2024). AI-Driven Innovations in Agriculture: Transforming Farming Practices and Outcomes.

Alzubaidi, L., Zhang, J., Humaidi, A. J., Al-Dujaili, A., Duan, Y., Al-Shamma, O. et al. (2021). Review of deep learning: concepts, CNN Architectures, Challenges, Applications, Future Directions. 8: 1–74.

Arshaghi, A., Ashourian, M., Ghabeli, L. J. M. T., and Applications. (2023). Potato Diseases Detection and Classification Using Deep Learning Methods. 82: 5725–5742.

Arya, S., and Singh, R. (2019). A Comparative study of CNN and AlexNet for detection of disease in Potato and Mango leaf. 2019 International conference on issues and challenges in intelligent computing techniques (ICICT). IEEE, 1–6.

Badidi, E. J. F. I. (2023). Edge AI for Early Detection of Chronic Diseases and the Spread of Infectious Diseases: Opportunities, Challenges, and Future Directions. 15: 370.

Balakrishna, K., Rao, M. J. I. J. O. C. V., and Processing, I. (2019). Tomato plant leaves disease classification using KNN and PNN. 9: 51–63.

Barbedo, J. G. A. J. D. (2019). A Review on the use of Unmanned Aerial Vehicles and Imaging Sensors for Monitoring and Assessing Plant Stresses. 3: 40.

Barburiceanu, S., Meza, S., Orza, B., Malutan, R., and Terebes, R. J. I. A. (2021). Convolutional neural networks for texture feature extraction. Applications to Leaf Disease Classification in Precision Agriculture. 9: 160085–160103.

Basandrai, A. K., Mehta, A., and Basandrai, D. J. I. P. (2023). Virulence structure of wheat powdery mildew pathogen, *Blumeria graminis* tritici: a review. 76: 21–45.

Basavaiah, J., and Arlene Anthony, A. J. W. P. C. (2020). Tomato Leaf Disease Classification Using Multiple Feature Extraction Techniques. 115: 633–651.

Birch, P. R., Bryan, G., Fenton, B., Gilroy, E. M., Hein, I., Jones, J. T. et al. (2012). Crops that feed the world 8: potato: are the trends of increased global production sustainable? 4: 477–508.

Borguini, R. G., and Ferraz Da Silva Torres, E. A. J. F. R. I. (2009). Tomatoes and Tomato Products as Dietary Sources of Antioxidants. 25: 313–325.

Chakravarthy, A. S., and Raman, S. (2020). Early blight identification in tomato leaves using deep learning. 2020 International conference on contemporary computing and applications (IC3A). IEEE, 154–158.

Chaudhary, A. K., Yadavb, J., Guptac, A. K., and Guptad, K. J. T. A. (2021). Integrated Disease Management of Early Blight (*Alternaria solani*) of Potato. 2: 77–81.

Chen, J., Chen, J., Zhang, D., Sun, Y., Nanehkaran, Y. A. J. C., and Agriculture, E. I. (2020). Using Deep Transfer Learning for Image-Based Plant Disease Identification. 173: 105393.

Chen, W., Wellings, C., Chen, X., Kang, Z., and Liu, T. J. M. P. P. (2014). Wheat stripe (yellow) rust caused by *Puccinia striiformis* f. sp. tritici. 15: 433–446.

Debnath, M., Prasad, G. B., and Bisen, P. S. (2010). Molecular diagnostics: promises and possibilities, Springer Science & Business Media.

Elbasi, E., Mostafa, N., Zaki, C., Alarnaout, Z., Topcu, A. E., and Saker, L. J. A. S. (2024). Optimizing Agricultural Data Analysis Techniques through AI-Powered Decision-Making Processes. 14: 8018.

Fatumo, D. E. (2023). Design and Development of Mobile-based e-Marketing Application for Communal Farmers in Eastern Cape. South Africa.

Francis, M., and Deisy, C. (2019). Disease detection and classification in agricultural plants using convolutional neural networks—a visual understanding. 2019 6th international conference on signal processing and integrated networks (SPIN), 2019. IEEE, 1063–1068.

Fratello, M., Tagliaferri, R. J. E. O. B., and Bioinformatics, C. B. A. O. (2018). Decision trees and random forests. 1.

Garrett, K., Alcalá-Briseño, R., Andersen, K., Buddenhagen, C., Choudhury, R., Fulton, J. et al. (2018). Network Analysis: A Systems Framework to Address Grand Challenges in Plant Pathology. 56: 559–580.

Govardhan, M., and Veena, M. (2019). Diagnosis of tomato plant diseases using random forest. 2019 Global Conference for Advancement in Technology (GCAT). IEEE, 1–5.

Guebsi, R., Mami, S., and Chokmani, K. J. D. (2024). Drones in Precision Agriculture: A Comprehensive Review of Applications, Technologies, and Challenges. 8: 686.

Helm, J. M., Swiergosz, A. M., Haeberle, H. S., Karnuta, J. M., Schaffer, J. L., Krebs, V. E. et al. (2020). Machine Learning and Artificial Intelligence: Definitions, Applications, and Future Directions. 13: 69–76.

Hossen, M. H., Mohibullah, M., Chowdhury, S. M., Ahmed, T., Acharjee, S., Panna, M. B. J. I. J. O. A. C. S. et al. (2022). Wheat Diseases Detection and Classification Using Convolutional Neural Network (CNN). 13.

Huerta-Espino, J., Singh, R., German, S., Mccallum, B., Park, R., Chen, W. Q. et al. (2011). Global status of wheat leaf rust caused by *Puccinia triticina*. 179: 143–160.

Huerta-Espino, J., Singh, R., Roelfs, A. P., Misra, J., Tewari, J., Deshmukh, S. et al. (2014). Rusts fungi of wheat. 27: 217–259.

Ireri, D., Belal, E., Okinda, C., Makange, N., and Ji, C. J. A. I. I. A. (2019). A Computer Vision System For Defect Discrimination And Grading In Tomatoes Using Machine Learning And Image Processing. 2: 28–37.

Jackulin, C., and Murugavalli, S. J. M. S. (2022). A Comprehensive Review on Detection of Plant Disease Using Machine Learning and Deep Learning Approaches. 24: 100441.

John, M. A., Bankole, I., Ajayi-Moses, O., Ijila, T., Jeje, O., and Lalit, P. J. A. J. O. P. S. (2023). Relevance of advanced plant disease detection techniques in disease and Pest Management for Ensuring Food Security and Their Implication: A review. 14: 1260–1295.

Kasteel, M., Ketelaar, T., and Govers, F. (2023). Fatal attraction: How Phytophthora zoospores find their host. Seminars in Cell & Developmental Biology. Elsevier, 13–21.

Khalid, M. M., Karan, O. J. I. J. O. M., Statistics, and Science, C. (2024). Deep Learning for Plant Disease Detection. 2: 75–84.

Konnik, M., Ahmadi, B., May, N., Favata, J., Shahbazi, Z., Shahbazmohamadi, S. et al. (2021). Training AI-based Feature Extraction Algorithms, for Micro CT images, Using Synthesized Data. 40: 1–13.

Kumar, A., and Vani, M. (2019). Image based tomato leaf disease detection. 2019 10th International Conference on Computing, Communication and Networking Technologies (ICCCNT). IEEE, 1–6.

Kumar, D., Kukreja, V. J. C., and Agriculture, E. I. (2024). Image segmentation, classification, and recognition methods for wheat diseases: Two Decades' Systematic Literature Review. 221: 109005.

Kumari, C. U., Prasad, S. J., and Mounika, G. (2019). Leaf disease detection: feature extraction with K-means clustering and classification with ANN. 2019 3rd international conference on computing methodologies and communication (ICCMC). IEEE, 1095–1098.

Liu, H., Wang, Y., Fan, W., Liu, X., Li, Y., Jain, S. et al. (2022). Trustworthy ai: A computational perspective. 14: 1–59.

Long, M., Hartley, M., Morris, R. J., and Brown, J. K. J. P. P. (2023). Classification of Wheat Diseases Using Deep Learning Networks With Field and Glasshouse Images. 72: 536–547.

Mahum, R., Munir, H., Mughal, Z.-U.-N., Awais, M., Sher Khan, F., Saqlain, M. et al. (2023). A Novel Framework for Potato Leaf Disease Detection using an Efficient Deep Learning Model. 29: 303–326.

Majeed, A., Siyar, S., and Sami, S. J. H. P. P. J. (2022). Late Blight of Potato: From the Great Irish Potato Famine to the Genomic Era–An Overview. 15: 1–9.

Manoharachary, C., and Kunwar, I. K. (2014_. Host–pathogen interaction, plant diseases, disease management strategies, and future challenges. Future Challenges in Crop Protection Against Fungal Pathogens. Springer.

Martinelli, F., Scalenghe, R., Davino, S., Panno, S., Scuderi, G., Ruisi, P. et al. (2015). Advanced Methods of Plant Disease Detection. A Review. 35: 1–25.

Mazumdar, P., Singh, P., Kethiravan, D., Ramathani, I., and Ramakrishnan, N. J. P. (2021). Late Blight in Tomato: Insights into the Pathogenesis of the Aggressive Pathogen *Phytophthora infestans* and Future Research Priorities. 253: 119.

Miller, S. A., Beed, F. D., and Harmon, C. L. J. A. R. O. P. (2009). Plant Disease Diagnostic Capabilities and Networks. 47: 15–38.

Mohamed, E. S., Belal, A., Abd-Elmabod, S. K., El-Shirbeny, M. A., Gad, A., Zahran, M. B. J. T. E. J. O. R. S. et al. (2021). Smart Farming for Improving Agricultural Management. 24: 971–981.

Mzoughi, O., and Yahiaoui, I. J. E. I. (2023). Deep Learning-Based Segmentation for Disease Identification. 75: 102000.

Nahiyoon, S. A., Ren, Z., Wei, P., Li, X., Li, X., Xu, J. et al. (2024). Recent development trends in plant protection UAVs: A Journey from Conventional Practices to Cutting-Edge Technologies—A Comprehensive Review. 8: 457.

Nayyer, M. A., Ahmad, M. F., Shamim, M., Lal, D., Srivastava, D., and Tripathi, V. (2021). Molecular approaches for the control of Septoria leaf spot in tomato. Biotic Stress Management in Tomato. Apple Academic Press.

Pandey, D. K. & Mishra, R. J. A. I. I. A. (2024). Towards sustainable agriculture: Harnessing AI for global food security.

Pankaj, R., Gupta, S., Azad, C., and Singh, R. (2020). Diseases of Tomato Crops and Their Management. Diseases of Fruits and Vegetable Crops. Apple Academic Press.

Patil, V. U., Vanishree, G., Pattanayak, D., Sharma, S., Bhardwaj, V., Singh, B. et al. (2017). Complete mitogenome mapping of potato late blight pathogen, *Phytophthora infestans* A2 Mating Type. 2: 90–91.

Rehman, S. U., Qiao, L., Shen, T., Hua, L., Li, H., Ahmad, Z. et al. (2024). Exploring the Frontier of Wheat Rust Resistance: Latest Approaches, Mechanisms, and Novel Insights. 13: 2502.

Ren, M., and Yi, P. H. J. S. R. (2022). Deep Learning Detection of Subtle Fractures using Staged Algorithms to Mimic Radiologist Search Pattern. 1–9.

Rhouma, A., Hajji-Hedfi, L., and Atallaoui, K. J. D.-L. S. (2024). Potato Late Blight: The Pathogen, The Menace, The Sustainable Control. 5: 37–51.

Riley, M. B., Williamson, M. R. & Maloy, O. J. T. P. H. I. (2016). Plant disease diagnosis. 10.

Sahoo, L., Mohapatra, D., Raghuvanshi, H. R., Kumar, S., Kaur, R., Chawla, R. et al. (2024). Transforming Agriculture through Artificial Intelligence: Advancements in Plant Disease Detection, Applications, and Challenges. 27: 381–388.

Sahu, A. K., Deep, H., Vishnoi, U., and Saraswat, M. (2023). Ayurvedic plant leaf detection using HOG feature descriptor and SVM classifier. The International Conference on Recent Innovations in Computing, . Springer, 357–370.

Sarker, I. H. J. S. C. S. (2021). Deep Learning: A Comprehensive Overview on Techniques, Taxonomy, Applications and Research Directions. 2: 420.

Shafi, U., Mumtaz, R., Qureshi, M. D. M., Mahmood, Z., Tanveer, S. K., Haq, I. U. et al. (2023). Embedded AI for Wheat Yellow Rust Infection Type Classification. 11: 23726–23738.

Shaheed, K., Qureshi, I., Abbas, F., Jabbar, S., Abbas, Q., Ahmad, H. et al. (2023). EfficientRMT-Net—An Efficient ResNet-50 and Vision Transformers Approach for Classifying Potato Plant Leaf Diseases. 23: 9516.

Siddiqua, A., Kabir, M., Ferdous, T., Ali, I. and Weston, L. (2022). Evaluating plant disease detection mobile applications: quality and limitations. Agronomy. 12: 1869. s Note: MDPI stays neutral with regard to jurisdictional claims in published ….

Singh, B. P., Singh, B., and Lal, M. K. J. I. J. O. I. H. (2022). Seven Decades of Potato Research in India: Achievements and Future Thrusts. 11: 158–183.

Skelsey, P., Humphris, S. N., Campbell, E. J., and Toth, I. K. J. P. O. (2018). Threat of establishment of non-indigenous potato blackleg and tuber soft rot pathogens in Great Britain under climate change. 13: e0205711.

Španić, V., Šunić, K., Duvnjak, J., Viljevac Vuletić, M., and Katanić, Z. J. C. A. F. M. I. C. F. T. Y. (2023). Wheat Breeding Efforts Toward Improvement of Grain Yield Related to the New Challenges for Photosynthesis Research. 87.

Tsedaley, B. J. J. B. A. H. (2014). Review on early blight (*Alternaria* spp.) of Potato Disease and its Management Options. 4: 191–199.

Vadivel, T., Suguna, R. J. A. A. S., Section B—Soil & Science, P. (2022). Retracted Article: Automatic Recognition of Tomato Leaf Disease Using Fast Enhanced Learning with Image Processing. 72: 312–324.

Vallad, G. E., Messelink, G., Smith, H. A. J. T., 2nd Ed., Euvelink, E., Ed. and Cab International: Boston, M., USA (2018). Crop Protection: Pest and Disease Management. 207–257.

Van Der Wolf, J. M., Acuña, I., De Boer, S. H., Brurberg, M. B., Cahill, G., Charkowski, A. O. et al. (2021). Diseases caused by Pectobacterium and Dickeya Species Around the World. 215–261.

Verma, N. V., Shukla, M., Kulkarni, R., Srivastava, K., Claudic, B., Savara, J. et al. (2022). Emerging extraction and diagnostic tools for detection of plant pathogens: recent trends, challenges, and future scope. 2: 858–881.

Wani, J. A., Sharma, S., Muzamil, M., Ahmed, S., Sharma, S., and Singh, S. J. A. O. C. M. I. E. (2022). Machine Learning and Deep Learning Based Computational Techniques in Automatic Agricultural Diseases Detection: Methodologies, Applications, And Challenges. 29: 641–677.

Woolhouse, M. J. P. T. O. T. R. S. B. B. S. (2011). How to make predictions about future infectious disease risks. 366: 2045–2054.

Xu, L., Cao, B., Zhao, F., Ning, S., Xu, P., Zhang, W. et al. (2023). Wheat Leaf Disease Identification Based on Deep Learning Algorithms. 123: 101940.

Xu, Y., Liu, X., Cao, X., Huang, C., Liu, E., Qian, S. et al. (2021). Artificial intelligence: A Powerful Paradigm for Scientific Research. 2.

Zhang, C., Marzougui, A., Sankaran, S. J. C., and Agriculture, E. I. (2020). High-resolution satellite imagery applications in crop phenotyping: An overview. 175: 105584.

CHAPTER 8

Utilizing AI for Remote Sensing and Monitoring of Crop Health

Prajakta Y. Pachorkar,[1,*] *Chaitanya S. Mogal,*[2] *Jagruti Chavan,*[3] *Chetan A. Patil,*[4] *Monali Sonwane,*[5] *Pranay Deore,*[5] *Harshali Deore,*[5] *Pratiksha Agare*[5] and *Minal Jagtap*[5]

1. Introduction

The preservation of biodiversity and natural environments is today a major concern for nature conservation. The introduction of non-native invasive species is one of the main dangers. Native flora and animals are harmed by these species, which spread swiftly. The importance of this topic is acknowledged by the United Nations (UN). The UN monitors the number of nations attempting to prevent and control invasive plant species (Indicator 15.8.1) and includes measures to control these species in its Sustainable Development Goals (Goal 15.8). We require up-to-date, comprehensive location data in order to comprehend and forecast the spread of invasive species. This issue poses a severe threat to biodiversity, not just in Hungary and Central Europe but globally as well. In addition to harming the ecosystem, invasive species also cost the agricultural industry money (Chytry et al., 2012). For the sake of both agriculture and nature preservation, it is crucial to map their paths, track their spread, and pinpoint impacted areas. Predicting which places may be impacted in the future is also necessary (Cosgrove and Loucks, 2015). Over the last 15 to 20 years, invasive species have been tracked using a variety of sensors on various platforms. For this, hyperspectral (HS) satellites such as Huan Jing-1 (HJ-1) and EO-1 Hyperion are helpful. Invasive species can also be efficiently mapped using sensors installed on

[1] Dept. of Microbiology, KRT Arts, BH Commerce and AM Science College, Nashik.
[2] COF, NAU, Navsari.
[3] School of Sciences, Yashwantrao Chavan Maharashtra Open University, Nashik.
[4] Dept. of Environment Science, KRT Arts, BH Commerce and AM Science College, Nashik.
[5] Dept. of Microbiology, KRT Arts, BH Commerce and AM Science College, Nashik.
* Corresponding author: prajaktapachorkar@gmail.com

DOI: 10.1201/9781003546702-8

aircraft using programs like CASI, AISA, APEX, AHS, and HySpex. In contrast to multispectral sensors that record a wider range of wavelengths, hyperspectral data offers comprehensive spectral information, making it more effective at differentiating between plant species (Papp et al., 2021). In conventional agriculture, agrochemicals known as plant protection products (PPPs) are widely employed to stop pests and diseases from reducing crop yields by up to 65% (Balafoutis et al., 2021). About 34% of the overall production costs in the 2016–2017 production year came from the use of PPPs in processed orange orchards in southwest Florida. In Florida, air-carrier ground sprayers are currently used for the majority of citrus PPP applications. Airblast spray treatments, however, pose a serious risk of spray drift, which can contaminate surrounding water surfaces, people, wildlife, and vegetation.

Innovative methods have been created to improve the effectiveness of PPP applications while lowering expenses in order to address this problem. Variable-rate technologies (VRT), which enable sprayer machines to modify the quantity of PPP delivered using data or sensors, are one such method. Three key components underpin the growing use of VRTs in agriculture: sensors to gather plant data, processing algorithms to calculate the necessary PPP quantity, and flow rate control. Numerous sensor technologies, like as light detection and ranging (LiDAR), ultrasonic, infrared, and machine vision sensors, are used to evaluate canopy structural aspects in VRT applications. Applications can be further optimised by machines with sensors and real-time processing capabilities, even though VRTs can use data that has already been gathered. However, because of the high data volumes and processing times required, real-time LiDAR sensor applications necessitate data downscaling preprocessing processes (Partel et al., 2021). Ultrasound, infrared, and video sensors have demonstrated some success for real-time field applications. Up to 37% less pesticide waste was produced when canopy analysis in a variable-rate sprayer was done using an ultrasound-based approach. However, the target must be perpendicular to the sensor for precise wave reflection, and ultrasonic sensors have a limited range. There has also been some success with a real-time infrared system for target detection in electrostatic spraying applications. The accuracy of plant classification provided by machine vision using inexpensive cameras is on par with laser radar sensors. To determine vineyard application needs for VRT uses, four machine vision algorithms were created, which might result in a 30% reduction in PPP usage. However, uncontrolled wind and lighting conditions reduce the effectiveness of machine vision sensors, while rain, dust, and fog increase noise levels (Mowla et al., 2023). Smart machines in agricultural settings require algorithms that can swiftly adjust to changing conditions because of the uncontrolled outdoor environment. A solution is provided by sensor fusion, which aggregates data from multiple sources to produce more trustworthy information. This approach uses sophisticated algorithms, including artificial intelligence (AI), to combine and process different types of data. By taking into account different agricultural field noises, AI algorithms improve data processing reliability and allow technologies to precisely detect crops, pests, and fruits while making production decisions (Mowla et al., 2023). AI has been used to create plant inventories, identify weeds, assess the quality of fruits and vegetables, diagnose illnesses and plant stress, and support crop classification and decision-making.

AI models are frequently utilised in the creation of intelligent agricultural equipment because of their versatility and dependability (Qazi et al., 2022). To overcome the primary drawbacks of the current variable-rate spraying technologies, such as their incapacity to differentiate trees from other objects, categorise the density or health of tree leaves, or adjust to changes in the environment A real-time smart sprayer sensing system was created and assessed in this study. Using citrus as a case study, this system applies LiDAR, machine vision, GPS, sensor fusion, and AI to tree crops (Partel et al., 2021). The sensor fusion system was designed to accomplish three tasks: (i) detecting the presence of an object; (ii) classifying the object to determine whether it is a tree (such as a citrus tree); (iii) estimating the height of the tree canopy and the density of its leaves; and (iii) detecting and counting the fruits on the identified tree canopy. Citrus orchards were used to test this sensing device, which was installed on an airblast sprayer (Partel et al., 2021). Global agricultural production systems are confronted with significant obstacles, including climate change, dwindling irrigation water supplies, increased production costs, and a shortage of agricultural workers (Turral et al., 2010). These problems have been made worse by the COVID-19 pandemic, which is now endangering global food supply systems and production. The sustainability of present and future food systems is threatened by these issues, which call for important innovations to solve them (Thompson and Scoones, 2009). The need to sustainably supply high-quality food is urgent given the world's population growth. Innovative technologies have long been used by agricultural researchers to incorporate them into farming methods. Understanding the various elements of agricultural systems and how they interact has been made much easier by crop simulation models (SS et al., 2024). Due to its capacity to use big data, which is becoming more widely available through Unmanned Aircraft Systems (UAS), Artificial Intelligence (AI) has become a potent instrument in agriculture in recent years. UAS offer a previously unheard-of chance to manage agricultural systems using advanced analytics, increasing their resilience and effectiveness (Jung et al., 2021). By providing a diagnostic tool that serves as an early warning system, remote sensing technologies allow the agricultural industry to take proactive measures to solve possible problems before they worsen and have a negative impact on crop output. Agricultural practitioners today have access to a variety of remote sensing options thanks to recent advancements in sensor technologies, data management, and analytics (Khanal et al., 2020). Because precision agricultural applications require high spatial, spectral, radiometric, and temporal resolutions, remote sensing systems that use information and communication technologies often generate large amounts of spectrum data (Sabir et al., 2024). Advanced data processing methods like machine learning, artificial intelligence, and big data analysis have been used to glean insightful information from this massive volume of data. This vast amount of data has also been processed, stored, and disseminated for PA applications using cloud computing technologies. Field crop, horticultural, viticultural, pasture management, and livestock management decision-making processes are supported globally by these advanced data collection and processing techniques (Sishodia et al., 2020). Satellite remote sensing has emerged as the main technique for local, regional, and worldwide crop monitoring since the 1970s. According to Wu et al. (2023), "crop monitoring" specifically refers to the observation of staple crops, as opposed to the more general phrase "agricultural monitoring," which includes livestock, horticulture, and aquaculture. Agroclimatic studies, crop condition and stress monitoring, and crop production forecasting are examples of common crop monitoring tasks. In order to provide early warnings of possible food shortages, some systems also include assessments of food security (Nakalembe et al., 2021). From the standpoint of remote sensing, agroclimatic analysis and food security assessments concentrate on comprehending the causes and effects, whereas crop monitoring mainly aims to evaluate crop growth status and final production (Thenkabail, 2024). New opportunities for fast, all-weather crop monitoring with high geographical and temporal resolutions have been made possible by the free availability of medium- to high-resolution satellite data. Cloud computing platforms like as Google Earth Engine (GEE), Microsoft AI for Earth, and Amazon have greatly enhanced the ability to process and extract information from satellite data. These platforms have eliminated earlier limitations on satellite data and processing capabilities in crop monitoring when paired with auxiliary data including weather, soil, and elevation information (Wu et al., 2022). Precision agriculture may accomplish a number of important goals by combining the Geographic Information System (GIS), Global Navigation Satellite System (GNSS), and remote sensing technologies: (1) surveying fields; (2) precisely administering pesticides, fertilisers, and irrigation management plans; and (3) closely monitoring crop health, soil moisture, pests, and diseases. By putting science-based decision-making procedures into practice, one can increase production and revenues, support environmental sustainability, raise crop quality, and eventually increase food security. With the growing use of yield monitoring technology, auto-guidance systems, and variable rate technology (VRT) in both developed and developing nations over the previous 20 years, precision agriculture has become more widely used worldwide in recent years (Zheng et al., 2021).

The multidisciplinary discipline of artificial intelligence (AI) aims to imitate human cognitive processes and behaviours, including learning and problem-solving, in robots. AI is currently being used by researchers and extension experts to address problems with agricultural productivity (Shaikh et al., 2022). By helping farmers choose the right crop varieties, implement better soil and nutrient management techniques, control pests and diseases, estimate crop production, and predict commodity prices, artificial intelligence (AI) can help farmers increase yields. To solve agricultural issues, it uses cutting-edge techniques like deep learning, robots, the Internet of Things (IoT), image processing, artificial neural networks, wireless sensor networks (WSN), and machine learning. These technologies help farmers make well-informed decisions by enabling real-time monitoring of farm parameters, including weather, temperature, water usage, and soil conditions. AI makes it easier to create intelligent farming techniques that increase yields and reduce losses (Javaid et al., 2023).

Using information from machine learning, AI-powered precision agriculture helps farmers with operations like irrigation, crop rotation, harvesting, crop selection, planting, and pest management. The premise that human intelligence may be explained in a way that allows a computer to reproduce and carry out tasks of various complexity forms the basis of artificial intelligence (AI) (Adinarayana et al., 2024). AI seeks to accomplish perception, thinking, and learning. It has a wide-ranging effect, as industries everywhere aim to automate certain operations with intelligent devices. This happens when the definition of human intellect is made machine-understandable. By automating a variety of operations, from easy to difficult, artificial intelligence (AI) in agriculture has the potential to greatly enhance the sector. The ultimate objective is for machines to be able to learn, reason, and observe in order to simplify a variety of jobs in a variety of industries (Javaid et al., 2023).

Artificial intelligence (AI) has revolutionized modern agriculture by seamlessly integrating with remote sensing technologies to enhance the accuracy, efficiency, and scalability of crop health monitoring. By combining data from multiple sensors such as LiDAR, infrared, hyperspectral, ultrasonic, and machine vision systems, AI enables real-time monitoring and data-driven decision-making across diverse agricultural landscapes. This multi-sensor fusion provides detailed insights into plant growth, soil conditions, and environmental factors, allowing for precise and timely interventions. AI further contributes to developing prescriptive solutions tailored to site-specific challenges such as pest management, irrigation scheduling, nutrient optimization, and disease prevention. These intelligent systems minimize the overuse of fertilizers, pesticides, and water resources, thus promoting sustainability and reducing environmental impact. Moreover, advanced remote sensing and AI-powered analytics facilitate early detection of potential threats, enabling farmers to take proactive measures against pests, diseases, and climatic stresses. Innovations in AI-based agricultural machinery—such as variable-rate sprayers, autonomous drones, and automated harvesting tools—are significantly improving operational efficiency, labor management, and yield quality. Collectively, these advancements highlight the transformative potential of AI-driven precision agriculture in fostering sustainable, resilient, and highly productive farming systems.

Objectives of Chapter

1. Investigate how artificial intelligence integrates with remote sensing technologies to enhance the accuracy, efficiency, and scale of monitoring crop health.
2. Assess the effectiveness of combining data from various sensors (e.g., LiDAR, infrared, hyperspectral, ultrasonic, and machine vision) for real-time monitoring and decision-making in agricultural practices.
3. Analyse the potential of AI in creating prescriptive solutions that address site-specific challenges such as pest control, irrigation, nutrient management, and disease prevention.
4. Evaluate how AI-driven precision agriculture technologies can minimize the overuse of fertilizers, pesticides, and water, promoting sustainable farming practices.
5. Study the effectiveness of advanced remote sensing tools and data analytics in providing early warnings about potential threats to crop health, such as diseases, pests, or environmental stresses.
6. Assess innovations in AI-based machinery, such as variable-rate sprayers and automated harvesting tools, and their impact on farming efficiency.

2. Types of Sensors for Smart farming

Traditional farming methods are being quickly advanced by innovation. The agricultural sector is changing as a result of technology including sensor technologies, unmanned aerial vehicles (UAVs), and satellite photography. Information technology is used in smart farming to maximise intricate agricultural processes. Its goal is to make intelligent use of the information gathered in order to improve solutions or solve problems. Almost every aspect of farming is included in smart farming. With the use of intelligent algorithms, farmers may respond to events by using portable devices, like smartphones and tablets, to monitor real-time data (such fertilisation, climate, and irrigation) and make well-informed judgements. Agricultural data is read and processed by a variety of sensors. The most popular sensors used in smart farming are shown below, along with details about each:

2.1 Water Content Sensor

This instrument calculates the proportion of water in the soil under test in relation to the total amount of soil. By monitoring variations in dielectric permittivity, it evaluates the soil's capacity to retain an electrical charge. The sensor's readings go from 0 (dry soil) to the saturation of the soil's porosity. The ratio of pore volume to total soil sample volume is known as porosity saturation. The sensor needs to be calibrated for various sites because measurements vary depending on the kind of soil (Elbasi et al., 2024).

2.2 Volumetric Water Content Sensor

This sensor determines the amount of water in the soil by measuring the water suction, which shows how hard it is for plant roots to draw water out. It calculates how much water is stored in the soil or how much irrigation is needed to keep the water levels at the appropriate levels (Ersin et al., 2022).

2.3 Electrical Conductivity Sensor

This sensor estimates the concentration of solutes in the soil, which can be detrimental to crops if it is excessively high. Usually caused by salt buildup from irrigation water, salinity around plant roots can result in long-term land degradation (Ersin et al., 2022).

2.4 pH Sensor

This sensor determines the acidity or alkalinity of the soil by measuring its pH values. The ideal pH range for soil is 6.0 to 7.0. A nutritional shortage is indicated by values outside of this range. Farmers must use acidic or alkaline fertilisers to change the pH in order to increase yield (Ersin et al., 2022).

2.5 Weed Seeker Sensor

This sensor detects and targets resistant weeds by using sophisticated optics and computational power. It accurately delivers herbicide to the weed by signalling an attached spray nozzle when it detects a weed. The sensor can target and spray only the weeds since it has an active light source and a selective spray sensor that detects chlorophyll. This technique optimises chemical application, minimises expenses, and minimises chemical usage by up to 90% (Ersin et al., 2022).

2.6 Temperature Sensor

This sensor is essential for identifying which crops can be grown in a field since it notifies users when the temperature deviates from the typical range. Water absorption and transpiration during photosynthesis are two critical plant growth processes that are influenced by soil temperature. Every crop has a range of temperatures that are essential for growth; if the temperature drops below this range, the growth-promoting enzymes stop working (Ersin et al., 2022).

2.7 Wind Speed Sensor

This sensor is crucial for tracking variations in wind speed patterns and directions since it monitors wind speed at a particular surface level. The crop determines the sensor's mounting height (Ersin et al., 2022). Figure 1 illustrates the many kinds of AI sensors used in agriculture (Jha et al., 2019).

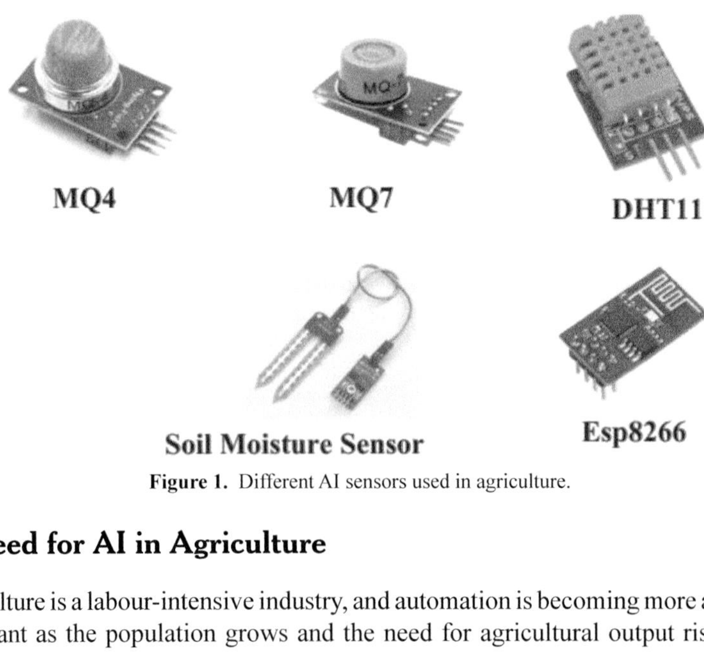

Figure 1. Different AI sensors used in agriculture.

3. Need for AI in Agriculture

Agriculture is a labour-intensive industry, and automation is becoming more and more important as the population grows and the need for agricultural output rises. With its components, technology, and applications, AI significantly helps farmers. Crop quality and supply are guaranteed by predictive analytics and sophisticated farm and crop management systems. Companies calculate acreage and track crop health in real time using satellite imagery and meteorological data. Businesses can assess tomato output and yield, forecast pricing, and detect pest and disease infestations by utilising big data, artificial intelligence, and machine learning technology. They can also give farmers advice on future pricing trends, pesticide use, crop varieties that will yield the highest profits, and demand levels. AI will be an effective tool for handling the increasing complexity of contemporary agriculture, greatly reducing the need for labour and resources. It is imperative that big businesses make investments in this area. As seen in Figure 2, a number of industries, including finance, transportation, healthcare, and agriculture, use AI technology to increase productivity and efficiency.

As the global population grows and urbanization accelerates, consumer behavior changes with rising disposable income. Farmers are under immense pressure to meet increasing demand and need methods to boost output. More people will need to be fed, and innovative farming solutions will be required due to the limited availability of fertile soil. Strategies must be developed to help farmers mitigate or manage risks. The global adoption of AI in agriculture presents an exciting opportunity. Many food producers currently struggle with risks and threats to their crops from pests and diseases, exacerbated by climate change, monoculture, and extensive pesticide use. These factors create new challenges for farmers. Farming, heavily dependent on natural forces, faces significant stress, amplified by unpredictable rainfall, labour shortages, and the annual need for higher yields. The agricultural sector will need to scale up massively in the coming years, doubling farm efficiency to meet targets. AI addresses these challenges by providing agricultural automation (Mohd et al., 2022).

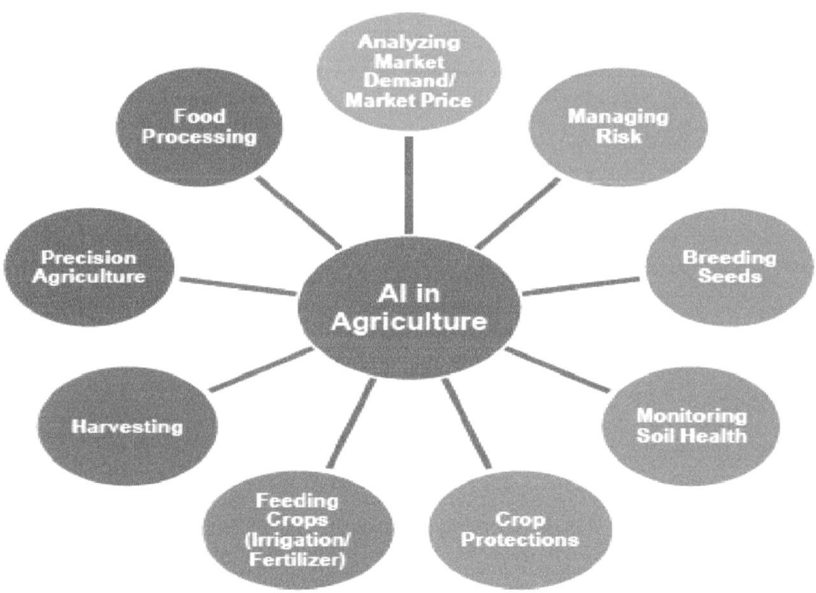

Figure 2. Need of AI in Agriculture (Source: /iiss.icar.gov.in/eMagazine).

4. Process of Artificial Intelligence in Agriculture

The quality of data accessible to AI determines its effectiveness in solving agricultural problems, and AI is seen as a promising technology for offering revolutionary agrarian solutions. However, obtaining the necessary information at the farmer level presents a significant challenge. Combining image classification techniques with remote and local sensing data can revolutionize the use of farm machinery and efficiency, especially in weed eradication, early disease diagnosis, crop harvesting, and grading. AI solutions enable continuous monitoring of high-value goods, which is crucial in horticultural practices that require comprehensive monitoring at all stages of plant growth. By utilizing precision agricultural software, soil sensors, soil analysis drones, or even smartphone photographs, AI systems can continuously monitor soil nutrient levels and compare them to historical data that have produced the highest yields for specific crops. AI can analyze the environmental impact of various fertilizer dosages to identify the optimal dose that maximizes production while minimizing negative effects, contributing to environmentally friendly farming. The rise in pollution and weather unpredictability over time has made it challenging for farmers to determine the best time to plant seeds, a problem where AI can provide significant assistance. AI can analyse how weather, seasonal sunlight, wind, and rain will affect crop planting cycles, aiding farmers in planning and scheduling seed planting. Advances in computer vision, mechatronics, AI, and ML have made it possible to use remote sensing technologies for identifying and managing plants, weeds, pests, and diseases. These technologies also offer opportunities to develop precise fertilization techniques. AI tools can help farmers reduce waste, improve product quality, and ensure quicker market access. Automated tractors, equipped with GPS technology, prepare the farm for planting and harvesting without a driver. Drones collect data, monitor farm and crop health from a distance, and relay this information back.

AI assists in pest control by identifying the central plant on the farm and determining the appropriate pesticides and their quantities, applying them using drone technology to save time. Predictive analytics can estimate precipitation and evapotranspiration, and when combined with soil samples and other data, ML models can provide insights into soil moisture, temperature, and overall condition. This enables farmers to irrigate their crops more efficiently, benefiting both their profit margins and the environment. Automated systems can continuously monitor farm conditions, leading to a less labour-intensive and more resource-efficient form of agriculture. Identifying and managing field variability ensures that crops receive precisely what they need, improving yields, fertilizer efficiency, and profitability. Precision agriculture enhances sustainability, environmental protection, productivity, and efficiency. Many businesses worldwide are adopting AI technologies to improve agricultural efficiency and also applying Expert system in agriculture which is shown in Table 1.

Numerous expert systems have been designed and implemented by various organizations and research institutions to address specific challenges within the agricultural sector. Certain crops are specifically targeted by the expert systems.

Table 1. AI Expert system in agriculture which is shown in Table 1 (Sharma, 2021).

System Name	Crop Name	Description
AGREX	Fruits, Vegetables, Paddy	In Kerala, The Centre for Informatics Research and Advancement (CIRA) has built an expert system program named as AGREX which can support the farmers to receive well-timed and trustworthy guidance. This expert program finds increased use in paddy, vegetables, postharvest technology, and fruits in the field of crop protection, fertilizer application, disease diagnostics, and irrigation scheduling.
TEAPEST	Tea	It is an object-oriented, rule-based expert system that can identify serious insect pests of tea and therefore suggest a suitable control mechanism.
JAFexpert	jute	Central Research Institute for Jute and Allied Fibres (ICAR-CRIJAF) developed a web-based expert system called JAFexpert. It is capable of providing data for management and accurate identification of injurious organisms and abiotic damages for jute and allied fibre crops.
AMPRAPALIKA	Mango	This expert system program is utilized for making a diagnosis of a specific disease of mango. The system's knowledge base includes information about the indicators and therapies of 14 mango tree illnesses that exist during the non-fruiting and fruiting seasons.
Rice-Crop Doctor	Rice	The National Institute for Agricultural Extension Management (MANAGE) developed this expert system program which is used to detect diseases and rice pests and then suggest the cure.
AGPEST	Wheat, Rice	This expert system is designed for the detection of diseases triggered by wheat plants and rice pests respectively. It also aids the decision support module with a collaborative console base UI for analysis made alongside the questions relating to some specific disease symptom.

The aforementioned description field talks about how the expert systems give the crops facilities. A number of industries have experimented with AI in the last ten years to enhance their operations. But the potential of AI to greatly improve agricultural decision-making has only lately come to light. AI technology can help farmers make better decisions that increase the productivity of their livestock and crops. The agricultural research and industry sectors use AI to evaluate human intelligence development through theories, techniques, algorithms, and applications. AI-based analytics are essential for data mining. AI employs machine learning (ML) methods to extract useful information from massive datasets because of the size and complexity of big data, which are beyond the capabilities of conventional data-processing systems. This aids in the comprehension of crop development patterns, possible diseases, the use of fertiliser or pesticides based on disease patterns, and the prediction of diseases based on the size, colour, or growth of leaves. Machines such as low-flying drones or small robots moving across the field can be equipped with a variety of sensors, such as feed, weight, soil, temperature, intensity, and several camera types (Mohd et al., 2022).

5. Artificial Intelligence-Monitored Agricultural Parameters

Given that agriculture is a labour-intensive industry by nature, it is not surprising that a labour shortage occurs frequently. Automation, however, offers a workable remedy. AI-based harvesting robots, self-driving tractors, intelligent irrigation systems, and automated fertilisation and spraying systems are a few examples. Assisting farmers in comprehending the overall AI system is one of the challenges facing software vendors. AI is used in agriculture to do activities including pest control, health monitoring, field harvesting, and deficit identification. Machine learning (ML) and artificial intelligence (AI) are taking the place of antiquated forecasting and intelligence techniques in the agricultural sector, increasing the sector's adaptability through the integration of cutting-edge technologies. Soil moisture and fertility can be monitored by biosensors, and raw data is gathered using a variety of techniques rather than relying on simple linear regression models. The ability of neural networks to forecast historical weather patterns with non-linear relationships is very helpful in figuring out when to plant vital crops like rice, wheat, and maize, which need a lot of rainfall to thrive.

Every day, farms produce hundreds of data points about weather, water use, soil conditions, and temperature. Real-time processing of this data by AI and ML models offers insightful information on the best time to plant seeds, which crops to choose, and which hybrid seeds to choose for increased yields. Precision agriculture is improved by AI technology, which raises harvest quality and accuracy overall. It facilitates the early identification of agricultural diseases, pests, and nutritional deficiencies. By using AI sensors to detect and target weeds, the right herbicides may be chosen and used less frequently, saving money. Many tech companies have created robots that monitor and precisely spray weeds using AI and computer vision. AI has impacted many industries, including marketing, sales, architecture, medical, and finance. In agriculture, AI, machine learning, and Internet of Things sensors are utilised to fertilise, irrigate, and safeguard crops.

To get the optimal yield, crops need to be irrigated at the appropriate times. Insufficient resources made it difficult for a small number of farmers to implement large-scale irrigation. The introduction of intelligent irrigation systems mainly consists of hardware implementation using IoT devices and an ML model to determine soil moisture. Drones, satellite imagery, remote sensing, and precision technologies are all used in ML's crop and soil monitoring systems. Traditionally labour-intensive jobs, such as harvesting crops, can now be completed more quickly and effectively by autonomous robots than by human workers. For the early and accurate detection of diseases, allergies, chemicals, and pollutants in food, plant and animal production systems, water, and soil, nano-based sensing mechanisms and smart sensors are being developed and tested.

AI is now able to gather and analyse the millions of data points produced daily by agricultural land. This solves long-standing issues by enabling farmers to make well-informed decisions instantly. The ability to forecast weather ahead of time aids farmers in determining when to plant their crops. AI can now assess hybrid seeds and their yields before planting, lowering the chance of unsuccessful harvests and boosting output. It has also resolved the problem of water utilisation. There is a labour shortage in farming as a result of people moving from rural to urban lifestyles. At first, farms needed a lot of seasonal labourers to gather crops, but as cities have grown, fewer people are available to work in agriculture. This scarcity is mitigated by AI bots. New agricultural technologies help gather information so farmers can keep an eye on and improve their crops. The fact that these systems are always being updated to account for changing environmental and ecosystem parameters shows how important technology improvements are to the modern agriculture industry. Adoption of innovative agricultural technologies contributes to the attainment of sustainability objectives. AI technology is boosting production and efficiency across a range of industries. AI in agriculture boosts productivity and lessens adverse environmental effects. In order to solve a variety of farming issues, the agriculture sector has embraced AI to optimise production and operation processes in food, biosystem engineering, and agriculture. Crop cultivation can benefit from the collection of data from sensors, drones, and satellites. AI can be used to analyse this data, giving farmers the ability to make better decisions. By analysing plant biomass and traits through phenotyping, artificial intelligence (AI) can find correlations between damaged crops and particular disease causes. Finding pests and contaminants in crops contributes to a higher total output. Agronomic, climatic, and hydrological balance are all impacted by agricultural water management. More efficient use of irrigation systems is made possible by advanced machine learning applications, such as daily, weekly, or monthly evapotranspiration estimates. Additionally, daily dew point temperature forecast aids in estimating evaporation and evapotranspiration. The enormous potential of AI will hasten disruption and fundamentally alter the way food is produced and consumed. AI will unleash the potential of new agricultural regions, generating new job opportunities as agriculture undergoes a digital revolution. AI is making controlled farming of both existing items like leafy greens and new ones like insects more feasible and economical. Farmers can overcome challenges like pests, water scarcity, and climate change with the aid of this technology. AI helps farmers make educated agricultural selections by using machine learning, deep learning, and other techniques to produce insightful predictions from vast volumes of data. AI is more accurate than humans in analysing large amounts of data. For jobs that used to need human labour, many farmers now use AI-powered robots. There is less workforce available for agricultural operations as a result of growing urbanisation and fewer individuals choosing to become farmers (Mohd et al., 2022).

6. Approach

For the purpose of spraying tree crops, such citrus, a 500-gallon back sprayer was equipped with a smart sensing system that included both hardware and software. Four sensors are used in this system to collect data: (i) a 2D LiDAR sensor to measure tree height and canopy leaf density; (ii) two USB cameras, one on each side, to take pictures for image classification (e.g., differentiating between trees, buildings, and people); (iii) a GPS sensor to provide location and tractor speed information; and (iv) two flow meters, one on each side, to track the rate of application. An embedded computer unit that manages the sprayer and provides a user interface processes the data that has been collected. Below is a thorough explanation of the smart sensing system.

6.1 Sprayer

The detecting system was used with a back sprayer that held 1892.71 L (500 gallons) (Power Blast 500 gallons, Chemical Containers, Florida, USA). The sprayer features 24 nozzles, 12 on the right and 12 on the left, and a PTO-driven fan with a diameter of 83.82 cm (33 in.). There are four zones on each side of the sprayer nozzles—a total of eight. Each zone is operated by an electric valve (ARAG HYEV-1, Italy) and features three nozzles. The direct-acting valves respond in 0.6 seconds throughout the course of a full cycle. The centrifugal pump in the sprayer runs between 150 and 250 psi and is likewise PTO-driven.

6.2 Intelligent Sensing Apparatus

Lidar A cheap, portable, 2D, 360-degree laser scanner with a scanning frequency of 10 Hz (600 rpm) and a maximum scan range of 10 m was used.

6.3 Vision of Machines

To capture images, two USB cameras (ELP USB130W01MT-DL36, Shenzhen, China) were utilised.

6.4 GPS Unit

In order to calculate speed and position in real time, a USB GPS module (Gowoops GPS module, Shenzhen, China) was utilised.

6.5 Flow Gauges

Before the valves, two electromagnetic flow meters (ARAG Orion 2, Italy) were positioned on either side of the sprayer.

6.6 Intelligent Controller Unit

An Arduino micro-controller (Arduino Mega2560, Italy) receives a signal from the Nvidia computer, which processes data in real-time. The nozzles are then turned on when the microcontroller turns on the relays that were requested to operate the electric valves of the sprayer.

6.7 Development of Software

The embedded computer's software was separated into two categories: user interface and data processing. Fruit identification and counting, picture tree categorisation (e.g., tree health status), and raw sensor data processing are all part of the data processing. With its various options, the user interface enables the user to customise the sprayer to suit his or her needs.

7. Applications of Remote Sensing in Accurate Agriculture

7.1 Connecting Agricultural Variables of Interest to Remote Sensing Observations

Crop features (such as morphological, biochemical, and physiological qualities), soil parameters (such as soil moisture, organic matter, pH, and drainage), and topography (such as elevation and slope) and their spatial and temporal fluctuations are important variables of interest in agriculture. Since remote sensing (RS) cannot directly measure these characteristics, they are usually inferred by combining ground-truth data and spectral measurements using empirical, mechanistic, or a mix of methods. For example, by comparing the spectral signatures of a target field with those of a well-fertilized reference plot that is comparable to the target field, or by combining RS-derived leaf area index and chlorophyll content with crop models, it is possible to empirically link nitrogen (N) stress in crops to RS observations. Similar to this, crop yield estimation using RS takes into account other factors like vegetation status, soil characteristics, and weather conditions (such as temperature, precipitation, and solar radiation) using empirical or mechanistic approaches. Crop geometry, leaf area indices, field circumstances early in the growing season, and picture spatial resolution must all be analysed in order to map crop emergence. Empirical methods are comparatively simple because they use only statistical methods to connect inputs to outputs. However, gathering a lot of data is necessary to improve model robustness. On the other hand, by taking into consideration a variety of biophysical processes, mechanistic models investigate cause-and-effect correlations between inputs and outputs. Frequently, these models depend on assumptions that might not always be true. The benefits and drawbacks of these strategies have been covered in detail in earlier research. Although the same characteristics can be described by both empirical and mechanistic approaches, one may be better suited for a given application than the other (Khanal et al., 2020).

7.2 Remote Sensing Data on Relevant Agriculture Variables

The attributes of sensors and their platforms, such as satellite orbit location, unmanned aerial system (UAS) orientation, and sensor characteristics, determine how helpful information from remote sensing (RS) is for agricultural decision-making. Temporal, spectral, and geographical resolutions are important characteristics of RS data. The term "spatial resolution" describes the size of an image's pixels, which affects the capacity to identify objects on the surface of the earth. When it comes to a sensor's capacity to resolve features in the electromagnetic regions (EMRs), "spectral resolution" refers to the quantity and magnitude of spectral sampling intervals. "Temporal resolution" shows the frequency of data collection. Furthermore, the platform of a sensor has an impact on air conditions and illumination during data collection, which can change the signal-to-noise ratio. Artefacts from measuring conditions and sensor features must be taken into consideration for precise variable detection and quantification in order to guarantee "error-free" inferences.

High temporal resolution (less than weekly) and high spatial resolution (less than decametric) imagery are generally used for precision agriculture (PA). Applications such as weed identification, crop counting, and crop growth stage identification benefit from high-resolution images. Crop water and nutrient stress fluctuation within a season can be captured with frequent photography at particular growth stages. The characterisation, identification, and classification of crop and soil attributes can be enhanced using hyperspectral photography, which records spectral signatures in multiple tiny spectral bands in EMRs. This is more difficult when using only a few broad spectral bands (multispectral). Compared to photography from visible and near-infrared regions, thermal and microwave imagery is especially helpful for identifying thermal abnormalities in crops and soils, such as water stress. The following part examines the temporal and spatial resolutions of RS data utilised in earlier research, assessing how well RS supports decision-making for different productive agricultural management activities. Pre-season planning, field preparation, planting, in-season monitoring, pest control spraying, harvesting, and post-harvest activities are many examples of these operations (Sami et al., 2020).

8. Applications of Artificial Intelligence In Agriculture

Numerous technological developments, such as data analytics, IoT applications, consultancy services, and the installation of cameras and sensors, are made possible by AI in agriculture. By assessing a variety of data sources, including weather, soil conditions, crop performance, and temperature, AI will get better at producing predicted insights. Examining agricultural hazards such as data reliability, safety and security concerns, interoperability problems, and unforeseen socio-ecological consequences of employing machine learning models to maximise yields are all part of this. Additionally, by quickly detecting plant illnesses and effectively administering agrochemicals, AI can improve crop management and yields.

8.1 Weather Forecasting

Weather forecasting and other important agricultural parameters, including crop cycles, groundwater levels, land quality, and plant disease detection, are made easier by AI technology. Soil and plant sensors, as well as multispectral photos from satellites or drones, make it easier to monitor crop health. By using more sophisticated unsupervised machine learning algorithms to analyse this data, AI solutions can boost productivity and lower crop loss. AI systems can also watch drone or security camera footage to identify unauthorised people, birds, or wild animals that could harm crops. In order to increase farmers' productivity, artificial intelligence (AI) in agriculture uses data and tools like soil sensors, smart drones, autonomous tractors, and other technology. AI is being used by many companies in the agriculture industry for predictive analytics and precision farming. The majority of agricultural start-ups have used AI-enabled techniques, which has greatly increased farming operations' productivity and results (Mohd et al., 2022).

8.2 Enabling Farmers to Attain Effective Outcomes

AI has the ability to revolutionise agriculture by giving farmers several advantages and increasing yields with less work. AI, however, is not a stand-alone technology; it facilitates the shift from conventional to creative farming methods. It's possible that many farmers have the misconception that artificial intelligence is only used in digital fields and are unaware of how it may help with real-world farming. AI uses cognitive IoT, which includes imaging, remote sensing, proximity sensing, and soil testing, to integrate growth-driven crops. AI can increase productivity by understanding, storing, and learning enormous volumes of data. AI can evaluate data and provide insightful recommendations for increasing crop productivity thanks to cognitive IoT. Hardware solutions maximise output by combining robotics and data collecting software to create the best fertiliser for certain crops (Mohd et al., 2022).

8.3 Promote Healthier Crop Yields

The agricultural sector is using AI technology more and more to manage data for farmers, monitor soil and growing conditions, control pests, produce healthier crops, ease workloads, and enhance other agriculture-related jobs along the food supply chain. Farmers struggle to decide when to plant seeds because of shifting weather patterns and increased pollution. Farmers can determine which crops to grow and when to sow them by combining weather forecasts with artificial intelligence (AI) to analyse weather conditions. AI makes crop selection easier and assists farmers in determining which crops yield the highest profits. Farmers can minimise operational errors and lower the risk of crop failure by using forecasting and predictive analytics. Additionally, AI collects data on plant growth, which helps produce crops that are resistant to disease and environmentally friendly. Additionally, AI systems are capable of reliably estimating missing nutrients and conducting chemical analyses of soil (Mohd et al., 2022).

8.4 Forecast Plant Ailments

AI is able to detect and even predict plant diseases, identify and eradicate weeds, and suggest effective pest management methods. AI assists in predicting the optimal agronomic product combinations, determining the most effective watering schedules, and scheduling fertiliser applications. AI can be used to automate harvesting and even predict when it will be most effective. The use of predictive analytics has the potential to revolutionise entire sectors. Farmers can process and collect more data using AI than they could without it. AI can be used by farmers to solve crucial problems including pricing forecasts, market demand analysis, and agricultural planting and harvesting schedules. Before delivering the fruit to the market, this can help growers sort it into stacks of different readiness levels. Field management uses high-quality photos from drones and copters to make forecasts in real time. In order to make forecasts during cultivation, feed and field maps are created, and regions where crops need more or less water, fertiliser, or pesticides are identified. Cognitive solutions give farmers useful information on seed types, infestation, weather forecasts, and soil conditions in a particular area (Mohd et al., 2022).

8.5 Monitoring of Crops and Soil

Crop and soil monitoring is a typical use of AI in agriculture. Data can be collected via drones, the Internet of Things (IoT), and satellite photos of fields. AI-based apps can then evaluate the data to find the best answers. AI applications use machine learning (ML) methods to make data processing rapid and simple while detecting diseases, plant pests, and defects in the soil. Apps for mobile agriculture can help farmers analyse crop data, keep track of their operations, keep an eye on weather patterns, and run their farms more effectively. The objective is to maximise the use of pesticides and herbicides to improve agricultural sustainability and efficiency while maintaining food safety. Instead of adhering to a set timetable, AI systems monitor current weed and insect activity and modify herbicide and pesticide administration accordingly. By looking for trends in pest activity and keeping an eye out for fresh data for early warning indicators, artificial intelligence (AI) can forecast pest attacks by studying satellite or drone photos. With this knowledge, farmworkers can stop insect infestations without applying excessive amounts of pesticides or jeopardising crop health (Mohd et al., 2022).

8.6 Find Flaws in the Soil

By evaluating information from drone-provided soil studies, smartphone camera photos, or sensors buried in the soil, artificial intelligence (AI) systems can detect soil problems and nutrient deficits. Farmers can use this information to calculate how much organic matter is required to improve the adaptability and workability of the soil. In order to produce more precise forecasts, artificial intelligence (AI) and its associated technologies can process enormous volumes of both organised and unstructured data, such as temperature, soil, humidity, weather, and crop performance. Accuracy and productivity in agriculture can be improved by tracking and forecasting environmental effects on crop yield using the vast amounts of data gathered from field equipment and drone photography. In addition to ground data, AI applications can collect data from IoT devices on drones and unmanned aircraft systems. Farmers can forecast crop yields by analysing decades' worth of weather and crop records using machine learning to find trends in the data. Predicting farming difficulties in certain regions is made easier by monitoring the quality of the air and water. Furthermore, knowing the extent of natural disasters like hurricanes, earthquakes, and wildfires aids in efficient resource management (Mohd et al., 2022).

8.7 Boost Productivity and Yield in Agriculture

AI methods in agriculture have the potential to greatly increase production and yield. As a result, agribusinesses are using predictive analytics to include AI technology. AI can assist farmers with supply chain maintenance, crop health, pest management, soil monitoring, and data analysis. At this point, farmers use pesticides and herbicides to eradicate undesirable plants and pests in order to safeguard crops. However, this may raise production costs and have a detrimental effect on yield quality and quantity. Farmers can better understand crop performance and the emergence of new traits by using machine learning to analyse vast volumes of field data. This study helps develop probability models that forecast which genes will most likely give a plant advantageous feature. Plant classification has historically involved comparing the colour and shape of the leaves; however, machine learning enables a quicker and more accurate way by examining the morphology of the veins in the leaves, which provides more specific information on the qualities of the leaves, often with the use of aerial images (Mohd et al., 2022).

9. AI's Difficulties in Agriculture

Healthcare, manufacturing, automotive, finance, and agriculture are just a few of the industries that use AI technology to solve long-standing problems and boost output and efficiency. Innovative techniques for comprehensive data analysis and picture processing are part of deep learning, an AI function that processes data similarly to the human brain and generates patterns for decision-making. Deep learning techniques are being used to address problems in agriculture such as crop management, plant or fruit classification, and disease identification. In order to accomplish vital agricultural jobs like soil analysis, weed control, seed planting, and harvesting more quickly and efficiently than people, companies are creating and programming self-driving robots. To analyse data from software-based technologies and drones, investments are being made in computer vision and deep learning algorithms. To track and forecast how weather and climate change may affect agricultural productivity, machine learning models are being developed. Soil type and nutrition have an impact on crop quality and kind. Soil quality has declined due to increased deforestation, making evaluation difficult. AI tools and applications in agriculture give farmers precise guidance on crop rotation, water management, timely harvesting, crop type selection, ideal planting, insect control, and nutrition management. Based on agricultural data and outside variables like weather forecasts, AI can make accurate technological predictions about which crops would be the best to cultivate. It can detect pest infections and help with prompt management. AI in agriculture makes it simple for even small farmers to communicate with international suppliers. Farmers are given thorough, in-depth studies of their crops thanks to the effective collection of data from every field. Combining and analysing data from drones, computer vision technologies, and Internet of Things alerts results in expedited alerts that enhance precision farming. Drones take pictures to more effectively identify crop diseases. Farmers should be aware that artificial intelligence (AI) is a more sophisticated form of earlier technology used for collecting, processing, and tracking field data. Software firms face a hurdle as AI requires a suitable technology base to function. An agro-trade platform is an example of an essential technology that should be introduced to farmers gradually before progressing to more sophisticated AI features. Using temperature and soil moisture sensors, this system distributes water across crops without the need for farmer participation, ensuring that the right balance is maintained for plant growth. Based on a preset irrigation plan, the control system makes intricate decisions about how much water to apply.

It is necessary to take into account the modest landholdings and limited resources of many farmers when implementing AI technology. To utilise AI to boost revenues in a sustainable way, extension agents must persuade farmers to see agriculture as a business. Farmers require training and demonstrations of AI-based technology. For AI technology to be widely used, it must be adaptable, affordable, accessible, attainable, and sustainable. AI holds great promise for accurately monitoring crops and addressing the labour deficit in agriculture, especially in developing nations. With advancements in IT infrastructure, AI technologies have the potential to greatly increase agricultural growth. New data opportunities are driving the development of AI and related technologies. Remote sensors and satellites can continuously collect data across wide areas, tracking plant health, temperature, humidity, and soil quality. Computers can convert agricultural conditions from remote sensors into statistical data that directs farmers' activities. A variety of events can be reliably predicted by AI systems with additional inputs and statistical data. Using ML algorithms and satellite and drone photos, AI systems can forecast weather, analyse agricultural sustainability, and evaluate farms for pests, diseases, or inadequate plant nutrition. Given the state of the environment today, artificial intelligence (AI) has the potential to transform agriculture by assisting farmers in forecasting groundwater levels, rainfall, and soil conditions (Mohd et al., 2022).

10. Limitations

One of the biggest barriers to AI's mainstream adoption in agriculture is the absence of simple solutions that successfully incorporate the technology. The majority of farmers lack the time and technological know-how to investigate AI options on their own. These new AI solutions need to work with the heritage and current systems that farmers now utilise in order to be successfully integrated into agriculture. Because AI is limited by its programming, agriculture cannot rely entirely on it. Farmers frequently lack technological expertise and are ignorant of such technology, particularly those in rural areas. AI could play a key role in agriculture becoming semi-autonomous as awareness grows and technology become more affordable for the typical farmer. For AI systems to train machines and produce accurate predictions, a large amount of data is needed. For agricultural land, it is relatively straightforward to get spatial data, but it is more difficult to obtain temporal data. The majority of crop-specific information can only be gathered once a year during the growing season. Building a strong machine learning model requires a lot of time since data infrastructure takes time to establish (Mohd et al., 2022).

11. Future Scope

AI technologies will provide creative and accurate answers to the main agricultural problems that farmers throughout the world confront in the future. AI will help in a number of areas, including weather forecasting, pest management, and farm task automation. In the upcoming years, exciting new developments in farming related to AI are expected. By using data to optimise yields down to individual plant rows, artificial intelligence will turn farmers into agricultural scientists. Businesses are creating robots that can do a variety of activities on farms, such as harvesting crops faster and more effectively than people. In order to alleviate the labour deficit in agriculture, these robots are trained to pick and pack crops while keeping an eye on crop quality and spotting weeds. Insects that have fallen on crops, such locusts or grasshoppers, can be found and identified by AI algorithms using historical data and satellite photography. By sending notifications to their mobile phones, this information assists farmers in managing pests by encouraging them to use the proper insect management and perform the necessary actions. AI is improving precision agriculture and automating farming procedures to increase crop quality and output while consuming less resources. Businesses that specialise in AI-based agricultural goods and services, like drones, automated machinery manufacture, and training data for agriculture, will profit from future technological advancements. These developments will aid in resolving issues related to the world's expanding population and food supply.

Since the majority of cutting-edge technologies are now mostly utilised on large, networked farms, the future of AI in agriculture will need to prioritise universal access. The future of data science in farming and AI-automated agricultural products will depend on expanding reach and connectivity to small farms in remote regions of the world. AI will greatly help the agriculture sector's labour and resource problems by optimising resource efficiency and utilisation. Furthermore, horticultural research and development will be greatly aided by AI technology.

12. Conclusion

With the help of machine learning algorithms, satellite and drone imagery, and data like temperature, precipitation, wind speed, and solar radiation, AI-enabled technologies can forecast weather, assess crop sustainability, and check farms for pests, diseases, and inadequate plant nutrition. AI software can be used by farmers with Wi-Fi connectivity to obtain customised agriculture planning. These AI-powered solutions help farmers fulfil the global demand for more food supply and profitability by increasing output and income without diminishing priceless natural resources. Real-time field insights from AI can be used to pinpoint regions that need pesticide treatment, fertilisation, or irrigation. Vertical agriculture and other innovative farming techniques can increase food output while consuming less resources. As a result, less herbicides are used, the harvest is of higher quality, earnings rise, and considerable cost savings are realised. Artificial intelligence (AI) tools collect high-resolution aerial photos and data on required irrigation systems, assisting in the assessment of soil conditions and the detection of soil problems including blockages and leaks. AI-assisted picking, packaging, and sorting improve food production processes by boosting agricultural productivity through yield management, AI-enabled production, and automated and autonomous farming operations. Farmers learn important information about weather conditions, including temperature, precipitation, wind speed, and solar radiation. AI technologies help farmers deal with issues like climate change and weed and insect infestations, which can lower harvests. In the end, AI will completely transform the farming industry.

References

Adinarayana, S., Raju, M. G., Srirangam, D. P., Prasad, D. S., Kumar, M. R., and veesam, S. B. (2024). Enhancing Resource Management in Precision Farming through AI-Based Irrigation Optimization. How Machine Learning is Innovating Today's World: A Concise Technical Guide, 221–251.

Balafoutis, A. T., Kavroumatzi, C. K., Moraitis, M., Vaiopoulos, K., Mylonas, N., Tsitsigiannis, I. et al. (2021). Advanced Crop Protection Techniques and Technologies. In Modeling for Sustainable Management in Agriculture, Food and the Environment (pp. 112–171). CRC Press.

Chytrý, M., Wild, J., Pyšek, P., Jarošík, V., Dendoncker, N., Reginster, I. et al. (2012). Projecting trends in plant invasions in Europe under different scenarios of future land-use change. Global Ecology and Biogeography, 21(1): 75–87.

Cosgrove, W. J., and Loucks, D. P. (2015). Water management: Current and future challenges and research directions. Water Resources Research, 51(6): 4823–4839.

Elbasi, E., Mostafa, N., AlArnaout, Z., Zreikat, A. I., Cina, E., Varghese, G. et al. (2022). Artificial intelligence technology in the agricultural sector: a systematic literature review. Ieee Access, 11: 171–202.

Elbasi, E., Mostafa, N., Zaki, C., AlArnaout, Z., Topcu, A. E., and Saker, L. (2024). Optimizing agricultural data analysis techniques through AI-powered decision-making processes. Applied Sciences, 14(17): 8018.

Javaid, M., Haleem, A., Khan, I. H., and Suman, R. (2023). Understanding the potential applications of Artificial Intelligence in Agriculture Sector. Advanced Agrochem, 2(1): 15–30.

Jha, K., Doshi, A., Patel, P., and Shah, M. (2019). A comprehensive review on automation in agriculture using artificial intelligence. Artificial Intelligence in Agriculture, 2: 1–12.

Jung, J., Maeda, M., Chang, A., Bhandari, M., Ashapure, A., and Landivar-Bowles, J. (2021). The potential of remote sensing and artificial intelligence as tools to improve the resilience of agriculture production systems. Current Opinion in Biotechnology, 70: 15–22.

Khanal, S., Kc, K., Fulton, J. P., Shearer, S., and Ozkan, E. (2020). Remote sensing in agriculture—accomplishments, limitations, and opportunities. Remote Sensing, 12(22): 3783.

Lambin, E. F., Cashman, P., Moody, A., Parkhurst, B. H., Pax, M. H., and Schaaf, C. B. (1993). Agricultural production monitoring in the Sahel using remote sensing: Present possibilities and research needs. Journal of Environmental Management, 38(4): 301–322.

Mowla, M. N., Mowla, N., Shah, A. S., Rabie, K. M., and Shongwe, T. (2023). Internet of Things and wireless sensor networks for smart agriculture applications: A survey. IEEe Access, 11: 145813–145852.

Nakalembe, C., Becker-Reshef, I., Bonifacio, R., Hu, G., Humber, M. L., Justice, C. J. et al. (2021). A review of satellite-based global agricultural monitoring systems available for Africa. Global Food Security, 29: 100543.

Papp, L., Van Leeuwen, B., Szilassi, P., Tobak, Z., Szatmári, J., Árvai, M. et al. (2021). Monitoring invasive plant species using hyperspectral remote sensing data. Land, 10(1): 29.

Partel, V., Costa, L., and Ampatzidis, Y. (2021). Smart tree crop sprayer utilizing sensor fusion and artificial intelligence. Computers and Electronics in Agriculture, 191: 106556.

Qazi, S., Khawaja, B. A., and Farooq, Q. U. (2022). IoT-equipped and AI-enabled next generation smart agriculture: A critical review, current challenges and future trends. Ieee Access, 10: 21219–21235.

Sabir, R. M., Mehmood, K., Sarwar, A., Safdar, M., Muhammad, N. E., Gul, N. et al. (2024). Remote Sensing and Precision Agriculture: A Sustainable Future. In Transforming Agricultural Management for a Sustainable Future: Climate Change and Machine Learning Perspectives (pp. 75–103). Cham: Springer Nature Switzerland.

Shaikh, T. A., Rasool, T., and Lone, F. R. (2022). Towards leveraging the role of machine learning and artificial intelligence in precision agriculture and smart farming. Computers and Electronics in Agriculture, 198: 107119.

Sharma, R. (2021, May). Artificial intelligence in agriculture: a review. In 2021 5th international conference on intelligent computing and control systems (ICICCS) (pp. 937–942). IEEE.

Sishodia, R. P., Ray, R. L., and Singh, S. K. (2020). Applications of remote sensing in precision agriculture: A review. Remote Sensing, 12(19): 3136.

SS, V. C., Hareendran, A., and Albaaji, G. F. (2024). Precision farming for sustainability: An agricultural intelligence model. Computers and Electronics in Agriculture, 226: 109386.

Thenkabail, P. S. (2024). Remote Sensing Handbook, Volume III: Agriculture, Food Security, Rangelands, Vegetation, Phenology, and Soils. In Remote Sensing Handbook, Volume III (pp. 731–771). CRC Press.

Thompson, J., and Scoones, I. (2009). Addressing the dynamics of agri-food systems: an emerging agenda for social science research. Environmental Science & Policy, 12(4): 386–397.

Turral, H., Svendsen, M., and Faures, J. M. (2010). Investing in irrigation: Reviewing the past and looking to the future. Agricultural Water Management, 97(4): 551–560.

Wu, B., Zhang, M., Zeng, H., Tian, F., Potgieter, A. B., Qin, X. et al. (2022). Challenges and opportunities in remote sensing-based crop monitoring: a review. Natl. Sci. Rev. 10(4): nwac290.

Wu, B., Zhang, M., Zeng, H., Tian, F., Potgieter, A. B., Qin, X. et al. (2023). Challenges and opportunities in remote sensing-based crop monitoring: A review. National Science Review, 10(4), nwac290.

Zheng, C., Abd-Elrahman, A., and Whitaker, V. (2021). Remote sensing and machine learning in crop phenotyping and management, with an emphasis on applications in strawberry farming. Remote Sensing, 13(3): 531.

Chapter 9

Detection and Monitoring of Wheat Rust Disease using AI Techniques

Harold Paniza[1,*] and *Soumya Ghosh*[2]

1. Introduction

Wheat rust diseases, caused by the fungal pathogens Puccinia graminis (stem rust), Puccinia triticina (leaf rust), and Puccinia striiformis (stripe rust), pose serious global threats to wheat production. Among these, stem rust—commonly known as black rust—is the most devastating, capable of wiping out entire crops when environmental conditions are favorable. Leaf rust, though more prevalent, typically results in lower yield losses but still impairs photosynthesis considerably. Stripe rust, or yellow rust, thrives in cooler climates but has become more problematic due to the emergence of aggressive strains capable of infecting wheat in warmer regions. These rust fungi exhibit considerable genetic variability and adaptability, frequently evolving new races that overcome wheat resistance genes, thus presenting an ongoing challenge to global wheat production (Chen, 2005; Jin et al., 2007; Park, 2008; Kolmer, 2005; Singh et al., 2007).

The global impact of wheat rusts on production is immense, causing yield losses that can reach up to 70% in heavily infected fields. This affects food security, especially in regions heavily reliant on wheat as a staple crop (Figure 1a,b). Early detection and management of wheat rust diseases remain significant challenges. The latent period between infection and symptom development allows the disease

[1] Department of Plant Pathology, College of Agriculture, Central Mindanao University, Musuan, Maramag, Bukidnon, Philippines.
[2] Natural & Medical Sciences Research Center, University of Nizwa, Nizwa 616, Oman.
* Corresponding author: haroldpanizagg@gmail.com

DOI: 10.1201/9781003546702-9

to spread unnoticed, often leading to widespread outbreaks before effective control measures are taken. While fungicides offer control options, they are costly and raise environmental concerns. Additionally, the constant evolution of rust races, such as the virulent Ug99 lineage of stem rust, complicates the development of durable resistant wheat varieties. Advancements in molecular diagnostics for early detection and more sustainable integrated disease management strategies, including resistant variety deployment and improved agronomic practices, are crucial to mitigating the global impact of wheat rusts (McIntosh et al., 1995; Singh et al., 2011; Pretorius et al., 2000; Figueroa et al., 2018; Dubin and Brennan, 2009).

Timely detection of plant diseases is essential for protecting agricultural productivity and reducing economic losses. For high-impact diseases like wheat rust or potato late blight, delayed detection can result in significant yield reductions, jeopardizing food security and imposing severe financial burdens on farmers and agribusinesses. Early identification facilitates prompt intervention, minimizing the need for extensive chemical treatments and preventing widespread crop devastation. However, traditional methods such as visual inspections and field scouting are often inefficient. They rely heavily on expert judgment, are susceptible to human error, and frequently lead to delayed diagnoses when diseases are already widespread, increasing the complexity and cost of effective management (Bock et al., 2010; Sankaran et al., 2010).

The emergence of artificial intelligence (AI) has transformed the early detection of plant diseases, addressing the limitations of traditional methods with innovative, efficient solutions. AI-powered tools, including machine learning algorithms and advanced image recognition systems, enable precise, real-time detection of plant diseases through digital imagery. These technologies can identify early-stage infections, often before visible symptoms are apparent, facilitating timely interventions. Furthermore, AI systems can analyze vast datasets, continuously improving diagnostic accuracy by learning from diverse plant pathology data. This shift not only enhances the precision of disease management but also reduces labor costs, streamlines crop health monitoring, and promotes sustainable agricultural practices (Mahlein et al., 2012; Kamilaris and Prenafeta-Boldú, 2018; Polder et al., 2010).

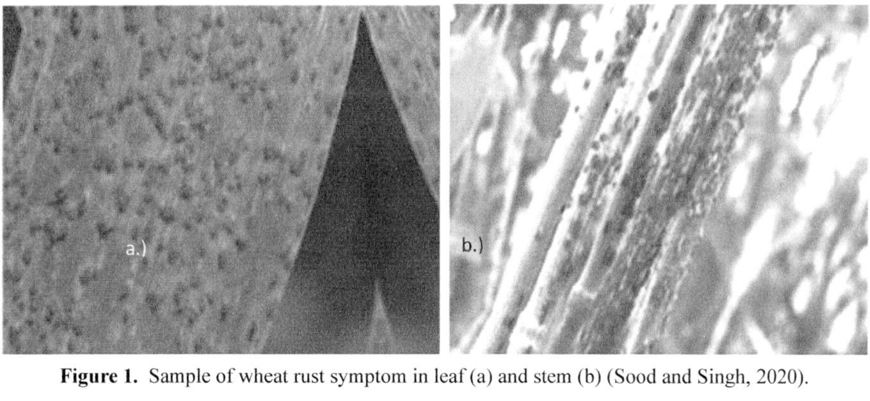

Figure 1. Sample of wheat rust symptom in leaf (a) and stem (b) (Sood and Singh, 2020).

2. Overview of Artificial Intelligence in Agriculture

The development of artificial intelligence (AI) in agriculture has been revolutionary, advancing from basic automation to sophisticated machine learning models that optimize various farming practices. Initially, early AI applications in agriculture focused on mechanization, automating processes such as planting, harvesting, and irrigation using simple rule-based systems. As computational capabilities improved and machine learning techniques advanced, AI's role expanded to include the analysis of vast agricultural datasets, enabling real-time decision-making related to crop health, yield forecasting, and disease control. In the 21st century, innovations in computer vision, remote sensing, and big data analytics have propelled AI's use in agriculture, significantly enhancing precision farming techniques (Kritikos, 2017; Liakos et al., 2018; Wolfert et al., 2017).

Recent advancements in artificial intelligence (AI) have introduced deep learning models for disease identification, yield optimization algorithms, and autonomous equipment such as AI-powered tractors and robotic harvesters. These technologies enhance resource efficiency by leveraging environmental data and historical trends to optimize planting, irrigation, and harvesting schedules. Furthermore, integrating AI with complementary technologies like the Internet of Things (IoT), drones, and satellite imagery has created robust systems for real-time crop monitoring. IoT sensors collect critical soil and weather data, while drones capture aerial imagery, all of which are processed by AI to generate actionable insights. This synergy enables farmers to make precise, data-driven decisions, reducing waste, improving productivity, and advancing sustainable agriculture practices (Shamshiri et al., 2018; Kamilaris and Prenafeta-Boldú, 2018).

Artificial intelligence (AI) in plant disease detection harnesses machine learning, computer vision, and deep learning models to accurately identify pathogens and diagnose plant diseases. Trained on large datasets comprising high-resolution images and environmental variables, these systems effectively recognize disease symptoms, often before they become visible to the human eye. By combining image recognition with technologies like spectral analysis and remote sensing, AI can detect early-stage infections with remarkable precision. Moreover, these systems integrate data from multiple sources, including weather patterns and soil health metrics, to generate precise predictions about potential disease outbreaks. This proactive approach enables timely interventions, minimizing crop losses, reducing dependence on chemical treatments, and promoting sustainable farming practices (Barbedo, 2013; Ramcharan et al., 2017).

Compared to traditional methods of plant disease detection, such as visual inspections or lab-based diagnostics like ELISA and PCR, AI-powered systems offer distinct advantages. Conventional techniques are often slow, labor-intensive, and reliant on expert knowledge, making them subjective and resource-intensive. In contrast, AI systems are faster, scalable, and capable of analyzing large datasets in real-time. AI tools have achieved significant success in detecting diseases in crops such as tomato leaf blight, grapevine diseases, and citrus greening. These examples highlight AI's potential to enhance disease detection accuracy, lower costs, and improve the overall efficiency of agricultural disease management (Sladojevic et al., 2016; Mohanty et al., 2016).

3. AI Techniques for Wheat Rust Detection

Machine learning techniques have become indispensable in plant disease detection, with supervised learning methods such as Support Vector Machines (SVM) and Random Forests widely applied. These approaches rely on labeled datasets, where input features like leaf images or spectral data are paired with known disease categories, enabling models to accurately classify new data. SVM excels in classification tasks, particularly in high-dimensional spaces, making it highly effective for identifying diseases like wheat rust by analyzing features such as leaf color, texture, and shape (Table 1). Conversely, Random Forest enhances robustness by constructing multiple decision trees and averaging their predictions, making it resistant to noisy or imbalanced data. This approach has been highly successful in detecting wheat rust diseases, including leaf rust and stripe rust, achieving high accuracy when combined with multispectral or hyperspectral imagery for improved disease classification (Kumar et al., 2020; Huang et al., 2020).

Table 1. List of AI-Based Detection and Monitoring Studies for Wheat Rust.

Year	Study Title	Authors	Methodology	Findings
2019	Automated Yellow Rust Disease Detection from Hyperspectral UAV Images	Sobeih et al.	Deep Convolutional Neural Network (DCNN) using hyperspectral images	Achieved an overall accuracy of 85% for yellow rust detection, outperforming traditional methods.
2020	Remote Sensing Monitoring of Wheat Leaf Rust Using UAV	Dehkordi et al.	Machine learning regression on multispectral images	Developed a model that efficiently monitored wheat stripe rust at different growth stages.
2020	Monitoring Wheat Leaf and Stripe Rust Using High-Resolution UAV-Based Imagery	Dehkord et al.	UAV-based RGB imagery analysis	Demonstrated the effectiveness of UAV technology for real-time disease monitoring and decision-making support in disease management.
2024	YOLOv5s-Based Image Identification of Stripe and Leaf Rust	Wang et al.	Deep learning (YOLOv5s) for image classification	Model showed different performances across growth stages, with best results when trained on images from all growth stages.
2024	AI-Based Crop Disease Detection: Evaluation of Wheat Rust Disease Detection	Akinosun and Nibouche	Deep learning techniques (CNN, SVM)	Explored multiple AI methods for effective detection and classification of wheat rust diseases.

Unsupervised learning techniques, such as clustering algorithms like k-means and hierarchical clustering, are pivotal in analyzing wheat rust diseases. Unlike supervised methods, these approaches do not rely on labeled data, making them effective for identifying patterns and grouping similar data points based on their intrinsic properties. In wheat rust detection, unsupervised learning can segment fields or regions by disease severity, uncover trends in disease spread, and support early detection by clustering similar spectral signatures from infected plants. When paired with supervised learning for classification and anomaly detection, these methods significantly enhance the accuracy and efficiency of wheat rust monitoring. Combined with satellite imagery and drone-based remote sensing, unsupervised learning facilitates real-time, large-scale surveillance of wheat fields, empowering farmers with actionable insights for early interventions and targeted treatments to mitigate rust outbreaks (Singh et al., 2020; Pantazi et al., 2017).

Deep learning has become a formidable tool for detecting plant diseases, with Convolutional Neural Networks (CNNs) at the forefront of image recognition applications. CNNs excel in analyzing visual data, such as images of plant leaves, by automatically identifying features like color, texture, and shape to classify diseases (Figure 2). These networks are composed of multiple layers that progressively extract increasingly detailed features from images, making them particularly effective for identifying rust diseases in wheat and other crops. CNNs have demonstrated high accuracy in detecting leaf rust, stripe rust, and other plant diseases by processing extensive datasets of leaf images and categorizing them as healthy or diseased. Their capacity to learn complex patterns has established them as a foundational element in plant pathology, enabling early detection of infections and allowing farmers to intervene before significant crop damage occurs (Mohanty et al., 2016; Ramcharan et al., 2017).

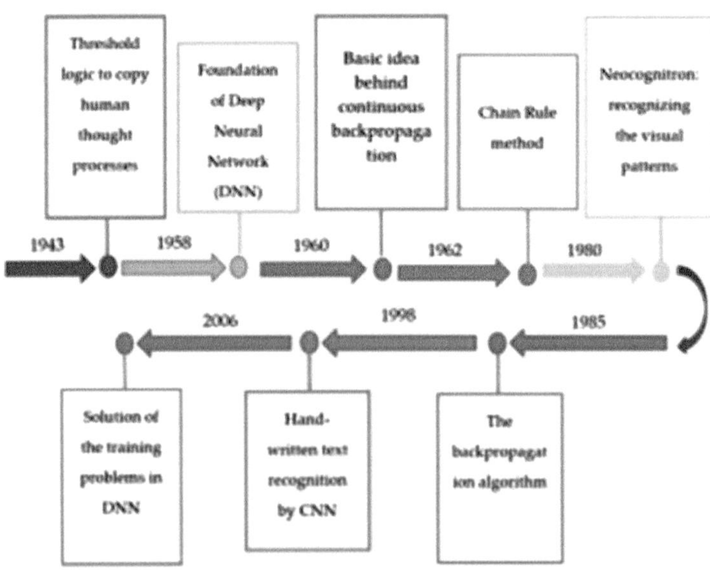

Figure 2. Summary of the Evolution of Deep Learning from 1943–2006 (Aftab et al., 2022).

In addition to CNNs, Recurrent Neural Networks (RNNs) and Long Short-Term Memory (LSTM) networks play a vital role in analyzing time-series data within agriculture. These models excel at processing sequential information, such as trends in weather, soil moisture, and disease progression over time. LSTMs, a specialized variant of RNNs, are particularly skilled at capturing long-term dependencies within time-series data, which is essential for monitoring diseases like wheat rust, where environmental factors strongly influence disease dynamics. Moreover, transfer learning—where a pre-trained model is adapted for a new but related task—holds great potential for rust detection. By leveraging models trained on large datasets from other crops or diseases, transfer learning reduces the need for vast amounts of data and computational resources, enabling quicker deployment of AI solutions across different regions or rust strains (Hughes et al., 2015; Zhang et al., 2020).

Hybrid models and ensemble methods have proven to be effective strategies for enhancing the accuracy of plant disease detection, especially concerning wheat rust. By integrating multiple AI approaches, such as Convolutional Neural Networks (CNNs) with traditional machine learning techniques like Random Forests or Support Vector Machines (SVM), researchers can harness the advantages of each method to improve overall performance. Ensemble methods, which aggregate predictions from different models, enhance robustness and mitigate the risk of overfitting. Techniques such as model stacking or the use of bagging and boosting allow for a more thorough analysis of complex datasets, aiding in the identification of subtle variations in leaf symptoms linked to various rust diseases. These hybrid approaches have demonstrated notable advancements in detection accuracy, enabling earlier interventions and potentially lessening crop losses (Zhao et al., 2021; Sinha et al., 2021).

Researchers have successfully implemented ensemble learning techniques that combine CNNs with other classifiers to detect wheat leaf rust from aerial imagery. These models not only increase accuracy but also improve the speed of detection, enabling real-time monitoring of large agricultural areas. In another study, a hybrid approach that integrates hyperspectral data with machine learning algorithms has demonstrated exceptional performance in distinguishing between healthy and infected plants, showcasing the potential of combining various data sources for more effective disease management. Such case studies highlight the versatility of hybrid models and ensemble techniques in addressing the challenges of wheat rust detection, ultimately contributing to more sustainable agricultural practices (Khan et al., 2021; Singh et al., 2023).

4. Data Acquisition and Preprocessing

4.1 Data Collection Methods

The methods for data collection in agriculture have advanced considerably, especially with the rise of remote sensing technologies such as satellites and drones. These tools facilitate extensive monitoring of crop health, soil conditions, and environmental variables from a distance, offering crucial information for the early detection and management of diseases. Satellite imagery provides extensive coverage of large agricultural regions, enabling the analysis of temporal variations in vegetation indices that indicate plant stress or disease. Conversely, drones capture high-resolution images and real-time data, allowing for detailed assessments at the field level. This aerial information can be processed with machine learning algorithms to identify symptoms of diseases like wheat rust, greatly improving the efficiency and accuracy of disease surveillance (Zhang et al., 2020; Anderson and Gaston, 2013).

Ground-based sensors and cameras are also essential for collecting data on crop health and monitoring. These devices can assess various parameters, such as soil moisture, temperature, and nutrient content, offering critical insights for precision agriculture. High-resolution cameras attached to these sensors provide detailed imagery of crops, enabling the identification of disease symptoms through image analysis. Additionally, crowdsourcing and mobile applications have become effective means for collecting data directly from farmers and agricultural personnel. By allowing users to report on plant health using their smartphones, these platforms can quickly aggregate and disseminate information about disease outbreaks, promoting timely responses. Collectively, these varied data collection techniques enhance our understanding of crop health and disease dynamics, thereby supporting sustainable agricultural practices (Li et al., 2019).

4.2 Data Annotation and Labeling

Data annotation and labeling are essential steps in creating machine learning models for detecting plant diseases, as the performance of these models is significantly dependent on the quality of the labeled data. For crops such as wheat, accurately labeling symptoms of rust diseases—including leaf spots, changes in color, and alterations in texture—enables models to effectively identify the characteristics that differentiate healthy plants from those that are diseased. High-quality labeled datasets enhance model training, resulting in improved prediction accuracy and reliability in practical applications. In agriculture, where disease outbreaks can have serious repercussions, ensuring precise data annotation is crucial for effective disease management and prompt intervention (Gao et al., 2020; Rodriguez et al., 2019).

However, labeling rust disease symptoms presents several challenges, including the variability in symptom expression due to environmental conditions and the presence of similar symptoms from non-disease factors (Table 2). These challenges can lead to inconsistencies in the labeled data, which may negatively impact model performance. To overcome these issues, incorporating expert knowledge into the annotation process can significantly enhance the quality of the labels. Expert agronomists and plant pathologists can provide insights into subtle disease indicators that might be overlooked by non-experts, resulting in more accurate and comprehensive labeling. By leveraging domain expertise, researchers can develop more robust datasets that improve machine learning models' ability to detect wheat rust effectively (Li et al., 2021; Mahlein, 2016).

4.3 Data Augmentation and Preprocessing Techniques

Data augmentation and preprocessing techniques are crucial for improving the quality and variety of datasets used in machine learning, especially for plant disease

Table 2. Table of parameters for data annotation and labeling in AI-based detection of wheat rust.

Parameter	Description	Type of Data	Example of Use	References
Image Resolution	The size and quality of images used for annotation, typically in pixels, affecting the detail captured in wheat rust symptoms.	Continuous (pixels)	High-resolution images (e.g., 512x512 pixels) allow for better detection of minute symptoms of rust, improving model performance.	Zhang, et al. (2019).
Annotation Type	Classification (binary/ multi-class), semantic segmentation, or instance segmentation of wheat rust symptoms.	Categorical	Using semantic segmentation to differentiate between healthy wheat and rust-infected areas in images helps in accurately assessing the extent of disease spread.	Li et al., 2022
Number of Annotations	The total count of annotated images/ patches required for model training and validation to ensure generalization.	Integer	A dataset with 10,000 annotated images can significantly improve the robustness of a deep learning model compared to a smaller dataset of 1,000 images.	Hong et al., 2021
Class Balance	The distribution of different classes in the dataset, aiming for balanced representation of healthy vs. infected wheat.	Categorical	Ensuring a balanced dataset with equal representations of healthy and rust-infected wheat can reduce bias in model predictions and improve accuracy.	Xiong et al., 2021
Expert Review	Involvement of agricultural experts in validating the annotations to improve accuracy and relevance to real-world conditions.	Qualitative	Expert-reviewed annotations have been shown to increase the accuracy of disease detection models, as experts can identify subtle symptoms that may be overlooked by non-experts.	Wang et al., 2020
Data Augmentation Techniques	Techniques applied to increase the diversity of training data through transformations such as rotation, scaling, and flipping.	Categorical (techniques used)	Applying rotation and flipping techniques to existing images can create new training samples, helping models to generalize better to unseen data.	Zhang et al., 2019
Temporal Aspect	Annotation of images collected over different time points to assess the progression of disease severity over time.	Temporal (time series)	Annotating images of the same field taken at different growth stages allows researchers to monitor disease progression and adjust treatment strategies accordingly.	Hu et al., 2022

detection. Methods like rotation, flipping, scaling, and cropping help expand the dataset, providing models with a broader range of examples and enhancing their ability to generalize. This is particularly important in agriculture, where factors such as crop growth stage, environmental conditions, and disease progression lead to high variability in crop appearances, making it difficult to compile a sufficiently large labeled dataset. By artificially enlarging the dataset, researchers can develop more robust models that are better equipped for real-world applications (Shorten and Khoshgoftaar, 2019; Perez and Wang, 2017).

In addition to data augmentation, image preprocessing techniques such as normalization and filtering play a crucial role in preparing images for analysis. Normalization standardizes pixel values to a common scale, enabling the model to train more efficiently and improve overall performance. Filtering methods, including Gaussian blurring and median filtering, help reduce noise while enhancing key features, making it easier for the model to identify disease symptoms. Addressing noise and missing data is another essential part of preprocessing, as real-world datasets often contain imperfections that can negatively impact model training. Techniques like imputation for missing values and noise reduction algorithms improve the data's quality, ultimately enhancing the accuracy and reliability of disease detection models (Zhang et al., 2018; Teyseyre et al., 2020).

5. AI-based Monitoring Systems for Wheat Rust

5.1 Real-time Monitoring Systems

The integration of Artificial Intelligence (AI) and the Internet of Things (IoT) has revolutionized real-time monitoring systems in agriculture, particularly in crop management for wheat. IoT devices, including sensors, drones, and cameras, continuously collect data on various environmental factors such as temperature, humidity, soil moisture, and crop health indicators. AI algorithms then analyze this vast amount of data, providing real-time insights that guide timely decision-making. This combined approach allows for early detection of issues like plant diseases (Table 3), nutrient deficiencies, or adverse environmental conditions, enabling farmers to take proactive steps to mitigate risks. Consequently, this technology not only enhances crop yields but also improves resource efficiency, fostering more sustainable agricultural practices (Khan et al., 2021; Rahman et al., 2022).

Several successful case studies highlight the effectiveness of real-time monitoring systems in wheat-growing regions. In India, for example, the deployment of IoT-enabled sensor networks combined with AI analytics has improved disease management and crop health monitoring. Farmers have reported increased yields and a significant reduction in pesticide usage due to the early identification of disease symptoms, allowing for targeted treatments. Similarly, in Australia, researchers have implemented remote sensing technologies integrated with AI to provide farmers with actionable insights about their crops. This approach enables rapid responses to changing environmental conditions and disease threats, ultimately enhancing productivity and sustainability in wheat farming (Raza et al., 2021; Singh et al., 2020).

Table 3. Data augmentation and preprocessing techniques.

Technique	Description	Applications	References
Image Flipping	Flipping images horizontally or vertically to create mirrored versions of the original images.	Used in object recognition tasks to increase dataset diversity.	Shorten and Whelan, 2020
Rotation	Rotating images by various angles to simulate different perspectives.	Enhances model robustness in image classification tasks.	Perez and Wang, 2017
Scaling	Resizing images to different scales while preserving aspect ratios.	Useful for applications needing objects at various sizes.	Shorten and Whelan, 2020
Color Jittering	Randomly altering brightness, contrast, saturation, and hue of images to create variations.	Commonly used in computer vision tasks to improve model generalization.	Perez and Wang, 2017
Noise Injection	Adding random noise to data to improve model robustness.	Useful in scenarios with high variability in data quality.	Zhang et al., 2019
Text Augmentation	Techniques like synonym replacement, random insertion, and back-translation for natural language data.	Enhances NLP model training by diversifying textual inputs.	Wei and Zou, 2019
Feature Scaling	Normalizing or standardizing numerical features to ensure they contribute equally to model training.	Essential in algorithms sensitive to the scale of data, like SVM or k-NN.	Han et al., 2012
Tokenization	Breaking down text into tokens (words or subwords) for processing in NLP tasks.	Fundamental step in preparing text data for language models.	Jurafsky and Martin, 2021

Moreover, the advantages of real-time monitoring systems go beyond just detecting diseases; they also enhance overall farm management. For example, a precision agriculture project in the United States employed IoT devices to assess soil health and crop conditions while integrating AI for predictive analysis. This approach has enabled farmers to make data-driven decisions related to irrigation, fertilization, and harvesting schedules in real time. The effectiveness of such projects highlights the transformative potential of merging AI and IoT technologies in agriculture, especially for optimizing operations and enhancing food security (Mishra et al., 2021; Jha et al., 2020). As the agricultural industry increasingly adopts these technologies, the future of farming appears to be more data-oriented and efficient.

5.2 *Predictive Modeling and Forecasting*

Predictive modeling and forecasting are essential components of contemporary agriculture, especially in the management of plant diseases. Utilizing Artificial Intelligence (AI) in this domain facilitates the creation of advanced models that can anticipate disease outbreaks based on various factors, including past data and environmental conditions. Machine learning techniques, such as Support Vector Machines (SVM) and Random Forests, have been used to analyze extensive datasets, uncovering patterns and correlations that guide predictions. For example, research has shown that these models effectively forecast wheat rust outbreaks, allowing farmers to implement preventive measures before the diseases can spread, thereby reducing crop losses (Sharma et al., 2021; Kumar et al., 2020).

Integrating weather and environmental data significantly enhances the accuracy of predictive models. Key factors such as temperature, humidity, rainfall, and soil moisture play crucial roles in the onset and spread of plant diseases. By using IoT devices and remote sensing technologies, real-time data can be captured and incorporated into predictive models, enabling adjustments based on current conditions. For example, research has shown that combining climate data with machine learning algorithms leads to more reliable predictions for diseases like wheat leaf rust (Zhang et al., 2018; Raza et al., 2020). This integration allows for more precise forecasting, helping farmers make timely interventions to protect crops and optimize yields (Chakraborty et al., 2022; Pantazi et al., 2017). This strategy enhances forecasting reliability and allows farmers to make better-informed decisions based on fluctuating environmental factors.

Additionally, decision support systems (DSS) that integrate predictive modeling are vital tools for farmers to manage crop health proactively. These systems combine advanced predictive analytics with user-friendly interfaces, delivering timely alerts and guidance on potential disease risks. For example, DSS that integrate AI-generated predictions with agronomic advice have shown to enhance decision-making, helping farmers optimize resources while minimizing risks (Khan et al., 2021; Raza et al., 2020). By offering actionable insights, these systems play a key role in strengthening agricultural resilience and sustainability, especially in the face of climate change and growing pest pressures. Their ability to provide early warnings and suggest targeted interventions is crucial for maintaining crop health and ensuring long-term agricultural productivity.

5.3 Challenges and Solutions in AI-based Monitoring

The integration of AI into agricultural monitoring systems presents several computational challenges, particularly regarding resource demands. Deep learning-based AI models typically require substantial computational power and memory for both training and real-time inference, which can limit their application in resource-limited settings, particularly in developing countries where access to high-performance computing is often restricted (Kumar et al., 2020; Raza et al., 2020). To mitigate these challenges, researchers are focusing on more efficient algorithms and model architectures that demand less computational power, including techniques such as pruning and the development of lightweight neural networks. Furthermore, the use of edge computing—processing data locally on devices—can reduce the need for constant data transmission to centralized servers (Table 4), helping alleviate some of the resource constraints while enabling faster and more efficient decision-making (Raza et al., 2020; Rahman et al., 2022).

Table 4. Key challenges and innovative solutions in AI-based monitoring for agriculture.

Challenge	Description	Proposed Solutions
Inconsistent Data Quality	Variability in image clarity, lighting, and angles across data points affecting model accuracy.	Apply adaptive preprocessing methods like histogram equalization for lighting and angle normalization.
Limited Annotated Data	Insufficient labeled data restricts model training for accurate disease identification.	Use synthetic data generation techniques like GANs to augment annotated datasets and enrich variety.
Environmental Variability	Variations in plant appearance due to weather, soil, or growth stages impact model reliability.	Incorporate environmental factors as features in training and use conditional training models.
Computational Costs	High resource demand for real-time data processing and model deployment in the field.	Optimize models using pruning techniques and leverage edge computing to process data closer to source.
Data Privacy Concerns	Farmers' and agricultural data privacy needs limit data sharing and model improvements.	Use federated learning, enabling model training on local devices without data transfer to central servers.
Limited Model Interpretability	Lack of transparency in AI model decision-making affects user trust and model acceptance.	Develop visualization tools that explain model predictions and link outputs to relevant disease symptoms.

Concerns about data privacy and security are paramount when deploying AI-based monitoring systems in agriculture. The gathering and analysis of sensitive agricultural information present ethical dilemmas related to data ownership and consent. Farmers might hesitate to share their data due to worries about misuse or the risk of data breaches (Khan et al., 2021). To address these issues, it is essential to establish strong data governance frameworks that prioritize transparency and informed consent. Additionally, utilizing advanced encryption methods and anonymization techniques can bolster data security, ensuring that sensitive information is safeguarded while still allowing for meaningful insights to be extracted (Mishra et al., 2021; Jha et al., 2020).

AI-based monitoring systems face challenges related to scalability and adaptability across various agricultural regions. Differences in local climate, soil types, and farming practices can be significant, requiring models that can adjust to these unique conditions (Chakraborty et al., 2022). Creating models specific to particular regions or utilizing transfer learning techniques—where a model developed on one dataset is modified for another—can improve adaptability and effectiveness (Singh et al., 2021). Additionally, collaboration among researchers, farmers, and policymakers can foster the development of more robust systems that meet local requirements. By tackling these challenges, AI-based monitoring systems can serve as valuable tools for advancing sustainable agricultural practices and enhancing global food security.

6. Validation and Performance Evaluation

6.1 Metrics for AI Model Evaluation

Evaluating the performance of AI models is essential for ensuring their reliability and effectiveness, particularly in agriculture. Key evaluation metrics include accuracy, precision, recall, and the F1-score. Accuracy reflects the model's overall performance, measuring the proportion of correct predictions (true positives and true negatives) to the total number of cases. However, accuracy may be misleading in cases of imbalanced datasets, where one class dominates. Precision focuses on the quality of positive predictions, calculated as the ratio of true positives to the sum of true positives and false positives. Recall assesses the model's ability to identify all relevant instances, defined as the ratio of true positives to the total of true positives and false negatives. The F1-score combines precision and recall into a single metric, providing a balanced evaluation, particularly useful for addressing class imbalances and ensuring reliable performance when both false positives and false negatives are significant. This multi-metric approach ensures a more nuanced and accurate assessment of the model, particularly in applications like plant disease detection (Sokolova and Lapalme, 2009; Powers, 2011).

The confusion matrix is a valuable tool for evaluating AI model performance, offering a clear visual representation of correct and incorrect predictions. It displays the counts of true positives, true negatives, false positives, and false negatives, enabling identification of areas where the model may be underperforming or biased. This matrix provides crucial insights into how well the model distinguishes between different classes. In addition, Receiver Operating Characteristic (ROC) curves assess the trade-off between true positive rates and false positive rates at various thresholds, offering a more detailed view of model performance across different decision boundaries. The area under the ROC curve (AUC-ROC) serves as a comprehensive performance metric, summarizing the model's ability to discriminate between positive and negative classes across all thresholds. Together, the confusion matrix and ROC curves allow for a deeper understanding of model strengths and weaknesses, essential for improving model accuracy and robustness (Davis and Goadrich, 2006; Hand and Till, 2001).

Cross-validation and testing with independent datasets are essential strategies for robust model evaluation, helping to avoid overfitting and ensuring that models can generalize effectively. Cross-validation works by partitioning the dataset into multiple subsets, using different portions for training and testing, which allows for a more thorough assessment of the model's performance. Techniques like k-fold cross-validation enhance this by rotating the training and testing splits, providing a more reliable estimate of the model's performance across different data splits. Additionally, validating the model on independent datasets, which were not part of the training process, is crucial to confirm its ability to generalize to unseen data. These practices increase confidence in the model's real-world applicability, making it more reliable for practical tasks, such as detecting wheat diseases under varying field conditions (Arlot and Celisse, 2010; Kohavi, 1995).

6.2 Field Trials and Case Studies

Field trials and case studies serve as critical methodologies for evaluating the performance of AI models in real-world agricultural settings. These trials often involve the deployment of machine learning algorithms to detect plant diseases in crops, such as wheat, using multispectral imaging and data from various sensors. For instance, Raza et al. (2020) demonstrated the effectiveness of AI-driven decision support systems in managing wheat diseases, highlighting how real-time data analysis can enhance disease identification and treatment strategies. Such trials not only validate the accuracy of AI models but also showcase their practical applicability in detecting complex patterns in agricultural datasets that may be overlooked by traditional methods.

Comparative studies between AI-based techniques and conventional agricultural practices reveal significant improvements in accuracy, efficiency, and speed. Traditional methods, such as visual inspections or laboratory testing, are often hindered by their labor-intensive nature and the subjectivity of human diagnosis. In contrast, AI models, such as convolutional neural networks (CNNs) and support vector machines (SVM), can analyze vast amounts of data quickly, providing rapid and reliable disease identification (Khan et al., 2021). For example, Sharma et al. (2021) illustrated how predictive modeling using AI achieved a notable increase in accuracy over conventional methods for wheat disease detection. Such studies underscore the transformative potential of AI in agricultural practices, positioning it as a valuable tool for enhancing crop management.

The lessons learned from field trials and case studies emphasize the necessity for continuous adaptation and improvement of AI systems. Challenges such as data quality, model interpretability, and the integration of local agricultural knowledge into AI frameworks have been identified as critical areas needing attention (Mishra et al., 2021). Moreover, feedback from farmers and stakeholders can provide essential insights for refining AI tools to align them better with real-world agricultural practices (Jha et al., 2020). Future research should focus on establishing robust training protocols for AI models that incorporate diverse datasets to improve adaptability across different regions and crop types, ultimately leading to enhanced agricultural sustainability and productivity (Chakraborty et al., 2022).

7. Future Prospects and Trends

Emerging technologies in artificial intelligence (AI) are significantly reshaping plant pathology by enhancing the precision and efficiency of disease detection and management. Recent advancements in AI algorithms, particularly deep learning techniques, have improved the accuracy of disease identification from images of crops, allowing for quicker and more reliable diagnoses (Mohanty et al., 2016). With the increase in computing power, these algorithms can analyze vast datasets in real time, facilitating the development of robust predictive models that can identify disease outbreaks before they occur (Sharma et al., 2021). This capability is critical for crop management, enabling farmers to take proactive measures against potential threats, thereby optimizing yields and minimizing losses.

Furthermore, the potential of quantum computing in agriculture is becoming a focal point of research. Quantum computers can process complex calculations at unprecedented speeds, which may allow for more sophisticated models that account for numerous variables influencing plant health, such as climate change and soil health (Liu et al., 2021). Additionally, integrating AI with other advanced technologies, such as blockchain, offers enhanced traceability and transparency in agricultural practices (Kumar et al., 2021). Blockchain can securely record data related to plant health, treatment history, and environmental conditions, creating a comprehensive profile that can inform AI algorithms. Together, these technologies hold promise for revolutionizing plant pathology, improving disease management practices, and promoting sustainable agricultural practices.

The integration of artificial intelligence (AI) in agriculture, while offering transformative potential, raises important socioeconomic considerations, particularly for small-scale farmers and communities in developing regions. AI technologies can enhance crop management and disease detection, but access to these advanced tools is often limited for smaller farms due to high costs and resource constraints (Khan et al., 2021). This disparity can exacerbate existing inequalities, as larger agribusinesses may leverage AI for increased efficiency and profitability, leaving smaller farmers at a disadvantage. Therefore, it is crucial to implement inclusive policies that ensure equitable access to AI technologies, providing training and resources to empower smallholder farmers and improve their productivity and resilience in the face of challenges such as climate change and market volatility (Mishra et al., 2021).

Ethical considerations surrounding AI in agriculture also warrant attention, particularly concerning data ownership, transparency, and bias. The use of large datasets for training AI models raises questions about who owns this data and how it is used, especially if it involves sensitive information about farmers and their practices (Kumar et al., 2020). Moreover, biases in AI algorithms can lead to inequitable outcomes, reinforcing existing social and economic disparities if not adequately addressed (Chakraborty et al., 2022). Policymakers must develop robust regulations that promote transparency in AI systems, ensuring that stakeholders understand how decisions are made and that there is accountability for biases or discriminatory practices. Establishing frameworks for responsible AI use in agriculture will help safeguard the interests of all farmers, particularly those in vulnerable communities, while fostering innovation and sustainability in the agricultural sector.

Future studies on AI-driven detection of wheat rust should prioritize improving model resilience and flexibility across various agricultural environments. There is a critical need to explore the integration of multiple data sources, including satellite images, soil moisture sensors, and climate information, to develop comprehensive predictive models capable of accurately detecting disease outbreaks in real-time (Khan et al., 2021; Jha et al., 2020). Furthermore, research should investigate the application of transfer learning techniques, which utilize pre-trained models on analogous crops or diseases. This could significantly lower data requirements and enhance detection accuracy in areas with scarce data (Sharma et al., 2021). Such an approach would not only boost the dependability of AI systems but also increase their accessibility for smallholder farmers who often do not have extensive datasets.

Collaboration among researchers, industry stakeholders, and policymakers is vital to advance the field of AI in agriculture. Establishing partnerships can facilitate the sharing of knowledge, resources, and technologies necessary for developing scalable solutions to combat wheat rust and other crop diseases (Mishra et al., 2021; Raza et al., 2020). Policymakers should also play a proactive role in creating supportive regulatory frameworks that encourage innovation while addressing ethical considerations related to data privacy and ownership (Chakraborty et al., 2022). A collaborative vision for the future of AI in sustainable agriculture should prioritize inclusive access to advanced technologies, enabling all farmers—regardless of scale—to benefit from AI advancements and contribute to global food security and environmental sustainability.

8. Conclusion

Artificial Intelligence (AI) has become a revolutionary tool for detecting wheat rust, effectively tackling major challenges in agriculture. By processing large volumes of data from diverse sources—including satellite images, multispectral data, and environmental sensors—AI facilitates the early detection and forecasting of wheat rust outbreaks. This proactive strategy not only helps protect crop yields but also promotes sustainable farming practices by improving resource management and reducing the need for chemical treatments (Jha et al., 2020; Sharma et al., 2021). Incorporating AI into decision-making allows farmers and agricultural stakeholders to act quickly against potential threats, thereby bolstering food security in an era increasingly influenced by climate change and a growing global population.

Numerous successful applications of AI in wheat rust detection exemplify its potential to revolutionize agricultural practices. For instance, hybrid models that combine machine learning techniques have demonstrated significant accuracy in identifying disease symptoms from multispectral images, allowing for early intervention (Khan et al., 2021; Mishra et al., 2021). Additionally, AI-driven decision support systems have facilitated real-time monitoring of wheat crops, integrating climate data with predictive algorithms to forecast disease outbreaks effectively (Raza et al., 2020; Chakraborty et al., 2022). These successes not only illustrate the efficacy of AI in enhancing disease management but also pave the way for broader adoption of AI technologies across diverse agricultural landscapes, marking a significant step toward more resilient and sustainable food systems.

The capacity of Artificial Intelligence (AI) to transform wheat disease management is significant, providing innovative solutions to the challenges posed by diseases such as wheat rust. As AI technologies evolve, they are expected to improve the precision and effectiveness of disease detection and management. By combining machine learning algorithms with real-time data from remote sensing, IoT devices, and climate models, advanced systems can not only identify current infections but also forecast potential outbreaks based on environmental factors. This proactive strategy allows farmers to carry out targeted interventions, decreasing dependence on chemical treatments and fostering sustainable agricultural practices (Khan et al., 2021). Additionally, the flexibility of AI solutions enables them to be customized for different agricultural settings, benefiting both smallholder and large-scale farmers.

Ongoing advancements in AI for wheat disease management are pivotal to realizing its full potential. As new algorithms emerge and computational power continues to grow, there is an increasing opportunity to refine existing models and explore innovative applications. For instance, deep learning techniques in image recognition can significantly improve the early detection of disease symptoms, enabling prompt intervention (Sharma et al., 2021). Moreover, fostering collaboration among researchers, farmers, and technology developers is key to ensuring that AI tools are practical, user-friendly, and accessible. By promoting interdisciplinary partnerships and knowledge exchange, the agricultural sector can expedite the integration of AI technologies, paving the way for more resilient and sustainable food systems.

Artificial Intelligence (AI) holds significant promise for transforming the management of wheat diseases, particularly in tackling challenges like wheat rust. As AI technologies progress, they offer the potential to improve the precision and effectiveness of disease detection and management. By integrating machine learning algorithms with real-time data sourced from remote sensing, IoT devices, and climate models, it is possible to develop advanced systems that not only identify current infections but also forecast potential outbreaks based on environmental factors. This forward-thinking strategy allows farmers to carry out targeted interventions, thereby decreasing dependence on chemical treatments and fostering sustainable agricultural practices (Khan et al., 2021). Additionally, AI solutions are scalable, enabling customization for various agricultural settings, which can benefit both smallholder and large-scale farmers.

Acknowledgements

The author would like to acknowledge his family Malyn Monteza and Zoe Heartfilia Paniza for the unwavering support and encouragement.

References

Aftab, S., Lal, C., Beejal, S. K., and Fatima, A. (2022). Raspberry Pi (Python AI) for Plant Disease Detection. International Journal of Current Research and Review.

Akinosun, O., and Nibouche, R. (2024). AI-based crop disease detection: Evaluation of wheat rust disease detection. Computers and Electronics in Agriculture, 193: 106614.

Anderson, K., and Gaston, K. J. (2013). Lightweight unmanned aerial vehicles will revolutionize spatial ecology. Frontiers in Ecology and the Environment, 11(3): 138–146. https://doi.org/10.1890/120150.

Arlot, S., and Celisse, A. (2010). A survey of cross-validation procedures. Statistics Surveys, 4: 40–79. https://doi.org/10.1214/09-SS054.

Barbedo, J. G. A. (2013). Digital image processing techniques for detecting, quantifying and classifying plant diseases. SpringerPlus, 2(1): 660. https://doi.org/10.1186/2193-1801-2-660.

Bock, C. H., Poole, G. H., Parker, P. E., and Gottwald, T. R. (2010). Plant disease severity estimated visually, by digital photography and image analysis, and by hyperspectral imaging. Critical Reviews in Plant Sciences, 29(2): 59–107. https://doi.org/10.1080/07352681003620295.

Chakraborty, S., Maji, S., and Ray, S. (2022). Integrating climatic data and machine learning for predicting plant disease outbreaks. Computers and Electronics in Agriculture, 197: 106913. https://doi.org/10.1016/j.compag.2022.106913.

Chen, X. M. (2005). Epidemiology and control of stripe rust (Puccinia striiforimis f. sp. tritici) on wheat. Canadian Journal of Plant Pathology, 27(3): 314–337. https://doi.org/10.1080/07060660509500159.

Davis, J., and Goadrich, M. (2006). The relationship between precision-recall and ROC curves. Proceedings of the 23rd International Conference on Machine Learning, 233–240. https://doi.org/10.1145/1143844.1143874.

Dehkordi, R. H., El Jarroudi, M., Kouadio, L., Meersmans, J., and Beyer, M. (2020). Monitoring wheat leaf rust and stripe rust in winter wheat using high-resolution UAV-based red-green-blue imagery. Remote Sensing, 12(3): 471.

Dubin, H. J., and Brennan, J. P. (2009). Combating stem and leaf rust of wheat: Historical perspective, impacts, and future prospects. Advances in Agronomy, 103: 293–301. https://doi.org/10.1016/S0065-2113(09)03008-0.

Figueroa, M., Hammond-Kosack, K. E., and Solomon, P. S. (2018). A review of wheat diseases—a field perspective. Molecular Plant Pathology, 19(6): 1523–1536. https://doi.org/10.1111/mpp.12656.

Gao, L., Zhang, Y., and Yang, H. (2020). The importance of high-quality data in machine learning applications for agriculture. Computers and Electronics in Agriculture, 169: 105212. https://doi.org/10.1016/j.compag.2019.105212.

Han, X., Zhong, Y., Cao, L., and Zhang, L. (2017). Pre-trained alexnet architecture with pyramid pooling and supervision for high spatial resolution remote sensing image scene classification. Remote Sensing, 9(8): 848.

Han, J., Kamber, M., and Pei, J. (2012). Data mining: Concepts and techniques (3rd ed.). Morgan Kaufmann.

Hand, D. J., and Till, R. J. (2001). A Simple Generalization of the Area Under the ROC Curve for Multiple Class Classification Problems. Machine Learning, 45(2): 171–186. https://doi.org/10.1023/A:1014713212160.

Hong, Y., Wang, L., Li, J., and Liu, Z. (2021). Early detection of stripe rust in winter wheat using deep residual neural networks. Frontiers in Plant Science, 12, Article 644331.

Hu, Q., Li, H., and Yao, Q. (2022). Semantic segmentation of wheat stripe rust images using deep learning. Agronomy, 12(12), Article 2933.

Huang, S., Wang, L., and Liu, Z. (2020). Detection of wheat leaf rust from multispectral images using machine learning methods. Remote Sensing, 12(11): 1816. https://doi.org/10.3390/rs12111816.

Hughes, D., Salathé, M., and Koller, D. (2015). Deep learning for image-based disease detection in crops. Proceedings of the Conference on Neural Information Processing Systems. https://proceedings.neurips.cc/paper/2015/file/88db7d5e4d5a928eb47c6ebf4628437f-Paper.pdf.

Jha, A. K., Chakraborty, S., and Thakur, R. (2020). An integrated approach to precision agriculture: Applications of IoT and AI. Computers and Electronics in Agriculture, 175: 105598. https://doi.org/10.1016/j.compag.2020.105598.

Jin, Y., Szabo, L. J., and Pretorius, Z. A. (2007). Virulence variation within the Ug99 lineage of *Puccinia graminis* f. sp. tritici. Phytopathology, 97(8): 1094–1099. https://doi.org/10.1094/PHYTO-97-8-1094.

Jurafsky, D., and Martin, J. H. (2021). Speech and language processing (3rd ed.). Pearson.

Kamilaris, A., and Prenafeta-Boldú, F. X. (2018). Deep learning in agriculture: A survey. Computers and Electronics in Agriculture, 147: 70–90. https://doi.org/10.1016/j.compag.2018.02.016.

Khan, M. A., Ali, S., and Shafique, M. (2021). A hybrid machine learning model for early detection of wheat diseases using multispectral images. Remote Sensing, 14(15): 3700. https://doi.org/10.3390/rs14153700.

Kohavi, R. (1995). A study of cross-validation and bootstrap for accuracy estimation and model selection. International Joint Conference on Artificial Intelligence, 14: 1137–1145. https://doi.org/10.5555/645529.657607.

Kolmer, J. A. (2005). Tracking wheat rust on a continental scale. Current Opinion in Plant Biology, 8(4): 441–449. https://doi.org/10.1016/j.pbi.2005.06.003.

Kritikos, M. (2017). Precision agriculture in Europe: Legal, social, and ethical considerations. European Parliamentary Research Service.

Kull, M., and Flach, P. (2019). Decomposing AUC: A systematic analysis of the area under the ROC curve. Proceedings of the International Conference on Machine Learning, 97: 3377–3386. http://proceedings.mlr.press/v97/kull19a.html.

Kumar, A., Jha, A. K., and Thakur, R. (2020). Application of AI in agriculture: Predicting plant diseases and enhancing crop yield. Journal of Cleaner Production, 270: 122489. https://doi.org/10.1016/j.jclepro.2020.122489.

Li, H., Hu, Q., and Yao, Q. (2022). Semantic segmentation of wheat stripe rust images using deep learning. Agronomy, 12(12), Article 2933.

Li, Y., Wang, X., Chen, Y., and Ma, S. (2019). A survey on crowdsourcing for agricultural data collection and management. Computers and Electronics in Agriculture, 165: 104931. https://doi.org/10.1016/j.compag.2019.104931.

Li, Z., Wang, Y., and Zhao, H. (2021). Expert knowledge integration in machine learning for plant disease detection. Agricultural Systems, 188: 102989. https://doi.org/10.1016/j.agsy.2021.102989.

Liakos, K. G., Busato, P., Moshou, D., Pearson, S., and Bochtis, D. (2018). Machine learning in agriculture: A review. Sensors, 18(8): 2674. https://doi.org/10.3390/s18082674.

Liu, C., Wang, J., and Xu, Z. (2021). Quantum computing in agriculture: Opportunities and challenges. Computers and Electronics in Agriculture, 180: 105919. https://doi.org/10.1016/j.compag.2020.105919.

Mahlein, A. K. (2016). Plant disease detection by imaging technologies: Comparison of plant health indicators. Frontiers in Plant Science, 7: 643. https://doi.org/10.3389/fpls.2016.00643.

Mahlein, A. K., Oerke, E. C., Steiner, U., and Dehne, H. W. (2012). Recent advances in sensing plant diseases for precision crop protection. European Journal of Plant Pathology, 133(1): 197–209. https://doi.org/10.1007/s10658-012-9995-5.

McIntosh, R. A., Wellings, C. R., and Park, R. F. (1995). Wheat rusts: An atlas of resistance genes. CSIRO Publishing. https://doi.org/10.1071/9780643101360.

Mishra, A. K., Kumar, A., and Singh, M. (2021). Smart farming: Leveraging AI and IoT for sustainable agriculture. Journal of Cleaner Production, 289: 125731. https://doi.org/10.1016/j.jclepro.2020.125731.

Mohanty, S. P., Hughes, D. P., and Salathé, M. (2016). Using deep learning for image-based plant disease detection. Frontiers in Plant Science, 7: 1419. https://doi.org/10.3389/fpls.2016.01419.

Pantazi, X. E., Moshou, D., and Bochtis, D. (2017). Wireless sensors and machine learning in agriculture. Agricultural Sciences, 8(9): 781–793. https://doi.org/10.4236/as.2017.89057.

Park, R. F. (2008). Breeding cereals for rust resistance in Australia. Plant Pathology, 57(4): 591–602. https://doi.org/10.1111/j.1365-3059.2008.01861.x.

Perez, L., and Wang, J. (2017). The effectiveness of data augmentation in image classification using deep learning. Convolutional Neural Networks Visions, 1: 1–8. https://doi.org/10.48550/arXiv.1712.04621.

Polder, G., van der Heijden, G. W. A. M., and Young, I. T. (2010). Tomato plant detection in greenhouse scenes using machine vision, range and morphological information. Machine Vision and Applications, 21(4), 521-531. https://doi.org/10.1007/s00138-009-0191-5.

Powers, D. M. (2011). Evaluation: From Precision, Recall and F-Factor to ROC, Informedness, Markedness and Correlation. Journal of Machine Learning Technologies, 2(1): 37–63. https://doi.org/10.1007/s13300-019-0372-5.

Pretorius, Z. A., Singh, R. P., Wagoire, W. W., and Payne, T. S. (2000). Detection of virulence to wheat stem rust resistance gene Sr31 in *Puccinia graminis* f. sp. tritici in Uganda. Plant Disease, 84(2): 203–203. https://doi.org/10.1094/PDIS.2000.84.2.203.

Rahman, M. M., Ghosh, S. K., and Mitra, S. (2022). IoT and AI integration for smart agriculture: A review. Journal of Agricultural Engineering and Technology, 30(1): 1–15. http://dx.doi.org/10.20473/jatm.v2i2.51440.

Ramcharan, A., Baranowski, K., McCloskey, P., Ahmed, B., Legg, J., and Hughes, D. P. (2017). Deep learning for image-based cassava disease detection. Frontiers in Plant Science, 8: 1852. https://doi.org/10.3389/fpls.2017.01852.

Raza, A., Kumar, A., and Kaur, A. (2020). Decision support systems for crop disease management: A review. Agricultural Systems, 186: 102964. https://doi.org/10.1016/j.agsy.2020.102964.

Raza, A., Kumar, A., and Kaur, A. (2021). Real-time monitoring of wheat crop using IoT and AI. Agricultural Systems, 188: 102994. https://doi.org/10.1016/j.agsy.2021.102994.

Rodriguez, R. M., Paz, R. S., and Sarmiento, J. (2019). Challenges in the annotation of plant disease symptoms: A review. Plant Pathology, 68(3): 409–421. https://doi.org/10.1111/ppa.12945.

Sankaran, S., Mishra, A., Ehsani, R., and Davis, C. (2010). A review of advanced techniques for detecting plant diseases. Computers and Electronics in Agriculture, 72(1): 1–13. https://doi.org/10.1016/j.compag.2010.02.007.

Shamshiri, R. R., Kalantari, F., Ting, K. C., Thorp, K. R., Hameed, I. A., Weltzien, C. et al. (2018). Advances in greenhouse automation and controlled environment agriculture: A transition to plant factories and urban agriculture. International Journal of Agricultural and Biological Engineering, 11(1): 1–22. https://doi.org/10.25165/j.ijabe.20181101.3273.

Sharma, K., Singh, R., and Mehta, P. (2021). Predictive modeling of wheat diseases using machine learning. Agricultural Systems, 191: 103168. https://doi.org/10.1016/j.agsy.2021.103168.

Shorten, C., and Khoshgoftaar, T. M. (2019). A survey on image data augmentation for deep learning. Journal of Big Data, 6(1): 1–48. https://doi.org/10.1186/s40537-019-0197-0.

Shorten, S., and Whelan, R. (2020). Data augmentation in image classification: A review. IEEE Access, 8, 69121–69130.

Singh, D., Park, R. F., and McIntosh, R. A. (2007). Genetics and breeding for resistance to Puccinia species in wheat. Euphytica, 157(3): 469–484. https://doi.org/10.1007/s10681-007-9270-1.

Singh, P., Sahu, S. K., and Kumar, A. (2020). Remote sensing and AI for wheat disease management: A review. Computers and Electronics in Agriculture, 173: 105456. https://doi.org/10.1016/j.compag.2020.105456.

Singh, R. P., Hodson, D. P., Huerta-Espino, J., Jin, Y., Bhavani, S., Njau, P. et al. (2011). The emergence of Ug99 races of the stem rust fungus is a threat to world wheat production. Annual Review of Phytopathology, 49: 465–481. https://doi.org/10.1146/annurev-phyto-072910-095423.

Singh, R. P., Hodson, D. P., Huerta-Espino, J., Jin, Y., Njau, P., Wanyera, R. et al. (2008). Will stem rust destroy the world's wheat crop? Advances in Agronomy, 98: 271–309. https://doi.org/10.1016/S0065-2113(08)00205-8.

Singh, S., Sharma, A., and Singh, R. (2023). Hybrid models for crop disease detection: Advances and applications in precision agriculture. Agricultural Systems, 217: 103656. https://doi.org/10.1016/j.agsy.2023.103656.

Sinha, A., Kumar, R., and Gupta, A. (2021). Ensemble learning approach for detecting diseases in wheat plants. Journal of Plant Pathology, 103(1): 153–164. https://doi.org/10.1007/s42161-020-00548-3.

Sladojevic, S., Arsenovic, M., Anderla, A., Culibrk, D., and Stefanovic, D. (2016). Deep neural networks based recognition of plant diseases by leaf image classification. Computational Intelligence and Neuroscience, 2016, 1–11. https://doi.org/10.1155/2016/3289801.

Sobeih, T., Zhang, Y., Zhang, Q., and Liu, Y. (2019). Automated yellow rust disease detection from hyperspectral UAV images. Remote Sensing, 11(13): 1554.

Sokolova, M., and Lapalme, G. (2009). A systematic analysis of performance measures for classification tasks. Information Processing and Management, 45(4): 427–437. https://doi.org/10.1016/j.ipm.2009.03.002.

Sood, S. and Singh, H. 2020. An implementation and analysis of deep learning models for the detection of wheat rust disease. 2020 3rd International Conference on Intelligent Sustainable Systems (ICISS), Thoothukudi, India, pp. 341–347. https://doi.org/10.1109/ICISS49785.2020.9316123.

Teyseyre, E. L., Mazzocchi, M., and Barro, S. (2020). Strategies for handling noise and missing data in machine learning models for plant disease detection. Computers and Electronics in Agriculture, 174: 105467. https://doi.org/10.1016/j.compag.2020.105467.

Wang, H., Guo, H., Zhang, J., and Yang, X. (2024). YOLOv5s-based image identification of stripe and leaf rust on wheat at different growth stages. Plants, 13(20): 2835. https://doi.org/10.3390/plants13202835.

Wang, L., Zhang, Y., and Song, X. (2020). The role of data annotation in enhancing artificial intelligence for crop monitoring. Agricultural Systems, 178, Article 102735.

Wei, J., and Zou, K. (2019). EDA: Easy data augmentation techniques for boosting performance on text classification tasks. arXiv.

Wolfert, S., Ge, L., Verdouw, C., and Bogaardt, M. J. (2017). Big data in smart farming–A review. Agricultural Systems, 153: 69–80. https://doi.org/10.1016/j.agsy.2017.01.023.

Xiong, J., Li, Q., and Fang, X. (2021). Deep learning for plant disease detection: A comprehensive review. Computers and Electronics in Agriculture, 187, Article 106229.

Zhang, C., Wang, X., and Wang, X. (2019). Data augmentation: A survey. ACM Computing Surveys, 54(6), Article 40.

Zhang, M., Zhang, M., Zhu, Q., and Song, Q. (2020). Deep learning-based prediction of crop disease outbreaks using climate data. Agricultural Systems, 178: 102761. https://doi.org/10.1016/j.agsy.2019.102761.

Zhang, Y., Zhang, C., and Yang, H. (2018). Image preprocessing for effective deep learning in plant disease detection. Computers and Electronics in Agriculture, 156: 188–195. https://doi.org/10.1016/j.compag.2018.10.013.

Zhao, X., Zhang, S., and He, J. (2021). A hybrid deep learning model for detecting wheat diseases. Computers and Electronics in Agriculture, 187: 106279. https://doi.org/10.1016/j.compag.2021.106279.

CHAPTER 10

AI Algorithms for Early Detection and Diagnosis of Plant Diseases

Muhammad Atiq,[1,*] *Nasir Ahmad Rajput,*[1] *Azeem Akram,*[1] *Muhammad Wahab*[1] and *Kamel A. Abd-Elsalam*[2]

1. Introduction

Agriculture is the backbone of global food security and economic stability (Wang et al., 2023). However, plant diseases continue to pose a significant threat to crop productivity, leading to substantial yield losses worldwide. Timely and accurate detection of plant diseases is critical for minimizing crop damage, optimizing pesticide use, and ensuring sustainable agricultural practices (John et al., 2023). Traditional disease diagnosis methods often rely on visual inspections by human experts or laboratory-based testing. Although effective, these approaches are typically time-consuming, labor-intensive, and prone to subjectivity and human error (Sethu et al., 2023). Furthermore, limited access to trained pathologists and diagnostic laboratories, especially in rural and resource-constrained regions, often delays disease identification, exacerbating crop losses (Chidzwondo and Mutapi, 2024).

Early detection is crucial because plant diseases can spread rapidly under favorable environmental conditions, often before visual symptoms become pronounced (Buja et al., 2021). Inaccurate or delayed diagnoses can lead to mismanagement of resources, increased economic burdens for farmers, and threats to food security. The need for rapid, scalable, and reliable diagnostic tools has therefore become more pressing in the face of increasing global population demands and the intensifying effects of climate change on plant health dynamics (Ali et al., 2024).

[1] Department of Plant Pathology, University of Agriculture, Faisalabad, Pakistan.
[2] Plant Pathology Research Institute, Agricultural Research Center, 12619, Giza, Egypt.
* Corresponding author: muhammad.atiq@uaf.edu.pk

DOI: 10.1201/9781003546702-10

In recent years, Artificial Intelligence (AI) has emerged as a transformative technology in plant disease management. AI systems, particularly those leveraging machine learning (ML) and deep learning (DL) techniques, are capable of analyzing complex datasets—including images, sensor outputs, and environmental data—to detect and diagnose diseases with remarkable speed and accuracy (Lepakshi, 2022). Unlike traditional approaches, AI-based models can learn from vast amounts of annotated data, identify subtle patterns invisible to the human eye, and continuously improve their performance over time (Ghiurău and Popescu, 2024). The integration of AI into plant disease diagnosis brings several key advantages. AI-powered systems offer scalability, enabling large-scale monitoring of agricultural fields with minimal human intervention (Sahoo et al., 2024). They facilitate precision agriculture by allowing site-specific disease management, reducing unnecessary pesticide use, and lowering production costs. Additionally, AI tools can be embedded in user-friendly smartphone applications and Internet of Things (IoT) devices, democratizing access to sophisticated diagnostic capabilities even for smallholder farmers (Kumar et al., 2025).

Despite its promise, the adoption of AI in plant disease diagnosis also introduces new challenges, such as ensuring model robustness across diverse crop varieties, environmental conditions, and disease stages (Delfani et al., 2024). Ethical considerations, including data privacy, algorithmic bias, and equitable access, must also be carefully addressed to maximize the societal benefits of these technologies (Hanna et al., 2024). Overall, AI-driven plant disease diagnosis represents a paradigm shift towards more proactive, efficient, and sustainable plant health management. As the field continues to evolve, interdisciplinary collaborations among plant pathologists, agronomists, computer scientists, and policymakers will be critical in fully realizing the potential of AI for global agricultural advancement. This chapter discusses the limitations of conventional plant disease diagnostic techniques, presents artificial intelligence algorithms, examines deep learning and CNNs' role, and explores AI integration with smartphones and IoT for real-time disease monitoring.

2. Deep Learning and CNNs in Plant Disease Identification

The application of deep learning, particularly Convolutional Neural Networks (CNNs), has emerged as a transformative approach in the field of plant disease identification. Traditional methods such as manual scouting and visual inspection are often time-consuming, subjective, and limited by the expertise of the observer. In contrast, deep learning techniques, especially CNNs, provide a scalable, accurate, and automated solution to detect plant diseases from digital images, enabling rapid and reliable crop health assessment (Ferentinos, 2018). CNNs are a type of deep neural network specifically designed to process visual data. In the context of plant disease identification, CNNs can learn and extract complex patterns from leaf images, such as texture, shape, and color variations, which are often subtle but indicative of disease symptoms (Ramcharan et al., 2017). The model automatically identifies these features during the training phase, eliminating the need for manual feature engineering (Too et al., 2019). Typically, a large dataset of labeled plant leaf images—healthy and diseased—is used to train the model. During training, the CNN learns to associate specific image patterns with corresponding disease categories using supervised learning techniques.

The architecture of a CNN usually includes convolutional layers, activation functions (like ReLU), pooling layers, and fully connected layers. The convolutional layers are responsible for feature extraction, while pooling layers reduce dimensionality and help prevent overfitting. The fully connected layers integrate the learned features to classify the image into one of the predefined categories. The final output layer usually uses a softmax activation function to predict the probability of each class, aiding in the final classification decision (Ferentinos, 2018). Model performance is typically evaluated using metrics such as accuracy, precision, recall, and F1-score. Accuracy provides an overall measure of the correct predictions, while precision and recall offer insights into how well the model handles imbalanced classes, such as rare diseases (Picon et al., 2019). The confusion matrix is also a helpful tool to visualize how well the model differentiates between disease classes. In addition, cross-validation is often used to assess the model's generalizability and avoid overfitting (Sladojevic et al., 2016).

Recent advancements have also explored transfer learning, where pre-trained models like VGG16, ResNet, or Inception are fine-tuned using plant disease datasets. This approach significantly improves accuracy and reduces training time, especially when the dataset is small (Mohanty et al., 2016). Furthermore, deep learning models can be deployed in mobile or edge devices, allowing farmers to capture images in the field and receive instant disease diagnoses (Figure 1). In conclusion, the integration of deep learning and CNNs into plant pathology is revolutionizing disease detection

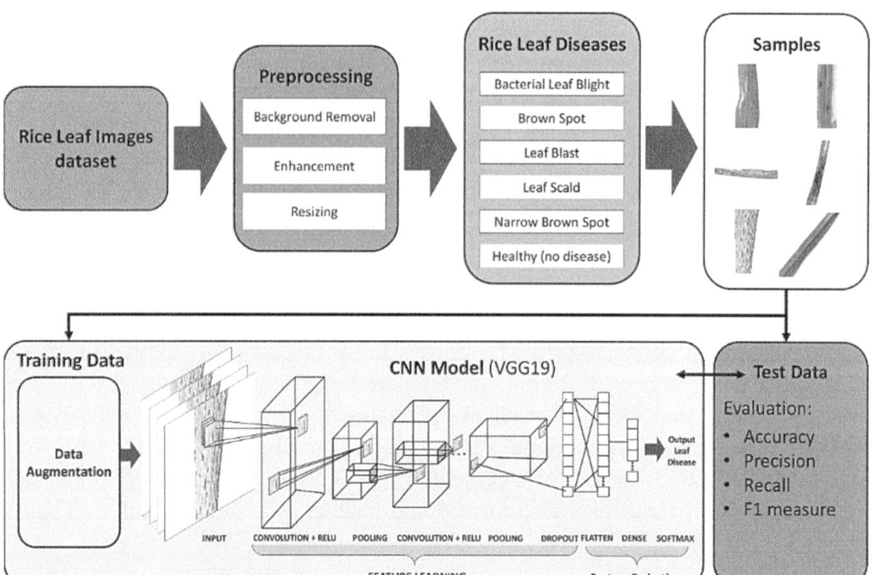

Figure 1. A technique based on transfer learning and deep convolutional neural networks (CNN) was proposed for the classification of leaf diseases (reprinted from Latif et al., 2022).

by offering fast, objective, and highly accurate identification methods. These technologies not only assist researchers and agricultural professionals but also empower farmers with practical tools to make timely disease management decisions, contributing to increased crop productivity and food security.

3. AI Integration in Mobile Apps, Sensor-Based Systems, and Real-Time Diagnostics

Artificial Intelligence (AI) is revolutionizing how plant diseases are detected and managed, especially through mobile applications, sensor-based systems, and real-time diagnostic tools used directly in the field (Aziz et al., 2025). These AI-powered technologies bridge the gap between research and practical agriculture, enabling farmers and agronomists to make informed decisions quickly and accurately (Titirmare et al., 2024). AI-enabled mobile apps have become essential tools for farmers and extension workers. Using computer vision and deep learning, these apps allow users to capture images of crops with smartphones and receive instant diagnoses. Convolutional Neural Networks (CNNs) are commonly employed to identify disease symptoms from leaf textures, colors, and patterns (Lu et al., 2021). Many apps are also integrated with geotagging, allowing for spatial disease monitoring and early warning systems across regions. Some examples include Plantix, Agrio, and Leaf Doctor, which provide real-time feedback, treatment suggestions, and historical disease data (Sarma et al., 2024). AI is also integrated into sensor-based networks that continuously monitor crop environments. These sensors collect data such as temperature, humidity, leaf moisture, soil conditions, and atmospheric CO_2 levels (Kaya, 2025). AI algorithms analyze this data to detect early signs of disease outbreak or predict the likelihood of disease spread. Machine learning models such as Decision Trees, Support Vector Machines (SVM), and Random Forests are often used to process this time-series data and identify patterns that precede infections (Santangelo et al., 2023). The systems are capable of sending alerts to farmers, enabling timely intervention and targeted pesticide application. AI-driven diagnostic devices are now being deployed in the field to provide real-time insights (Figure 2). These may include portable imaging systems, drone-mounted cameras or IoT-enabled hand-held devices (Khan, 2024). Combined with AI algorithms, these tools can assess plant health, detect nutrient deficiencies, and classify multiple disease types on-the-spot. For instance, edge AI devices allow for computation on the device itself, reducing reliance on cloud connectivity and ensuring faster responses in remote locations. This approach ensures disease containment at an early stage, reducing crop loss and minimizing the need for blanket chemical treatments. Together, these AI-powered tools form a comprehensive digital ecosystem that empowers farmers to shift from reactive to proactive disease management. Their integration into everyday farming practices marks a significant step toward sustainable, precision agriculture (Yadav and Yadav, 2025).

Figure 2. The AI-powered mobile application detects plant diseases using image recognition and geotagging, providing real-time feedback and management recommendations. A sensor-based system monitors plant health parameters, integrating multi-sensor arrays for continuous data collection. AI models on edge devices process data for real-time alerts, risk assessment, and control strategies. A system integrates AI, IoT, and big data for smart plant health management (created via https://ideogram.ai).

4. GLCM-Based Texture Feature Extraction and Its Integration with AI Algorithms for Plant Disease Diagnosis

The emergence of artificial intelligence (AI) and machine learning (ML) has revolutionized agricultural diagnostics, particularly in the early detection and classification of plant diseases (Ahmad and Yadav, 2023). Among the various techniques used to support automated decision-making, the Grey Level Co-occurrence Matrix (GLCM) has proven to be a highly effective statistical tool for texture analysis and feature extraction. GLCM is a second-order statistical method that quantifies how often a pair of pixel intensities (i, j) occur in a specified spatial relationship within an image. It constructs a two-dimensional square matrix of order N, where N is the number of grey levels present in the image (Liu et al., 2018). The matrix captures spatial dependencies by calculating the frequency of co-occurrences at specific directions (0°, 45°, 90°, and 135°) and distances (usually d = 1), which enables modeling of image texture (Zahid et al., 2019; Radovanovic and Dukanovic, 2020). For instance, consider an image composed of eight grey levels ranging from 0 to 7. In such a matrix, if grey levels 1 and 5 appear next to each other at 0° with a distance of one pixel, this co-occurrence is recorded and contributes to the value at row 1, column 5 in the GLCM. If this pairing occurs twice, the matrix value becomes 2 at that location (Panchal et al., 2019; Iqbal et al., 2018). From the GLCM, numerous statistical features can be derived commonly referred to as Haralick features that are essential for training machine learning models to distinguish between healthy and infected plant tissues. These features include contrast ($\sum_i \sum_j (i - j)^2 \cdot P(i, j)$), correlation ($\sum_i \sum_j ((i - \mu_i)(j - \mu_j) \cdot P(i, j)) / (\sigma_i \cdot \sigma_j)$), energy or angular second moment ($\sum_i \sum_j (P(i, j))^2$), and homogeneity ($\sum_i \sum_j P(i, j) / (1 + |i - j|)$). Additional descriptors include entropy ($-\sum_i \sum_j P(i, j) \cdot \log(P(i, j))$), dissimilarity ($\sum_i \sum_j |i - j| \cdot P(i, j)$), and inverse difference moment ($\sum_i \sum_j P(i, j) / (1 + (i - j)^2)$). Mean is computed as ($\mu = \sum_i i \cdot P(i)$), variance as ($\sigma^2 = \sum_i (i - \mu)^2 \cdot P(i)$), skewness as ($\sum_i ((i - \mu)^3 \cdot P(i)) / \sigma^3$), and kurtosis as ($\sum_i ((i - \mu)^4 \cdot P(i)) / \sigma^4$). Other textural descriptors such as smoothness ($1 - 1 / (1 + \sigma^2)$) and root mean square ($RMS = \sqrt{(\sum_i \sum_j P(i, j)^2)}$) further enhance the discriminative power of feature sets (Hufkens et al., 2019; Arsenovic et al., 2019). These GLCM-based features are widely used as input parameters for both conventional ML classifiers such as support vector machines (SVM), random forests (RF), and k-nearest neighbors (KNN) as well as deep learning architectures like convolutional neural networks (CNNs). The integration of such features enables the development of automated and precise systems for plant disease detection, allowing AI algorithms to detect complex texture patterns associated with biotic stress. In many cases, combining GLCM with deep learning improves model performance by complementing pixel-based learning with handcrafted texture descriptors (Figure 3). As a result, GLCM serves as a bridge between classical statistical texture modeling and contemporary AI-driven approaches, significantly enhancing the efficiency, accuracy, and scalability of plant disease diagnosis tools used in precision agriculture.

4	1	0	7	0	2
3	1	1	5	3	1
0	3	4	2	7	6
0	1	5	2	0	1
3	6	3	1	5	3

Figure 3. Illustration showing eight discrete grey levels utilized in the construction of a grey level co-occurrence matrix (GLCM). The grid presents pixel intensity values ranging from 0 to 7, with specific grey levels (1 and 5) highlighted to demonstrate a co-occurrence relationship at an orientation of 0° and a distance of 1. These spatial relationships serve as the foundation for generating the GLCM, which facilitates the extraction of statistical texture features for image-based classification and early plant disease diagnosis.

5. Successful Implementations and Global Case Studies

Artificial Intelligence (AI) is revolutionizing plant disease diagnostics globally, and Pakistan has increasingly embraced this transformation through public and private sector initiatives (Sajal, 2024). In recent years, research institutes and Agri-tech startups across the country have initiated AI-based systems to address crop health challenges especially for staple crops such as wheat, cotton, and rice, which are critical to national food security and export economy (Jain et al., 2024). One of the most promising projects has emerged from the National Agricultural Research Centre (NARC) and affiliated institutions, where machine learning models have been developed for the early detection of wheat rust (Singla et al., 2024). These models use image recognition techniques trained on locally sourced datasets, enabling farmers to identify rust at an early stage, thereby minimizing yield loss and reducing the indiscriminate use of fungicides (Mohanty et al., 2016).

Meanwhile, Pakistan's thriving Agri-tech startup ecosystem has also contributed significantly. Pak Agri Market and Zarai Digital, for instance, are integrating AI-driven diagnostic features in mobile apps that allow farmers to upload images of diseased plants (Hasan et al., 2022). The backend system uses convolutional neural networks (CNNs) to classify the disease and recommend targeted treatments. These apps are designed with low-bandwidth environments in mind, making them suitable for rural Pakistan (Too et al., 2019). Furthermore, Punjab Information Technology Board (PITB) has implemented pilot projects incorporating AI in agricultural extension services. By combining satellite imagery, remote sensing, and AI-based forecasting, they have launched early warning systems for pest and disease outbreaks particularly in cotton-growing regions of South Punjab, where pest infestations have historically caused significant economic damage World Bank. (2021).

In collaboration with international partners, including Chinese and European research centers, Pakistan has also begun using unmanned aerial vehicles (UAVs) coupled with AI algorithms to survey large-scale agricultural zones. These systems can detect disease stress patterns across vast areas in real-time, facilitating rapid responses and promoting precision agriculture practices (Pakistan Ministry, 2023). These case studies demonstrate Pakistan's commitment to integrating AI in agricultural systems for enhanced food security, environmental sustainability, and economic resilience. While still in the growth phase, these efforts highlight the country's potential as a regional leader in smart AI-driven agriculture.

6. Limitations and Ethical Considerations in AI Deployment

The deployment of artificial intelligence solutions for plant disease diagnosis faces several boundaries alongside ethical questions regarding their usage. Data quality and availability represent the main hurdle that must be resolved (Garrido et al., 2024). AI models succeed only when they access extensive diverse labeled datasets which are properly marked. Agri-cultural data remains scattered throughout various regions and presents inconsistent information leading to prediction inaccuracies and reduced model application potential (Kamilaris and Prenafeta-Boldú, 2018). Insufficient data creates a major challenge during detection of rare or new plant diseases because early disease identification becomes impossible (Barbedo, 2016). Algorithmic bias becomes a critical issue because models develop mistaken conclusions from training data that is unbalanced or from biased samples (Fazelpour and Danks, 2021). Such biases in the model output system inadvertently marginalize specific types of farmers because the algorithms may show preferences for select crops or regional practices and farming systems (Gwelo, 2025). Biased information has the potential to make existing agricultural resource distribution gaps worse (Bronson et al., 2021). AI system implementation for the agricultural industry creates ethical challenges regarding how farmers give consent and their rights to their data along with their privacy (Mark, 2019). Farmers usually remain uninformed about the ways their data gets utilized and they typically have limited capability to establish equitable terms (McCown, 2002). The use of proprietary algorithms managed by big

Agri-tech companies generates ethical challenges regarding accountability in addition to transparency matters and possible wrong uses of analytical insights (Carbonell, 2016).

AI tools meet limited accessibility as well as usability standards for agricultural workers with smallholdings and few resources (Singh et al., 2024). The adoption of AI tools faces obstacles in rural areas due to high expenses and minimal digital expertise of residents and inadequate basic infrastructure (Mhlanga, 2021). The digital gap threatens to build two separate agricultural sectors because well-funded farms receive all technological benefits while others remain unconnected (Photos, 2023). The lack of comprehensive regulatory frameworks together with appropriate policies creates barriers for deploying AI at higher levels of responsibility (De Almeida et al., 2021). The current basic framework of policies falls short of following technological advances which creates blind spots mainly in areas of liability standards and ethical addressing procedures (Calo, 2015). Society requires quick establishment of inclusive governance structures which combine different sectors to fill regulatory gaps while enabling access to responsible innovation (Stilgoe et al., 2020). AI has strong potential to modify plant disease management however both its limits and ethical implications need thorough consideration (Tzachor et al., 2022). The needs for joint collaboration between technologists, policymakers, farmers, and ethicists exist to ensure the development of sustainable, inclusive and fair AI innovations (Bronson, 2018).

7. Conclusion

Artificial Intelligence has demonstrated substantial potential in revolutionizing plant disease detection and diagnosis. Through the application of machine learning, deep learning, and data-driven technologies, faster and more accurate identification of diseases has become achievable across diverse agricultural contexts. Despite significant progress, certain limitations persist. These include variability in dataset quality, performance inconsistencies under diverse environmental conditions, algorithmic bias, and limited accessibility in resource-poor settings. Ethical concerns and policy frameworks also require careful consideration to ensure fair and responsible deployment of AI technologies in agriculture. Ongoing advancements in AI models, sensor integration, and computational power, coupled with multidisciplinary collaboration, are expected to further enhance the reliability and scalability of AI-driven disease diagnosis. Adoption of these technologies holds promise for more resilient, sustainable, and productive agricultural systems globally.

References

Ahmed, I., and Yadav, P. K. (2023). Plant disease detection using machine learning approaches. Expert Systems, 40(5): e13136.

Ali, F., Rehman, A., Hameed, A., Sarfraz, S., Rajput, N. A., and Atiq, M. (2024). Climate change impact on plant pathogen emergence: artificial intelligence (AI) approach. In Plant quarantine challenges under climate change anxiety (pp. 281–303). Cham: Springer Nature Switzerland.

Arsenovic, M., Karanovic, M., Sladojevic, S., Anderla, A., and Stefanovic, D. (2019). Solving current limitations of deep learning based approaches for plant disease detection. Symmetry (Basel), 11(7): 939.

Aziz, D., Rafiq, S., Saini, P., Ahad, I., Gonal, B., Rehman, S. A. et al. (2025). Remote sensing and artificial intelligence: revolutionizing pest management in agriculture. Frontiers in Sustainable Food Systems, 9: 1551460.

Barbedo, J. G. A. (2016). A review on the main challenges in automatic plant disease identification based on visible range images. Biosystems Engineering, 144: 52–60.

Bronson, K. (2018). Smart farming: including rights holders for responsible agricultural innovation. Technology Innovation Management Review, 8(2): 7–14.

Bronson, K., Rotz, S., and D'Alessandro, A. (2021). The human impact of data bias and the digital agricultural revolution. In Handbook on the human impact of agriculture (pp. 119–137). Edward Elgar Publishing.

Buja, I., Sabella, E., Monteduro, A. G., Chiriacò, M. S., De Bellis, L., Luvisi, A. et al. (2021). Advances in plant disease detection and monitoring: From traditional assays to in-field diagnostics. Sensors, 21(6): 2129.

Calo, R. (2015). Robotics and the Lessons of Cyberlaw. Calif. L. Rev., 103: 513.

Carbonell, I. M. (2016). The ethics of big data in big agriculture. Internet Policy Review, 5(1): 1–13.

Chidzwondo, F., and Mutapi, F. (2024). Challenge of diagnosing acute infections in poor resource settings in Africa. AAS Open Research, 4: 28.

De Almeida, P. G. R., dos Santos, C. D., and Farias, J. S. (2021). Artificial intelligence regulation: a framework for governance. Ethics and Information Technology, 23(3): 505–525.

Delfani, P., Thuraga, V., Banerjee, B., and Chawade, A. (2024). Integrative approaches in modern agriculture: IoT, ML and AI for disease forecasting amidst climate change. Precision Agriculture, 25(5): 2589–2613.

Fazelpour, S., and Danks, D. (2021). Algorithmic bias: Senses, sources, solutions. Philosophy Compass, 16(8): e12760.

Ferentinos, K. P. (2018). Deep learning models for plant disease detection and diagnosis. Computers and Electronics in Agriculture, 145: 311–318.

Ghiurău, D., and Popescu, D. E. (2024). Distinguishing Reality from AI: Approaches for Detecting Synthetic Content. Computers, 14(1): 1.

González-Rodríguez, V. E., Izquierdo-Bueno, I., Cantoral, J. M., Carbú, M., and Garrido, C. (2024). Artificial intelligence: a promising tool for application in phytopathology. Horticulturae, 10(3): 197.

Gwelo, F. A. (2025). Algorithmic Bias and Farmers' Autonomy in AI-Driven Agricultural Marketing and Supply Chains. In AI and Machine Learning Applications in Supply Chains and Marketing (pp. 313–344). IGI Global.

Hanna, M., Pantanowitz, L., Jackson, B., Palmer, O., Visweswaran, S., Pantanowitz, J. et al. (2024). Ethical and Bias considerations in artificial intelligence (AI)/machine learning. Modern Pathology, p. 100686.

Hasan, S. K., Kabir, M. R., Kabir, M. R., Islam, M. R., Akhter, M. J., and Moury, J. Y. (2022). Proximate composition, minerals, phytochemicals, and functional activities of jujube fruits grown in Bangladesh. Journal of Agriculture and Food Research, 8: 100302.

Hufkens, K., Melaas, E. K., Mann, M. L., Foster, T., Ceballos, F., Robles, M. et al. (2019). Monitoring crop phenology using a smartphone based near-surface remote sensing approach. Agricultural and Forest Meteorology, 265: 327–337.

Iqbal, Z., Khan, M. A., Sharif, M., Shah, J. H., ur Rehman, M. H., and Javed, K. (2018). An automated detection and classification of citrus plant diseases using image processing techniques: A review. Computers and Electronics in Agriculture, 153(4): 12–32.

Jain, R., Ashok, A., and Kumar, V. (2024). Transforming Agriculture: AI-driven Solutions from Emerging Agri-Tech Ventures. Agribusiness Incubation for Promoting Agripreneurship through Startups, 158.

John, M.A., Bankole, I., Ajayi-Moses, O., Ijila, T., Jeje, T. and Lalit, P. (2023). Relevance of advanced plant disease detection techniques in disease and Pest Management for Ensuring Food Security and Their Implication: A review. American Journal of Plant Sciences, 14(11): 1260–1295.

Kamilaris, A., and Prenafeta-Boldú, F. X. (2018). Deep learning in agriculture: A survey. Computers and Electronics in Agriculture, 147: 70–90.

Kaya, C. (2025). Intelligent Environmental Control in Plant Factories: Integrating Sensors, Automation, and AI for Optimal Crop Production. Food and Energy Security, 14(1): e70026.

Khan, A. O. R., Islam, S. M., Sarkar, A., Islam, T., Paul, R., and Bari, M. S. (2024). Real-time predictive health monitoring using AI-driven wearable sensors: Enhancing early detection and personalized interventions in chronic disease management. International Journal for Multidisciplinary Research.

Kumar, A., Nanjundan, P., and Thomas, L. (2025). Democratising Intelligent Farming Solutions to Develop Sustainable Agricultural Practices. In AI in Agriculture for Sustainable and Economic Management (pp. 53–66). CRC Press.

Latif, G., Abdelhamid, S. E., Mallouhy, R. E., Alghazo, J., and Kazimi, Z. A. (2022). Deep learning utilization in agriculture: Detection of rice plant diseases using an improved CNN model. Plants, 11(17): 2230.

Lepakshi, V. A. (2022). Machine learning and deep learning based AI tools for development of diagnostic tools. In Computational approaches for novel therapeutic and diagnostic designing to mitigate SARS-CoV-2 infection (pp. 399–420). Academic Press.

Liu, B., Shi, Y., Duan, Y., and Wu, W. (2018). UAV-based crops classification with joint features from orthoimage and DSM data. The International Archives of the Photogrammetry, Remote Sensing and Spatial Information Sciences (ISPRS Archives), 42(3): 1023–1028. https://doi.org/10.5194/isprs-archives-XLII-3-1023-2018.

Lu, J., Tan, L., and Jiang, H. (2021). Review on convolutional neural network (CNN) applied to plant leaf disease classification. Agriculture, 11(8): 707.

Mark, R. (2019). Ethics of using AI and big data in agriculture: The case of a large agriculture multinational. The ORBIT Journal, 2(2): 1–27.

McCown, R. L. (2002). Changing systems for supporting farmers' decisions: problems, paradigms, and prospects. Agricultural Systems, 74(1): 179–220.

Mhlanga, D. (2021). Artificial intelligence in the industry 4.0, and its impact on poverty, innovation, infrastructure development, and the sustainable development goals: Lessons from emerging economies?. Sustainability, 13(11): 5788.

Mohanty, S. P., Hughes, D. P., and Salathé, M. (2016). Using deep learning for image-based plant disease detection. Frontiers in plant science, 7, 215232.

Pakistan Ministry of National Food Security & Research. (2023). CPEC Agricultural Modernization Report. Government of Pakistan.[URL: mnfsr.gov.pk/cpec-agri] (Official report)

Panchal, P., Raman, V. C., and Mantri, S. (2019). Plant diseases detection and classification using machine learning models. In 2019 4th International Conference on Computational Systems and Information Technology for Sustainable Solution (CSITSS) (Vol. 4, pp. 1–6). IEEE.

Photos, A. P. (2023). Bridging the Gap: Accelerating Technology Adoption for Sustainable Food Production.

Picon, A., Alvarez-Gila, A., Seitz, M., Ortiz-Barredo, A., Echazarra, J., and Johannes, A. (2019). Deep convolutional neural networks for mobile capture device-based crop disease classification in the wild. Computers and Electronics in Agriculture, 161: 280–290.

PITB & World Bank. (2021). Smart Agriculture Punjab: AI for Pest Forecasting. Punjab Government Technical Report. [URL: pitb.gov.pk/smart_agri] (Official PITB project page).

Radovanovic, D., and Dukanovic, S. (2020). Image-based plant disease detection: A comparison of deep learning and classical machine learning algorithms. In 2020 24th International Conference on Information Technology (IT) (pp. 1–4). IEEE.

Ramcharan, A., Baranowski, K., McCloskey, P., Ahmed, B., Legg, J., and Hughes, D. P. (2017). Deep learning for image-based cassava disease detection. Frontiers in Plant Science, 8: 1852.

Sahoo, L., Mohapatra, D., Raghuvanshi, H. R., Kumar, S., Kaur, R., Chawla, R. et al. (2024). Transforming Agriculture through Artificial Intelligence: Advancements in Plant Disease Detection, Applications, and Challenges. Journal of Advances in Biology & Biotechnology, 27(5): 381–388.

Sajal, I. (2024). Exploring The Impact of AI in Disease Detection and Analyzing the Landscape of Health Infrastructure in Pakistan (Doctoral dissertation, School of Social Sciences & Humanities, S3H-NUST).

Santangelo, O. E., Gentile, V., Pizzo, S., Giordano, D., and Cedrone, F. (2023). Machine learning and prediction of infectious diseases: a systematic review. Machine Learning and Knowledge Extraction, 5(1): 175–198.

Sarma, H. H., Das, B. C., Deka, T., Rahman, S., Medhi, M., and Kakoti, M. (2024). Data-driven agriculture: Software innovations for enhanced soil health, crop nutrients, disease detection, weather forecasting, and fertilizer optimization in agriculture. Journal of Advances in Biology and Biotechnology, 27(8): 878–896.

Sethu, M., Kotla, B., Russell, D., Madadi, M., Titu, N. A., Coble, J. B. et al. (2023). Application of artificial intelligence in detection and mitigation of human factor errors in nuclear power plants: a review. Nuclear Technology, 209(3): 276–294.

Singh, N., Wang'ombe, J., Okanga, N., Zelenska, T., Repishti, J., Mishra, S. et al. (2024). Farmer. Chat: Scaling AI-Powered Agricultural Services for Smallholder Farmers. arXiv preprint arXiv: 2409.08916.

Singla, A., Nehra, A., Joshi, K., Kumar, A., Tuteja, N., Varshney, R. K. et al. (2024). Exploration of machine learning approaches for automated crop disease detection. Current Plant Biology, 100382.

Sladojevic, S., Arsenovic, M., Anderla, A., Culibrk, D., and Stefanovic, D. (2016). Deep neural networks based recognition of plant diseases by leaf image classification. Computational Intelligence and Neuroscience, 2016(1): 3289801.

Stilgoe, J., Owen, R., and Macnaghten, P. (2020). Developing a framework for responsible innovation. In The ethics of nanotechnology, geoengineering, and clean energy (pp. 347–359). Routledge.

Titirmare, S., Margal, P. B., Gupta, S., and Kumar, D. (2024). AI-powered predictive analytics for crop yield optimization. In Agriculture 4.0 (pp. 89–110). CRC Press.

Too, E. C., Yujian, L., Njuki, S., and Yingchun, L. (2019). A comparative study of fine-tuning deep learning models for plant disease identification. Computers and Electronics in Agriculture, 161: 272–279.

Tzachor, A., Devare, M., King, B., Avin, S., and Ó h Éigeartaigh, S. (2022). Responsible artificial intelligence in agriculture requires systemic understandings of risks and externalities. Nature Machine Intelligence, 4(2): 104–109.

Wang, X., Ma, L., Yan, S., Chen, X., and Growe, A. (2023). Trade for food security: The stability of global agricultural trade networks. Foods, 12(2): 271.

Yadav, A., and Yadav, K. (2025). Portable solutions for plant pathogen diagnostics: development, usage, and future potential. Frontiers in Microbiology, 16: 1516723.

Zahid, A., Abbas, H. T., Ren, A., Zoha, A., Heidari, H., Shah, S. A. et al. (2019). Machine learning driven non-invasive approach of water content estimation in living plant leaves using terahertz waves. Plant Methods, 15(1): 1–13. https://doi.org/10.1186/s13007-019-0522-9.

Chapter 11

The Role of AI in Plant Diseases Detection and Their Impact on Crop Yields

Heba Yousef, Kamel A. Abd-Elsalam* and *Salah M. Abdel-Momen*

1. Introduction to AI in Agriculture

The term artificial intelligence (AI) was initially used by John McCarthy at the Dartmouth Conference, where he proposed that machines could simulate human learning and intelligence. Since then, AI has become a central field in computer science, revolutionizing industries such as manufacturing, healthcare, education, and finance by solving complex problems beyond human capabilities. Although there is no universally accepted definition, AI is generally described as systems that simulate human-like cognition, capable of tasks like learning, reasoning, and problem-solving. The Turing Test, devised by Alan Turing, remains a landmark approach to assessing machine intelligence, requiring proficiency in areas like natural language processing and machine learning (Warwick, 2013).

Recent developments in AI focus on creating autonomous technologies that are capable of learning from huge amounts of data to perform tasks traditionally requiring human intelligence. AI's capabilities in automation, prediction, and decision-making are transforming industries, with significant potential in areas like agriculture. AI technologies help manage agricultural challenges such as crop management, soil health, pest control, and climate adaptation by processing and analyzing complex information from multiple sources, such as satellite photography and sensors (Mana et al., 2024).

In agriculture, diseases of plants carried on by pathogens including bacteria, fungus, and viruses pose a major threat, reducing global crop yields by about

Plant Pathology Research Institute, Agricultural Research Center (ARC), 12619 Giza, Egypt.
* Corresponding author: heba.yousef@live.com, https://orcid.org/0000-0003-3314-8856

DOI: 10.1201/9781003546702-11

30% (Gai and Wang, 2024). These diseases are impacted by a number of aspects, including as plant type, environmental circumstances, and human practices. Climate change, changing farming practices, and reduced biodiversity exacerbate the spread and impact of plant diseases. Traditional control methods using agrochemicals and cultivating disease-resistant varieties are often ineffective due to pathogen adaptation and the environmental damage caused by chemicals (Fones et al., 2020).

AI presents a promising solution to challenges in agriculture, particularly in managing plant diseases, optimizing crop yields, and adapting to climate change. By leveraging automation, machine learning, and data analysis, AI enhances disease detection, monitoring, and management, improving prediction accuracy and resource optimization for more sustainable farming practices (Mana et al., 2024). However, forecasting how climate change may affect plant diseases is complex, involving multiple interrelated components including environmental circumstances and pathogen behavior (Fones et al., 2020). With climate change altering pest and pathogen dynamics, AI tools are increasingly seen as critical for developing strategies to protect crops and ensure food security. As AI technologies continue to evolve, it is anticipated that they will be essential to address plant disease management challenges and protecting the world's food systems from the growing impacts of climate change (Taddeo and Floridi, 2018).

2. AI Technologies for the Detection of Plant Diseases

Specifically, the identification and control of plant diseases is being revolutionized by the convergence of Internet of Things (IoT), artificial intelligence (AI), and unmanned aerial vehicles (UAVs). These technologies improve disease reporting accuracy and efficiency, giving farmers access to timely information that is essential for reducing crop losses.

Even though agriculture is crucial to the world's food security, younger generations are becoming less interested in farming. Many young people are moving to cities in order to avoid the difficulties that come with working in agriculture, including unstable finances, hard labor, and the growing impact of climate change on crop quality. In addition to changes in host resistance and pathogen activity, climate change presents significant risks and makes the already challenging work of controlling diseases more challenging (Smith et al., 2020). This problem is made worse by an over reliance on pesticides, which might result in the emergence of resistant diseases. This highlights the urgent need for creative solutions to preserve plant productivity and health.

In this instance, it is essential to identify plant diseases accurately and promptly. The timely identification of plant diseases is among the fundamental tenets of precision farming, since it is essential for addressing the long-term development of pathogenic resistance and avoiding resource waste (Smith et al., 2020). Plant illnesses can be difficult to identify, though; some may not show obvious signs at first, and by the time they do, it can be too late to act effectively. The fact that misdiagnoses can result in large agricultural losses emphasizes the significance of having qualified experts who can correctly diagnose plant diseases.

Automated solutions utilizing computer vision advances are being developed to help with the detecting procedure. By using visual signs to identify plant diseases, these technologies can help both inexperienced gardeners and seasoned experts make accurate diagnoses of plant health problems (Hassan et al., 2021). These automated technologies can expedite and enhance the accuracy of the identification process by employing sophisticated imaging techniques and machine learning algorithms.

Conventional manual detection techniques are time-consuming and prone to mistakes, which can cause significant production losses and the quick spread of illnesses. This emphasizes the need for cutting-edge technologies like both Machine Learning (ML) and Deep Learning (DL), which automate the monitoring of vast agricultural regions and enable plant disease early detection (Shoaib et al., 2023). As an example, plants can be categorized using machine learning approaches according to a number of attributes, including shape, variety, maturity, and health condition. According to research, these systems can improve diagnosis accuracy by efficiently identifying diseases by analyzing large image collections. In order to train models for disease identification, one study used a feature dataset of 600 photos, including 50 photos for each variety of plant.

In plant pathology, technologies utilizing artificial intelligence (AI) are gradually being used to detect diseases and defects in plants. The technique by which automated detection systems work is as follows: sensors placed in agricultural fields take pictures of plants, which are subsequently processed and divided into several categories to produce information for machine learning algorithms. By predicting if a leaf is healthy or diseased, these algorithms allow for more effective disease management techniques. Deep learning algorithms are popular because they can automatically extract pertinent details from photos. This ability has greatly improved the accuracy of diagnosing a variety of plant diseases, such as cucumber mosaic and foliar diseases in crops like bananas and chili plants (Hassan et al., 2021).

Furthermore, IoT technology supports these programs by providing real-time data to farmers from agricultural fields, enabling them to remotely monitor environmental conditions. Proactive reactions to possible disease threats are made possible by this skill (Kumar and Kurian, 2023). For example, an IoT-based predictive model created by Liu et al. (2021) demonstrated how IoT might be used to transform disease management techniques by predicting disease outbreaks with an extraordinary 91% accuracy rate.

The integration of IoT, AI, and UAV technology improves resource use and encourages sustainable farming methods. A more adaptable agricultural sector in light of ongoing global warming and changing demographic patterns may be achieved by farmers incorporating these cutting-edge technologies to maximize output while reducing environmental consequences (Smith et al., 2020).

In conclusion, implementing cutting-edge technology like IoT and AI is essential to enhancing agricultural disease management and detection. These developments not only address the pressing issues facing modern agriculture, but they also pave the way for a sustainable future in food production. The field of plant pathology and agriculture in general has advanced significantly with the potential for automated methods to improve accuracy and lower manpower demands. This technological convergence could revolutionize farming methods, providing sustainability and food security in a world growing more complicated by the day.

3. The Fundamentals of Automated Plant Disease Detection

Plant diseases are a major cause of crop loss in the agricultural sector, which is being further taxed by the combined consequences of climate change as well as the growing need for food worldwide. Approximately 20 to 40 % of global crop yields are lost annually due to diseases, according to the Food and Agriculture Organization (FAO), underscoring the critical need for efficient disease management techniques (FAO, 2018). A ground-breaking method that uses technology to raise the speed and precision of identifying diseases is automated plant diagnosis of disease.

Traditional methods of disease detection occasionally rely on time-consuming hand examinations, which can be unreliable, particularly in expansive agricultural environments. On the other hand, automated systems continually monitor plant health and detect disease symptoms early on by using high-resolution imagery, real-time data analytics and machine learning algorithms. This change in technology not only speeds up diagnosis but also improves intervention precision, which results in enhanced crop care and decreased losses.

An important step toward sustainable farming methods is the use of automated plant disease detection systems as precision agriculture develops. These technologies improve productivity and promote the more general objectives of protecting the environment and food security by enabling prompt reactions to possible diseases. Global agricultural expansion is severely restricted by plant diseases, which cause enormous yearly financial losses (Mitra, 2021). Advances in machine learning technologies have consequently made the study of plant disease identification an essential component of the identification of patterns and the advancement of modern agriculture (Sanida et al., 2023).

Furthermore, advanced applications including production of crops prediction, weed detection and processing, and plant pathogens detection have been made possible by the widespread AI applications in agriculture (Albattah et al., 2022). Effective crop harvesting, avoiding growth problems, and lowering the need for pesticides—all depend on early plant disease identification and management. This promotes more sustainable farming methods by reducing the harm that pesticide use causes to the environment (Perveen et al., 2023). In conclusion, one significant invention that improves agricultural sustainability and resilience in the face of persistent difficulties is automated plant disease detection. The plant disease prediction framework with preset procedures is shown in Figure 1.

3.1 Acquisition of Plant Images

This stage involves taking pertinent pictures of the object in order to classify it using automated methods. An image is a set of binary data that may be manipulated and analyzed on a computer. Images in this area are taken with high-resolution digital cameras.

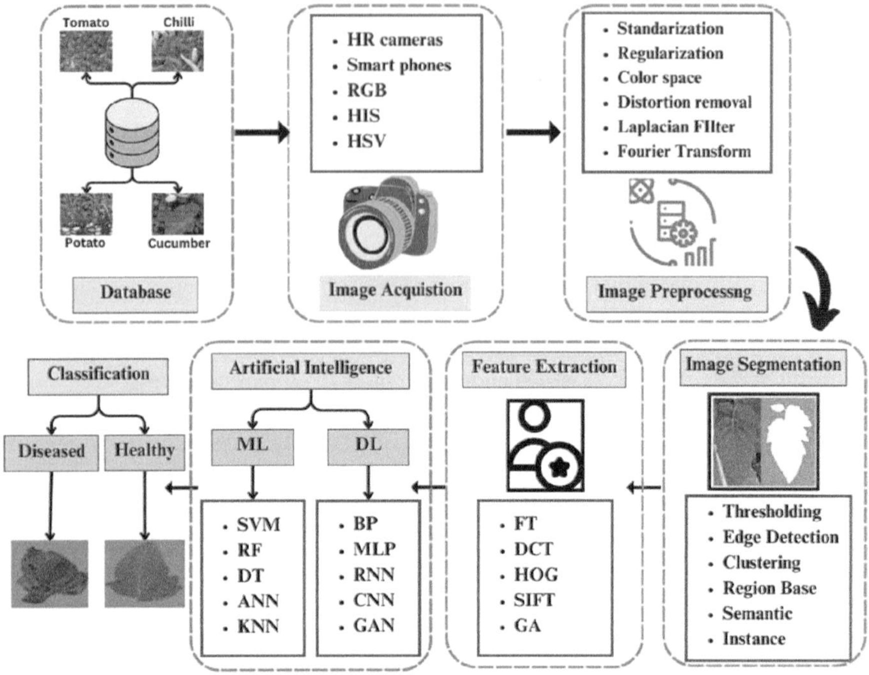

Figure 1. A plant disease prediction system that includes all the necessary steps (Reprinted from Jafar et al., 2024).

The ability to record image samples in a variety of supported formats, including jpg, png, tif, and more, has demonstrated the value of smartphones. Once all necessary photos have been taken, they are delivered to the picture preprocessing step so they can be edited before being used. The use of image-enhancing techniques is necessary if the gathered photos do not meet the processing specifications (Basavaiah and Anthony, 2020). The process of acquiring images is essential for a precise classification of diseases. The obtained images are crucial to the overall framework's effectiveness. These photos are used to train machine learning algorithms.

3.2 *Image Segmentation*

In agricultural research, segmentation is a basic approach that involves carefully decomposing a picture into its component elements. The primary goal is to carry out a more thorough analysis of every object in order to extract useful aspects that could improve our comprehension and knowledge. The retrieved traits can be used to distinguish between areas that are unaffected and those that are diseased. In order to extract several potentially useful features, it is essential to segment the preprocessed pictures in order to categorize sick leaves (Jafar et al., 2024). Conventional methods for segmenting the provided images, like clustering, region-based, thresholding, and edge detection all require knowledge of mathematics and the processing of images. Among the most effective segmentations techniques is thresholding, which divides images according to values of pixel intensity. It is extensively employed in many different possibilities, including remote sensing, detection, and classification.

3.3 Feature Extraction

In the field of agriculture, feature extraction is the procedure of identifying characteristics in unprocessed data. Properties of shape, color, and texture are the input image feature descriptors. It is essential for categorizing tasks. According to Basavaiah and Anthony (2020), a key method in machine learning is feature engineering, which entails transforming raw information into a set of relevant and practical features. The dataset is supplied as input for this stage, which determines the health of the plants. An image's fundamental elements include its color, texture, morphology, and other associated attributes. Morphological characteristics work better than other characteristics at identifying the damaged area on a leaf. Often utilized color qualities include Gabor texture and color moments. A variety of techniques are available to obtain these properties, including the color R moment (Al Rahhal et al., 2016), the color correlogram (Huang et al., 1997), the color histogram (Sugimura et al., 2015), and others. Potential additions to the texture include entropy, variation, homogeneity, and contrast. Texture feature utilization has been found to produce better results in the setting of plant disease diagnosis challenges.

3.4 Artificial Intelligence

Artificial intelligence (AI) in agricultural research is becoming more and more popular, particularly in the fields of plant disease classification and identification. This procedure begins with classification, which entails grouping data into classes. Detecting and classifying plant leaves more especially, separating healthy from diseased specimens is of special interest to us in this context. To execute, one must be familiar with ML and DL's classification and detection techniques.

The application of machine learning and deep learning to plant disease identification has grown in popularity and demonstrated encouraging outcomes in correctly diagnosing plant diseases from digital photos. Plant disease diagnostics has made substantial use of traditional machine learning methods including feature extraction and classification. These techniques allow a classifier to distinguish between healthy and unhealthy plants by using factors collected from pictures, such as color, texture, and form. Although these techniques have been widely used to detect diseases like rust, leaf blotch, and powdery mildew, in addition to symptoms caused due to abiotic stressors like drought and inadequate nutrients, they are not highly efficient at accurately identifying subtle disease symptoms or detecting diseases in their early stages (Shoaib et al., 2023). They also have trouble processing high-resolution and complicated pictures.

Plant disease identification has recently been approached using deep learning (DL) approaches which involve deep belief networks (DBNs) and convolutional neural networks (CNNs) (Pacal et al., 2024). By using these techniques, a network is trained to recognize the underlying characteristics of the images, making it possible to identify subtle disease symptoms that may be undetected by conventional image processing techniques (Khan et al., 2021). High-resolution photos are appropriate for DL models because they are capable of handling large and complex images. However, these approaches might not be appropriate for invisible diseases and require a significant amount of labeled training data.

3.5 Machine Learning Algorithms

AI includes machine learning as a subset. Computers can learn from experience according to machine learning (ML) (Shoaib et al., 2023). Regarding the ML field, we currently discover a number of subtypes, each of which is appropriate for a distinct learning scenario. Giving the data entered into the system and the corresponding target values that are expected from the data are known as supervised learning. The obvious goal is to get information and construct a connection that will permit the system to forecast outcomes according to the inputs. This entails using labeled data to train algorithms that categorize leaves into categories based on plant diseases. Conversely, unsupervised learning utilizes a distinct strategy.

In this instance, data is sent to the system without clear input-output specifications. Its goal is to search the data for hidden connections or trends (Attri et al., 2023). While controlled learning involves labeled data, and independent learning involves unlabeled data, semi-supervised learning occurs. In machine learning, it's also critical to distinguish between classification and regression tasks because they yield distinct kinds of output data. In classification tasks, inputs are arranged into classes and qualitative results are obtained. Plant leaf disease classification into discrete categories is one use of the classification (Shoaib et al., 2023). Regression tasks, in contrast, attempt to calculate values with input information and deal with numerical results. Among the often utilized methods include naive Bayes, artificial neural networks, support vector machines, k-Nearest Neighbors (k-NN), decision trees, random forests, linear discriminant analysis, and linear regression.

3.6 Deep Learning Models

Deep learning (DL), a branch of machine learning and artificial intelligence, has had major effects on fields like natural language processing, object recognition, and picture classification. DL gets away with the demand for intensive artificial feature engineering by applying neural networks to feature selection on their own. Creating abstract, high-level features by combining low-level data, it increases precision and adaptability in tasks like target identification and image recognition. There are two distinct periods in the evolution of DL: the first, which took place between 1943 and 1998, and the second, which took place between 2006 and till now (Lavecchia, 2019). The first stage produced ground-breaking innovations including the chain rule, backpropagation, Neocognitron, and frameworks like LeNet for handwritten text recognition. Deep belief networks (DBN), CNN, auto encoders, and other modern methods and structures emerged during the second stage of DL. They can be applied to a number of industries, including marketing, banking, text recognition, earthquake prediction, self-driving cars, healthcare, and image recognition.

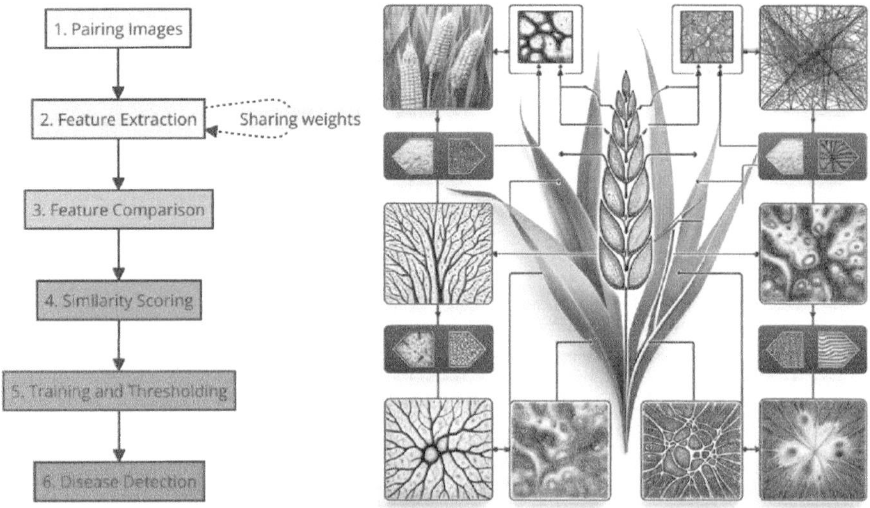

Figure 2. Siamese convolutional neural networks (SNNs) for plant disease detection from an image dataset are shown in this graph diagram (reprinted from González-Rodríguez et al., 2024).

The Siamese convolutional neural network (SNN), a recent development of convolutional neural networks (CNNs), was presented by Narain et al. (2022) to detect disease of tomato leaves. By comparing input pairings, SNNs improve a detection system, which makes them ideal for differentiation tasks (Figure 2). They assess picture pair similarities, in contrast to conventional CNNs, and are able to classify diseases with high accuracy even when there is limited labeled data and infraclass heterogeneity (González-Rodríguez et al., 2024).

4. Ethics in AI-Powered Agriculture

For AI agriculture to have ethical regulations that balance innovation and accountability, policy development is essential (Figure 3). Governments and organizations must develop frameworks that encourage technology growth while addressing ethical issues as AI changes the agricultural sector. Among these are rules to safeguard farmers' rights, maintain data privacy, and promote sustainability (Kshetri, 2021). Collaboration between various stakeholders, including farmers, technologists, and ethicists, is necessary for effective policy formation in order to jointly establish fair and useful policies (Ritchie and Roser, 2024). By fostering trust and making rules more relevant, this cooperative approach guarantees the equitable and responsible application of AI's advantages in agriculture.

4.1 Data Security and Privacy

Since data is necessary for planning and policies for agricultural growth and development, farmers' data should be gathered and used responsibly, protecting privacy and preventing misuse. Big data science is crucial to enhancing agriculture now more than ever because of population expansion and climate change, which

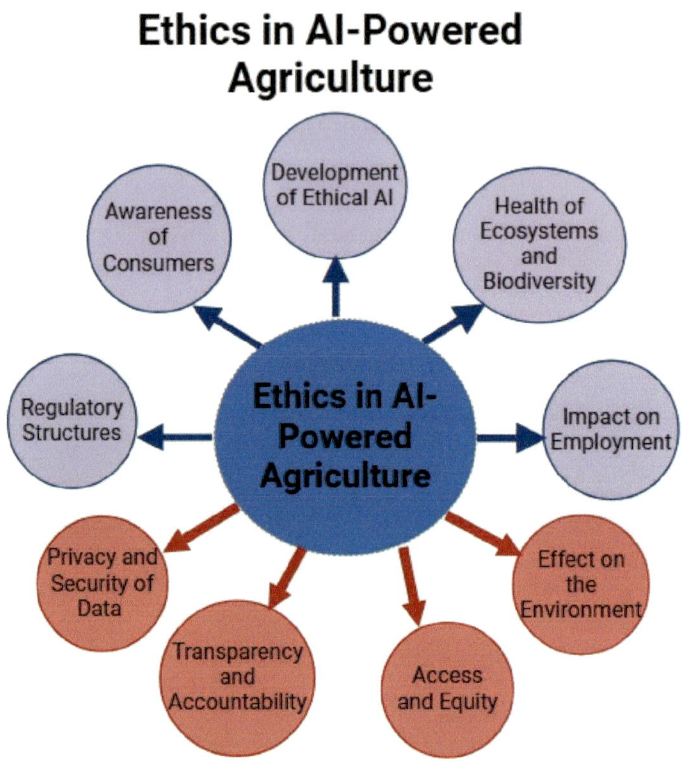

Figure 3. Ethics guidelines for agricultural artificial intelligence.

are driving up demand for sustainable strategies. Modern civilization has new opportunities due to big data, especially in agricultural research, which helps farmers make better decisions about how to produce crops. By combining big data technology with high-performance computing, farmers can support multiple agricultural areas and increase productivity through data-intensive decision-making (Tantalaki et al., 2019).

4.2 Transparency and Accountability

Building confidence in AI-powered agricultural systems requires transparency in the ways that data is obtained, kept, and used. A more social and efficient agricultural environment can be produced by stakeholders by creating uniform standards and keeping lines of communication open. Since knowing data management procedures motivates farmers to interact with technology and contribute their information, this openness is essential to fostering trust among farmers, technology providers, and other stakeholders. Clear rules should outline the kinds of data being gathered, including weather patterns, crop yields, and soil health, as well as the techniques used, such as sensors, surveys, and satellite imaging. Farmers must be made aware of the goal of data gathering and its intended application. Stakeholders should also understand how and where their data is held, as well as the safety measures taken to prevent breaches, like access controls, encryption, and cloud storage (Greenfield et al., 2022). Data utilization must also be transparent, including who can access the data, how it can be shared with outside parties (such as government agencies or researchers), and whether it will be used for purposes other than those initially planned, like market research or policymaking.

4.3 Access and Equity

Maintaining access to AI technologies for smallholder farmers and those in emerging nations is essential to halting the growth of agricultural inequality. The digital divide is a serious problem since farmers like these may fall behind those who have more money if they do not have access to cutting-edge equipment and resources (Singh et al., 2024). Designing AI solutions with price in mind is crucial to resolving this challenge. Advanced technologies will only help wealthy farmers if they continue to be unaffordable, which will exacerbate already-existing inequality. The agriculture industry should promote inclusivity and enable all farmers to use AI for increased production and sustainability by emphasizing cost-effective solutions and offering resources and training. In addition to improving people's quality of life, this strategy helps vulnerable areas achieve greater economic growth and food security (Davis, 2016).

4.4 The impact on the Environment

The application of AI in agriculture should be encouraged in order to limit environmental damage and lessen dependency on dangerous pesticides. In order to reduce the need for synthetic inputs, artificial intelligence can help farmers make informed decisions about fertilization, pest control, and crop management by using predictive modeling and advanced analytics. (Williams et al., 2023). Furthermore, the ethical application of artificial intelligence requires handling resources like land and water as efficiently and responsibly as possible. In addition to increasing agricultural productivity now, this strategy protects their availability for generations to come. The agriculture industry may contribute to a more resilient food chain, address urgent environmental problems, and advance ecological health by incorporating AI into sustainable practices.

4.5 Impact on Employment

Significant worries regarding job displacement in traditional farming roles are raised by the growth of automation in agriculture through AI (Smith and Johnson, 2014). As human workers are replaced by robots and algorithms, many may lose their jobs, raising serious concerns about retraining and assistance for those impacted. In order to solve this problem, a balance must be struck between using AI technology to supplement human labor and avoiding complete replacement (Brown, 2021). Rather than totally replacing workers' jobs, the industry may foster a more collaborative atmosphere by emphasizing how AI can help people and increase productivity, for example, by providing data-driven insights or precision agriculture technologies. To ensure that the shift to an AI-driven agricultural landscape is not only effective but also inclusive and supportive of the workforce, it is important to invest in education and skill development programs that will give workers the tools they need to adjust to new technology (Chen et al., 2024).

4.6 Health of Ecosystems and Biodiversity

Although AI-driven agricultural methods are frequently intended to increase productivity and efficiency, they run the risk of encouraging monocultures, which endanger biodiversity and undermine ecosystem resilience. Over-reliance on single-crop systems can reduce genetic diversity and increase short-term yields while simplifying management and increasing ecosystem susceptibility to pests, diseases, and climate change. Applications of AI must so actively promote ecosystem health and take into account the larger ecological effects of farming techniques. Agro ecological concepts can be integrated with AI to support a variety of agricultural systems and preserve important ecosystem services including pollination, soil health, and water conservation. In the end, a well-rounded strategy that preserves biodiversity not only maintains agricultural output but also cultivates an environment that is sufficiently strong to resist future difficulties (Wilson, 2023).

4.7 Development of Ethical AI

Algorithm bias is a serious issue when developing AI systems for agriculture since it might result in unfair solutions that disproportionately impact particular farming practices. Designing AI systems to reduce bias while taking into consideration the varied settings and requirements of different agricultural stakeholders is essential to ensuring justice and inclusion. According to Smith et al. (2023), this entails carrying out comprehensive testing and validation across different social classes and farming practices in order to prevent unexpected results. Accountability is also important; developers and businesses must take responsibility for the decisions their AI systems make and the consequences those choices have on the environment and communities. Stakeholders can build trust and guarantee that AI technologies contribute favorably to agricultural sustainability and equity, which will ultimately benefit all farmers, by putting in place clear accountability frameworks that include transparency in algorithm design and decision-making processes (Davis, 2022).

4.8 Awareness of Consumers

Food production must be transparent in order to enable consumers to make knowledgeable decisions about the goods they purchase. Customers should be made aware of AI's impact on food quality and safety as it continues to improve supply chain efficiency and optimize crop yields in agriculture. By giving consumers comprehensive information on how AI is used in agriculture and food production, businesses can increase consumer trust and contribution. Furthermore, consumers may comprehend the ethical implications of their purchases, particularly those pertaining to sustainability, labor practices, and environmental impact, thanks to clear labeling, which is essential to this transparency. A more ethical food system will result from such labeling since it not only assists consumers in making decisions that are consistent with their values but also motivates producers to embrace more accountable and transparent processes (Greenfield, 2023).

4.9 Regulatory Structures

To maintain ethical behaviors in AI agriculture and strike a balance between innovation and accountability, policy development is crucial. Governments and institutions must develop frameworks that address ethical issues while promoting technical growth as AI technologies continue to transform the agricultural sector (Smith et al., 2021). Setting rules that safeguard data privacy, advance sustainability, and defend the rights of farmers is a component of the process (Kshetri, 2021). Furthermore, cooperation between various stakeholders, such as farmers, technologists, and ethicists, is necessary for the formulation of effective policies. These groups can jointly develop policies that are realistic, fair, and considerate of the intricacies of contemporary agriculture by combining diverse viewpoints. In addition to improving the efficacy and relevance of laws, this cooperative approach fosters stakeholder confidence and guarantees that the advantages of AI in agriculture are distributed fairly and responsibly (Ritchie and Roser, 2024). In AI-powered agriculture, ethical issues are complicated yet essential for ensuring that technology works for farmers, consumers, and the environment. A proactive strategy may reduce risks, encourage sustainable behaviors, and help realize AI's maximum potential.

5. Limitations of AI in Plant Disease Detection

Although the application of artificial intelligence has demonstrated considerable promise for detecting agricultural diseases, there are lots of limitations that need to be resolved to improve its efficacy. A significant challenge is obtaining the variety and quality of data needed to train AI algorithms; since many systems rely on datasets that do not accurately reflect the heterogeneity inherent in actual agricultural circumstances, which could result in inaccurate disease identification (Hammond et al., 2020). Furthermore, it may be difficult for AI models to differentiate between similar symptoms of other diseases, leading to incorrect diagnosis and treatment suggestions (González-Rodríguez et al., 2024). Future research on advancing AI in disease detection should concentrate on combining data from multiple sources, including genomic, environmental, and meteorological data, to increase the models' durability and accuracy. According to Zhang et al. (2024), incorporating cutting-edge machine learning techniques, like ensemble methods and transfer learning, may enhance diagnostic accuracy in a variety of agricultural scenarios. The possibility for artificial intelligence to enhance crop health as well as food safety by resolving these issues and investigating novel strategies for managing diseases in agriculture.

5.1 Analysis of Background and Noise

Images of plant diseases include extraneous noise, multicolored backdrops, and extra features like roots, grass, soil, etc. Finding and separating these components is essential. One technique for removing polluted areas from the pictures taken is segmentation. The suggested automatic system must remove unnecessary elements from the image and isolate only the desired portion in order to efficiently identify diseased plants in the fields in real time (Jafar et al., 2024).

5.2 Factors Affecting the Acquisition of Images

Image acquisition in agricultural applications, especially with respect to plant disease diagnosis, is influenced by a number of aspects. It can be challenging to obtain a comparable image due to many elements such as environmental variables, moisture content, and light intensity. The performance of the system is significantly impacted by the tools used for image acquisition. The type of sample-taking device, the amount of moisture, the time of day, and the intensity of the light—all play a role in how accurate forecasts are. To get the best image quality in a variety of circumstances it is important to control these different factors (Jafar et al., 2024).

5.3 Identification and Isolation of Disease Symptoms

Agricultural research relies heavily on digital image processing, especially when it comes to recognizing and distinguishing identical signs of different diseases. To perform better, it is essential to separate the symptoms of diseases with identical characteristics. This endeavor, however, becomes difficult when many diseases have similar environmental variables and symptoms (Farhood et al., 2022).

5.4 The Impact of Dataset Size on Model Performance

In machine learning, dataset size has a big impact on model performance; bigger datasets allow for improved learning and generalization, capturing complex patterns and lowering the danger of overfitting. On the other hand, small datasets might make learning more difficult, increase variation, and result in inaccurate predictions, especially when it comes to regression and classification analyses. Data quality is essential; noisy or skewed data can impair performance, even while growing dataset size is advantageous. Together with the advantages of huge datasets, there are drawbacks that must be considered. Larger datasets also demand more computational power, requiring advanced equipment and effective algorithms (Bailly et al., 2022).

5.5 Unbalanced Data for Various Kinds of Diseases

Unbalanced patterns in the training dataset pose a challenge to automated detection methods. Diseases must be represented by comparable-sized samples, both infected and healthy, to prevent discrimination and maintain an equal and unbiased dataset for accurate analysis and prediction (Chen et al., 2024).

5.6 Concurrent Multiple Diseases

Malnutrition, various illnesses, Malnutrition, pests, and many diseases can all coexist in a single image; it is sometimes accurate to believe that each image represents a single disease. This makes it more difficult to identify and monitor a particular disease, and the way symptoms appear can differ depending on the exact geographic area. Consequently, when evaluating photos, it is essential to take these elements into perspective (Barbedo, 2016).

5.7 Disease with Similar Symptoms

The similarity of symptoms and patterns makes it difficult to identify diseases in agriculture. When conducting research, scientists usually go into the visible spectrum. Although it adds complexity, expense, and difficulties, including infrared spectral bands may aid in illness differentiation. Existing approaches might still be prone to mistakes, but these novel approaches might lessen the need for large datasets and the possibility of mistakes in farming techniques (Jafar et al., 2024).

6. Challenges of AI in Plant Disease Detection

Agricultural disease detection using AI provides a lot of promise, but before it can be widely adopted and shown to be successful, a number of challenges must be resolved. It is necessary to comprehend these challenges to enable develop trustworthy AI applications that can successfully support disease identification (Sahoo et al., 2024).

6.1 Differences in Image Quality and Lighting

Changes in lighting and image quality can have a significant impact on AI model performance, leading to inaccurate classifications and reduced disease detection performance (Wang et al., 2022).

6.2 Phases of Disease Development and Similarities Across Classes

Different stages of disease progression may show similar visual indicators, which makes it challenging for AI models to reliably differentiate between classes (Kumar et al., 2023).

6.3 Restricted Access to Various and Annotated Datasets

The lack of diverse and well-annotated datasets makes it challenging to train robust AI models and hinders their ability to generalize across a wide range of settings and conditions (Rather et al., 2024).

6.4 Difficulties with Computation in Managing Large Datasets

Because of the significant computational resources needed to process large-scale datasets, numerous companies may find it challenging to use AI solutions (Selmy et al., 2023).

6.5 Disease Identification in the Field in Real Time

Variable conditions and the requirement for quick processing and analysis make it difficult to achieve real-time diseases diagnosis in natural environments (Naveed et al., 2024).

6.6 Influence of Environmental Conditions on the Efficiency of Disease Detection

Humidity, temperature, and soil conditions are examples of environmental parameters that can impact the precision of AI-based disease diagnosis, making model training more difficult (Wang et al., 2021).

6.7 Merging Several Modal Data Sources

It is challenging but essential to combine data from multiple sources, including pictures, sensor readings, and environmental data, in order to increase the precision of disease diagnosis (Zhang et al., 2020).

6.8 Constructing Strong and Accessible Models to Address Social and Ethical Issues

Reliable and interpretable artificial intelligence models are critically needed to solve ethical problems as well as develop user trust (Gilpin et al., 2018).

6.9 Climate Change's Effects on Plant Diseases

Plant disease dynamics may change as an effect of climate change, creating AI systems that can adjust to changing environmental factors and disease trends critical (Harrison et al., 2024).

6.10 Incorporating IoT Devices, Drones, and Mobile Applications

Climate change can alter the dynamics of plant diseases, necessitating AI systems that adapt to evolving conditions and disease patterns (Harrison et al., 2024).

6.11 Innovations in Transfer Learning, Domain Adaptation, and Data Enhancement

While data augmentation and transfer learning can improve model performance, there are challenges in ensuring successful application in a variety of industries (Shorten and Khoshgoftaar, 2019).

6.12 Utilizing Crowdsourcing Platforms and Employing Citizen Scientists

Although using crowdsourcing platforms and using citizen scientists can improve collecting data, they may also raise worries over the accuracy and dependability of the information (Fritz et al., 2019).

7. The Future of AI in Plant Protection

7.1 AI-Powered Solutions for Integrated Smart Plant Protection

AI is transforming plant protection by making it possible to create networked smart systems that better monitor and preserve crops. These AI-powered services employ complex algorithms to identify diseases and pests quickly, enabling quick treatment. For instance, artificial intelligence (AI) can evaluate sensor and satellite imagery data to forecast outbreaks and optimize pesticide use, minimizing environmental effect and enhancing crop health (Zhang et al., 2022). The agriculture industry can solve important environmental problems, advance ecological health, and help create a more sustainable food system by incorporating AI into sustainable operations.

By developing networked systems for early disease and pest detection, optimizing pesticide use, and enhancing crop health, artificial intelligence is revolutionizing plant protection. Using cutting-edge computer vision models like YOLOv5 and AlexNet, studies in this field concentrate on AI models for a range of applications, such as fruit detection and disease diagnosis. Interestingly, the EADD-YOLO model performs quite well, with a detection accuracy of 95.5% and a speed of 625 FPS. AI algorithms and several mobile apps are available for diagnosing plant diseases (Table 1). Furthermore, numerous trials have proven AI approaches for non-visual tasks, such as protein site prediction, pest density estimation, and safe data-sharing techniques (Zhang et al., 2022). These developments highlight AI's enormous potential in intelligent plant protection and improve resilient as well as sustainable farming methods (Sawant et al., 2023).

Table 1. The roles of the various AI models used in plant pathology.

AI Model	Function
Convolutional Neural Networks (CNN)	Disease detection and classification from plant images.
Recurrent Neural Networks (RNN)	Disease progression modeling and time-series analysis.
Support Vector Machines (SVM)	Classification of diseases and identification of patterns.
Random Forests	Evaluation and forecasting of disease risk.
Deep Belief Networks (DBN)	Identification of diseases and feature extraction.
Long Short-Term Memory (LSTM)	Disease forecasting and temporal modeling.
Generative Adversarial Networks (GAN)	Creation of synthetic data for disease modeling.
Decision Trees	Diagnosis and decision-making of diseases based on rules.
Gaussian Processes	Analysis of disease propagation using Bayesian modeling.
K-nearest Neighbors (KNN)	Analysis of similarity and grouping of diseases.

7.2 Dynamic Fruit Detection and Counting in Real Time

Real-time fruit detection and counting has been greatly improved by recent developments in AI, such as better models like YOLOv5s, which are particularly helpful in large-scale orchards. To increase accuracy and efficiency while spotting fruits against complicated backdrops, these AI models use advanced techniques including feature concatenation and attention processes. Farmers can get accurate yield estimates by integrating drones or mobile cameras for real-time photograph capturing. This allows for more efficient harvest planning and resource allocation while reducing waste (Zhang et al., 2022).

This technology is a major advancement in automating agricultural processes and increasing production, even though it faces difficulties such as changing fruit conditions and the requirement for smooth connection with current farm systems. The effective operation of intelligent agricultural tools, especially when it comes to automated crop harvesting, depends on real-time object detection. On edge devices, which frequently have constrained computing power, ensuring the precision and speed of model recognition and detection is still a major challenge.

The architecture of application models must be optimized for stationary agricultural landscapes, even though the diversity of things is lower than in certain public datasets, such as the COCO dataset. Nevertheless, the complicated backgrounds found in agricultural environments, where leaves and fruits frequently overlap, provide difficulties for model precision and detection speed.

Deep learning methods are increasingly being applied to a range of agricultural applications, such as ripeness detection and yield prediction, fruit quality assessment, and autonomous picking, as smart agriculture continues to advance quickly. To predict crop yields by identifying fruit and improving the quality of incoming crops by tracking their growth conditions, several researchers have used deep learning over the past few decades (Dhaka et al., 2021).

7.3 Smart Clustering for Pest Management

An inventive method that uses AI to improve real-time monitoring and management of agricultural pests is adaptive clustering for pest detection. These models are very effective in a variety of agricultural contexts because they make use of complex algorithms that can learn and adjust to diverse environmental factors. The models can precisely detect pest infestations quickly by dynamically combining information from several sources, such as satellite imagery and sensors (Wang et al., 2024).

The capacity of adaptive clustering to continuously learn from fresh data is one of its main advantages. These AI models maintain excellent accuracy by updating their clustering algorithms in response to changes in the environment and pest behavior. This flexibility indicates that the pest detection system will continue to function well despite changing climatic conditions and agricultural landscapes (Wang et al., 2024).

When insect infestations are detected early, prompt and focused responses are possible. Because of its accuracy, fewer chemical pesticides are used, which lowers expenses and reduces the harm that excessive use of pesticides causes on the environment. For example, adaptive clustering models can assist farmers in identifying pest-affected areas within a field so they can apply pesticides only where necessary instead of all over the crop (Wang et al., 2024).

Furthermore, by integrating data from several sources, such as soil health indicators and weather patterns, these systems support methods for sustainable agriculture. Making better decisions is made possible by this holistic approach, which offers a thorough grasp of the variables affecting pest populations (Wang et al., 2024).

Recent research has shown that adaptive clustering works well in practical situations, such as Wang et al., (2024) study. Important improvements in pest detection timeliness and accuracy were demonstrated by their research, which improved crop health and yield results. The potential of adaptive clustering to transform pest management and boost overall agricultural productivity is demonstrated by these developments.

8. The Impact of AI on Sustainable Agriculture

8.1 Precision Farming

By analyzing data from several sources, including sensors, satellite imaging, and weather forecasts, artificial intelligence (AI) assists farmers in optimizing resource use. This reduces waste and environmental effect by enabling targeted applications of herbicides, fertilizers, and water (Adewusi et al., 2024). Precision agriculture gathers and analyzes enormous amounts of sensor data, drones, satellites, and agricultural equipment using cutting-edge technologies and artificial intelligence algorithms. Through this connection, farmers may maximize crop output, use resources optimally, and make data-driven decisions while reducing their negative effects on the environment. Controlling and observing crops, collecting and assessing data, Automated farming, weed and insect control, and automated agriculture are a few examples of AI applications in this industry (Kumar et al., 2024).

8.2 Crop Monitoring

Efficient crop health monitoring and rapid disease detection are essential for optimizing yields and reducing losses in the constantly shifting field of agriculture. In crop health management, smart sensing technologies have become revolutionary, transforming farmers' field monitoring practices and facilitating the early identification of possible hazards (Paul et al., 2022). Proactive treatments and focused utilization of resources are made possible by these cutting-edge technologies, which offer real-time information about the physiological and environmental aspects of crops. By continuously monitoring important attributes including plant health, nutritional status, and conditions in the environment, smart sensor devices offer crucial data for enhancing crops management strategies.

Additionally, the integration of machine learning algorithms and the early diagnosis of diseases using spectrum analysis offer valuable devices for preventing the spread of infections and establishing early disease management approaches into action. Drones and image technologies driven by AI further improve real-time crop health monitoring by spotting problems like pest infestations or nutrient deficits early enough. This allows for rapid action to minimize pesticide use and crop loss. Farmers' approaches to monitoring crop health, preventing disease, and using sustainable farming methods are being revolutionized by this fast and accurate monitoring, which provides efficient resource allocation and maximizes crop management techniques (Kumar et al., 2021).

8.3 Soil Management

Agriculture depends significantly on soil management, which includes actions to preserve and enhance soil health for the best possible crop yield and growth. This includes evaluating and modifying the nutrient levels in the soil, enhancing its structure, and controlling hazardous substances. Effective soil management guarantees agricultural sustainability as the cornerstone of the agricultural sector (Zhang et al., 2024). In agriculture, artificial intelligence (AI) technology that resembles human intellect is being utilized more and more to improve soil management.

AI can forecast important indications like nutrient content, pH value, and moisture levels by using machine learning algorithms to examine enormous quantities of data on soil health. This enables AI to offer scientific recommendations for fertilization and irrigation. According to Paul et al. (2022), this encourages soil fertility, enhances crop yield and quality, lowers soil erosion and pollution, and supports sustainable agricultural growth. Furthermore, high-precision soil monitoring and management are made possible by the combination of AI, remote sensing, and drone technology, which helps to detect problems like soil erosion, pests, and diseases early on. Long-term agricultural sustainability is promoted by this proactive approach, which safeguards the ecosystem and soil resources (Agrawal and Arafat, 2024).

8.4 Climate Adaptation

As a major worldwide concern, climate change requires proactive adaptation measures for preserving infrastructure, businesses, and communities. AI has become a potent instrument for creating novel approaches to climate change adaptation, identifying high-risk regions, forecasting natural disasters, and creating successful reaction plans. Early warning systems and climate modeling driven by AI provide vital assistance, facilitating prompt responses and improving community stability. To guarantee inclusion and transparency, however, ethical issues and biases in AI systems need to be addressed (Griffin et al., 2023).

Through the identification of sensitive areas and the optimization of resource usage to reduce emissions, AI-enabled solutions provide substantial benefits for adaptation to changes in the climate (Bag et al., 2023). Machine learning algorithms, for instance, create prediction models for future climate scenarios, while AI-powered sensors and drones allow for possibility to manage and monitor the consequences of severe weather in real time. AI also helps in the creation of energy-efficient infrastructure, which decreases carbon footprints and increases resilience (Zhang et al., 2020). Despite these advantages, issues like data quality, privacy, and unbalanced accessibility to technologies must be resolved if AI is successfully and fully utilized in adapting to climate change (Brown et al., 2023).

8.5 Optimization of the Supply Chain

Blockchain, IoT, and cloud computing are examples of Industry 4.0 technologies that are driving the world's progressive transformation to a digital future. The capacity of artificial intelligence (AI) to replicate human talents and improve problem-solving with increased precision and speed makes it stand out among these (Dirican, 2015). Even though the field of artificial intelligence (AI) is not novel, recent technological advancements have expanded its applications across a range of industries, including supply chain management (SCM) (Jarrahi, 2018).

In order to improve functionality, many businesses are switching from remote monitoring to highly autonomous AI-based systems as AI technology becomes a substantial competitive advantage. Business research and supply chain management, two domains known to gain greatly from AI applications, are among those that are being impacted by the expanding scholarly conversation on AI (Soni et al., 2023).

The role of AI to SCM has to be investigated, despite the strong interest from researchers and practitioners (Jarrahi, 2018; Nishant et al., 2020). AI contributes to sustainability by improving supply chain efficiency through demand prediction, logistics optimization, food waste reduction, and on-time delivery to customers (Davis, 2022).

8.6 Automated Farming

AI is revolutionizing farming methods by increasing productivity, sustainability, and efficiency when combined with agricultural robotics. Using accurate machine learning and computer vision algorithms, AI-powered robots effectively carry out tasks including harvesting, weeding, sowing and planting, avoiding crop damage and the need for pesticides. These autonomous field-navigating robots increase output, save expenses, and help with shortages of workers.

As the cornerstone of human civilization, agriculture is currently confronted with issues such as resource scarcity, population increase, and climate change. AI tools that maximize crop management and distribution of resources comprise of computer vision, deep learning, and robotics. Real-time insights regarding crop circumstances, variations in the weather, and the state of the soil are provided by artificial intelligence systems that evaluate data from drones, sensors, and satellites. This reduces resource waste and allows for sound decision-making (Dirican, 2015).

Through image identification and data analysis, artificial intelligence (AI) improves disease control and pest management by enabling early identification and response, which lowers the need for chemical pesticides. Automated irrigation systems and labor-intensive jobs are further revolutionized by autonomous robots, which provide optimal water use and support sustainable water controlling. Artificial intelligence and the Internet of Things (IoT) have been combined to create "smart farms," where a whole agricultural ecosystem is created using AI algorithms and networked devices (Jarrahi, 2018).

For optimal use of AI in agriculture, however, difficulties like data security, ethical issues, and small-scale farmers' accessibility must be resolved. By precisely planting, weeding, and harvesting, robotics and AI-powered equipment increase productivity while lowering the need for workers and chemical resources (García, 2023).

8.7 Data-Informed Decision Making

Agricultural decision-making is being revolutionized by automated forecasting and machine learning, which give farmers practical insights from large datasets. Farmers may forecast crop yields, keep updated on insect incidents, and manage resources as efficiently as possible by using predictive analytics, which uses statistical algorithms and historical data to forecast outcomes in the future. By detecting hidden patterns in data, a subfield of artificial intelligence ,namely, machine learning improves predictive analytics. For example, the best times to grow and harvest crops can be predicted by machine learning algorithms using variables like soil composition, weather trends, and past crop performance. This helps farmers to make well-informed decisions that minimize waste and resource usage while optimizing yields (Sharma et al., 2020).

Additionally, these technological advancements enable continuous monitoring in real time and modifications, sending out notifications for possible disease outbreaks or unfavorable weather patterns, allowing for proactive crop protection measures. Farmers may increase sustainability, lower costs, and boost production by using machine learning and predictive analytics into farm management techniques. This will ultimately make agricultural systems more adaptable to global problems. More sustainable practices and higher yields can result from farmers utilizing AI analytics to make intelligent choices based on evidence from the past and present (Miller, 2021).

8.8 Supports for Biodiversity

The provision of vital ecosystem services like clean water, pollination, food, and medicine depends on biodiversity, which includes all life forms and their interactions. However, short-term, unsustainable practices have reduced biodiversity and contributed to the global extinction problem. The threat to native ecosystems and species is increasing as a result of the impacts of global warming and the estimated 2.4 billion additional human beings by 2050. The urgent need for efficient policies that align with The Framework for Global Biodiversity after 2020 is demonstrated by inability to reach 20 Aichi Biodiversity Targets (Silvestro et al., 2022).

Since the 1960s, the focus of conservation efforts has shifted from protecting nature alone to acknowledging the complex interactions between biodiversity and people. As a result, systematic conservation planning and spatial preservation assigning have been developed. As natural environments and human activities increasingly overlap, it is critical to preserve intact ecosystems, such as tropical rainforests. Although they frequently fail to take climate impacts and temporal variations into consideration, tools such as Marxan helps in conservation planning optimization by finding protected areas that are cost-effective. By tracking species on farms, artificial intelligence (AI) tools improve assessments of biodiversity and direct policies that preserve ecosystems while increasing agricultural output (Watts et al., 2021).

AI is revolutionizing sustainable agriculture by improving crop management, facilitating more effective use of resources, and fostering climate change resilience. Technology has a growing potential to help develop a more sustainable food system as it develops further (Johnson, 2024).

9. Conclusions

AI in agriculture offers a number of opportunities to transform traditional farming into a reliable and efficient system. Artificial intelligence (AI) tools such as the Internet of Things (IoT) and machine learning increase agricultural sustainability and productivity by automating disease diagnosis and enhancing crop management. Nonetheless, there are still challenges, particularly concerning data quality, the need for substantial computing power, and issues with real-time disease diagnosis. To advance AI-driven agriculture ethically, it is imperative to protect data privacy, facilitate accessibility for smallholder farmers, and create clear legislative frameworks. The development of reliable AI models that function in various environmental settings and easily interface with other digital tools should be the main goal of future research. AI could improve food safety worldwide and have an impact on sustainable agriculture and environmental preservation with sustained innovation and effective assistance from the government.

References

Adewusi, A. O., Asuzu, O. F., Olorunsogo, T., Iwuanyanwu, C., Adaga, E., and Daraojimba, D. O. (2024). AI in precision agriculture: A review of technologies for sustainable farming practices. World Journal of Advanced Research and Reviews, 21(1): 2276–2285.

Agrawal, J., and Arafat, M. Y. (2024). Transforming Farming: A Review of AI-Powered UAV Technologies in Precision Agriculture. Drones, 8(11): 664. https://doi.org/10.3390/drones8110664.

Agrios, G. N. (2005). Plant pathology. Elsevier.

Al Rahhal, M. M., Bazi, Y., AlHichri, H., Alajlan, N., Melgani, F., and Yager, R. R. (2016). Deep learning approach for active classification of electrocardiogram signals. Information Sciences, 345: 340–354.

Albattah, W., Nawaz, M., Javed, A., Masood, M., and Albahli, S. (2022). A novel deep learning method for detection and classification of plant diseases. Complex & Intelligent Systems, 1–18.

Alengebawy, A., Abdelkhalek, S. T., Qureshi, S. R., and Wang, M. Q. (2021). Heavy metals and pesticides toxicity in agricultural soil and plants: Ecological risks and human health implications. Toxics, 9(3): 42.

Attri, I., Awasthi, L. K., Sharma, T. P., and Rathee, P. (2023). A review of deep learning techniques used in agriculture. Ecological Informatics, 102217.

Bag, S., Rahman, M. S., Rogers, H., Srivastava, G., and Pretorius, J. H. C. (2023). Climate change adaptation and disaster risk reduction in the garment industry supply chain network. Transportation Research Part E: Logistics and Transportation Review, 171: 103031.

Bailly, A., Blanc, C., Francis, É., Guillotin, T., Jamal, F., Wakim, B., and Roy, P. (2022). Effects of dataset size and interactions on the prediction performance of logistic regression and deep learning models. Computer Methods and Programs in Biomedicine, 213: 106504.

Bannerjee, G., Sarkar, U., Das, S., and Ghosh, I. (2018). Artificial intelligence in agriculture: A literature survey. International Journal of Scientific Research in computer Science applications and Management Studies, 7(3): 1–6.

Barbedo, J. G. A. (2016). Expert systems applied to plant disease diagnosis: survey and critical view. IEEE Latin America Transactions, 14(4): 1910and1922.

Basavaiah, J., and Arlene Anthony, A. (2020). Tomato leaf disease classification using multiple feature extraction techniques. Wireless Personal Communications, 115(1): 633–651.

Belsky, J., and Joshi, N. K. (2020). Effects of fungicide and herbicide chemical exposure on Apis and non-Apis bees in agricultural landscape. Frontiers in Environmental Science, 8: 81.

Brown, H., Butterfield, S., Davies, J., Dodd, S., and Morris, A. (2023). Exploring the association between health, local area characteristics and climate action plans in the UK: Cross-sectional analysis using administrative data from 2018 and a citizen science ranking of climate action plans from 2021. PLOS Climate, 2(9): e0000166.

Brown, P. (2021). Education, technology and the future of work in the fourth industrial revolution. Digital Futures of Work Research Programme, Working Paper, 2.

Chakraborty, S., and Newton, A. C. (2011). Climate change, plant diseases and food security: an overview. Plant Pathology, 60(1): 2–14.

Chaloner, T. M., Gurr, S. J., and Bebber, D. P. (2020). The global burden of plant disease tracks crop yields under climate change. bioRxiv, 2020-04.

Chen, W., Yang, K., Yu, Z., Shi, Y., and Chen, C. L. (2024). A survey on imbalanced learning: latest research, applications and future directions. Artificial Intelligence Review, 57(6): 1–51.

Davis, E. (2016). Sustainable development in Vietnam: The interconnectedness of climate change, socio-economic development, land use, and food security. Pursuit-The Journal of Undergraduate Research at the University of Tennessee, 7(1): 11.

Davis, K. F., Downs, S., and Gephart, J. A. (2021). Towards food supply chain resilience to environmental shocks. Nature Food, 2(1): 54–65.

Davis, P., Zarowiecki, M., Arnaboldi, V., Becerra, A., Cain, S., Chan, J. et al. (2022). WormBase in 2022—data, processes, and tools for analyzing *Caenorhabditis elegans*. Genetics, 220(4): iyac003.

Delgado-Baquerizo, M., Guerra, C. A., Cano-Díaz, C., Egidi, E., Wang, J. T., Eisenhauer, N. et al. (2020). The proportion of soil-borne pathogens increases with warming at the global scale. Nature Climate Change, 10(6): 550–554.

Dhaka, V. S., Meena, S. V., Rani, G., Sinwar, D., Ijaz, M. F., and Woźniak, M. (2021). A survey of deep convolutional neural networks applied for prediction of plant leaf diseases. Sensors, 21(14): 4749.

Dirican, C. (2015). The impacts of robotics, artificial intelligence on business and economics. Procedia-Social and Behavioral Sciences, 195: 564–573.

Fao, F. A. O. S. T. A. T. (2018). Food and agriculture organization of the United Nations. Rome, URL: http://faostat. fao. org, 403-403.

Farhood, A. N., Merhij, M. Y., and Al-Fatlawi, Z. H. (2022). Drought stress effects on resistant gene expression, growth, and yield traits of wheat (*Triticum aestivum* L.).

Fitt, B. D., McCartney, H. A., and West, J. S. (2006). Dispersal of foliar plant pathogens: mechanisms, gradients and spatial patterns. The Epidemiology of Plant Diseases, 159–192.

Fones, H. N., Bebber, D. P., Chaloner, T. M., Kay, W. T., Steinberg, G., and Gurr, S. J. (2020). Threats to global food security from emerging fungal and oomycete crop pathogens. Nature Food, 1(6): 332–342.

Fritz, S., See, L., Carlson, T., Haklay, M., Oliver, J. L., Fraisl, D. et al. (2019). Citizen science and the United Nations sustainable development goals. Nature Sustainability, 2(10): 922–930.

Gai, Y., and Wang, H. (2024). Plant Disease: A Growing Threat to Global Food Security. Agronomy, 14(8): 1615.

García, R., Aguilar, J., Toro, M., Pérez, N., Pinto, A., and Rodríguez, P. (2023). Autonomic computing in a beef-production process for Precision Livestock Farming. Journal of Industrial Information Integration, 31: 100425.

Gilpin, L. H., Bau, D., Yuan, B. Z., Bajwa, A., Specter, M., and Kagal, L. (2018, October). Explaining explanations: An overview of interpretability of machine learning. In 2018 IEEE 5th International Conference on data science and advanced analytics (DSAA) (pp. 80–89). IEEE.

González-Rodríguez, V. E., Izquierdo-Bueno, I., Cantoral, J. M., Carbú, M., and Garrido, C. (2024). Artificial intelligence: A promising tool for application in phytopathology. Horticulturae, 10(3): 197.

Greenfield, P., Slavich, E., Jamieson, W., and Dencheva, N. (2022). The Advanced Scientific Data Format (ASDF): An Update. In SciPy (pp. 1–6).

Greenfield, R. (2023). Catering to the experts: food marketing and health professionals in the early 20th century. Journal of Historical Research in Marketing, 15(3): 147–170.

Griffin, C., Wreford, A., and Cradock-Henry, N. A. (2023). 'As a farmer you've just got to learn to cope': Understanding dairy farmers' perceptions of climate change and adaptation decisions in the lower South Island of Aotearoa-New Zealand. Journal of Rural Studies, 98: 147–158.

Hammond, J., Rosenblum, N., Breseman, D., Gorman, L., Manners, R., van Wijk, M. T. et al. (2020). Towards actionable farm typologies: Scaling adoption of agricultural inputs in Rwanda. Agricultural Systems, 183: 102857.

Harrison, S., Franklin, J., Hernandez, R. R., Ikegami, M., Safford, H. D., and Thorne, J. H. (2024). Climate change and California's terrestrial biodiversity. Proceedings of the National Academy of Sciences, 121(32): e2310074121.

Hassan, S. A. E., Akbar, S., Gull, S., Rehman, A., and Alaska, H. (2021, April). Deep learning-based automatic detection of central serous retinopathy using optical coherence tomographic images. In 2021 1st International Conference on Artificial Intelligence and Data Analytics (CAIDA) (pp. 206–211). IEEE.

Huang, J., Kumar, S. R., Mitra, M., Zhu, W. J., and Zabih, R. (1997, June). Image indexing using color correlograms. In Proceedings of IEEE computer society conference on Computer Vision and Pattern Recognition (pp. 762–768). IEEE.

Jafar, A., Bibi, N., Naqvi, R. A., Sadeghi-Niaraki, A., and Jeong, D. (2024). Revolutionizing agriculture with artificial intelligence: plant disease detection methods, applications, and their limitations. Frontiers in Plant Science, 15: 1356260.

Jarrahi, M. H. (2018). Artificial intelligence and the future of work: Human-AI symbiosis in organizational decision making. Business Horizons, 61(4): 577–586.

Johnson, D. (2024). Food security, the agriculture value chain, and digital transformation: The case of Jamaica's agricultural business information system (ABIS). Technology in Society, 77: 102523.

Johnson, M., Albizri, A., Harfouche, A., and Fosso-Wamba, S. (2022). Integrating human knowledge into artificial intelligence for complex and ill-structured problems: Informed artificial intelligence. International Journal of Information Management, 64: 102479.

Khan, I. H., Liu, H., Li, W., Cao, A., Wang, X., Liu, H. et al. (2021). Early detection of powdery mildew disease and accurate quantification of its severity using hyperspectral images in wheat. Remote Sensing, 13(18): 3612.

Kshetri, N. (2021). Blockchain and sustainable supply chain management in developing countries. International Journal of Information Management, 60: 102376.

Kumar, K. A., Reddy, G. O., Masilamani, P., Turkar, S. Y., and Sandeep, P. (2021). Integrated drought monitoring index: A tool to monitor agricultural drought by using time-series datasets of space-based earth observation satellites. Advances in Space Research, 67(1): 298–315.

Kumar, R., Farooq, M., and Qureshi, M. (2024). Advancing precision agriculture through artificial intelligence: Exploring the future of cultivation. In A Biologist s Guide to Artificial Intelligence (pp. 151–165). Academic Press.

Kumar, T. S., and Kurian, C. P. (2023). Real-time data based thermal comfort prediction leading to temperature setpoint control. Journal of Ambient Intelligence and Humanized Computing, 14(9): 12049–12060.

Kumar, Y., Koul, A., Singla, R., and Ijaz, M. F. (2023). Artificial intelligence in disease diagnosis: a systematic literature review, synthesizing framework and future research agenda. Journal of Ambient Intelligence and Humanized Computing, 14(7): 8459–8486.

Lavecchia, A. (2019). Deep learning in drug discovery: opportunities, challenges and future prospects. Drug Discovery Today, 24(10): 2017–2032.

Liu, L., Guo, X., Liu, W., and Lee, C. (2021). Recent progress in the energy harvesting technology—from self-powered sensors to self-sustained IoT, and new applications. Nanomaterials, 11(11): 2975.

Mana, A. A., Allouhi, A., Hamrani, A., Rahman, S., el Jamaoui, I., and Jayachandran, K. (2024). Sustainable AI-Based Production Agriculture: Exploring AI Applications and Implications in Agricultural Practices. Smart Agricultural Technology, 100416.

McCarthy, J., Minsky, M. L., Rochester, N., and Shannon, C. E. (2006). A proposal for the dartmouth summer research project on artificial intelligence, august 31, 1955. AI magazine, 27(4): 12-12.

Mesías-Ruiz, G. A., Pérez-Ortiz, M., Dorado, J., De Castro, A. I., and Peña, J. M. (2023). Boosting precision crop protection towards agriculture 5.0 via machine learning and emerging technologies: A contextual review. Frontiers in Plant Science, 14: 1143326.

Miller, M. (2021). Identifying critical thresholds for resilient regional food flows: A case study from the US upper midwest. Frontiers in Sustainable Food Systems, 5: 684159.

Mitra, D. (2021). Emerging plant diseases: research status and challenges. Emerging trends in plant pathology, 1–17.

Narain, T., Sahu, P., and Singh, A. P. (2022, September). Plant Disease Classification Using Siamese Convolutional Neural Network. In International Conference on Innovative Computing and Communications: Proceedings of ICICC 2022, Volume 2 (pp. 57–65). Singapore: Springer Nature Singapore.

Nath, C. P., Kumar, N., Das, K., Hazra, K. K., Praharaj, C. S., and Singh, N. P. (2021). Impact of variable tillage based residue management and legume based cropping for seven years on enzymes activity, soil quality index and crop productivity in rice ecology. Environmental and Sustainability Indicators, 10: 100107.

Naveed, M., Majeed, M., Jabeen, K., Hanif, N., Naveed, R., Saleem, S. et al. (2024). Plant Disease Diagnosis with Artificial Intelligence (AI). In Microbial Data Intelligence and Computational Techniques for Sustainable Computing (pp. 217–234). Singapore: Springer Nature Singapore.

Nishant, R., Kennedy, M., and Corbett, J. (2020). Artificial intelligence for sustainability: Challenges, opportunities, and a research agenda. International Journal of Information Management, 53: 102104.

Pacal, I., Kunduracioglu, I., Alma, M. H., Deveci, M., Kadry, S., Nedoma, J. et al. (2024). A systematic review of deep learning techniques for plant diseases. Artificial Intelligence Review, 57(11): 304.

Parveen, A., Anjum, M., Mahmood, S., Nawaz, R., and Khalid, A. (2023). Environmental implication and cancer risk assessment of residual pollutants in cotton crop: a case study of Multan District, Pakistan. Environmental Monitoring and Assessment, 195(9): 1100.

Paul, K., Chatterjee, S. S., Pai, P., Varshney, A., Juikar, S., Prasad, V. et al. (2022). Viable smart sensors and their application in data driven agriculture. Computers and Electronics in Agriculture, 198: 107096.

Rather, I. H., Kumar, S., and Gandomi, A. H. (2024). Breaking the data barrier: a review of deep learning techniques for democratizing AI with small datasets. Artificial Intelligence Review, 57(9): 226.

Ritchie, H., and Roser, M. (2024). Smallholders produce one-third of the world's food, less than half of what many headlines claim. Our World in Data.

Sahoo, P., Sharma, S. K., Saha, S., Jain, D., and Mondal, S. (2024). A multistage framework for respiratory disease detection and assessing severity in chest X-ray images. Scientific Reports, 14(1): 12380.

Sanida, T., Sideris, A., Sanida, M. V., and Dasygenis, M. (2023). Tomato leaf disease identification via two–stage transfer learning approach. Smart Agricultural Technology, 5: 100275.

Sawant, S. B., Raju, R. S., Jyothi, G. B. N., and Behera, L. (2023). Biotica Research Today.

Selmy, H. A., Mohamed, H. K., and Medhat, W. (2023). Big data analytics deep learning techniques and applications: A survey. Information Systems, 102318.

Sharma, R., Kamble, S. S., Gunasekaran, A., Kumar, V., and Kumar, A. (2020). A systematic literature review on machine learning applications for sustainable agriculture supply chain performance. Computers & Operations Research, 119: 104926.

Shoaib, M., Shah, B., Ei-Sappagh, S., Ali, A., Ullah, A., Alenezi, F. et al. (2023). An advanced deep learning models-based plant disease detection: A review of recent research. Frontiers in Plant Science, 14: 1158933.

Shorten, C., and Khoshgoftaar, T. M. (2019). A survey on image data augmentation for deep learning. Journal of Big Data, 6(1): 1–48.

Silvestro, D., Goria, S., Sterner, T., and Antonelli, A. (2022). Improving biodiversity protection through artificial intelligence. Nature Sustainability, 5(5): 415–424.

Singh, M. M., Rastogi, P., Verma, S., Singh, R. P., and Tiwari, P. (2024). A Review on Cutting-Edge Deep Learning Models in the Detection and Classification of Plant Diseases. Available at SSRN 4804223.

Smith, A., and Anderson, J. (2014). AI, Robotics, and the Future of Jobs. Pew Research Center, 6: 51.

Smith, D., and Johnson, S. (2024). AI-Powered Automation: Impacts on Workforce Dynamics and Economic Growth. MZ Computing Journal, 5(1).

Smith, M., Lal, P., and Vedwan, N. (2023). Motivations underlying farmers' management decisions and willingness to adopt sustainable practices: A case study of the northeastern United States. Journal of Rural Studies, 103: 103138.

Smith, V., Wesseler, J. H., and Zilberman, D. (2021). New plant breeding technologies: An assessment of the political economy of the regulatory environment and implications for sustainability. Sustainability, 13(7): 3687.

Smith, W., Grant, B., Qi, Z., He, W., Qian, B., Jing, Q. et al. (2020). Towards an improved methodology for modelling climate change impacts on cropping systems in cool climates. Science of the Total Environment, 728: 138845.

Snoeijer, B. T., Burger, M., Sun, S., Dobson, R. J., and Folarin, A. A. (2021). Measuring the effect of Non-Pharmaceutical Interventions (NPIs) on mobility during the COVID-19 pandemic using global mobility data. NPJ Digital Medicine, 4(1): 81.

Soni, P., Dave, V., and Repalle, N. B. (2023, December). AI Based Performance Boost in Solar PV Fuel Cell Hybrids. In International Conference on Renewable Energy, Green Computing, and Sustainable Development (pp. 3–16). Cham: Springer Nature Switzerland.

Sugimura, D., Mikami, T., Yamashita, H., and Hamamoto, T. (2015). Enhancing color images of extremely low light scenes based on RGB/NIR images acquisition with different exposure times. IEEE Transactions on Image Processing, 24(11): 3586–3597.

Taddeo, M., and Floridi, L. (2018). How AI can be a force for good. Science, 361(6404): 751–752.

Tantalaki, N., Souravlas, S., and Roumeliotis, M. (2019). Data-driven decision making in precision agriculture: The rise of big data in agricultural systems. Journal of Agricultural & Food Information, 20(4): 344–380.

Větrovský, T., Kohout, P., Kopecký, M., Machac, A., Man, M., Bahnmann, B. D. et al. (2019). A meta-analysis of global fungal distribution reveals climate-driven patterns. Nature Communications, 10(1): 5142.

Wang, J., Chen, Y., Huang, J., Jiang, X., and Wan, K. (2024). Leveraging machine learning for advancing insect pest control: A bibliometric analysis. Journal of Applied Entomology.

Wang, T., Zhang, T., Zhang, B., Ouyang, H., Chen, D., Chen, Q. et al. (2022). Pretraining is all you need for image-to-image translation. arXiv preprint arXiv:2205.12952.

Wang, X., Liu, J., and Zhu, X. (2021). Early real-time detection algorithm of tomato diseases and pests in the natural environment. Plant Methods, 17(1): 43.

Warwick, K. (2013). Artificial intelligence: the basics. Routledge.

Watts, N., Amann, M., Arnell, N., Ayeb-Karlsson, S., Beagley, J., Belesova, K. et al. (2021). The 2020 report of The Lancet Countdown on health and climate change: responding to converging crises. The lancet, 397(10269): 129–170.

Williams, M. J., Sikder, M. N. K., Wang, P., Gorentala, N., Gurrapu, S., and Batarseh, F. A. (2023). The application of artificial intelligence assurance in precision farming and agricultural economics. In AI Assurance (pp. 501–529). Academic Press.

Wilson, M. (2023). Constructing measures: An item response modeling approach. Routledge.

Zhang, J., Yin, Z., Chen, P., and Nichele, S. (2020). Emotion recognition using multi-modal data and machine learning techniques: A tutorial and review. Information Fusion, 59: 103–126.

Zhang, L., Ling, J., and Lin, M. (2022). Artificial intelligence in renewable energy: A comprehensive bibliometric analysis. Energy Reports, 8: 14072–14088.

Zhang, T., Cai, Y., Zhuang, P., and Li, J. (2024). Remotely sensed crop disease monitoring by machine learning algorithms: A review. Unmanned Systems, 12(01): 161–171.

CHAPTER 12

Precision Spraying and Targeted Pesticide Application

Subhan Ali, Akhtar Hameed,[1,*] *Abdul Rehman,*[2] *Rana Binyamin,*[1] *Hafiz Muhammad Usman Aslam,*[1,3] *Muhammad Waqar Alam*[4] and *Rabia Rasheed*[1]

1. Introduction

Precision agriculture is a recent concept of managing agricultural production by employing technology to practices that enhance crop productivity and minimize the effects of the agriculturalactivity on the environment. It employs different tools and processes, for instance, GPS technique, sensors as well as data metric to obtain information on crops growth, soil health and pest attacks. The data collected is later analyzed and utilized to allocate resources, manage crops, and other interventions that are deemed necessary. It relies on proper assessments of spatial variability in a short scale of resolution in order to inform farmers regarding resource allocation and pest control (Zanin et al., 2022). Most agricultural prices fluctuate in response to market forces of supply and demand. Reducing production costs without sacrificing the production of crops or raising the activity's risk is the only option for farmers to become more profitable. Precision farming aims to reduce environmental impact while improving production and profitability through increased machine and input efficiency (Schiefer and Dillon, 2015). Using field-based approaches, precision agriculture improves agricultural production management. Accurately assessing spatial differences at short resolution scales is required (Jurado-Expósito et al., 2019; Zanin et al., 2022) (Figure 1).

[1] Institute of Plant Protection, MNS University of Agriculture Multan 61000, Pakistan.
[2] Department of Plant Pathology, University of Agriculture Faisalabad 38000, Pakistan.
[3] Department of Plant Pathology, San Luis Valley Research Center, Colorado State University, Fort Collins, CO, United States.
[4] Department of Botany, University of Okara 53600, Okara Pakistan.
* Corresponding author: akhtar.hameed@mnsuam.edu.pk

DOI: 10.1201/9781003546702-12

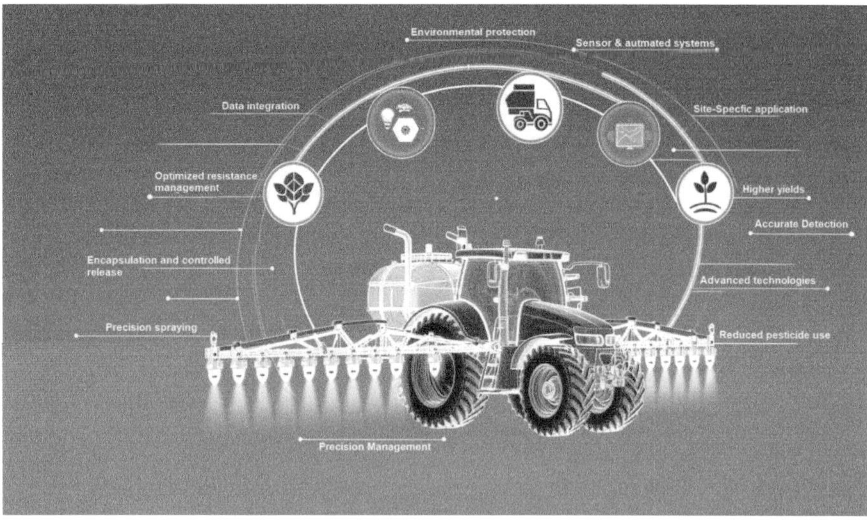

Figure 1. Precision spraying application in agriculture enhancing yields and environmental protection with reduced pesticide.

Throughout crop management, pesticides will inevitably leak out, especially through spray drift, which has resulted in a number of adverse results: high risk of exposure and inhalation for livestock as well as individuals (Feltracco et al., 2022); environmental contamination of surface water and other living things (Carazo-Rojas et al., 2018; Tauchnitz et al., 2020); inadvertent damage to nearby plants and crops (Marrs and Frost, 1997; Marrs et al., 1989); and excessive toxic residues in agricultural food and products (Damalas and Eleftherohorinos, 2011; Xun et al., 2023).

The use of chemical pesticides with their hazardous properties should be reduced by 50% by 2030, according to two pesticide reduction targets that have been proposed as part of the Farm to Fork Strategy in a European Green Deal. These targets are in response to the growing public concern about the environmental contamination and health risks caused by agrochemical applications (European Commission, 2020). These challenging objectives have been approached from a variety of aspects and with a range of approaches, including introducing new tools and technologies to maximize the application of spray, enhancing end-user education, and raising the level of new technology and precision farming adoption (Xun et al., 2023).

1.1 Importance of Precision Spraying and Targeted Pesticide Application

Precision agriculture involves advanced technologies for precision spraying and targeted pesticide application, ensuring minimal environmental contact and minimal contact with the environment. It has been found that precision spraying is capable of reducing the pesticide usage up to 70% while ensuring the 100% removal of weeds (Zanin et al., 2022). There are several drawbacks of conventional pesticide application, which include chemical over-reliance, causing environmental pollution,

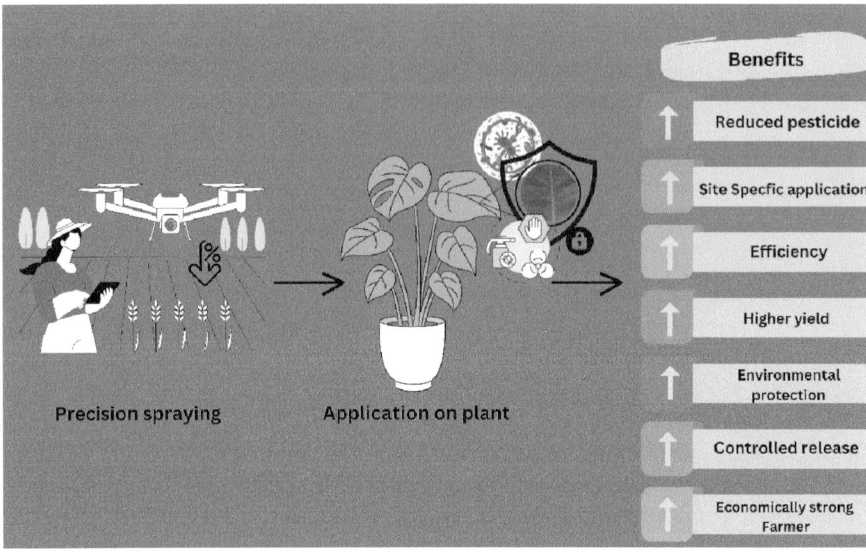

Figure 2. Importance of precision spraying and targeted pesticide application in agriculture.

adverse impacts on human health and leading to pest pesticide resistance. On the other hand, precision spraying is a method of pesticide application in specific areas or on specific plants only thereby limiting the quantity of the chemicals used and hence limits harm to non-target species and the environment (Batte and Ehsani, 2006). It has gained importance especially in contemporary farming because it assists in dealing with the emerging issues of toxicity of pesticides to the environment and human health (Figure 2).

Crop quality and yield may be significantly affected by crop protection, which is a crucial step in crop management. The primary technique for controlling weeds, diseases and pests is still the application of chemical pesticides, even if the use of biological agents has increased. However, in reality pesticide applications, a significant amount—between 30 and 50 percent—of the applied plant protection products (PPPs) are lost to non-target sites due to run-off and spray drift loss because of growers' ongoing lack of knowledge and training for enhancing and optimizing the spray application (Xun et al., 2023).

1.1.1 Historical context and evolution

Precision spraying has a rich history dating back to the early 20th century. The first sprayer was used in the USA in the 1850s to protect potatoes from the Colorado Beetle. This early device had a knapsack tank but no pump, so the spray was fed by gravity to two "sprinklers." A pump on the knapsack was introduced later, and the use of pesticides became more widespread (Matthews, 2018b).

The first major advance in precision spraying came in early 20th century through the introduction of air-blast sprayers. These sprayers utilized powerful air jets to propel the pesticide spray up as well as into the crop canopy, which made it easier and more effective to apply the pesticide (Mahmud et al., 2021). Since 1920, most of the work done in site-specific agriculture has focused on nutrient management of crops (Franzen and Mulla, 2015). From this innovation, air assisted spraying became more popular among farmers since they understood the importance of improving on this method

In the subsequent decades more enhancements of the design and technology of sprayer were made to increase the level of accuracy of pesticide application. Boom sprayers and other new techniques including mist blowers, handgun, etc., were invented in the Europe which were adopted by growers from all across the world (Wang et al., 2022). These sprayers enabled the pesticides to be applied in large surfaces with less effects on the environment since they were precise in their application. In the recent two to three decades, the enhanced modern sprayers that include the tunnel sprayer, tower sprayer and the precision sprayer have boosted the prospect of precision spraying (Mahmud et al., 2021). These advanced systems incorporate a number of sensors and automatons for the accurate and appropriate spraying of pesticide, reducing the impacts of off-target deposition and drift.

One of the technological advancement in precision spraying has been the use of GPS and GIS (Mahmud et al., 2021; Franzen and Mulla, 2015). They can help the farmers to locate their fields and monitor the path of their sprayers so that they spray pesticides on the right areas only. It has also contributed to a dramatic decrease in the volumes of pesticides applied, better protection of crops, and fewer effects on the environment. Another significant advancement in precision spraying has been the incorporation of drones and unmanned aerial vehicles (UAVs) (Mulla and Khosla, 2016). These technologies have been in use in Japan since mid 1990s to apply sprays on rice, a method that is now popular in China. In trials in the USA, drones have been evaluated for spraying vines. But with nearly 100l/ha there is an issue with frequent re-filling required due to low payload capacity of drones (Matthews, 2018a; Matthews, 2018b).

1.2 Relevance of Nanotechnology in Agriculture

1.2.1 Introduction to nanofertilizers

Nanotechnology has been considered as a potential technology in the agricultural field in the aspect of contributing a superior technique for pesticide spraying. The application of nanotechnology in pesticide formulation has been aimed at increasing the efficacy, specificity and environmentally friendly nature of these compounds (Kim et al., 2018). Nanotechnology is the capacity to manipulate structures that are between 1 and 100 nanometers in size. In the case of pesticides, nanotechnology results in the development of nano-formulations which is comprised of one or more nanomaterials. Advantages of these nano-formulations over conventional pesticide formulation embraces better dissolution, higher bio-availability and targeted delivery to the pests/tissues of the targeted plant (Kah et al., 2013). By regulating the size, morphology, and surface chemistry of the nanomaterials, the researchers are well placed to develop more effective and less hazardous nano-pesticides.

The difficulties associated with applying conventional pesticides can be used to demonstrate the significance of adopting nanotechnology in pesticide application. This could be due to reasons such as low solubility, prompt biodegradation, other side effects, which imply that higher quantities of pesticides are used in the process which therefore leads to pollution. Such limitations can be overcome by the nano formulations because of the increased stability and the bioavailability of the pesticides used, small quantities of pesticides, and very low impact on the non-target organisms (Kah and Hofmann, 2014). This integration could mean more efficient pest control, low levels of pesticide residues on the food and the environment and slow rate of emergence of resistance in the target pest species.

1.3 Integration of Nanotechnology in Pesticide Application

Nano-formulations can be applied through modern techniques which include UAV and ground applicator equipped with sensors and GPS. These precision spraying systems can identify and spray only the precise pest or weed, thereby minimizing the total quantity of pesticide used and off-target exposure rates (Khot et al., 2012). Also, pesticides in nano-formulations can be formulated in a manner that it releases the active ingredients a the right time and at the correct location where the target organisms are found.

Several nano-formulations of pesticides have been studied by researchers which include nano-emulsions, nano-suspensions, and nano-capsules. These formulations have been found to be effective in enhancing efficacy, decreasing toxicity and increasing the biodegradability of the formulations (Kah et al., 2013). For instance, Kah et al. (2014) found that nano-formulations of imidacloprid insecticide yielded better results in controlling aphids than the conventional formulations despite the reduced application rate of between 50%.

2. Fundamentals of Precision Spraying

Precision spraying methods have come out as the hope towards achieving sustainable agriculture in the face of increasing global population demanding for more food while not compromising the environment (Chowdhury et al., 2023). Through the efforts of agricultural engineers, crop protection strategies have been stepped up through the use of advanced tools and better strategies in transforming blanket spraying methods to strategic and efficient programs. Swapping from conventional spraying reduces costs and pollutes the environment since the implementation of precision spraying directs the inputs into the required areas. Precision spraying, which aims at achieving high degrees of accuracy and efficiency in the use of resources, can significantly reshape crop protection and facilitate the improvement of environmental standards (Mahmud et al., 2021). Not only does precision spraying have environmental benefits, but farmers might save money on it as well. Pesticides, herbicides, and other related inputs may be used as carefully as possible so that farmers can minimize the costs of using chemicals and maximize the efficient use of resources. Precision spraying also enhances crop protection, and this leads to higher yields, better product quality, and increased market access and competitiveness. Farmers are enabled to move to a more sustainable and notably more profitable type of farming with the help of new sensor technology, automation, variable rate application systems, and site-specific spraying (Zanin et al., 2022).

The site specific pesticide application principle involves determining the spatial distribution of the crop problem that is being addressed (by weed, disease, crop canopy density or insects). Site-specific weed control is more prevalent and may be carried out using two main approaches: real-time mapping and application and weed mapping, and the generation of prescription map having herbicide volume rates (Zanin et al., 2022). Precision spraying is among the features of modern farming that contributes to productivity while the world is facing the necessity to produce more food with regards to the sustainable usage of resources.

2.1 Advantages Over Traditional Spraying Methods

The effectiveness and efficiency of the application are significantly affected by the spraying mechanism, which produces and delivers spray droplets containing active substances to the targeted plant. As of now, Europe uses three primary types of spraying systems for tree crops: high, efficient and traditional production systems, which are compatible with BMPs. The traditional method is most commonly used with traditional airblast sprayers, usually running at a high rate of airflow/pesticide volume without adequate control or crop architecture modification. This spraying technique is commonly used by growers where they overspray insecticides with high chances of drift and runoff (Balsari et al., 2002; Salyani et al., 2013). The sprayer is carefully adjusted and fine-tuned to deliver the desired spraying system that conforms to best management practices based on the type of tree crop targeted, the prevailing environmental conditions, and the use of application technology (Balsari et al., 2011). Reducing the spray drift when comparing the optimization to the traditional system is another finding (Balsari et al., 2002; Doruchowski et al., 2009). The precise spraying system has been designed and put into use, thanks to advancements in sensing technology. Based on the extensive information gathered about the target vegetation volume, weather, and pest and disease infections, this sophisticated technology can accurately change the spray volume and/or air flow rate in real-time to accomplish site-specific management (Balsari et al., 2008; Khot et al., 2012; Gil et al., 2013; Escolà et al., 2013; Chen et al., 2013). The precision spraying system shows a 34–88% potential savings of spray volume compared with the conventional system while still achieving similar or even better deposition quality in the target vegetation (Llorens et al., 2010; Chen et al., 2011; Campos et al., 2019; Xun et al., 2023).

2.2 Technological Components

Precision agriculture depends on the application of technology in improving the effectiveness of application of pesticides. Two significant subtopics are Global Positioning System (GPS) and Geographic Information System (GIS) technologies, and sensors and automated systems. These technologies facilitate effective management and supervision of the agricultural activities thus minimizing wastage and enhancing productivity (Table 1).

Table 1. Technological Components in Precision Spraying and Targeted Pesticide Application.

Component	Function	Mechanism	Benefits	Example	References
GPS and GIS Technologies	Accurate field mapping and navigation	Provides precise location data for equipment navigation and field mapping.	Minimizes over-application, reduces environmental impact.	Mapping fields to identify pest-prone areas	Loudjani (2014)
GPS	Location data provision	Utilizes satellites to determine exact positions on the earth's surface.	Enhances precision in pesticide application.	Tractor navigation for spraying	Loudjani (2014)
GIS	Spatial data analysis	Analyzes and visualizes spatial data to create detailed maps of fields.	Guides targeted application of pesticides, improves resource management.	Field maps showing soil moisture levels	Mulla (2013)
Sensors and Automated Systems	Real-time condition monitoring and adjustment	Detects environmental and crop conditions to inform application rates dynamically.	Reduces waste, ensures precise pesticide application.	Optical sensors detecting weed density	Sudduth et al. (2008)
Optical Sensors	Environmental condition detection	Uses light-based technology to measure soil moisture, nutrient levels, and pest presence.	Enables selective herbicide application, reduces waste.	Identifying areas needing more nutrients	Shanahan et al. (2008)
Automated Systems	Dynamic application adjustment	Links to sensors to automatically adjust pesticide application rates based on real-time data.	Ensures exact amount of pesticide application, enhances efficiency.	Automated sprayers adjusting for pest density	Bongiovanni and Lowenberg-Deboer (2004)
Variable Rate Technology (VRT)	Variable pesticide dosage application	Adjusts pesticide dosage according to the specific needs of different field zones.	Reduces waste, improves application precision.	Spraying higher doses in areas with higher pest concentration	Bongiovanni and Lowenberg-Deboer (2004)
UAVs and Ground Robots	Autonomous navigation and application	Navigates fields autonomously and applies pesticides with high precision.	Reduces labor costs, improves application accuracy.	Drones spraying pesticides on crops	Perz and Wronowski (2019)

2.2.1 *GPS and GIS technologies*

GPS and GIS technologies are critical in precision agriculture. GPS gives accurate location information which is important in mapping of fields and navigation of agricultural equipment with a lot of precision. This makes it easier for farmers to spray the pests only on the affected areas thus avoiding over-spraying which results in polluting the environment (Loudjani, 2014).

On the other hand, while GPS provides means for navigation, GIS provides tools for analyzing and displaying data gathered on different geographical scales. Farmers can map out their fields and produce different shades to show differences in soil type, moisture content and pest infection. These maps help in the selective use of pesticides where each part of the field requires the treatment to be applied. The application of GPS and GIS has brought about major advancements in precision farming, particularly in resource management and crop yields (Mulla, 2013).

2.2.2 *Sensors and automated systems*

Precision spraying is facilitated by the use of sensors as these are instrumental in furnishing real-time information about the different environmental and crop conditions. They can for example, measure the moisture content of the soil, nutrient concentration, and pests' presence. For example, optical sensors can determine weed species and weed density to apply exactly the necessary amount of herbicide (Shanahan et al., 2008).

In many cases, these sensors are automated systems that change the amount and frequency of pesticide application according to the data collected. This automation helps to apply only the right amount of pesticide required avoiding wastage of pesticide and improved productivity. Additional advanced systems may embrace usage of variable rate technology (VRT) which calculates the precise concentration of the pesticide to be applied depending on the level of pest pressure in the various field sections (Bongiovanni and Lowenberg-Deboer, 2004).

Sensors with automated systems also assist UAVs and ground robots advantageous for precision spraying and the development of smart agriculture. Some can self-propel themselves on the fields and calibrate the amount of pesticides to apply, doing so with precision and ease. Recognizing that UAVs could potentially lower labor intensity and increase the efficiency of the dispersed pesticide, the study delves into the applicability of UAV towards precision farming (Perz and Wronowski, 2019).

2.3 Types of Precision Sprayers

There are several categories of precision sprayers based on the method of operation as well as the control system, including the aerial and ground-based sprayers, the manual and automatic systems (Table 2).

2.3.1 *Aerial vs. ground-based sprayers*

Aerial sprayers that utilize UAVs and manned aircraft implement precision agriculture that involves the large area mapping using GPS and GIS for spraying uniform patterns (Zhang and Kovacs, 2012). For instance, drones have garnered

Table 2. Types of Precision Sprayers in Agriculture.

Type	Function	Mechanism	Benefits	Disadvantages	Example	Citations
Aerial Sprayers	Rapid, large-area pesticide application	Utilize drones or manned aircraft equipped with GPS and GIS for precise application patterns.	Cover large areas quickly, ideal for difficult terrains, high accuracy.	- Typically more expensive to operate - May not be as precise as ground-based systems - Can be affected by weather conditions	Drones spraying pesticides on tall crops	Zhang and Kovacs (2012); Jin and Tang (2010)
Ground-based Sprayers	Close-range, precise application	Mounted on tractors or autonomous vehicles, equipped with sensors and VRT for real-time adjustments.	Ensures precise application, reduces waste, adapts to specific field conditions.	- Limited to areas accessible by ground vehicles - May require more time to cover large areas	Tractors with VRT systems for variable pesticide application	Batte and Ehsani (2006); Sudduth et al. (2005)
Manual Sprayers	Direct, adaptable pesticide application	Operated by human labor, allowing for observation and immediate adjustment.	Cost-effective, adaptable for small-scale farms, precise for individual plants.	- Labor-intensive and time-consuming - May not provide consistent application rates - Limited precision and control	Backpack sprayers in vineyards and orchards	Popp et al. (2013)
Automated Systems	Efficient, large-scale pesticide application	Use of GPS, GIS, sensors, and robotic technologies for uniform and precise application.	Reduces labor costs, minimizes human error, enhances efficiency.	- Typically more expensive to purchase and maintain - May require specialized training for operation - Can be affected by technical issues or malfunctions	Tractor-mounted sprayers with VRT, drones for targeted spraying	Zhang and Kovacs (2012); Sudduth et al. (2005); Sishodia et al. (2020)

much attention in precision agriculture because of their operation at relatively low altitudes and spatial positioning of the application, especially in variable rate (Jin and Tang, 2010).

In contrast, ground-based sprayers can be towed by tractors or autonomous ground equipment. It is especially useful for the close-range application, and can be fitted with soil and crop condition monitors. The ground-based systems may use VRT, which determines an appropriate application rate for the area of the field (Batte and Ehsani, 2006). This has the advantage of making sure that every patch in the field is given the right amount of pesticide or fertilizer hence making it efficient and friendly to the environment (Sudduth et al., 2005).

2.3.2 *Manual vs. automated systems*

Mechanically operated hand sprayers are preferred in certain agricultural sectors because they offer flexibility in the amount of human labor required. Such systems may also be cheaper as well as easier to maintain than the fully automated systems. However, manual sprayers make it easy to observe when in application and make changes as needed which may be suitable for small-scale farmers or particular situations (Berenstein and Edan, 2017). For instance, backpack sprayers are preferred in vineyard and orchards where the spray solution must be applied to specific plants. But, manual sprayers are tiresome and less effective, especially for large area applications (Popp et al., 2013).

Automated systems are electronic-based systems that improve efficiency and effectiveness in the application of pesticides. Such equipment includes; tractors equipped with sprayers, robotic sprayers and drones that are fitted with GPS, GIS and different sensors. Mechanized sprayers can apply pesticides uniformly and accurately on large fields, thereby minimizing cost and human interference (Zhang et al., 2018). For example, VRT equipped with the tractor mounted sprayers update the application rates of pesticides using real time data of soil and crop sensors, thus reducing the environmental contamination (Sudduth et al., 2005). In implementing crop spraying, drones with high intensity camera and GPS can be used to access difficult areas hard to reach by ground machinery (Sishodia et al., 2020).

2.3.3 *Advantages and disadvantages of different types of precision sprayers*

Aerial broadcasters, including drone models, are fast in applying the pesticide and can reach rough surfaces allowing high accuracy in pesticide distribution. But they entail higher fixed costs, legal requirements and limited carrying capacity (Zhang and Kovacs, 2012). The ground-based sprayers suitable for local application and may feature real-time sensors but are characterized by slow speeds of coverage and are restricted at the hard to reach or lofty crop fields (Batte and Ehsani, 2006). The use of manual sprayers is cheap and enables direct monitoring, and instant modification since it is mostly used on small holder farms. However, they are capital saving but not as efficient for massive production (Popp et al., 2013). On the other hand, automated systems are efficient and accurate thus minimizing the cost that would be incurred in hiring people to do the job as well as minimizing errors that may be made by people. These systems are more expensive to install and need technical

Table 3. Advantages and Disadvantages of Different Types of Precision Sprayers.

Type	Pros	Cons
Aerial vs. Ground-based Sprayers		
Aerial Sprayers	- Cover large areas quickly	- Higher initial cost
	- Access difficult terrains easily	- Requires regulatory approval and training
	- High precision in pesticide application	- Limited payload capacity
Ground-based Sprayers	- Suitable for close-range applications	- Slower coverage speed
	- Can be equipped with various sensors for real-time monitoring	- Limited access to uneven or tall crop fields
	- Typically have a larger payload capacity	- Higher risk of soil compaction
Manual vs. Automated Systems		
Manual Sprayers	- Lower cost and simpler to maintain	- Labor-intensive
	- Allows for direct observation and immediate adjustment	- Less efficient for large-scale operations
	- Useful for small-scale farms and specific scenarios	- Higher risk of human error
Automated Systems	- High efficiency and precision	- Higher initial investment
	- Reduces labor costs and minimizes human error	- Requires technical knowledge for operation and maintenance
	- Capable of real-time data collection and adjustment	- Potential for technical malfunctions

expertise before they can be operated and maintained and there are usually dangers of technical breakdowns of some parts of the system (Shamshiri et al., 2018).

3. Targeted Pesticide Application

Targeted pesticide application which is also referred to as precision spraying is one of the most basic techniques in current agricultural systems whose goal is to apply pesticides in the right quantity, at the right time and in the correct area. This is in contrast with the conventional broadcast spraying where the pesticides are applied over the whole field regardless of the pest or the weeds distribution (Meshram et al., 2022).

Precision spraying has three major goals, including increasing the effectiveness of pesticide application, minimizing the amount of chemicals that finds its way to the environment, and decreasing input expenses for the farmer by achieving the aforementioned goals (Chowdhury et al., 2023). Research has evidenced that traditional techniques of spraying can cause up to 50% of the insecticides to fail to reach the intended crop or plant or tree or whatever targeted and is wasted out into the surroundings (Mahmud et al., 2021). This is not only a waste of resources, but also leads to potential dangers of water and soil pollution, exposure to non-target species.

Precision spraying technologies are more accurate and effective and involve the use of sensors, computer statistics, and gate systems that assess the presence of pests or weeds and sprays only when necessary. This can mean less pesticide use between 30–70% less than conventional spraying but at the same time enhancing the pest control outcome (Zanin et al., 2022). For instance, a study revealed that in a case when a new precision spraying system was implemented, it became possible to reduce the application of herbicides by 60% while maintaining effective control of weeds in soybean farms (Yang et al., 2016). Perhaps one of the first articles with focus on the problem of real-time, site-specific herbicide application proposed using electronic sensors for the detection of the weeds (Biller et al., 1997). When the reflectance properties of plants become different from those of soil, the sensors are able to identify green color of the soil from the plants. Dammer and Wartenberg (2007) demonstrated that compared to conventional uniform spraying, site-specific weed management reduced weed presence by 12.7% to 40.9% (Zanin et al., 2022).

In recent years, deep learning algorithms have made significant progress in the field of target recognition, enabling precise identification of plants in agricultural fields' complicated surroundings (Li et al., 2022). A research study has developed a machine vision-based system that integrates deep learning algorithms with spraying technology to enable precision, real-time targeted pesticide application in field environments. First, a suggested overall system structure was made up of a pressure-stabilized pesticide supply module, an electronically controlled spray module, and an image capture and identification module. Next, the model is refined and lightened by substituting the attention mechanism and the backbone network with the target detection model YOLOv5s. This led to the construction of a grille decision control method for solenoid valve group on-off, and the selection of frequent malignant weeds as objects for data collection and model training. Field tests were carried out at various speeds once the hardware system and detection model on the electric spray bar sprayer were finally deployed. According to the experimental findings, the enhanced method increases FPS by 18.16% while reducing the model size to 53.57% of the original model with minimal effect on mAP accuracy (Li et al., 2022).

The study developed a vision-guided mobile robot platform for precision spraying in tobacco fields, which uses deep learning-based YOLO models to recognize and identify tobacco plants, achieving an F 1-score of 87.2% and a frame per second rate of 67. The platform also introduced a novel disturbance-based pressure and flow control method, resulting in up to 60% reduction of agrochemicals compared to traditional methods (Nasir et al., 2023).

3.1 Mechanisms and Methods

Advanced technologies are utilized in target detection and selective application techniques to minimize pesticide use, reduce environmental impact, and enhance crop yield (Figure 3).

3.1.1 Target detection and identification

Precision pesticide application has been developed based on the target detection and identification with the aid of remote sensing, machine learning, image processing,

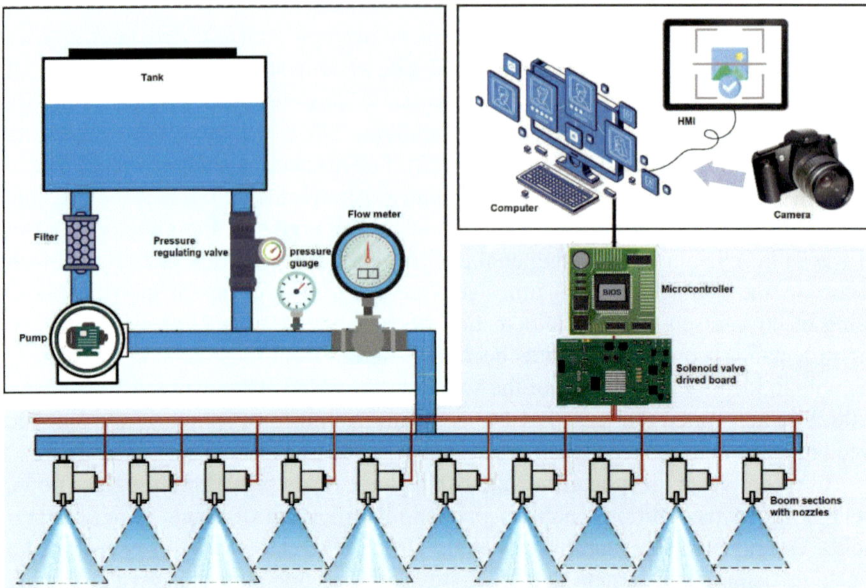

Figure 3. The real-time target spaying system.

etc. for detecting the infected regions (Song et al., 2015). Several applications of remote sensing in plant pathology include multispectral and hyperspectral imaging that help in disease and pests identification. These imaging methods gather data in various spectral bands to evaluate changes in the health status of plants suggesting the presence of pests or diseases. For instance, hyperspectral imaging can differentiate between the changes in the coloration of the leaves, water content as well as the structure of the cell; this will ensure that the farmer is fore-warned of pest infestation at an early stage (Garcia-Ruiz et al., 2013).

AI-based methods are used systematically and become incorporated into detecting targets with the help of remote sensing data. This way these models are trained with large databases of images tagged accordingly allowing the algorithms to identify pests and diseases with a high degree of accuracy. According to the comparison of the deep learning methods, it is possible to state that one of the most suitable methods for the differentiation of types of plant stress and definition of the affected zones are Convolutional Neural Networks (CNNs) (Kamilaris and Prenafeta-Boldú, 2018). However, target detection also employs image processing as part of their process or step. Remote sensing data of high spatial resolution received from drones or satellite are used to gain some important information of the crops' condition. The pest infestations can be then quantified and classified using methods like object based image analysis (OBIA) and vegetation indices like NDMI. It is these techniques that assists in tracking and regulating the infestation in real time and provides information for the application of pesticide (López-Granados, 2011).

3.1.2 Selective application techniques

Selective application techniques efficiently apply pesticides to specific areas in a field, based on target detection and identification, thereby reducing environmental impact and costs. One of the most used selective application techniques is variable rate technology (VRT). VRT systems have the activity which permits the modification of the pesticide application rate based on data received from the sensors and GPS. These systems can regulate the quantity of concentration across the different field areas to use only the necessary quantity of pesticide on the fields. This accuracy also reduces the amounts of chemicals used and the likelihood of over application of the chemicals (Bongiovanni and Lowenberg-DeBoer, 2004).

Spot spraying is another method of selective application another method of selective application. This is the process of using sensors and cameras in order to detect the weed or pest infected plant in real time. This is achieved in a manner that the sprayer orients itself to the targeted plant without damaging other plants in the garden. It assists in reducing the use of pesticides and it also aids in sparing the desirable insects and plants within that area from the impact. Other advancements in the efficiency of working of robots have also played a role in the efficiency of spot spraying for use in vast lánd areas (Slaughter et al., 2008).

Patch spraying differs from spot spraying as it is made on a slightly larger area of the intended surface. Patch spraying refers to the process of spraying only the affected part of the crops that is affected by insects or diseases. This technique is preferred where pest intensity is not the same in all the areas of the field (Gerhards and Oebel, 2006). The selective application of pesticides has been applied in large fields reviewing pests through drones and UAVs through their high-resolution cameras and sensors. These drone can spray the affected region without wastage of time and efforts applied in manual spraying (Zhang and Kovacs, 2012).

3.2 Technologies Involved

3.2.1 Drones and UAVs

Drone and UAV systems are valuable in precision agriculture as they offer high-detailed pictures and targeted pesticide spraying in addition to the ground techniques such as cameras, multispectral sensors, and GPS. They can operate in extensive areas that are otherwise unattainable and apply the pesticide directly to the target species, this conserves resources and the environment (Zhang and Kovacs, 2012). Some of these are more efficiency and lesser costs on labor when crop spraying is done by drones. Thus far, pesticide drones have been proved to disperse formulations more uniformly and use lesser chemicals in comparison to conventional techniques.

By use of multispectral or hyperspectral images taken from the drones, it was possible to identify various problems in plant health such as nutrient and pest, as well disease. These images are helpful in order to get high accurate crop-maps and to establish regions that deserve attention (García-Ruiz et al., 2013). Furthermore, thermographic cameras operating in drones may also be useful for determining soil moisture, a water deficit in plants, and the optimization of pesticide spraying (Sishodia et al., 2020). For example, hyper spectral imaging, which is an unmanned aerial system that gives a map of the field conditions that may identify stressing fields, pest, or disease early stage. These sensors release crucial data that is useful in decision making and resource utilization in spraying (Zhang et al., 2018).

Modern development like the integration of machine learning algorithms in the manufacture of drones prompted its use. AI models can be employed for identifying jumbo chunks of aerial imagery for pest hotspot and spraying patterns (Zhang et al., 2020). For instance, the recognition algorithms like the CNNs can help to discern between the normal and affected crops to reduce the use of fungicides and hence minimize transportation and have little impact on the environment (Kamilaris and Prenafeta-Boldú, 2018).

Drones also have an added benefit of duration of pesticide application as they applied fast and probably function under other climate situations. That kind of capability may be more helpful in being able to respond quickly to new pest events or disease occurrence to prevent other occurrences due to resulting early action (Lost Filho et al., 2020). As for volumetric usage, it is expected that more development will be witnessed regarding the application of drones in practical precision agriculture as a result of advanced sensor, AI and autopilot technologies (Manfreda et al., 2018).

3.2.2 *Robotics and AI*

Automation is now in vogue of precision agriculture through Robotics & AI to reduce large amount of man power involvement and also increases the accuracy of pesticide application. Self-driving mechanism equipped with various sensors fitted with intelligent algorithms are capable of moving across the field and locating the specific locations of pest infestations and apply pesticides selectively. Some examples include robotic sprayers that can roam through fields spraying herbicides for weed control or insecticides for pest control as they detect weeds, pests and diseases using computer vision and machine learning. These robots are intended to spray only the targeted areas making the use of chemicals low and protecting the environment (Slaughter et al., 2008). The application of AI results in these robots to gain insight from the surrounding living environment to be able to determine the best decision making that in turn optimizes their efficiency and effectiveness (Shamshiri et al., 2018). The machine learning skills enables robotic instruments to maneuver through the crops using computations solving and searching for targets that require pesticide and apply it correctly. Automated sprayers reduce drift and thus improve the consistency of the application and eliminate errors that would normally be made by people, therefore decreasing the amounts of chemicals that affect other areas (Patel, 2016).

AI technologies in the form of Deep learning and AI are used to analyze the data collected by sensors of robots and drones. These algorithms can be of assistance in identifying the pattern that relates to the crop health and the places that require the pesticide sprays to be delivered. For example, it is possible to use AI models to diagnose diseases in crops at a stage where early signs are not visible in plants as they are at the time the initial outward signs are observed (Kamilaris and Prenafeta-Boldú, 2018). Precision agriculture with robotics is not only a ground-based platform but an aerial platform. Autonomous drones are also other aerial robots that would require the incorporation of AI into their system. These systems are able to plan and execute the spraying operations without any human intervention while changing the flying paths and spraying rates based on information collected by the sensing instruments mounted in the aircrafts (Lost Filho et al., 2020). This level of automation also minimizes the level of interference from human beings and increases the standard as well as efficacy of pesticide application.

Cobots which are also referred to as collaborative robots are another new technology adopted in precision agriculture. These robots act as assistants to human operators, boosting the competence of their operations. For example, cobotics can be used for crop scouting, data gathering and spot application of pesticides in a manner that leaves more strategic planning to the farmer (Shamshiri et al., 2018).

3.2.3 Sensor-based measurement technologies for precision spraying

The many real-time sensor applications for variable-rate spraying, such as infrared, stereo vision, ultrasonic, and laser scanners, have been examined (Zhang et al., 2018).

Camera sensor-based technologies

Camera-based sensing tries to mimic human perception to provide information or inputs based on processed image data to spraying systems needed for canopy characterization-based pesticide applications. Pre-processing, segmentation, feature extraction, and classification are all included in image processing. The segmentation of the desired region (such as diseases, canopy, and stressors) from the crop photos is made easier with the use of image pre-processing. According to Mahmud et al. (2021) it entails picture rotation, normalization, resizing, color alteration, denoising, contrast enhancement, and so forth. Color transformation is among the most crucial of these for segmentation. Cheng et al. (2001) provide a thorough explanation, including the benefits and drawbacks of several color spaces utilized in machine vision systems. Garcia-Mateos et al. (2015) provided a detailed presentation on the conversion of several color spaces, such as channels and transformation function, from RGB to be utilized in target detection. When various image processing methods were readily available, the use of camera sensors for precise spraying in tree fruit orchards was started. Table 4 lists the applications of camera sensors for variable rate spraying.

Ultrasonic sensor-based technologies

The sensing technology used to measure the tree canopy parameter for precision spraying is ultrasonic sensor systems. These are used to monitor the target information during spraying because they are accurate, affordable, and lightweight (Balsari et al., 2008; Zaman and Salyani, 2004). Sensors created waves in the direction of the targets and then received the reflected waves back from them to identify the item. The ultrasonic sensors can both detect the presence of the targets and determine their distance since the time difference between the delivered sound wave and the received sound wave is measurable. Depending on the evaluation of the strength of the reflection, the characteristics and configurations of the target are gathered (NDK, 2020). An amplifier, an output signal generator, and an ultrasonic transceiver make up the ultrasonic sensing system. Various transceivers, like Prowave 400EP14D (Hočevar et al., 2011) and Prowave 400EP250 (Stajnko et al., 2012), have been successfully used in agricultural applications. A selective amplifier with a 40-dB gain and 40 kHz frequency was used for target spraying in an orchard. Ultrasonic sensors control spraying using sound waves to assess canopy characteristics and manage spray operation. Table 4 provides an overview of site-specific spraying applications.

In the first stage, Giles et al. (1987) assessed the effectiveness of ultrasonic sensors in orchard spraying. The system developed adjusted the sprayer's flow rate according to the canopy size variations detected by the ultrasonic sensors. Spray saves for apples and peaches were from 36% to 52% and 28% to 35%, respectively.

Laser sensor-based technologies

Over the past few decades, laser sensing, an emerging technology primarily utilized for plant phenotyping and remote sensing applications, has demonstrated potential in the detection of tree canopy parameters (such as canopy volume and target distance) (Mahmud et al., 2021). The main benefit of employing a laser sensor for canopy volume estimate over other types of sensors is that its electromagnetic signal may pass through the plant canopy and provide details about the inner and exterior structures of the tree, as well as canopy density (Berk et al., 2016). Through a lens, the laser beams direct visible laser light in the direction of the target. A receiving point of convergence on the sensor concentrates the light that is reflected diffusely from the exterior of the object by the laser, generating a pinpoint of light on the direct imager. The distance to the sensor and the angle from the horizontal show how many recognized points the laser bars get for each yield from the item sliced. Using a laser sensor, point cloud data collection is the first step in the sensing process. The ground information is also included in the raw data and has to be eliminated by designating a zone of interest. Tree canopy points are sorted using laser scanner sensors, which are more accurate than ultrasonic sensors in gathering data and specifications about trees and fruit, with promising results compared to hand measurements (Berk et al., 2020; Mahmud and He, 2020; Mahmud et al., 2021). Rosell et al. (2009) obtained three-dimensional (3D) structural features of the trees in the orchard using a laser scanning technology. In a single test, the study used a 3D LiDAR sensor with a maximum scanning angle of 180 and a 5-mm standard deviation at a range of up to 8 m, and it obtained an accuracy of $\pm$15 mm. Because LiDAR-based laser-sensing technologies provide great precision and accuracy in real-time measurements, they have been tried in fruit orchards during the past ten years (Mahmud et al., 2021). Table 4 provides an overview of the uses of laser sensors for site-specific spraying. Prior to the 2010s, laser-sensing technology was thoroughly researched for the creation of sprayers; nevertheless, it was not until the 2010s that it was effectively incorporated into a precision sprayer (Chen et al., 2012).

3.3 Environmental and Economic Benefits

The elements of precision spraying consist of several main areas, including environmental and economic factors. These benefits are important to explain the opportunities of using precision spraying technologies in the framework of agriculture.

Environmental benefits

Precision spraying has enormous advantages when it comes to the protection of the environment since it minimizes the use of pesticides in farming. This reduction is

Table 4. Summary of Sensor-Based Canopy Measurement Technologies for Precision Spraying.

Type	Crops	Sensors	Detected	Accuracy	Chemical Saving	Limitations	References
Camera sensor	Apple	RGB camera	Tree canopy	-	23% (0.96 L min 1 flow rate decrease)	The variety in tree size and shape was not considered when calculating the amount of savings.	Ho˘cevar et al. (2010)
	Orange and Grapefruit	RGB camera	Tree canopy	-	22% to 45%	Savings were only possible when tiny trees were sprayed	Thériault et al. (2001)
	Grapevine	Multispectral Camera	Powdery mildew disease	85% - 100% of the affected region was found	Approximately 65% to 85%	The developed system's false-positive rate ranged from 5% to 20%.	Oberti et al. (2016)
Ultrasonic Sensor	Apple	-	Tree canopies	Error averages of +0.53 cm and +5.11 cm on field and laboratory scales, respectively	not spray	An increase in unpredictability in the field dramatically decreased the sensor's accuracy.	Escolà et al. (2011)
		-	Contour of the tree canopy	-	Approximately 20.2% per nozzle	The savings varied by the size and training of orchard trees	Hočevar et al. (2011)
		-	Tree canopy volume	-	Approximately 58%	The vegetation gap width was found from 0.35 to 1.20 meters	Balsari, and Tamagnone (1998)
	Citrus	-	Tree canopy	-	30% of saving in time	Only handgun sprayer used	Moltó et al. (2000)
	Grape	-	Tree row volume	R2 of 0.66 was reported for TRV measurement	Approximately 58%	Not examines the effects of ground speed	Llorens et aL (2010)

Table 4 contd....

...Table 4 contd.

Type	Crops	Sensors	Detected	Accuracy	Chemical Saving	Limitations	References
Laser Sensor-Based Technologies	Apple	-	Tree canopies		60–67%	Only compared the economics of variable-rate and constant-rate sprayers	Manandhar et al. (2020)
	Apple	-	Canopy volume		Approximately 46%	Not analyze different stages of growth	Li et al. (2018)
	Apple	-	Tree canopy foliage volume		Two-year average of 60.5% by volume	Only small canopies trees were examined	Chen et al. (2019)
	Crab-apple	-	Tree size, shape, and leaf density		19.86% + 3.0%	Only conducted in nursery field conditions.	Zhu et al. (2017)
	Peach	-	Tree canopy volume		50%, 40%, and 13% at bloom, pit hardening, and final swell, respectively	Spray coverage was limited at the final swell	Asaei et al. (2019)

achieved by an improved efficiency in the precision of the application of the pesticide over the area, through a controller, which means a lesser quantity of pesticide is used. For instance, a study conducted on precision spraying of specialty crops revealed that when using precision spraying equipment, the amount of pesticide released was 10 to 35% less than that from conventional spraying tools (Allmendinger et al., 2022). This reduction in the use of pesticides can also be regarded as beneficial that pesticide has minimal chances of finding its way into water supplies and make them toxic.

Economic benefits

The benefits that are tied to precision spraying are numerous and possibly holds significant value in supporting farmers. This cost can therefore be managed by the farmers applying measures that would ensure the quantity to be used is reduced. Second, precision spraying may also have the ability to reduce the cost of labor and fuel as compared with conventional spraying. For instance, in a study done on the effects of precision guidance in agricultural economy, it was realized that the use of UAS system of precision spraying could save about 44% of spraying material than could be roughly sprayed (Tona et al., 2018). However, precision spraying can also improve the utilization efficiency of matters concerning farming activities; farmers are able to spray for longer durations and gain higher speeds which implies efficiency together with profitability.

4. Integration of Nanotechnology

4.1 Nanofertilizers and Nano-pesticides

The areas of application of nanotechnology in the agricultural products are the nanofertilizers and the nano-pesticides. Nanofertilizers that are manufactured in the nanoscale could improve nutrient use efficiency and nutrient uptake by plants and reduce losses relative to normal fertilizers. For example, Dimkpa et al. (2020) reported that ZnO nanoparticles improved wheat grain yield by up to 25% compared to bulk Zn fertilizer.

Likewise, nano-pesticides are pesticides that contain nanomaterials and that improve the efficiency, stability, and selectivity of the compounds (Kah and Hofmann, 2014). This has implication in terms of solubility enhancement and controlled release of the active ingredients hence reducing the application rates and environmental effects. According to Kah et al. (2018) review, another advantage is that nano form pesticides can be applied at a 90% lower concentration than conventional pesticides but the effectiveness in pest control might even be enhanced.

Some of the characteristics include high surface-to-volume ratio, specific physicochemical properties that can be tailored, and the ability to transverse biological barriers, which afford nanomaterials the potential to increase agricultural productivity and sustainability (Dimkpa and Bindraban, 2018). These technologies are still in their formative stages but they could change the manner in which crops are fertilized and protected against pests.

4.2 Benefits of using Nanomaterials

Nanomaterials have several advantages for precision spraying in the agricultural industry. One of the major uses is enhancement in the effectiveness of the delivery of agrochemicals such as pesticides and fertilizers. Traditional spraying techniques are inefficient, which allows only 10 to 30 per cent of pesticides to reach the target areas. However, nanoformulation does help in enhancing the bioavailability and durability of active substances and in extending the release time period of the same (Duhan et al., 2017; Shang et al., 2019). For instance, nanoparticle encapsulation on pesticide reduces leaching losses, degradation, and evaporation (Duhan et al., 2017).

Nanomaterials also enable the proper and selective utilization of the agrochemicals during the agricultural process. High area of contact and rapid uptake are preferred for specific deposition in some parts of the plant or in pests (Shang et al., 2019). Such levels of precision assist in reducing the quantities that are required, and at the same time, also raising the level of productivity. According to one study, it is mentioned that the nanoparticle based pesticides were found to contain just only 10% of the active ingredient of the existing pesticide formulations (Duhan et al., 2017).

In addition, the movement of the transferred agrochemicals could also be enhanced in terms of stability and mobility through nanomaterials. New pesticides come as formulations that have nanomaterials used to improve the stability of the pesticides and their ability to get into the plant tissues and the pests themselves (Shang et al., 2019). This in turn leads to increase in efficiency and the extent of pollution of the environment is also reduced due to control of pesticide usage.

Nano-materials are small particles to make effective biocides with high pest efficacy but low impact of ecosystems. These insecticides or fungicides are useful for controlling pests and diseases in crops. The encapsulated material helps the pesticides get to the intended area, thus it is much safer for the environment (Duhan et al., 2017; Shang et al., 2019).

4.3 Precision Spraying and Application of Nano-Materials

Precision spraying systems also employ nano-materials in a bid to enhance efficiency or productivity or output. Perhaps, an area that would be more compelling to explore is the enhancement of more stable and selective nano-coated pesticide formulations (Kah et al., 2018). For instance, Dimkpa et al. (2020) examines that the nano-coated pesticide cut down the use of the pesticide by 80%–90% and still protect their crops against pests. The second pertinent implementation of nanomaterials in precision spraying is that it can be applied to create nano-sensors. These sensors help in identifying and supervising many situations about the tractors, crops, temperature, moisture and health of crops and also help farmers in decision making of spraying (Shaikh et al., 2021). For instance, in a study conducted by Zia et al. (2022) they found that in identification of pests and diseases affecting crops, nano-sensors can be employed hence there would not be general spraying of pesticides on the fields.

The application of nanomaterial in precision spraying also means possibilities to develop new generation and environmentally friendly spraying equipment. For example, it is easier to form thin films using nanomaterials so as to alter the amount of water and chemicals used during spraying (Butt, 2022). It is therefore crucial to have awareness about thin film coating and how it can help to attain good quality and reasonably low costs. The following methods give a successful combination of quality and cost which could make it an effective tool in many industrial societies.

4.4 Application Methods

4.4.1 Encapsulation and controlled release

The development of nanotechnology provides new strategies for formulations and release of pesticide with improved utilization of the pesticides. It can also be made to protect active agents and deliver them to target pests or diseases and at the same time avoid degradation of the agents (Kah et al., 2013). It will also help in minimizing on the amount of pesticide required while at the same time increasing the effectiveness of the pesticide. One such approach is to encapsulate the pesticides within polymeric nanoparticles such as chitosan for chitosan for PLGA. Such nanoparticles have the ability to release the active ingredients after a certain time interval depending on pH, temperature, or even enzymatic activity (Campos et al., 2014). For instance, Campos et al. (2019) analyzed the release characteristics of PLGA nanoparticles containing the insecticide imidacloprid where 80 percent of the active ingredient was detected within 30 days. The other approach is the conjugation of pesticides into lipid based nano-carriers such as liposome or solid lipid nanoparticles (SLNs). These nano-carriers can protect the active ingredients from degradation and also increase their accumulation into target organisms (Campos et al., 2019). According to Campos et al. (2019) the tebuconazole-encapsulated SLNs were more pathogenic to *B. cinerea* than free fungicide.

4.4.2 Improved adherence and absorption

Nanotechnology also improves the sticking and penetrating capacity of pesticides on the plant cuticles and within target pests. The surface chemistry of nanoparticles designed to have a desired electrical charge or hydrophobicity to enhance adhesion to plant tissues or the intended pests (Kah et al., 2013). This results in longer duration that the pesticide stays on the plant surface hence requiring less frequent application. Also, nanoparticles can help the active ingredients to enter the target organism which may be an insect or a fungus by releasing the active ingredients through interaction with cell membranes or cuticles (Campos et al., 2014). Campos et al. (2014) provided evidence that chitosan nanoparticles containing imidacloprid had higher insecticidal efficacy against *Aphis gossypii* with an LC50 of 0. 02 mg/L and 0. 05 mg/L for the free insecticide that was tested. Rathore et al. (2024) utilized nanoparticles to target pesticides in cotton crops, resulting in a 30% reduction in pesticide usage and a 25% increase in cotton yield compared to traditional methods.

4.4.3 Successful implementation in various crops

Precision spraying technologies have been successfully utilized in many crops and enhanced agricultural productivity. For instance, the use of drones in vineyards entails precision spraying which results in efficient pest control and low pesticide use. A similar study by Matese and Di Gennaro (2015) on the use of UAVs for Italian vineyards established that accurate application reduced the total quantity of chemicals and paved way for improved grapes quality. Multispectral cameras on drones can help spray pesticides where needed, reducing pest and disease pressure and increasing crop yields. According to Smith et al. (2020), VRT systems that can be used for field zoning cut pesticide consumption by 20% while raising yields by 15% against standard spraying. According to a survey conducted in Brazil, precision spraying and targeted pesticide application reduced pesticide usage by 79% in soybean and by 53.7% in maize, with an additional reduction of 36.6% reported in certain application stages or pesticide types. This makes soybean farming less destructive and less likely to release chemicals into the atmosphere. The technology achieved a cost reduction of 2.3 times lower than the cost associated with pesticide application over the entire area using a conventional sprayer (Zanin et al., 2022). A robotic sprayer for precision agriculture with tobacco recognition in real-time obtained an F1-score of 87. 2% and a frame per second rate of 67 using YOLOv5n model which was trained on real field data.

It adapted a new pressure and flow control mechanism primarily for handling any undesirable pressure changes, thus realized a 60% reduction in agrochemicals as compared to broadcast spraying (Nasir et al., 2023). Some research indicate that pesticides formulated in nanoemulsion have an application rate reduced up to 30–50% lower than those of conventional coatings, but the efficiency of the treatment is either preserved or improved. This is because the nanodroplets are much smaller, with sizes ranging from 20-200 nm, which can easily penetrate plant tissues better than larger droplets. In addition, the slow release of the pesticide at the nanoemulsion interface will enhance the activity of the pesticide and reduce the frequency of application (Yadav et al., 2023).

Other techniques include using nanoparticle carriers to transport the pesticides. For instance, chitosan nanoparticles have been used to encapsulate and to control the release of insecticides and fungicides (Shang et al., 2019). These nanocarriers can be made to act selectively on specific plant tissues or pests hence reducing negative side effects. There is up to 80% reduction of pesticide use by employing the nanoparticle-based formulation, but the pest control efficacy is enhanced (Ibrahim et al., 2024). Comparison between conventional practices and nanotechnology application for pesticide indicated more merits of latter. In conventional practices, 40–70% nitrogen, 80–90% phosphorus, and 50–70% potassium are leached into the environment, leading to pollution. On the other hand, losses that occur as a result of using microtechnology based approaches can be minimized to as low as 10% by using nanotechnology based approaches (Duhan et al., 2017).

Conventional pest control measures entail aerial spraying of pesticides over entire fields, even if there are differences in pest populations or crop health. This can also result in over utilization, high costs, and pollution in form of chemical leaching and ground water pollution (Popp et al., 2013). Precision spraying, in contrast, utilizes information technologies that identify spots that require treatment. For instance, Zhang and Kovacs (2012) conducted a research where they evaluated aerial and ground spraying in corn fields. The overall findings indicated that the use of drones in spraying the pesticides cut the use by a third but at the same time enhanced the effectiveness of pest control. The use of this targeted application helped to keep the chemicals away from areas that would harm other important insects and the environment.

5. Environmental Impact and Sustainability

5.1 Reduction in Chemical Usage

Precision spraying decreases the demand for pesticide application in agriculture as it is used when required, less chemicals as opposed to normal methods (Chowdhury et al., 2021). Studies have shown that precision spraying reduces pesticide use to 30% while achieving 100% efficacy on weeds (Zanin et al., 2022). It is beneficial in the following ways: it sparingly uses chemicals and it is environmental friendly. It also cuts down the risk of polluting soil water and other unpopular species through lowering the amount of pesticides used on the fields. This in turn has the desired outcome on the quality of our ecosystems and their corresponding levels of bio-diversity. Furthermore, lower consumption of pesticide has an impact in production and release of greenhouse gases, and will go a long way in reducing its impact on the carbon footprint of the agricultural activities (Chowdhury et al., 2021).

Precision spraying also reduce the rate of build up of resistance by the target organisms and as such, prolongs the effective shelf life of existing pesticides. Besides, precision spraying can also aid in reducing the shelf-life of the pesticides through right and effective application reducing on the need for the development and registration of more pesticides. It is also important to note the following economic advantages of precision spraying: By spending less amount of money on pesticides, the farmer is able to make realistic profits. For instance in one study, precision spraying technology was established to have slashed the common cost by 56%. 16% than that of the conventional total area spraying technique (Zanin et al., 2022).

5.2 Minimizing Runoff and Residue

Precision spraying minimizes pesticide runoff and residue, reducing environmental impact. Spray drift, the airborne movement of droplets, can cause pollution and food contamination, especially in aquatic ecosystems (Tudi et al., 2021; Berger et al., 2022). To mitigate this issue, precision spraying techniques can be employed. For instance, unmanned aerial vehicles (UAVs) with fine and very fine droplet attachments have been used in a number of commercial spray applications, enabling water-saving and very effective pesticide delivery. Nevertheless, there hasn't been enough discussion of the drift possibility related to UAV applications, particularly for tiny droplets produced by rotating disk nozzles, which might cause issues with the environment and regulations (Tudi et al., 2021).

Another approach is to optimize spray application methods. A reduction in the amount of pesticide used can be achieved by better informing farm operators about new low-risk, low-active-ingredient pesticides that are available. These products must be correctly combined, used as needed, and applied to the intended location.

A realistic short-term aim would be to improve on-farm pesticide selection and mixing, which might result in a 5–40% reduction in pesticide use, depending on the farm and particular crop (Berger et al., 2022). Longer term, better pesticide product scheduling may result from enhanced disease and insect models as well as sensing tools, thereby eliminating the need for some treatments. Depending on the cropping system, eliminating a single spray from the operation can cut the amount of pesticide used by 5–15%. Additionally, integrating remote sensing tools into spray equipment allows for targeted application, reducing the amount of pesticide released into the environment (Berger et al., 2022).

5.3 Improved Targeting Reducing Environmental Contamination

Precision spraying and targeted pesticide application strategies are crucial for reducing environmental contamination by minimizing pesticide release and their impact on non-target organisms and ecosystems (Gooding, 2024). Precision spraying and targeted application technologies have made significant strides in recent years. For example, self-driving automobile technology, image analyzation, and use of artificial intelligence has improved specific application of chemicals that are sprayed on vegetable crops in the western US (Giles, 2024; Gooding, 2024). These technologies help the farmers to use the pesticides at the right places and in the right proportions so that they do not affect the environment.

One more significant aspect of enhanced targeting is variable rate application and spot spraying. Variable rate applications apply the material across the field but the amount applied vary depending on conditions at the particular area. This approach could reduce the amount of pesticides used by approximately 50-70% compared to full-field application (Holtermark et al., 2024; Gooding, 2024). The other common technique is known as spot treatment whereby the pesticides are applied only to those parts of the field which is infested with weeds or pests hence using a lesser quantity of pesticides.

Precision spraying and application technologies are also likely to minimize spray drift which is a primary cause of environmental pollution. Spray drift refers to the situation whereby pesticide moves from the intended area to other adjacent areas or water sources due to factors such as wind. Some control measures include precision spraying and targeted application technologies that reduce the possibility of spray drift through the application of smaller and more precise amounts of pesticides (Van Boheemen et al., 2024; Gooding, 2024).

Apart from the fact that better targeting aids in preventing the polluting of the environment, from the economic perspective, it benefits farmers. By applying pesticides at the spot, the farmers are able to reduce costs because they apply pesticides only in the affected areas and do not harm other crops by applying excess pesticides. This approach may also help farmers maintain their certifications on environmental sustainability, which is helpful in the agricultural industry (Kempenaar et al., 2024; Gooding, 2024).

5.4 Long-term Sustainability

Precision spraying and selective use of pesticides are viable in the long run for the sustained health of the soil as well as the regulatory and ethical considerations.

5.4.1 Impact on soil health and biodiversity

Precision spraying and targeted pesticide application reduces environmental impact by minimizing pesticide use, soil pollution, and potential life-threatening risks (Kempenaar et al., 2024). It also has the added advantage of preserving soil integrity by preventing soil erosion and runoff that causes loss of arable land (Anastasiou et al., 2023; Gooding, 2024).

In addition, the precision spraying can be optimized to avoid the use of pesticides at times when they are likely to affect various organisms more such as during flowering or fruiting stage. This way of thinking not only leads to a decrease in harming the environment but also contributes to keeping the balance of species on earth. Wageningen Plant Research revealed that precision orchard sprayers deliver the spray deposition at the leaf canopy level up to about 20% more compared to the standard cross-flow fan orchard sprayers reducing the ground deposition underneath of trees by 30 % to 50 %. The improved precision spraying technique seen above can easily arise and decrease the overall effects of pesticide on the environment (FutureLearn, 2024).

5.4.2 Regulatory and ethical considerations

Government rules and regulations are important in shaping the future of precision spraying and application of pesticides. Current policies like the Green Deal by the European Union seeks to have the utilization of PPP by half the current levels within the next ten years (European Commission, 2020). To this end, regulatory authorities should offer policy and guidelines on the development and application of precision spraying systems. Clearly, there is a lack of consistent risk assessments and labels for precision spraying systems. This entails formulating reliable risk assessment frameworks that can capture the effects on the environment of these systems (Gregory, 2024; Gooding, 2024). Also, regulatory authorities should see to it that precision spraying systems are integrated and developed to reduce the likelihood of drift and contamination (Chapple et al., 2024; Gooding, 2024).

Pressing to this issue, the existing legislation does not sufficiently recognize or evaluate precision applications as regards to PPP authorization. This lack of recognition prevents the popularization of precision spraying and targeted application of pesticides, which could reduce the negative impact on the environment and can make pest control more environmentally friendly (Chowdhury et al., 2021). Ethical issues are also put into consideration. For example, precision spraying and targeted pesticide application are questionable in terms of their effects on non-target organisms as well as on the systematized consequences of these methods on ecosystems. Non-dietary risk assessor Alex Gregory also calls for better risk assessment models and equipment standardization when it comes to utilization of these techniques (Gooding, 2024).

The efficacy of using precision spraying and targeted pesticide application for sustainable agriculture is evident in these studies. For instance, the Dutch government's FRUIT4.0 project provided a precision orchard sprayer capable of task map-based spraying, which improved the spray deposition at leaf canopy and decreased the ground deposition under trees (FutureLearn, 2024). In another study conducted by ILVO Living Lab Agrifood Technology, real-time and task-map precision spray applications were evaluated for controlling weeds and diseases on vegetable row crops, demonstrating improved efficiency and reduced pesticide usage. The goals of this project were to quantify pesticide saving opportunities and to estimate the percentage decrease in runoff at various crop developmental stages (Willekens et al., 2023).

6. Challenges and Future Directions

6.1 Technical Challenges

There are many technical barriers restricting the development of precision spraying, one of which is high technology intensity. A major drawback of the current technology for detection is that the algorithms may be slightly off from what can be observed by a normal human being or even an experienced agronomist. For example, Syngenta trials revealed that the digital imagery used in the observation of various problems in the field was less precise as compared to human observation. Such a gap can create disparities and ineffectiveness of precision spraying with a great level of accuracy (Thomas, 2024).

Another complexity is the ability to incorporate digital mapping interfaces into application equipment. The interaction between these systems is not always seamless, which may further hinder the achievement of intended outcomes. For instance in Syngenta trials the precision potential was significantly affected by the inability of controlling the existing sprayers which could only be controlled in boom sections. This has raised the need for better technologies that can overcome these challenges (Thomas, 2024).

The incorporation of such sensors on the sprayers to pick targets on the crop and on or off the nozzles brings noble control. However, the software and algorithms that underpin this technology remain to be sufficiently robust and replicable when used in the field. For instance, in Syngenta trials in green-on-green optical spot spraying, the sprayer technology, and nozzle switching were effective; however, the software and algorithms offered questionable outcomes (Thomas, 2024).

The use of precision spraying also have some limitation that tends to do with the cost and the nature of the technology. The amount of investment involved in implementing precision spraying can be large and as a result, the process of improving vine estates may be slow. For example, in a research done on the factors influencing adoption of precision spraying in winegrowing farms, the author noted that economic conditions affect adoption of precision technologies (Bonicelli et al., 2010).

Concerning the identified technical issues, there is a special emphasis on creating "cost-effective" smart sprayers. However, a major challenge lies in the environmental qualification of equipment and practices to implement incentives for the replacement of old sprayers and to enhance the adoption of more effective spraying techniques (Bonicelli et al., 2010). Furthermore, for the improvements that are efficient at a national or a regional level, dedicated incentive systems need to be developed, although this is beyond the scope of the present work (Tudi et al., 2021).

Precision spraying involves several steps such as the use of monitoring and detection methods, application of precise spraying technology, and choice of the right treatment protocols. Nonetheless, the efficient translation of precision spraying as a technology into normal usage is hampered by the knowledge gap and the availability of cost-effective solutions (Bonicelli et al., 2010; Tudi et al., 2021).

6.2 Research and Development Needs

6.2.1 *Areas requiring further investigation*

R&D is a very elaborate area of study and practice that involves an understanding of numerous aspects, such as customer requirements, technological possibilities, and organizational objectives. One of the special topics to be explored is the significance of the R & D function for innovation. A World Economic Forum survey found that 80% of executives see R&D as critical to business innovation; however, only 40% of firms have an R&D blueprint (Church and Dwight, 2024). This means that there is a disparity between the perceived importance of R&D and the actual level of R&D investment.

One such area that needs further research is the environmental effects of R & D expenditure. The changing world culture towards sustainable practices means that consumers want products and services that are environmentally conscious. According to the Harvard Business Review, 75% of customers are willing to spend more on sustainable goods, yet only 20% of businesses possess a concrete sustainable plan (Satyendra, 2022). This indicates that there is need for companies to consider venturing into R&D in sustainable development.

6.2.2 *Innovations on the horizon*

Energy and technology industries could be quite risky, especially when conducting R & D but there are several innovations that are possible in the future. An area that is steadily attracting interest is AI application with focus on the R&D field. A study conducted by McKinsey done by the McKinsey Global Institute seems to indicate that AI has an approximate of 30% of productivity improvement but only 10% of companies use the advance in their R&D processes (McKinsey and Company, 2023).

Another area that is under focus is the application of the blockchain in the research and development process. A study by the World Economic Forum also suggested that the use of blockchain can boost effectiveness in research and development by as much as 20%; however, it also indicated that 95 percent of organizations still do not integrate blockchain into their research and development initiatives (World Economic Forum, 2017).

6.3 Future Prospects

Precision spraying is a crucial aspect of modern agriculture, especially in tree fruit orchards, where efficient and targeted pesticide application is essential to minimize environmental impact and optimize crop yields. Advancements in sensor technologies and computer-based control systems have significantly improved the accuracy and efficiency of precision spraying. Variable-rate spraying have adopted camera sensors, ultrasonic sensors, and LiDAR sensors to determine canopy geometry, density, LAI and LAD with improved accuracy. Variable rate treatment using satellite mapping is one of the emerging technologies in the precision spraying of fertilizers and pesticides among growers. Connectivity with other autonomous systems is said to transform the industry through the elimination of manpower costs and enhanced accurate treatments. Real time detection with the use of ultrasonic detectors and subsequent spot spraying can be used effectively in high value crops where selective herbicides cannot be applied. Parameters and delivery systems will be also controlled by Computer-based control systems in the future of precision spraying. Machine learning and Artificial intelligence can add more efficiency to the precision spraying systems and predict the spraying pattern with better accuracy and make real-time changes in it. The combination of sensors and algorithms can help generate the next generation of precision sprayers, focusing more on the variations in the canopy shape, density, and LAI.

7. Conclusion

The focus on precision spraying and targeted application has been identified as the key aspect of contemporary agriculture and is universally recognized as possessing significant potential in relation to minimizing the use of chemicals and their potential effects on the surroundings. In the field of precision spraying, the use of nanotechnology can enhance the performance of these technologies further. Precision spraying is the method of applying pesticide in very specific basins using GPS, GIS and other remote sensing technologies on the field. This method has been proved to reduce the chemical application by 40 percent and still achieve 100 percent weed control. The targeted application type is the application of computerized control system for the purpose of regulating several factors and coming up with the treatment maps for the specific field.

Nanotechnology is also expected to bring major changes to the agricultural sector due to more precise application of pesticides and fertilizers. The incorporation of nanoparticles can enhance the stability and effectiveness of these products, minimizing their quantities and environmental consequences. The fusion of nanotechnology in techniques used in precision spraying is likely to increase the efficacy of these modes.

As nanotechnology advances in the agricultural sector, the authors assert that there will be further improvement and development of precision spraying and precise application. Advancements in new and improved materials and technology will also enhance the delivery of pesticides and fertilizers in a more precise form such that minimal amounts will be used with consideration towards the environment. Precision spraying and targeted application in agriculture can significantly impact global food security by reducing waste and environmental impact. As the world's population grows, sustainable practices are crucial for future generations. The integration of nanotechnology in precision spraying can enhance efficiency and effectiveness, driving innovation and improvement in agriculture.

References

Allmendinger, A., Spaeth, M., Saile, M., Peteinatos, G. G., and Gerhards, R. (2022). Precision chemical weed management strategies: A review and a design of a new CNN-based modular spot sprayer. Agronomy, 12(7): 1620

Anastasiou, E., Fountas, S., Voulgaraki, M., Psiroukis, V., Koutsiaras, M., Kriezi, O. et al. (2023). Precision farming technologies for crop protection: A meta-analysis. Smart Agricultural Technology, 5: 100323.

Asaei, H., Jafari, A., and Loghavi, M. (2019). Site-specific orchard sprayer equipped with machine vision for chemical usage management. Computers and Electronics in Agriculture, 162: 431–439.

Balsari, P., and Tamagnone, M. (1998). An ultrasonic airblast sprayer. In International Conference on Agricultural Engineering AgEng'98 (pp. Paper-98). GBR.

Balsari, P., Doruchowski, G., Marucco, P., Tamagnone, M., Van de Zande, J. C., and Wenneker, M. (2008). A system for adjusting the spray application to the target characteristics. Agricultural Engineering International: the CIGR Ejournal, 8(002): 11.

Balsari, P., Marucco, P., and Oggero, G. (2002). Spray applications in Italian apple orchards: Target coverage, ground losses and drift. In 2002 ASAE Annual Meeting (p. Paper Number: 02–1002). American Society of Agricultural and Biological Engineers.

Balsari, P., Marucco, P., Doruchowski, G., Ophoff, H., and Roettele, M. (2011). Best management practice to reduce spray drift. TOPPS Prowadis Projects Sponsored by European Crop Protection Association ECPA: Brusselsm Belgium.

Batte, M. T., and Ehsani, M. R. (2006). The economics of precision guidance with auto-boom control for farmer-owned agricultural sprayers. Computers and Electronics in Agriculture, 53(1): 28–44.

Berenstein, R., and Edan, Y. (2017). Automatic adjustable spraying device for site-specific agricultural application. IEEE Transactions on Automation Science and Engineering, 15(2): 641–650.

Berger, L. T., Doll, D., Schwitzky, E., Lavelle, K., Skalsky, M., and Spinelli, F. (2022, September). Precision agriculture – an enabler for pesticide use reduction. EIP-AGRI Focus Group Sustainable ways to reduce pesticides in pome and stone fruit production. Retrieved from https://ec.europa.eu/eip/agriculture/sites/default/files/minipaper_1_220908_finaldraft.pdf.

Berk, P., Hocevar, M., Stajnko, D., and Belsak, A. (2016). Development of alternative plant protection product application techniques in orchards, based on measurement sensing systems: A review. Computers and Electronics in Agriculture, 124: 273–288.

Berk, P., Stajnko, D., Belsak, A., and Hocevar, M. (2020). Digital evaluation of leaf area of an individual tree canopy in the apple orchard using the LIDAR measurement system. Computers and Electronics in Agriculture, 169: 105158.

Biller, R. H., Hollstein, C., and Sommer, C. (1997). Precision application of herbicides by use of optoeletronic sensors. Precis. Agric. 2: 451–458.

Bongiovanni, R., and Lowenberg-DeBoer, J. (2004). Precision agriculture and sustainability. Precision Agriculture, 5: 359–387.

Bonicelli, B., Naud, O., Rousset, S., Sinfort, C., De Rudnicki, V., Lescot, J. M. et al. (2010, September). The challenge for precision spraying. In AgEng 2010: International Conference on Agricultural Engineering (pp. 11-p).

Butt, M. A. (2022). Thin-film coating methods: A successful marriage of high-quality and cost-effectiveness—A brief exploration. Coatings, 12(8): 1115.

Campos, E. V., de Oliveira, J. L., Pascoli, M., de Lima, R., and Fraceto, L. F. (2016). Nanoparticles for insect pest control: current status and prospects. Frontiers in Chemistry, 4, 61.

Campos, E. V., Proença, P. L., Oliveira, J. L., Bakshi, M., Abhilash, P. C., and Fraceto, L. F. (2019). Use of botanical insecticides for sustainable agriculture: Future perspectives. Ecological Indicators, 105: 483–495.

Campos, J., Llop, J., Gallart, M., García-Ruiz, F., Gras, A., Salcedo, R., and Gil, E. (2019). Development of canopy vigour maps using UAV for site-specific management during vineyard spraying process. Precision Agriculture, 20(6): 1136–1156.

Carazo-Rojas, E., Pérez-Rojas, G., Pérez-Villanueva, M., Chinchilla-Soto, C., Chin-Pampillo, J. S., AguilarMora, P. et al. (2018). Pesticide monitoring and ecotoxicological risk assessment in surface water bodies and sediments of a tropical agro-ecosystem. Environmental Pollution, 241: 800–809. https://doi. org/10.1016/j.envpol.2018.06.020.

Chapple, A. C., Douzals, J. P., and Doruchowski, G. (2024). Development of a digital label: DriftRadar as a case study. In Proceedings of the International Conference on Precision Agriculture (pp. 1–8).

Chen, L., Wallhead, M., Zhu, H., and Fulcher, A. (2019). Control of insects and diseases with intelligent variable-rate sprayers in ornamental nurseries. Journal of Environmental Horticulture, 37(3): 90–100.

Chen, Y., Zhu, H., and Ozkan, H. E. (2011). Development of LIDAR-guided sprayer to synchronize spray outputs with canopy structures. American Society of Agricultural and Biological Engineers Annual International Meeting 2011 ASABE 2011, 1(11): 11–24. https://doi.org/10.13031/2013.37206.

Chen, Y., Zhu, H., and Ozkan, H. E. (2012). Development of a variable-rate sprayer with laser scanning sensor to synchronize spray outputs to tree structures. Transactions of the ASABE, 55(3): 773–781.

Chen, Y., Zhu, H., Ozkan, H. E., Derksen, R. C., and Krause, C. R. (2013). Spray drift and off-target loss reductions with a precision air-assisted sprayer. Transactions of the ASABE, 56(6): 1273–1281. https://doi. org/10.13031/trans.56.10173.

Cheng, H. D., Jiang, X. H., Sun, Y., and Wang, J. (2001). Color image segmentation: advances and prospects. Pattern recognition, 34(12): 2259–2281.

Chowdhury, M., Kushwah, A., and Kumar, P. (2021). Precision Spraying: Enhancing Crop Protection and Environmental Sustainability.

Church and Dwight. (2024). What Skills Are Required for Research and Development? Retrieved from https://careers.churchdwight.com/winning-together/research-and-development/what-skills-are-required-for-research-development/.

Damalas, C. A., and Eleftherohorinos, I. G. (2011). Pesticide exposure, safety issues, and risk assessment indicators. International Journal of Environmental Research and Public Health, 8(5): 1402–1419. https:// doi.org/10.3390/ijerph8051402.

Dammer, K. and Wartenberg, G. (2007). Sensor-based weed detection and application of variable herbicide rates in real-time. Crop Prot., 26: 270–277. https://doi.org/10.1016/j.cropro.2005.08.018.

Dimkpa, C. O., and Bindraban, P. S. (2018). Nanofertilizers: New products for the industry? Journal of Agricultural and Food Chemistry, 66(26): 6462–6473. https://doi.org/10.1021/acs.jafc.7b02150.

Dimkpa, C. O., Latta, D. E., McLean, J. E., Britt, D. W., Boyanov, M. I., and Anderson, A. J. (2020). Fate of CuO and ZnO nano- and microparticles in the plant environment. Environmental Science and Technology, 47(9): 4734–4742. https://doi.org/10.1021/es304736y.

Doruchowski, G., Swiechowski, W., Holownicki, R., and Godyn, A. (2009). Environmentally-dependent application system (EDAS) for safer spray application in fruit growing. The Journal of Horticultural Science and Biotechnology, 84(6): 107–112. https://doi.org/10.1080/14620316.2009.11512605.

Duhan, J.S., Kumar, R., Kumar, N., Kaur, P., Nehra, K. and Duhan, S. (2017). Nanotechnology: The new perspective in precision agriculture. Biotechnology Reports, 15: 11–23.

Escolà, A., Planas, S., Rosell, J. R., Pomar, J., Camp, F., Solanelles, F. et al. (2011). Performance of an ultrasonic ranging sensor in apple tree canopies. Sensors, 11(3): 2459–2477.

Escolà, A., Rosell-Polo, J. R., Planas, S., Gil, E., Pomar, J., Camp, F. et al. (2013). Variable rate sprayer. Part 1 - Orchard prototype: Design, implementation and validation. Computers and Electronics in Agriculture, 95: 122–135. https://doi.org/10.1016/j.compag.2013.02.004.

European Commission. (2020). The EU's Green Deal. Retrieved from https://ec.europa.eu/clima/policies/green-deal_en.htm.

Feltracco, M., Barbaro, E., Maule, F., Bortolini, M., Gabrieli, J., De Blasi, F. et al. (2022). Airborne polar pesticides in rural and mountain sites of North-Eastern Italy: An emerging air quality issue. Environmental Pollution, 308(January): 119657. https://doi.org/10.1016/j.envpol.2022.119657.

Franzen, D. and Mulla, D. (2015). A history of precision agriculture. Precision agriculture technology for crop farming, pp. 1–20.

FutureLearn. (2024). Precision Farming: Environmental Considerations. Retrieved from https://www.futurelearn.com/info/courses/innovation-in-arable-farming/0/steps/164402.

García-Mateos, G., Hernández-Hernández, J. L., Escarabajal-Henarejos, D., Jaén-Terrones, S., and Molina-Martínez, J. M. (2015). Study and comparison of color models for automatic image analysis in irrigation management applications. Agricultural water management, 151: 158–166.

Garcia-Ruiz, F., Sankaran, S., Maja, J. M., Lee, W. S., Rasmussen, J., and Ehsani, R. (2013). Comparison of two aerial imaging platforms for identification of Huanglongbing-infected citrus trees. Computers and Electronics in Agriculture, 91: 106–115.

Gerhards, R., and Oebel, H. (2006). Practical experiences with a system for site-specific weed control in arable crops using real-time image analysis and GPS-controlled patch spraying. Weed Research, 46(3): 185–193.

Gil, E., Llorens, J., Llop, J., Fàbregas, X., Escolà, A., & Rosell-Polo, J. R. (2013). Variable rate sprayer. Part 2 - vineyard prototype: Design, implementation, and validation. Computers and Electronics in Agriculture, 95: 136–150.

Giles, D. K. (2024). Targeted spray weed control in specialty crops: Historical and current performance and issues. In Targeted Application, Spray Technology and Authorisation Workshop (pp. 1–12).

Giles, D. K., Delwiche, M. J., and Dodd, R. B. (1987). Control of orchard spraying based on electronic sensing of target characteristics. Transactions of the ASAE, 30(6):1624–1636.

Gooding, M. (2024). TASAW-FINAL-DIGITAL. Retrieved from https://www.aab.org.uk/wp-content/uploads/2024/02/TASAW-FINAL-DIGITAL.pdf.

Gregory, A. (2024). A non-dietary risk assessors' perspective of precision application and the challenge of conservative assumptions. In Proceedings of the International Conference on Precision Agriculture (pp. 1–8).

Hočevar, M., Širok, B., Jejčič, V., Godeša, T., Lešnik, M., and Stajnko, D. (2010). Design and testing of an automated system for targeted spraying in orchards/Konzeption und Erprobung eines automatisierten Systems für die gezielte Applikation von Pflanzenschutzmitteln in Obstanlagen. Journal of Plant Diseases and Protection, 71–79.

Hočevar, V. J. T. G. M., Lešnik, B. Š. A. M. M., and Stajnko, A. Š. D. (2011). Design and testing of an ultrasound system for targeted spraying in orchards. Strojniski Vestnik/Journal of Mechanical Engineering, 57(7-8): 587–598.

Holterman, H. J., Van Boheemen, K., and Huijsmans, J. F. M. (2024). Modelling spray drift in precision spray applications. In Targeted Application, Spray Technology and Authorisation Workshop (pp. 13–24).

Huang, B., Chen, F., Shen, Y., Qian, K., Wang, Y., Sun, C. et al. (2018). Advances in targeted pesticides with environmentally responsive controlled release by nanotechnology. Nanomaterials, 8(2): 102.

Ibrahim, M. M., Said, M. A., and Ilbaş, A. I. (2024). Harnessing nanotechnology for precision agricultural advancements. American Journal of Biomedical Science and Research, 22(3), DOI: 10.34297/AJBSR.2024.22.002960.

Jin, J., and Tang, L. (2010). Optimal coverage path planning for arable farming on 2D surfaces. Transactions of the ASABE, 53(1): 283–295.

Jurado-Expósito, M., Castro, A. I., Torres-Sánchez, J., Jiménez-Brenes, F. M. and López-Granados, F. (2019). *Papaver rhoeas* L. mapping with cokriging using UAV imagery. Precis. Agric. 20: 1045–1067. https://doi.org/10.1007/s11119-019-09635-z.

Kah, M., and Hofmann, T. (2014). Nanopesticide research: Current trends and future priorities. Environment International, 63: 224–235. https://doi.org/10.1016/j.envint.2013.11.015.

Kah, M., Beulke, S., Tiede, K., and Hofmann, T. (2013). Nanopesticides: state of knowledge, environmental fate, and exposure modeling. Critical Reviews in Environmental Science and Technology, 43(16): 1823–1867.

Kah, M., Kookana, R. S., Gogos, A., and Bucheli, T. D. (2018). A critical evaluation of nanopesticides and nanofertilizers against their conventional counterparts. Nature Nanotechnology, 13(8): 677–684. https://doi.org/10.1038/s41565-018-0131-1.

Kamilaris, A., and Prenafeta-Boldú, F. X. (2018). Deep learning in agriculture: A survey. Computers and Electronics in Agriculture, 147: 70–90.

Karimzadeh, R., Hejazi, M. J., Helali, H., Iranipour, S., and Mohammadi, S. A. (2011). Assessing the impact of site-specificspraying on control of *Eurygaster integriceps* (Hemiptera: Scutelleridae) damage and natural enemies. Precision Agriculture, 12: 576–593.

Kempenaar, C., Van Boheemen, K., Tielien, J., Sijbrandij, F., Michielsen, J. M., Veldhuisen, B. et al. (2024). Precision dosing and spraying in the Dutch National Fieldlab Precision agriculture and Farm of the Future. In Targeted Application, Spray Technology and Authorisation Workshop (pp. 37–48).

Khot, L. R., Ehsani, R., Albrigo, G., Larbi, P. A., Landers, A., Campoy, J. et al. (2012). Airassisted sprayer adapted for precision horticulture: Spray patterns and deposition assessments in small-sized citrus canopies. Biosystems Engineering, 113(1): 76–85. https://doi.org/10.1016/j.biosystemseng.2012.06.008.

Khot, L. R., Sankaran, S., Maja, J. M., Ehsani, R., and Schuster, E. W. (2012). Applications of nanomaterials in agricultural production and crop protection: A review. Crop Protection, 35: 64–70. https://doi.org/10.1016/j.cropro.2012.01.007.

Kim, D. Y., Kadam, A., Shinde, S., Saratale, R. G., Patra, J., and Ghodake, G. (2018). Recent developments in nanotechnology transforming the agricultural sector: a transition replete with opportunities. Journal of the Science of Food and Agriculture, 98(3): 849–864.

Li, H., Guo, C., Yang, Z., Chai, J., Shi, Y., Liu, J. et al. (2022). Design of field real-time target spraying system based on improved yolov5. Frontiers in Plant Science, 13: 1072631.

Li, L., He, X., Song, J., Liu, Y., Zeng, A., Liu, Y. et al. (2018). Design and experiment of variable rate orchard sprayer based on laser scanning sensor. International Journal of Agricultural and Biological Engineering, 11(1): 101–108.

Llorens, J., Gil, E., Llop, J., and Escolà, A. (2010). Variable rate dosing in precision viticulture: Use of electronic devices to improve application efficiency. Crop Protection, 29(3): 239–248. https://doi. org/10.1016/j.cropro.2009.12.022.

López-Granados, F. (2011). Weed detection for site-specific weed management: mapping and real-time approaches. Weed Research, 51(1): 1–11.

Iost Filho, F. H., Heldens, W. B., Kong, Z., and de Lange, E. S. (2020). Drones: innovative technology for use in precision pest management. Journal of Economic Entomology, 113(1): 1–25.

Loudjani, P. (2014). Precision Agriculture: An Opportunity for EU-Farmers–Potential Support with the CAP 2014–2020.

Mahmud, M. S., and He, L. (2020). Measuring tree canopy density using a lidar-guided system for precision spraying. In 2020 ASABE Annual International Virtual Meeting (p. 1). American Society of Agricultural and Biological Engineers.

Mahmud, M. S., Zahid, A., He, L., and Martin, P. (2021). Opportunities and possibilities of developing an advanced precision spraying system for tree fruits. Sensors, 21(9): 3262.

Manandhar, A., Zhu, H., Ozkan, E., and Shah, A. (2020). Techno-economic impacts of using a laser-guided variable-rate spraying system to retrofit conventional constant-rate sprayers. Precision Agriculture, 21(5): 1156–1171. https://doi.org/10.1007/s11119-020-09712-8.

Manfreda, S., McCabe, M. F., Miller, P. E., Lucas, R., Pajuelo Madrigal, V., Mallinis, G. et al. (2018). On the use of unmanned aerial systems for environmental monitoring. Remote Sensing, 10(4): 641.

Marrs, R. H., and Frost, A. J. (1997). A microcosm approach to the detection of the effects of herbicide spray drift in plant communities. Journal of Environmental Management, 50(4): 369–388. https://doi. org/10.1006/jema.1996.9984.

Marrs, R. H., Williams, C. T., Frost, A. J., and Plant, R. A. (1989). Assessment of the effects of herbicide spray drift on a range of plant species of conservation interest. Environmental Pollution, 59(1): 71–86. https:// doi.org/10.1016/0269-7491(89)90022-5.

Matese, A., and Filippo Di Gennaro, S. (2015). Technology in precision viticulture: A state of the art review. International journal of wine research, 69–81.

Matthews, G. (2018b). A history of pesticides and challenges for the future. Agrisciences. Retrieved from https://www.soci.org/news/agrisciences/a-history-of-pesticides-and-challenges-for-the-future.

Matthews, G. A. (2018a). A history of pesticides. Cabi.

McKinsey and Company. (2023). The Economic Potential of Generative AI: The Next Productivity Frontier. Retrieved from https://www.mckinsey.com/capabilities/mckinsey-digital/our-insights/the-economic-potential-of-generative-ai-the-next-productivity-frontier.

Meshram, A. T., Vanalkar, A. V., Kalambe, K. B., and Badar, A. M. (2022). Pesticide spraying robot for precision agriculture: A categorical literature review and future trends. Journal of Field Robotics, 39(2): 153–171.

Moltó, E., Martín, B., and Gutiérrez, A. (2000). Pm—Power and machinery: Design and testing of an automatic machine for spraying at a constant distance from the tree canopy. Journal of Agricultural Engineering Research, 77(4): 379–384.

Mulla, D. and Khosla, R. (2016). Historical evolution and recent advances in precision farming. Soil-Specific Farming Precision Agriculture, pp. 1–35.

Mulla, D. J. (2013). Twenty five years of remote sensing in precision agriculture: Key advances and remaining knowledge gaps. Biosystems Engineering, 114(4): 358–371.

Nasir, F.E., Tufail, M., Haris, M., Iqbal, J., Khan, S., and Khan, M. T. (2023). Precision agricultural robotic sprayer with real-time Tobacco recognition and spraying system based on deep learning. Plos One, 18(3): e0283801.

NDK. (2020). Available online: https://www.ndk.com/en/sensor/ultrasonic/index.html.

Oberti, R., Marchi, M., Tirelli, P., Calcante, A., Iriti, M., Tona, E. et al. (2016). Selective spraying of grapevines for disease control using a modular agricultural robot. Biosystems Engineering, 146: 203–215.

Patel, M. K. (2016). Technological improvements in electrostatic spraying and its impact to agriculture during the last decadeand future research perspectives–A review. Engineering in Agriculture, Environment and Food, 9(1): 92–100.

Perz, R., and Wronowski, K. (2019). UAV application for precision agriculture. Aircraft Engineering and Aerospace Technology, 91(2): 257–263.

Plant Health Engineering Division, NIPHM. (2024). Workshops and extension officers need better coordination and cooperation to reduce the number of sick equipment. https://niphm.gov.in/Recruitments/ASO-PHE-Manual-NIPHM-03102013.pdf.

Popp, J., Pető, K., and Nagy, J. (2013). Pesticide productivity and food security. A review. Agronomy for Sustainable Development, 33: 243–255.

Rathore, A., Hasan, W., NM, R., Pujar, K., Singh, R., Panotra, N. et al. (2024). Nanotech for Crop Protection: Utilizing Nanoparticles for Targeted Pesticide Delivery. Uttar Pradesh Journal of Zoology, 45(6): 46–71.

Rosell, J. R., Llorens, J., Sanz, R., Arnó, J., Ribes-Dasi, M., Masip, J. et al. (2009). Obtaining the three-dimensional structure of tree orchards from remote 2D terrestrial LIDAR scanning. Agricultural and Forest Meteorology, 149(9): 1505–1515.

Salyani, M., Miller, D. R., Farooq, M., and Sweeb, R. D. (2013). Effects of sprayer operating parameters on airborne drift from citrus air-carrier sprayers. Agricultural Engineering International: CIGR Journal, 15(1): 27–36.

Satyendra. (2022). Research and Development and the Needed Skills Requirement. Ispat Guru. https://www.ispatguru.com/research-and-development-and-the-needed-skills-requirement/.

Schiefer, J. and Dillon, C. (2015). Te economic and environmental impacts of precision agriculture and interactions with agro-environmental policy. Precis. Agric. 16: 46–61. https://doi.org/10.1007/s11119-014-9382-5.

Shaikh, A., Meroliya, H., Dagade-Gadale, S., and Waghmode, S. (2021). Applications of Nanotechnology in Precision Agriculture: A review. Res. Rev. Biotechnol. Biosci, 8: 105–117.

Shamshiri, R., Weltzien, C., Hameed, I. A., J Yule, I., E Grift, T., Balasundram, S. K. et al. (2018). Research and development in agricultural robotics: A perspective of digital farming.

Shanahan, J. F., Kitchen, N. R., Raun, W. R., and Schepers, J. S. (2008). Responsive in-season nitrogen management for cereals. Computers and Electronics in Agriculture, 61(1): 51–62.

Shang, Y., Hasan, M. K., Ahammed, G. J., Li, M., Yin, H., and Zhou, J. (2019). Applications of nanotechnology in plant growth and crop protection: a review. Molecules, 24(14): 2558.

Sishodia, R. P., Ray, R. L., and Singh, S. K. (2020). Applications of remote sensing in precision agriculture: A review. Remote sensing, 12(19): 3136.

Slaughter, D. C., Giles, D. K., and Downey, D. (2008). Autonomous robotic weed control systems: A review. Computers and Electronics in Agriculture, 61(1): 63–78.

Smith, D. R., Harmel, R. D., and Haney, R. L. (2020). Long-term agro-economic and environmental assessment of adaptive nutrient management on cropland fields with established structural conservation practices. Journal of Soil and Water Conservation, 75(3): 416–425.

Song, Y., Sun, H., Li, M., and Zhang, Q. (2015). Technology application of smart spray in agriculture: A review. Intelligent Automation and Soft Computing, 21(3): 319–333.

Stajnko, D., Berk, P., Lešnik, M., Jejčič, V., Lakota, M., Štrancar, A. et al. (2012). Programmable ultrasonic sensing system for targeted spraying in orchards. Sensors, 12(11): 15500–15519.

Sudduth, K. A., Kitchen, N. R., Wiebold, W. J., Batchelor, W. D., Bollero, G. A., Bullock, D. G. et al. (2005). Relating apparent electrical conductivity to soil properties across the north-central USA. Computers and Electronics In Agriculture, 46(1-3): 263–283.

Tauchnitz, N., Kurzius, F., Rupp, H., Schmidt, G., Hauser, B., Schrödter, M. et al. (2020). Assessment of pesticide inputs into surface waters by agricultural and urban sources - A case study in the Querne/ Weida catchment, central Germany. Environmental Pollution, 267. https://doi.org/10.1016/j.envpol.2020.115186.

Thériault, R., Salyani, M., and Panneton, B. (2001). Spray distribution and recovery in citrus application with a recycling sprayer. Transactions of the ASAE, 44(5): 1083.

Thomas, J. (2024). Precision future for pesticide application. Direct Driller. https://directdriller.com/precision-future-for-pesticide-application/[1].

Tona, E., Calcante, A., and Oberti, R. (2018). The profitability of precision spraying on specialty crops: a technical–economic analysis of protection equipment at increasing technological levels. Precision Agriculture, 19: 606–629.

Tudi, M., Daniel Ruan, H., Wang, L., Lyu, J., Sadler, R., Connell, D. et al. (2021). Agriculture development, pesticide application and its impact on the environment. International Journal of Environmental Research and Public Health, 18(3): 1112.

Van Boheemen, K., Huijsmans, J. F. M., and Holterman, H. J. (2024). Precision agriculture and crop protection: Definitions and the relation between precision applications and the authorisation procedure of PPPs. In Targeted Application, Spray Technology and Authorisation Workshop (pp. 25–36).

Wang, S., Xu, T.,and Li, X. (2022). Development Status and Perspectives of Crop Protection Machinery and Techniques for Vegetables. Horticulturae, 8(2): 166.

Willekens, A., Wieme, J., Van De Vijver, R., Van Beek, J., Windels, J., Nuyttens, D. et al. (2023). Real-time and task map precision spray applications for weeds and diseases: Use cases developed at ILVO Living Lab Agrifood Technology. In Targeted Application, Spray Technology and Authorisation Workshop (p. 23).

World Economic Forum. (2017). Realizing the Potential of Blockchain: A Multistakeholder Approach to the Stewardship of Blockchain and Cryptocurrencies. Retrieved from https://www3.weforum.org/docs/WEF_Realizing_Potential_Blockchain.pdf.

Xun, L., Campos, J., Salas, B., Fabregas, F. X., Zhu, H., and Gil, E. (2023). Advanced spraying systems to improve pesticide saving and reduce spray drift for apple orchards. Precision Agriculture, 24(4): 1526–1546.

Yadav, A., Yadav, K., Ahmad, R., and Abd-Elsalam, K. A. (2023). Emerging frontiers in nanotechnology for precision agriculture: Advancements, hurdles and prospects. Agrochemicals, 2(2): 220–256.

Yang, C., Sui, R., and Lee, W. S. (2016). Precision agriculture in large-scale mechanized farming. Precision Agriculture Technology for crop farming, 177–204.

Zaman, Q. U., and Salyani, M. (2004). Effects of foliage density and ground speed on ultrasonic measurement of citrus tree volume. Applied Engineering in Agriculture, 20(2): 173–178.

Zanin, A. R. A., Neves, D. C., Teodoro, L. P. R., da Silva Júnior, C. A., da Silva, S. P., Teodoro, P. E. et al. (2022). Reduction of pesticide application via real-time precision spraying. Scientific Reports, 12(1): 5638.

Zhang, C., and Kovacs, J. M. (2012). The application of small unmanned aerial systems for precision agriculture: a review. Precision Agriculture, 13: 693–712.

Zhang, Z., Wang, X., Lai, Q., and Zhang, Z. (2018). Review of variable-rate sprayer applications based on real-time sensortechnologies. Automation in Agriculture–Securing Food Supplies for Future Generations. Intech, 53–79.

Zhu, H., Liu, H., Shen, Y., Liu, H., and Zondag, R. H. (2017). Spray deposition inside multiple-row nursery trees with a laser-guided sprayer. Journal of Environmental Horticulture, 35(1): 13–23.

Zia, R., Taj, A., Younis, S., Bukhari, S. Z., Latif, F., Feroz, Y. et al. (2022). Application of nanosensors for pesticide detection. In Nanosensors for Smart Agriculture (pp. 259–302). Elsevier.

CHAPTER 13

New Concepts in Artificial Intelligence for Automated Weed Control

Kamel A. Abd-Elsalam,[1,*] *Heba Yousef,*[1] *Rawan K. Hassan,*[2] *Toka E. Abdelkhalek*[2] and *Hassan Almoammar*[3]

1. Introduction

The relentless challenge of weed management in agriculture has entered a revolutionary phase with the integration of AI technologies. Weeds, which have traditionally been controlled through labor-intensive manual removal or indiscriminate herbicide application, are now being targeted with surgical precision thanks to breakthroughs in machine learning, robotics, and computer vision (Khang et al., 2023). This chapter delves into the cutting-edge AI methodologies that are transforming weed control from a blanket approach to a sophisticated, automated, and sustainable practice (Zheng et al., 2024). Recent advancements in deep learning, particularly CNNs and transformer-based architectures, have enabled real-time, high-accuracy weed detection and classification even in complex, occluded field environments (Wang, 2022). These AI models, trained on vast datasets of multispectral and hyperspectral imagery, can distinguish between crops and weeds with remarkable precision, enabling targeted interventions (Yaoning et al., 2021). When combined with autonomous ground robots or drone-based systems, this technology facilitates precise mechanical removal or spot spraying, dramatically reducing herbicide usage and minimizing environmental impact (Du et al., 2019). Beyond simple detection, novel AI paradigms

[1] Plant Pathology Research Institute, Agricultural Research Center, Giza 12619, Egypt.
[2] Biotechnology English Program, Faculty of Agriculture, Cairo University, Giza 12613, Egypt.
[3] Wellness and Preventive Medicine Institute, Health Sector, King Abdulaziz City for Science and Technology (KACST), Riyadh, Saudi Arabia.
* Corresponding author: kamelabdelsalam@gmail.com

DOI: 10.1201/9781003546702-13

are being applied to weed management (Fengle et al., 2010). Reinforcement learning algorithms optimize decision-making for robotic weeders, allowing them to adapt to dynamic field conditions. Generative adversarial networks (GANs) are being used to synthesize training data, improving model robustness across diverse agricultural settings (Felizardo et al., 2016). Meanwhile, edge AI implementations bring real-time processing capabilities directly to field equipment, eliminating reliance on cloud connectivity and enabling scalable solutions for large farms (Sabzi et al., 2020). This chapter systematically explores these innovations, presenting case studies of successful deployments while also addressing critical challenges: the need for large, annotated datasets; model generalizability across different crops and geographies; energy efficiency constraints in edge devices; and economic barriers to adoption. By bridging the gap between AI research and agricultural practice, these technologies promise not only to enhance weed control but to redefine sustainable farming in the 21st century ushering in an era of precision agriculture where AI-driven automation maximizes productivity while minimizing ecological harm. The discussion extends to future directions, including the integration of digital twins for weed prediction, swarm robotics for coordinated weeding operations, and explainable AI (XAI) to build trust in autonomous systems. Through this comprehensive examination, the chapter provides both a technological roadmap and a critical assessment of how AI can meet one of agriculture's most persistent challenges in an era of climate change and growing food demand.

2. Fundamentals of Weed Control

Agricultural operations have experienced a transformation through autonomous vehicles and robots which boost operational efficiency and minimize labor expenses while increasing agricultural production. Autonomous vehicles utilize GPS and LIDAR and cameras as advanced sensors to execute planting and spraying tasks and harvesting duties in nearly automated conditions (Korres et al., 2019). Robots have become principal for precise agricultural activities such as pruning along with weeding and fruit harvesting through systems equipped with machine vision for detailed detection capabilities (Zimdahl et al., 2024). These technologies help create sustainable practices because they reduce herbicide usage and enable farmers to implement regenerative systems. The improvement of irrigation systems has successfully tackled problems of water management. Surface irrigation spreads water by gravity whereas localized systems distribute water by utilizing a pressurized network system. Modern automated irrigation technology utilizes timers alongside soil sensors and plant stress detectors to guarantee targeted water delivery (Harrison et al., 2017; Gao et al., 2024). They facilitate enhanced agricultural productivity while reducing waste. Insula's self-operating systems encounter practical challenges because of substantial project costs and technological obstacles faced by fundamental farmers. Fleet-based robotic systems alongside modular architectural solutions deliver improved scalability and affordability to agricultural operations. These technologies unite to transform agriculture into a more sustainable system which provides greater production rates (Scavo and Mauromicale, 2020).

2.1 Traditional Methods and Limitations

The accuracy of physical weed control methods through flame and laser weeding techniques is achieved but implementation faces multiple obstacles. The gas emissions from flame weeding operations create environmental problems (Riemens et al., 2022). Laser weeding operates on renewable energy to precisely destroy small weeds despite its inability to control mature weeds and its safety hazards to human health and exposure to fire risks (Vasileiou et al., 2024; Monteiro and Santos, 2022). The mechanical weeding robots display high efficiency but they create soil compaction and encounter problems with hilly surfaces which leads to reduced accuracy (Jitsuyama and Ichikawa, 2011). Sensor-based recognition systems require additional developments to achieve higher operational effectiveness during complex environment operations (Rakhmatulin and Andreasen, 2020). The use of bioherbicides presents environmental benefits but these products demonstrate short shelf life and reduced effectiveness combined with challenging extraction methods which increase safety risks and implementation costs (Andreasen et al., 2022). Future research needs to develop effective bioherbicide formulations through surfactant incorporation or nanoformulation methods along with identifying fresh plant-derived compounds (Fernández et al., 2018; Kaab et al., 2020). Chemical herbicides operate swiftly while being affordable yet create environmental damages to the soil and microorganisms while polluting groundwater (Morra et al., 2018). Herbicide reduction requires precise application techniques and the execution of various weed control strategies (Motmainna et al., 2021). Research-driven innovation represents a vital solution for achieving both sustainable and efficient weed management in agricultural environments (Machleb et al., 2020). AI in Automated Weed Control highlighting various aspects of its implementation is shown in Figure 1. It covers AI technologies, including machine learning algorithms and different learning approaches. Targeted weed management focuses on precise weed detection and variable rate herbicide application. Robotic weeding systems involve autonomous robots and mechanical weeding. Chemical weed control is emphasized through selective herbicide spraying and smart sprayer technology, though it is mentioned twice. Barriers & challenges include high costs and a lack of annotated weed datasets. Future directions explore advancements like swarm robotics, explainable AI (XAI), digital twin technology, and AI integration with smart farming.

2.2 Chemical Methods

Herbicide-based weed control serves as the primary weed management approach in present-day farming, though it creates environmental damage while producing weeds that adapt to herbicides. Agricultural workers discovered 2,4-D in the 1940s, thus creating herbicides that enabled farmers to reduce labor and achieve higher production efficiencies. Current herbicides are grouped according to chemical composition and timing of application before planting or after germination along with how they work and their formulation characteristics and target selection criteria. The different mechanisms of weed control through herbicides include blocking photosynthesis processes as well as interrupting amino acid synthesis and preventing

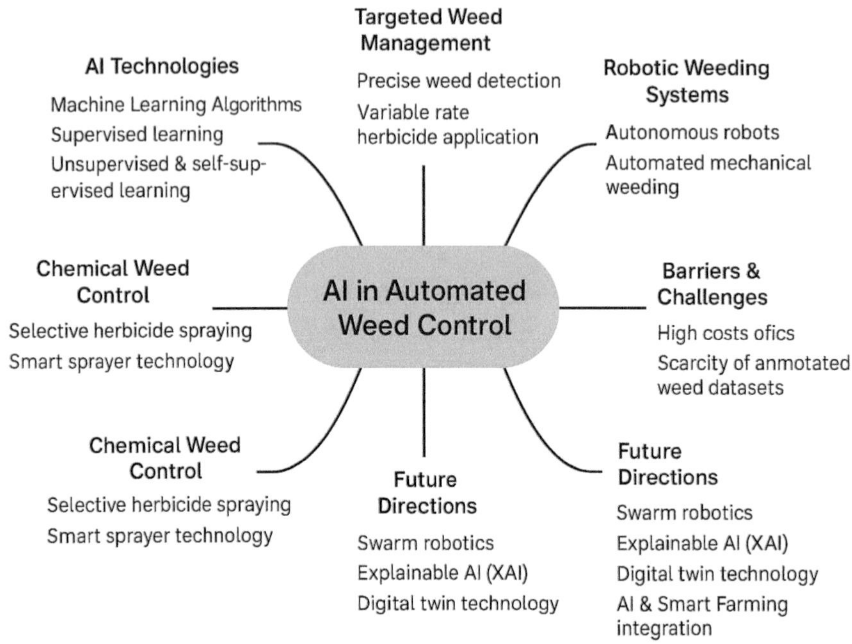

Figure 1. Automated weed control using AI involves the application of artificial intelligence and robotics to detect and manage weeds in agricultural environments. This technology is designed to improve farming productivity and decrease the dependency on chemical weed killers.

cell development (Marble et al., 2015). Herbicide overuse has developed resistance in 262 weed species across the world where 510 species-action site combinations are now impacted (Gao et al., 2024). The impact of environment factors including temperature and soil moisture content together with humidity directly affects how effectively herbicides work. Wind-driven spray drift affects proper herbicide application by combining factors of velocity with dropping size and application methods. Herbicides cause adverse impacts on population ecology for beneficial insects and all forms of soil-dwelling organisms. Research shows that ladybird beetle mortality rates reach 80% after exposure to 2,4-D and earthworms die upon contacting high concentration doses (Benaragama et al., 2016; Casagrande et al., 2016). Studies have established that herbicide formulations containing Glyphosate modify the intestinal bacterial population of honeybees and demonstrate ecological consequences. Precision chemical weed control presents a long-term solution which directs herbicides at weeds to minimize overall herbicide consumption while supporting crop yields and floral diversity (Cook et al., 2018). When precision technologies support alternative weed management approaches, they help protect the environment by reducing both chemical resistance and ecological damage. Intensive practices like crop rotation together with biological control methods and bioherbicide development must be utilized to decrease dependence on chemical herbicides. Improvements in herbicide formulations together with advanced application techniques work to reduce the impact of chemicals on unintended targets and environmental pollutants

(Kumar et al., 2025; Anderegg et al., 2023). Addressing herbicide resistance requires a multifaceted approach, including monitoring resistance patterns, diversifying weed control methods, and promoting integrated weed management (IWM) practices (Olorunmaiye et al., 2025; Harrison and Loux, 2017). Collaborative efforts among researchers, policymakers, and farmers are crucial for developing and implementing sustainable solutions. By balancing efficiency with environmental stewardship, modern agriculture can achieve effective weed control while preserving ecosystem health and ensuring long-term productivity (Gangstad, 2018).

2.3 Mechanical Methods

When farmers use mechanical methods for weed control their crops sustain accidental harm that leads to stunted growth and diminished resistance to pathogens and secondary infections resulting in yield losses. The damage from mechanical injuries to sugar beet roots creates openings that allow the diseases Fusarium alongside Phytophthora root rot to enter the plant system (Shaner et al., 2014). Eliminating weeds near the crop rows remains vital since they compete excessively for resources and decrease plant health. The development of mechanical weed control accuracy requires combining traditional practices with modern sensor systems and machine vision and high-precision navigation systems (Mesnage et al., 2021). Ultrasonic sensors together with LiDAR sensors serve as effective sensors for this specific purpose. The distance measurement for ultrasonic sensors works by using sound waves and LiDAR sensors establish distance calculations through light beam phase comparison. Ultrasonic sensors prove economical for their use within 100 mm to 10 m distance range and show reliable performance, yet LiDAR delivers more precise results for identifying features by detecting chlorophyll reflection (Chacko et al., 2021). Extensive agricultural fields can be inspected for weeds by these identification and classification sensors. The analysis capability of machine vision technology helps precision mechanical weed control through automatic identification of weeds from captured images. The FarmWise autonomous weeding robots use deep learning to analyze images of plants in order to perform weed removal tasks. The machine vision system implemented by F. Poulsen Engineering's Robovator separates weeds from crops through visual analysis which triggers the tine rake to change its operational actions depending on the identified plant species (Pantović et al., 2023). Future developments in machine vision technology will facilitate its effective operation in various cropping environments, including row configurations under 20 cm. Mechanical weeding robots need GNSS (Global Navigation Satellite System) as their core positioning system because it regulates their function with data from satellite signals. The major Global Navigation Satellite System services include GPS from the U.S. in addition to GLONASS managed by Russia as well as Galileo managed by the EU and Beidou operated by China (Hussain et al., 2021). GNSS satellites produce positioning information which might not reach required levels especially during signal disruptions (Almberg et al., 2018). The need for continuous improvements in navigation technology becomes essential to support precision agriculture systems.

3. AI Technologies for Weed Control

Conventional methods of mechanical weed control harm agricultural crops by causing injuries that restrict development while making them more susceptible to pathogens thus leading to secondary infections and decreased crop production. Sugar beets with mechanical injuries become more prone to develop Fusarium and Phytophthora root rot (Shaner et al., 2014; Partel et al., 2019). Intra-row weed control proves essential because these competitors attract valuable resources that jeopardize the condition of crops. The combination of traditional mechanical methods with advanced sensor technologies like ultrasonic and LiDAR sensors (Mesnage et al., 2021) enhances precision in weed control (Gerhards et al., 2022). Ultrasonic sensors use sound waves to determine distance, but LiDAR sensors reach higher accuracy by using laser beams. The chlorophyll-containing survey technology of LiDAR proves greatest at weed identification and categorization in extensive agricultural land areas through its ability to reflect near-infrared light waves (Chacko et al., 2021). The position calculation and weed detection functionality of precision mechanical weed control systems operates through machine vision technology which processes images to identify potential targets (Priyadarshini et al., 2024). Deep learning enables FarmWise autonomous robots to detect weeds using their technology effectively and remove them from agricultural fields automatically. Denmark-based F. Poulsen Engineering designed the Robovator adopting machine vision technology that integrates with a tine rake system which adjusts mechanical activities according to crop and weed presence across different growth periods (Pantović et al., 2023). Machine vision technology will evolve to address various cultivation environments that extend to the handling of sparse row patterns. Mechanical weeding robots need dependable navigation systems such as Global Navigation Satellite Systems (GNSS) to access their positioning data (Su and W. H., 2020). The accuracy issues and signal interruption problems provide challenges for specific operational tasks (Dasgupta et al., 2020; Amend et al., 2019).

3.1 Machine Learning Algorithms

Machine learning plays a crucial role in decision-making by uncovering rules and patterns within large datasets, particularly in agriculture. It is primarily divided into two types: unsupervised learning and supervised learning (Muchhadiya et al., 2024). Unsupervised learning, such as cluster analysis (CA), identifies patterns in unlabeled data without human intervention. In contrast, supervised learning, which includes methods like discriminant analysis (DA), learns to map inputs to outputs based on labeled examples (Priyadarshini et al., 2024). This approach facilitates easier system training by using example input-output pairs rather than manual programming. Popular supervised learning algorithms for weed detection in precision agriculture include convolutional neural networks (CNN), artificial neural networks (ANN), support vector machines (SVM), and soft independent modeling of class analogy (SIMCA) (Corceiro et al., 2023). Various sensing techniques enhance machine learning applications; satellite imagery provides a broad overview of weed distribution, while near-ground methods offer real-time precision (Mulla and D. J., 2013). Remotely sensed aerial scans can create prescription maps for targeted pesticide applications, although they require high-quality data and are not suited for real-time use (Hamuda et al., 2017; Hamuda et al., 2017). Research is advancing toward ground-based robotic sensors capable of precise weed identification by analyzing plant characteristics through visual data at specific wavelengths or 3D point clouds. These characteristics can include fruit quantity, trunk geometry, and plant stress indicators (Kool et al., 2023). Modern sensing technologies, categorized into imaging sensors (like smart cameras) and non-imaging sensors (such as lasers), provide enhanced spatial and temporal accuracy in agricultural applications (Ghaly et al., 2022).

3.2 Computer Vision and Remote Sensing

Imaging sensors, which utilize a range of imaging technologies including RGB cameras, near infrared, and hyperspectral imagery, are crucial for achieving accurate and focused weed management while reducing the need for pesticides and laborious tasks (Juwono et al., 2023). An important benefit of image sensors in robotic weeding is their capacity to offer instantaneous information on the distribution of crops and weeds in the cultivation area. Through the acquisition of intricate field pictures, these sensors enable precise discrimination and categorization of crops and weeds with exceptional precision (Shanmugam et al., 2020). Subsequently, algorithms powered by machine learning analyze these data to identify certain weed species and determine the most suitable responses, such as selectively applying pesticides or physically removing the weeds. Utilizing cameras in computerized weeding machines presents distinct advantages compared to conventional weed management approaches. First and foremost, the capacity to accurately focus on eliminating weeds while protecting crops leads to substantial financial savings and improved productivity (Gerhards et al., 2022). By selectively spraying herbicides exclusively in areas where weeds are prevalent, farmers may decrease their pesticide consumption, resulting in monetary benefits and limiting the ecological consequences of farming. RGB cameras are widely used due to their cost-effectiveness and ability to capture high-resolution color images in the visible spectrum, making them suitable for basic plant differentiation tasks. However, they are limited in detecting plant stress or characteristics beyond the visible range (Burkart et al., 2018). Near-infrared (NIR) cameras, on the other hand, are better in detecting chlorophyll levels and plant water content, but they are more expensive and lack the color differentiation of RGB cameras. Spectral and multispectral cameras go a step further, capturing data across multiple wavelengths, allowing for an advanced analysis of plant characteristics such as nutrient deficiencies or disease, making them ideal for precision agriculture. However, their high cost and the complexity of processing large datasets make them less accessible for real-time applications (Lu et al., 2020). Thermal cameras detect heat signatures, enabling the identification of plant stress and moisture levels, and they perform well in low light or nighttime conditions, although thermal imaging alone may not provide enough detail for accurate plant–weed differentiation and is sensitive to environmental factors such as wind and surface reflectivity (Gené et al., 2019). Therefore, each camera type presents a trade-off between cost, data richness, and the complexity of integration, and their selection depends on the specific requirements of the robotic weeding system. It should be noted that the independent functionality of robotic weeding systems coupled with image sensors enables continuous weed surveillance and control, guaranteeing prompt interventions and minimizing the likelihood of weed infection. The following are discussions of the most prevalent sensor systems mounted on weeding robots (Nguyen et al., 2016).

3.2.1 Image acquisition technologies

The implementation of robotic weeding requires technological solutions to improve automated weed control systems because they encounter multiple technical barriers. The main difficulty stems from sensing system longevity and dependability using RGB cameras as these devices have trouble distinguishing between crop and weed populations across diverse agricultural areas affected by different light conditions and variations in plant cover and terrain (Putra et al., 2020). The detection capabilities of optical sensors for weeds become complicated by heterogeneous soil conditions that show dissimilar traits such as color and structure along with moisture content (Costa et al., 2020; Gracia et al., 2018) and these capabilities also struggle against obstructing vegetation and young weed growth. The integration of multiple sensing systems including LIDAR and both ultrasonic detectors and thermal cameras leads to improved sensing accuracy. Sensor fusion algorithms act as essential components to integrate different sensor streams into precise weed-crop differentiation systems (Endres et al., 2013). Temperature and humidity changes sensors can automatically calibrate their performance for reliable operation across dynamic environments (de et al., 2019). Robotic weeding systems need better navigation systems to be truly effective. The precision of robotic maneuvering improves through GPS positioning systems connected with RTK and IMUs whereas LIDAR and ultrasonic sensors implemented in obstacle detection systems enable robots to navigate diverse terrains efficiently (Gašparović et al., 2020). The requirement for real-time data processing faces challenges because AI algorithms used for weed recognition have high computational requirements. Cloud-based systems serve agricultural farms by conducting quick data processing and model distribution among fields (Rasmussen et al., 2019). Equipment used for sensing must demonstrate resistance against harsh agricultural conditions. The processing equipment needs reliable protective packaging which safeguards sensors against environmental factors such as dust along with dirt and humidity (Hassanein et al., 2018; Hashemi et al., 2022). The establishment of solutions against these obstacles will create effective robotic weeding platforms delivering sustainable alongside precise weed management capabilities.

3.2.2 Data processing techniques

Precision agriculture relies on technologies that combine sensors, information systems, and informed management to optimize crop productivity and to reduce the environmental impact. Nowadays, precision agriculture has a broad range of applications, and it is employed in different agricultural contexts including pests' control, fertilization, and irrigation, sowing and harvesting (Rai et al., 2023).

Precision agriculture can be effectively applied to IWM also. In the last decade, precision agriculture has rapidly advanced because of technological innovations in the areas of sensors, computer hardware nanotechnology, unmanned vehicles systems and robots that may allow for specific identification of weeds that are present in the feld (Verhoeven and G., 2008). Unmanned aerial vehicles (UAV) are one of the most successful technologies applied in precision agriculture. Unmanned Vehicles systems are mobile Aerial (UAV) or Terrestrial (UTV) platforms that provide numerous advantages. UAV remote sensing has several advantages compared to satellite remote sensing: (1) UAV remote sensing can be equipped with specific sensors for observation, as required. (2) UAV remote sensing can observe targets at any time period allowed by weather and environmental conditions. (3) UAV remote sensing can set a repeatable flight route, to achieve multiple target observations from a set altitude and angle. (4) The image sensor mounted on the UAV is closer to the target, and the image resolution obtained by observation is higher (Jung et al., 2020). These characteristics have not only allowed the remote sensing community to produce new techniques in land cover/land use and change detection based on remote sensing satellite data in the past, but have also contributed to the growth of forest remote sensing, precision agriculture remote sensing, and other research directions. Forestry remote sensing is a relatively small aspect of satellite remote sensing. Multi-spectral and hyper-spectral data related to tree photosynthesis are among the various types of remote sensing satellite data. However, the spatial resolution of remote sensing satellite data is limited and cannot meet the requirements for specific tree types and disease identification. Additionally, remote sensing satellites cannot provide the high-accuracy elevation information necessary for forestry structure studies (Putra et al., 2017).

3.3 Robotic Weeding Technologies

In this section, the various approaches in the development of weeding robots are analysed. The reviewed research has been divided into two broad categories, i.e., robots which use chemical and non-chemical weed control methods. Chemical weed control methods involve the application of herbicides. Non-chemical weed control, on the other hand, involves mainly the use of mechanical weeding tools and laser-based tools (Sakamoto et al., 2012). In each of the two categories, the papers have been examined according to their focus, namely (a) locomotion and navigation, (b) sensing, and (c) weeding tool.

3.3.1 Spraying robots

Agricultural robots employ specific navigation and locomotion systems which match their targeted missions and crop growth areas by traveling through straight lines through row crops. A low-cost spraying robot operating in narrow rows depends on vision systems to detect weeds before applying herbicides to flaxseed fields because it combines self-charging capabilities (Zhu et al., 2020). A robot traverses predefined paths in sugar beet fields by adjusting herbicide distribution according to weed dimensions and adjustment factors related to its movement speeds to maximize pesticide use efficiency (Meshram et al., 2022). The recognition of both weeds and crops through vision technologies allows for precise herbicide application as a method to reduce environmental damage (Lochan et al., 2024). A modular robot having camera technology regulates multiple spraying nozzles for selective herbicide distribution around finger millet which operates with three-second precision insect identification (Ghafar et al., 2023). The vision-based system implemented for onion fields demonstrated weed detection proficiency reaching 97% success rate (Cantelli et al., 2019). A four-wheeled robotic system navigates cotton fields while precisely delivering herbicide applications to weed locations with an efficiency rate higher than 97%. The Asterix robot identifies carrot seed lines with its front-mounted camera then releases properly regulated herbicide doses when spotting weeds (Hejazipoor et al., 2021). The machines function to apply chemicals selectively which leads to decreased herbicide consumption and protects agricultural plants (Chaitanya et al., 2020). The two-wheeled intra-row robot demonstrated a remarkable performance by removing weeds effectively while damaging less than 1.95% of the crops in maize fields as well as 0.82% crop damage in Chinese cabbage fields (Wang et al., 2022). The technological developments confirm precision agriculture's potential to boost operational efficiency and environmental sustainability.

4. Data Collection and Analysis

Weed control with Artificial Intelligence (AI) depends on sophisticated data collection systems which gather agricultural records from different sources including crop management and soil records (Warren and Marz, 2015). The daily operations of a farm including tilling and planting and chemical treatments need to be registered in Farm Management Information Systems (FMIS) using manual data entry. Plant and soil samples go through examinations to learn their characteristics as well as to identify existing weeds (Talaviya et al., 2020). The small scale of data produced in precision agriculture practices leads to substantial volume when consolidated across entire agricultural operations. Distributed programs like DSSAT and ICASA utilize relational databases to efficiently process data while traditional storage models have stored operational data efficiently (Chang et al., 2008). Modern Big Data requirements pose challenges that traditional relational databases struggle to resolve when it comes to rapid query management and distributed database storage needs. The issues require alternative solutions such as NoSQL databases and distributed systems including Google's Bigtable to address them (Hunt et al., 1994; White et al., 2013).

4.1 Red-Green-Blue (RGB) Sensors

The agricultural sector is adopting Red-Green-Blue (RGB) sensors known as visual spectrum sensors because they provide affordability with versatility (Whitehead and Hugenholtz, 2014). The combination of red, green, and blue channels in sensors enables image capture for subsequent vegetation index calculations specifically the Green/Red Vegetation Index (GRVI), Greenness Index (GI), and Excessive Greenness (ExG) as described by Motohka et al. (2010) and Xue and Su (2017). The application of RGB sensors in machine learning supports three key functions: object recognition and phenology studies and pathology detection (Crimaldi et al., 2020; Pfanz et al., 2018) A standard RGB image procedure from Unmanned Aerial Vehicles encompasses pre-flight planning as well as image capturing and subsequent post-processing stages (Whitehead & Hugenholtz, 2014). The first step of UAV imagery collection involves defining the research area together with selecting flight altitude along with adhering to local governing authorities. Data acquisition needs sufficient flow to avoid errors when collecting images. Post-processing of RGB sensors lacks the need for radiometric calibration because it enables direct use of acquired data and orthomosaic generation (Torres-Sánchez et al., 2014). Machine learning algorithms require extensive datasets to conduct their training and testing process according to Bah et al. (2018). Researchers find available plant datasets through repositories such as PlantVillage (Hughes & Salathe, 2016) and also generate new datasets from the ground up (Sa et al., 2018). The classification of weeds and plant health assessment through RGB sensors remains effective despite their vulnerability to lighting variations (Hassler & Baysal-Gurel, 2019).

4.2 Multispectral and Hyperspectral Imaging

Remote sensing operations in agriculture benefit from RGB, multispectral and hyperspectral sensors (Table 1) because they collect electromagnetic spectrum data from a wider range than RGB sensors do. The multiple bands available in UAV system

Table 1. Comparison of RGB, Multispectral, and Hyperspectral Sensor.

Feature	RGB Sensors	Multispectral Sensors	Hyperspectral Sensors
Radiometric Bands	3 (Red, Green, Blue)	Typically 4–10	Hundreds to Thousands
Calibration	Generally not required	Radiometric and atmospheric correction required	Complex radiometric calibration required
Data Processing	Orthomosaic generation, vegetation index calculation	Image rectification, georeferencing, stacking	Quality assessment, narrowband index calculation
Applications	Object recognition, phenology, pathology detection	Plant disease detection, field quality assessment	Specific field characteristic detection
Cost and Complexity	Low cost, simple workflow	Moderate cost, moderately complex workflow	High cost, complex workflow
Advantages	Readily available, low cost, and simple to use	Broader range of vegetation indices can be calculated.	Can detect a specific field characteristic, early stage plant disease
Disadvantages	Only minimal spectral information using three significant bands	Lower resolution compared to RGB sensors, Multi-lenses nature of the sensors in phase 3 leads the data collected suffer from the parallax problem	High cost and weight can limit the UAV systems payload
References	(Whitehead and Hugenholtz, 2014)	(Cottrell., 2024)	(Brook and Ben-Dor, 2015)

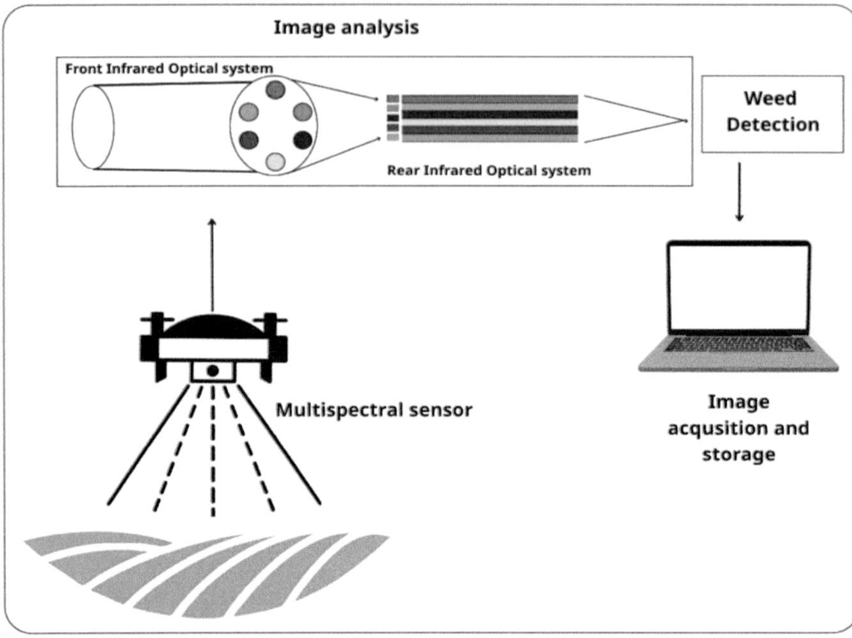

Figure 2. Weed detection system utilizing a multispectral sensor mounted on a drone, which captures and analyzes images through front and rear infrared optical systems for effective image acquisition and storage.

sensors known as multispectral sensors enable the calculation of different vegetation indices (Figure 2). Active multispectral sensors need both radiometric calibration and atmospheric correction that are supported by downwelling irradiance sensors and calibrated reflectance panels according to Cottrell (2024). The necessary condition for obtaining high-resolution images from multispectral sensors includes flying at low altitudes while maintaining appropriate image overlap because these sensors possess lower resolution capabilities than RGB sensors (Laliberte et al., 2011). The collected data experiences parallax errors which need image rectification along with georeferencing steps to create calibrated multispectral orthomosaics for identifying plant diseases during early stages and measuring soil water content (Mazzia et al., 2020). Field characteristics remain detectable through the use of hyperspectral sensors which acquire both hundreds and thousands of narrow electromagnetic bands. UAV utilization of hyperspectral sensors is restricted because the equipment maintains higher costs and weighs more than both RGB and multispectral systems despite price reductions. The process of radiometric calibration demands sophisticated methods that derive mainly from manned aircraft per Brook and Ben-Dor (2015). When developing hyperspectral imaging plans, one must decide on both spatial resolution and temporal resolution since the data output becomes more extensive. The post-processing steps enable the calculation of narrowband indices including chlorophyll absorption ratio index (CARI) together with photochemical reflectance index (PRI) (Adao et al., 2017).

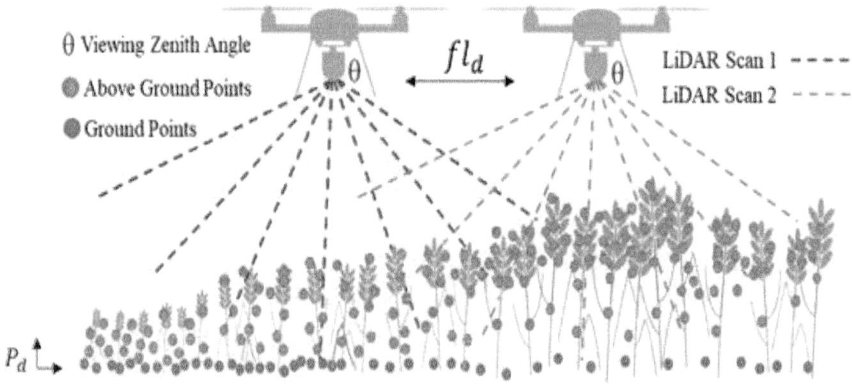

Figure 3. LiDAR system, showing the interaction between various angles and distances in the process of data collection from a landscape (Farhan et al., 2024).

4.3 3D imaging scanning LiDAR system

Three-dimensional imaging technology innovations have brought substantial improvements to weed management through better evaluation of both weed population distribution patterns across space and time. The accuracy of plant models from 3D imaging exceeds traditional 2D imaging because depth data removes the limitations posed by shadows and overlapping foliage. The technology enables weed mapping in real time while enabling targeted weed removal and invasive weed escape assessment, which reduces herbicide usage and increases the sustainability of the environment (Andújar et al., 2016; Smith et al., 2018).

Light Detection and Ranging (LiDAR) along with photogrammetry are the leading 3D imaging methods that exist. A high-resolution LiDAR system delivers precise point clouds at sub-centimeter resolution but exists outside farmers' reach due to its expensive nature (Andújar et al., 2013; Paturkar et al., 2020). Photogrammetry creates three-dimensional models from drone or ground vehicle obtained two-dimensional images through the affordable process of Structure-from-Motion (SfM) (Figure 3). Photogrammetric measurements today are highly sensitive to environmental conditions, particularly wind-related factors (Dandrifosse et al., 2020; Armean et al., 2021).

The technology of 3D imaging enables real-time weed patch mapping to detect developing infestations, which leads to more appropriate herbicide dosing and application times (Wu et al., 2020). Through the use of this method, researchers can detect plants that survive the late season, which reduces seedbank longevity while inhibiting new generations of resistant species (Kutugata et al., 2021). Automated weed control systems monitor weeds according to specifications of species and dimension and geographic position, which enhances precision and decreases operational expenditures (Li and Tang, 2018). Future adaptive weed management strategies' emergence is possible because of advances in AI combined with sensor fusion, which enhances photogrammetry's operational reliability under windy conditions (Zhang et al., 2016).

4.4 Real-Time Data Processing

The real-time analysis of data brings essential advantages to oil palm plantation weed management because weeds reduce crop yield output. The use of chemical herbicides leads to environmental damage and mechanical control damages beneficial life forms while eroding the soil structure. Such solutions need to be implemented according to Kudsk and Streibig (2003) and Rakhmatulin et al. (2021). Real-time weed identification and management receives effective support from Machine Learning (ML) and Artificial Intelligence (AI). Noise reduction methods enable automatic weed detection activities by acquiring data followed by labeling steps and model development processes. Mobile application integration of ML-based systems enables farmers to perform real-time and non-real-time prediction of weed categories thus improving operational efficiency while promoting environmental sustainability by minimizing harmful chemical usage (Su, 2020). Growers who employ real-time data collection methods can establish weed distribution patterns which helps them execute targeted control measures before the season begins. Early-season proactive interventions through this strategy stop late-season weed escapes while minimizing seedbank survival and preventing herbicide-resistant weed evolution. Artificial intelligence in automated systems optimizes the precision of weed removal through spraying and mechanical techniques which reduces production expenses while preserving environmental health (Kudsk and Streibig, 2003; Su, 2020).

4.4.1 Edge computing solutions

With edge computing technology farmers can process their real-time data where it is generated near Internet of Things devices and sensors in agricultural areas to enhance weed management. The system cuts down time delays and helps users make quick decisions which are vital for accurate weed control. Edge computing processes data at its entry point to decentralize cloud operations which reduces network congestion and optimizes the speed of weed detection measures (Zhang et al., 2020). The combination of edge computing with AI and machine learning algorithms enables weeds to be correctly identified through local site sensor analysis for accurate spatial distribution assessments. The processing power of edge-based systems analyzes drone imagery instantly to find weed infestations which leads to focused herbicide application or mechanical weed removal techniques that minimize environmental harm (Verma et al., 2023; Rai et al., 2019). Edge computing enables autonomous weed control systems to function by outfitting robots with intelligent sensors which identify weeds through multiple characteristics. The system proves advantageous for remote farming locations where internet connectivity remains restricted by allowing farmers to conduct weed management successfully (Kohzadi et al., 2019).

4.4.2 Cloud-based data analytics

The analysis of data within cloud platforms delivers major transformations to weed management practices in agricultural settings particularly in specialty crop cultivation systems. UAVs with various sensors attached to them enable streamlined pest and disease detection by automating the process (Panda et al., 2010). This approach decreases expenses and offers reduced manual labor time (Panda et al., 2010). The Agroview application functions through cloud and AI technologies to analyze the considerable data produced by UAVs (Ampatzidis et al., 2020). Users can utilize Agroview as an innovative system to conduct plant detection while also counting them and mapping their location and performing dimensional assessments and health assessments of their plants through the creation of health maps. Agroview demonstrates its usefulness in citrus tree monitoring by achieving extremely precise results for tree identification (MAPE 2.3%) and generating correct measurements of tree height (MAPE 4.5%) and canopy size (MAPE 12.9%) (Ampatzidis et al., 2020; Maillard and Gomes, 2016). Cloud computing systems including Amazon Web Services (AWS) make available scalable infrastructure solutions through which users can process data using specialized computing instances (Buyya et al., 2010). This architecture supports parallel processing, crucial for timely decision-making in weed management. Cloud analytics combined with UAV technology enables sustained crop health monitoring along with rapid detection of diseases and pests and exact measurement of horticultural practices. Remote sensing techniques enable growers to perform detailed and time-sensitive agricultural assessments which boost their weed control performance and improve sustainable operations (Das et al., 2018). The sharing of farmer and researcher data through cloud-based applications enables consortium development to advance efficient weed management practices. These emerging technologies show promise to enhance agricultural yield and decrease dependency on dangerous chemicals according to Shakoor et al. (2017). Cloud-based data analytics provides an essential advancement for precision agriculture because it optimizes weed management and supports sustainability of ecosystems (Singh, 2018).

4.5 Data Integration and Management

Today's weed control program implements artificial intelligence and machine learning applications by integrating different management systems and data repositories. Precision weed management occurs when farmers unite satellite imagery with drones and both airborne and station-based sensors and database field reports (Williamson et al., 2023; Harfouche et al., 2020). Broad scale weed distribution patterns become visible through satellite imagery yet drones along with ground sensors detect precise weed locations in combination with real-time environmental parameters to boost targeted control method performance (Williamson et al., 2023; Wang et al., 2020). Modern learning algorithms utilizing SVMs and CNNs operate on combined datasets to identify plant species between weeds and crops with excellent precision. The inclusion of weather data enables these systems to forecast weed development patterns and determine proper times for herbicide use (Wang et al., 2020). Cloud-based solutions like Cropin Cloud enable farmers to observe weed patterns and track crop wellness through real-time data organization which supports purpose-driven action planning (Albert and Singhal, 2024; Potamitis et al., 2015). Farmers can use predictive models that analyze environmental factors and historical records to anticipate upcoming weed infestations which allows them to take preventive actions by applying pre-emergence herbicides or modifying planting density (Williamson et al., 2023). The available data enables farmers to select sustainable weed control approaches which bring both efficiency and sustainability to their practices.

4.5.1 Interoperability with Farm Management Systems

Modern agricultural operations need Farm Management Systems (FMS) interoperability to maximize the effectiveness of their weed control operations. Farmers can accomplish better productivity by integrating weed detection information into their Farm Management Systems through seamless data integration (Schmaltz et al., 2022). Current digital solutions fail to cooperate with one another which results in isolated data databases that prevent efficient data sharing and analytical processes (Bökle et al., 2022; Saiz-Rubio and Rovira-Más, 2020). The purpose of Farm Management Information Systems (FMIS) is to support farmers in their planning needs and document their field activities alongside fieldwork assessment (Burlacu et al., 2014). Through this process farmers can decrease agricultural production expenses while maintaining compliance standards. The maximum performance of FMIS depends on merging weed detection technologies' datasets including drone-generated imagery and sensor networks alongside crop statistics and meteorological predictions (Saiz-Rubio and Rovira-Más, 2020). The implementation of interoperability enables parallel observation of farm operations that enable exact weed control and incremental herbicide use scheduling. The integration system supports precise regulatory documentation through clear documentation practices (Bökle et al., 2022). Interoperability enables farm operations to link farmers with contractors and enables digital farming solutions to reach small and medium-sized agricultural enterprises (Saiz-Rubio and Rovira-Más, 2020). Data security and ownership protection takes precedence during integration processes since farm owners need their information protected (Meola, 2020). The agricultural sector will improve sustainable and efficient weed management practices through solutions that resolve interoperability and security issues.

4.5.2 Data Visualization Tools

Modem weed management relies on data visualization tools because they convert sophisticated datasets into efficient monitoring systems which track weed distribution patterns alongside density measurements. These tools make the evaluation process simpler by using charts and graphs and maps to display weed patterns thus helping weed management teams make precise control decisions (Kumari et al., 2024). The Microsoft Power BI system integrates numerous datasets consisting of weed density alongside environmental factors to present such information using interactive dashboards. The research in Taif Saudi Arabia demonstrated that weed densities reached their peak at specific altitude points thus contributing to site-specific management practices that decrease herbicide usage (Majrashi et al., 2024). Programs that display advanced visual data help detect weeds through their compatibility with AI machine learning platforms including PyTorch and TensorFlow. This weed classification technology reaches high accuracy levels up to 91% while generating maps of spatial distribution for focused interventions (Haq et al., 2023; Blank et al., 2023). The combination of weed data with soil composition information and weather conditions through visualization tools enables precise prediction of infestations with optimized resource allocation. These tools benefit farming operations of every scale through their interactive features which enable detailed zooming abilities and comprehensive data analysis (Kumari et al., 2024).

5. Challenges and Barriers to Adoption

Current adoption of AI for weed control in agriculture encounters many technological as well as financial and social obstacles that restrict its broad usage. Accurate results must compete with quick response times which presents itself as a primary technological obstacle. Time-sensitive decisions from weed control algorithms become delayed because complex processing algorithms create delays in activities like herbicide spraying and mechanical weeding (Teal et al., 1989). Large data management plays a vital role since unorganized or insufficient datasets result in flawed predictions. Numerous stakeholders should cooperate to create centralized data centers while refining data synthesis methods (Ahngar et al., 2022). The adoption of AI systems for farms demands specific adjustments to match regional characteristics alongside crop types and operational patterns and presents resource costs. The integration of IoT devices with AI platforms faces challenges because standard protocols are lacking thus becoming an obstacle to adoption (Villa-Henriksen et al., 2020). Business economics create substantial barriers to implementation. High initial costs for hardware, such as drones and robotic weeders, and software for data processing and analysis limit accessibility, especially for small-scale farmers (Bosche et al., 2018). Internet connectivity costs alongside regular maintenance expenses create financial burdens for rural and developing regions. The high technology costs prevent smallholder farmers, who already function with tight profit margins, from accessing these solutions which widens accessibility gaps (Subeesh and Mehta, 2021). AI offers reduced herbicide costs and decreased manpower requirements yet initial costs for implementing this technology do not always return profit values specifically when applied toward low-value crops. The combination of government support through subsidies and NGO-supplied microloans and rental-based solutions created by Subeesh and Mehta (2021) makes AI technology more available for wider implementation. Farmers resist adopting new technology because of their cultural habits and they lack sufficient technical abilities. Most farmers lack knowledge about AI and fear that AI systems might replace them at their jobs. These barriers can be overcome through interfaces designed for easy use as well as training that happens locally combined with practical demonstrations.

6. Future Trends

Artificial Intelligence (AI) will drive the evolution of weed control through sustainable precise methods that solve agricultural problems in the future. Artificial Intelligence delivers precise weed management through machine learning and computer vision systems together with robotic implementation thereby reducing both herbicide usage and environmental consequences. Robotic weeders use sensors together with GPS systems to automatically detect weeds before extracting them while eliminating 95% of herbicides (Su et al., 2020). Weed maps generated by high-resolution drones create GPS-accurate weed locations for targeted intervention and laser weeding systems deliver precise weed removal at submillimeter scale without damaging crops or the ground (Slaughter et al., 2008). Predictive modeling utilizes collective historical data sets and real-time information to generate infestation forecasts which enables optimized resource deployments (Shobila and Mood, 2014). Through the use of AI alongside IoT devices there is enhanced automation of irrigation and planting and harvesting processes which leads to improved efficiency. The combination of scalable technologies and collaborative initiatives with AI-based weed control will create a sustainable future despite existing barriers related to costs and accessibility.

7. Conclusion

AI enables a revolutionary change in weed management through its integration which provides modern agriculture with advanced solutions to persisting problems. Modern agriculture encounters insufficient management through traditional methods which rely on man-powered removal operations combined with arbitrary herbicide spraying. Through machine learning collaborative with computer vision and robotics AI develops precise eco-friendly weed control methods which reduce herbicide usages while improving yields. CNNs and transformer models deliver advanced weed detection capability which surpasses field challenges according to the findings presented in this chapter. A combination of ground robots and drones taking advantage of these technologies enables precise mechanical weeding and targeted spraying operations that decrease chemical usage thus protecting environmental quality. Reinforcement learning enables robotic systems to optimize their adaptive decision-making abilities which improves their operational efficiency and dynamic environmental responsiveness in agricultural areas. AI-driven solutions require systematic solutions to address the requirement of extensive annotated datasets as well as the need for model generalizability across different crops and geographies and the economic barriers which stand in the way. The development of technology frameworks requires research scientists to organize partnerships with farmers and governmental officials to support innovative technology dissemination. Field implementations together with case studies show the concrete advantages AI delivers for both herbicide reduction and operational efficiency improvement. The future holds promise through emerging technologies such as swarm robotics and explainable AI (XAI) designing coordinated weeding operations while establishing trust in autonomous systems. Through digital twin applications and predictive modeling together with real-time decision systems, precision agriculture becomes more advanced leading to proactive weed prevention techniques that prevent infestations from worsening.

References

Adão, T., Hruška, J., Pádua, L., Bessa, J., Peres, E., Morais, R. et al. (2017). Hyperspectral imaging: A review on UAV-based sensors, data processing and applications for agriculture and forestry. Remote sensing, 9(11): 1110.

Ahngar, T. A., Bahar, F. A., Singh, L., Bhat, M. A., Singh, P., Mahdi, S. S. et al. (2022). Artificial intelligence in agriculture, applications, benefits and challenges: A review. The Pharma Innovation Journal, 11(3): 1407–1414.

Albert, J. T., and Singhal, J. K. (2024). Artificial Intelligence and Machine Learning in Pest and Weed Management for Sustainable Agriculture. International Journal of Sciences and Innovation Engineering, 1(2): 19–28.

Almberg, K. S., Turyk, M. E., Jones, R. M., Rankin, K., Freels, S., and Stayner, L. T. (2018). Atrazine contamination of drinking water and adverse birth outcomes in community water systems with elevated atrazine in Ohio, 2006–2008. International journal of environmental research and public health, 15(9): 1889.

Amend, S., Brandt, D., Di Marco, D., Dipper, T., Gässler, G., Höferlin, M. et al. (2019). Weed management of the future. KI-Künstliche Intelligenz, 33(4): 411–415.

Ampatzidis, Y., Partel, V., and Costa, L. (2020). Agroview: Cloud-based application to process, analyze and visualize UAV-collected data for precision agriculture applications utilizing artificial intelligence. Computers and Electronics in Agriculture, 174: 105457.

Anderegg, J., Tschurr, F., Kirchgessner, N., Treier, S., Schmucki, M., Streit, B. et al. (2023). On-farm evaluation of UAV-based aerial imagery for season-long weed monitoring under contrasting management and pedoclimatic conditions in wheat. Computers and Electronics in Agriculture, 204: 107558.

Andreasen, C., Scholle, K., and Saberi, M. (2022). Laser weeding with small autonomous vehicles: Friends or foes?. Frontiers in Agronomy, 4: 841086.

Andújar, D., Dorado, J., Fernández-Quintanilla, C., and Ribeiro, A. (2016). An approach to the use of depth cameras for weed volume estimation. Sensors, 16: 972.

Andújar, D., Escolà, A., Rosell-Polo, J. R., Fernández-Quintanilla C., Dorado, J. (2013). Potential of a terrestrial LiDAR-based system to characterize weed vegetation in maize crops. Comput. Electron. Agric. 92: 11–15.

Armean, Z. M., Low, T., and Hancock, N. (2021) Automatic leaf segmentation and overlapping leaf separation using stereo vision. Array, 12: 100099.

Bah, M. D., Hafane, A., and Canals, R. (2018). Deep learning with unsupervised data labeling for weed detection in line crops in UAV images. Remote Sensing, 10(11): 1690.

Benaragama, D., Shirtliffe, S. J., Gossen, B. D., Brandt, S. A., Lemke, R., Johnson, E. N. et al. (2016). Long-term weed dynamics and crop yields under diverse crop rotations in organic and conventional cropping systems in the Canadian prairies. Field Crops Research, 196: 357–367.

Blank, L., Rozenberg, G., and Gafni, R. (2023): Spatial and temporal aspects of weeds distribution within agricultural fields—a review. – Crop Protection 106300.

Bökle, S., Paraforos, D. S., Reiser, D., and Griepentrog, H. W. (2022). Conceptual framework of a decentral digital farming system for resilient and safe data management. Smart Agricultural Technology, 2: 100039.

Bosche, A., Crawford, D., Jackson, D., Schallehn, M., and Schorling, C. (2018). Unlocking opportunities in the Internet of Things. Bain & Company.

Brook, A., and Ben-Dor, E. (2015). Supervised vicarious calibration (SVC) of multi-source hyperspectral remote-sensing data. Remote Sensing, 7(5): 6196–6223.

Burkart, A., Hecht, V. L., Kraska, T., and Rascher, U. (2018). Phenological analysis of unmanned aerial vehicle-based time series of barley imagery with high temporal resolution. Precision Agriculture, 19: 134–146.

Burlacu, G., Costa, R., Sarraipa, J., Jardim-Golcalves, R., and Popescu, D. (2014, April). A conceptual model of farm management information system for decision support. In Doctoral conference on computing, electrical and industrial systems (pp. 47–54). Berlin, Heidelberg: Springer Berlin Heidelberg.

Buyya, R., Broberg, J., and Goscinski, A. M. (Eds.). 2010. Cloud computing: Principles and Paradigms, vol. 87, John Wiley & Sons.

Cantelli, L., Bonaccorso, F., Longo, D., Melita, C. D., Schillaci, G., and Muscato, G. (2019). A small versatile electrical robot for autonomous spraying in agriculture. AgriEngineering, 1(3): 391–402.

Casagrande, M., Peigné, J., Payet, V., Mäder, P., Sans, F. X., Blanco-Moreno, J. M. et al. (2016). Organic farmers' motivations and challenges for adopting conservation agriculture in Europe. Organic Agriculture, 6: 281–295.

Chacko, S. R., Raj, S. K., and Krishnasree, R. K. (2021). Integrated weed management in vegetables. Journal of Pharmacognosy and Phytochemistry, 10(2): 2694–2700.

Chaitanya, P., Kotte, D., Srinath, A., and Kalyan, K. B. (2020). Development of smart pesticide spraying robot. International Journal of Recent Technology and Engineering, 8(5): 2193–2202.

Chang, F., Dean, J., Ghemawat, S., Hsieh, W. C., Wallach, D. A., Burrows, M. et al. (2008). Bigtable: A distributed storage system for structured data. ACM Transactions on Computer Systems (TOCS), 26(2): 1–26.

Cook, S., Lacoste, M., Evans, F., Ridout, M., Gibberd, M., and Oberthür, T. (2018). An on-farm experimental philosophy for farmer-centric digital innovation. In 14th International Conference on Precision Agriculture.

Corceiro, A., Alibabaei, K., Assunção, E., Gaspar, P. D., and Pereira, N. (2023). Methods for detecting and classifying weeds, diseases and fruits using AI to improve the sustainability of agricultural crops: a review. Processes, 11(4): 1263.

Costa, L., Nunes, L., and Ampatzidis, Y. (2020). A new visible band index (vNDVI) for estimating NDVI values on RGB images utilizing genetic algorithms. Computers and Electronics in Agriculture, 172: 105334.

Cottrell, B., Kalacska, M., Arroyo-Mora, J. P., Lucanus, O., Inamdar, D., Løke, T. et al. (2024). Limitations of a Multispectral UAV sensor for satellite validation and mapping complex vegetation. Remote Sensing, 16(13): 2463.

Crimaldi, M., Cristiano, V., De Vivo, A., Isernia, M., Ivanov, P., and Sarghini, F. (2020). Neural network algorithms for real time plant diseases detection using UAVs. pp. 827–835. *In:* Coppola, A., Di Renzo, G. C.. Altieri, G., and Antonio, P. D. (Eds.). Innov. Biosyst. Eng. Sustain. Agric. For Food Prod.

Dandrifosse, S., Bouvry, A., Leemans, V., Dumont, B., and Mercatoris, B. (2020). Imaging

Das, U., Pathak, P., Meena, M., and Mallikarjun, N. (2018). Precision farming a promising technology in horticulture: a review. Int J Pure Appl Biosci, 6(1): 1596–1606.

Dasgupta, I., Saha, J., Venkatasubbu, P., and Ramasubramanian, P. (2020). AI crop predictor and weed detector using wireless technologies: A smart application for farmers. Arabian Journal for Science and Engineering, 45: 11115–11127.

de Lima, D. C., Saqui, D., Ataky, S., Jorge, L. A. D. C., Ferreira, E. J., and Saito, J. H. (2019). Estimating agriculture nir images from aerial rgb data. pp. 562–574. *In:* Computational Science–ICCS 2019: 19th International Conference, Faro, Portugal, June 12–14, 2019, Proceedings, Part I 19. Springer International Publishing.

Du, J., Yu, M. Q., Chen, Y. P., Li, M., Cheng, C. Q., and Yin, X. (2019). Development of an automatic speed control system for highclearance sprayer. J. China Agr. Univ., 24(12): 104–110.

Endres, F., Hess, J., Sturm, J., Cremers, D., and Burgard, W. (2013). 3-D mapping with an RGB-D camera. IEEE transactions on robotics, 30(1): 177–187.

Esposito, M., Crimaldi, M., Cirillo, V., Sarghini, F., and Maggio, A. (2021). Drone and sensor technology for sustainable weed management: A review. Chemical and Biological Technologies in Agriculture, 8: 1–11. Write a small text for this outline.

Etienne, A., and Saraswat, D. (2019). Machine learning approaches to automate weed detection by UAV based sensors. In Autonomous air and ground sensing systems for agricultural optimization and phenotyping IV (p. 110080R). International Society for Optics and Photonics.

Farhan, S. M., Yin, J., Chen, Z., and Memon, M. S. (2024). A comprehensive review of LiDAR applications in crop management for precision agriculture. Sensors (Basel, Switzerland), 24(16): 5409.

Felizardo, K. R., Mercaldi, H. V., Cruvinel, P. E., Oliveira, V. A., and Steward, B. L. (2016). Modeling and model validation of a chemical injection sprayer system. Applied Engineering in Agriculture, 32(3): 285.

Fengle, L., Xiaohui, Z., and Weiwei, M. (2010). Current development status of foreign big-scale protection machinery and application technology. Journal of Agricultural Mechanization Research, 3: 246–248.

Fernández-Quintanilla, C., Peña, J. M., Andújar, D., Dorado, J., Ribeiro, A., and López-Granados, F. (2018). Is the current state of the art of weed monitoring suitable for site-specific weed management in arable crops?. Weed Research, 58(4): 259–272.

Gangstad, E. O. (2018). Weed Control Methods for Rights of Way Management. CRC Press.

Gao, W. T., and Su, W. H. (2024). Weed management methods for herbaceous field crops: A review. Agronomy, 14(3): 486.

Gašparović, M., Zrinjski, M., Barković, Đ., and Radočaj, D. (2020). An automatic method for weed mapping in oat fields based on UAV imagery. Computers and Electronics in Agriculture, 173: 105385.

Gené-Mola, J., Vilaplana, V., Rosell-Polo, J. R., Morros, J. R., Ruiz-Hidalgo, J., and Gregorio, E. (2019). Multi-modal deep learning for Fuji apple detection using RGB-D cameras and their radiometric capabilities. Computers and Electronics in Agriculture, 162: 689–698.

Gerhards, R., Andujar Sanchez, D., Hamouz, P., Peteinatos, G. G., Christensen, S., and Fernandez-Quintanilla, C. (2022). Advances in site-specific weed management in agriculture—A review. Weed Research, 62(2): 123–133.

Ghafar, A. S. A., Hajjaj, S. S. H., Gsangaya, K. R., Sultan, M. T. H., Mail, M. F., and Hua, L. S. (2023). Design and development of a robot for spraying fertilizers and pesticides for agriculture. Materials Today: Proceedings, 81: 242–248.

Ghaly, A. E., and Ibrahim, M. M. (2022). Mechanization of weed management in sugar beet. pp. 327–367. *In:* Sugar Beet Cultivation, Management and Processing. Singapore: Springer Nature Singapore.

Gracia-Romero, A., Vergara-Díaz, O., Thierfelder, C., Cairns, J. E., Kefauver, S. C., and Araus, J. L. (2018). Phenotyping conservation agriculture management effects on ground and aerial remote sensing assessments of maize hybrids performance in Zimbabwe. Remote Sensing, 10(2): 349.

Hamuda, E., Mc Ginley, B., Glavin, M., and Jones, E. (2017). Automatic crop detection under field conditions using the HSV colour space and morphological operations. Computers and Electronics in Agriculture, 133: 97–107.

Haq, S. I. U., Tahir, M. N., and Lan, Y. (2023). Weed detection in wheat crops using image analysis and artificial intelligence (AI). Applied Sciences, 13(15): 8840.

Harfouche, A. L., Jacobson, D. A., Kainer, D., Romero, J. C., Harfouche, A. H., Mugnozza, G.S. et al. (2019). Accelerating climate resilient plant breeding by applying next-generation artificial intelligence. Trends in Biotechnology, 37(11): 1217–1235.

Harrison, S. K., and Loux, M. M. (2017). Chemical weed management. pp. 101–153. *In:* Handbook of weed management systems. Routledge.

Hashemi-Beni, L., Gebrehiwot, A., Karimoddini, A., Shahbazi, A., and Dorbu, F. (2022). Deep convolutional neural networks for weeds and crops discrimination from UAS imagery. Frontiers in Remote Sensing, 3: 755939.

Hassanein, M., and El-Sheimy, N. (2018). An efficient weed detection procedure using low-cost UAV imagery system for precision agriculture applications. The International Archives of the Photogrammetry, Remote Sensing and Spatial Information Sciences, 42: 181–187.

Hassler, S. C., and Baysal-Gurel, F. (2019). Unmanned aircraft system (UAS) technology and applications in agriculture. Agronomy, 9(10): 618.

Hejazipoor, H., Massah, J., Soryani, M., Vakilian, K. A., and Chegini, G. (2021). An intelligent spraying robot based on plant bulk volume. Computers and Electronics in Agriculture, 180: 105859.

Hughes, D. P., and Salathe, M. (2016). An open access repository of images on plant health to enable the development of mobile disease diagnostics. ArXiv151108060 Cs.

Hunt, L. A., Â Jones J. W., Â Hoogenboom G. Â et al. (1994). Input and output file structures for crop simulation models. *In*: Uhlir, P. F., and G. C. Carter (Eds.). Crop Modeling and Related Environmental Data: A Focus on Applications for Arid and Semiarid Regions in Developing Countries. 35â€“72. CODATA Commission on Global Change.

Hussain, M. I., Abideen, Z., Danish, S., Asghar, M. A., and Iqbal, K. (2021). Integrated weed management for sustainable agriculture. Sustainable Agriculture Reviews, 52: 367–393.

Jitsuyama, Y., and Ichikawa, S. (2011). Possible weed establishment control by applying cryogens to fields before snowfalls. Weed Technology, 25(3): 454–458.

Jung, C., Zhou, K., and Feng, J. (2020). Fusionnet: Multispectral fusion of RGB and NIR images using two stage convolutional neural networks. IEEE Access, 8: 23912–23919.
Juwono, F. H., Wong, W. K., Verma, S., Shekhawat, N., Lease, B. A., and Apriono, C. (2023). Machine learning for weed–plant discrimination in agriculture 5.0: An in-depth review. Artificial Intelligence in Agriculture, 10: 13–25.
Kaab, S. B., Rebey, I. B., Hanafi, M., Hammi, K. M., Smaoui, A., Fauconnier, M. L. et al. (2020). Screening of Tunisian plant extracts for herbicidal activity and formulation of a bioherbicide based on *Cynara cardunculus*. South African Journal of Botany, 128: 67–76.
Khang, A., Rath, K. C., Panda, S., Sree, P. K., and Panda, S. K. (2023). Revolutionizing agriculture: exploring advanced technologies for plant protection in the agriculture sector. pp. 1–22. *In:* Handbook of Research on AI-Equipped IoT Applications in High-Tech Agriculture. IGI Global.
Kohzadi, S., Shahmoradi, B., Ghaderi, E., Loqmani, H., and Maleki, A. (2019). Concentration, source, and potential human health risk of heavy metals in the commonly consumed medicinal plants. Biological Trace Element Research, 187: 41–50.
Kool, J., de Jonge, E., Nieuwenhuizen, A., and Braam, H. (2023). Green on Green weed detection: Finding weeds in a soybean crop in Brazilian fields with the Rometron WEED-IT sensor: intermediary report (No. WPR-10.18174/649472). Wageningen Plant Research.
Korres, N. E., Burgos, N. R., and Duke, S. O. (2019). Weed control.
Kudsk, P., and Streibig, J. S. (2003). Herbicides and two edge-sword. Weed Res., 43, 90–102. [Google Scholar] [CrossRef].
Kumar, A., Sharma, M. J., Mahajan, A., and Kumar, J. (2024). Principles and Practices For Modern Agronomy. Academic Guru Publishing House.
Kumar, M. K., Singh, A., Pandey, M. S., and Mehra, M. R. (2025). A Practical Manual on Fundamentals of Agronomy.
Kumari, S., Pandey, G. K., and Tiwari, S. (2024). Data Visualization Techniques in Smart Agriculture Implementation. pp. 122–159. *In:* AI Applications for Business, Medical, and Agricultural Sustainability. IGI Global.
Kutugata, M., Hu, C., Sapkota, B., and Bagavathiannan, M. (2021) Seed rain potential in late-season weed escapes can be estimated using remote sensing. Weed Sci., 69: 653–659.
Laliberte, A. S., Goforth, M. A., Steele, C. M., and Rango, A. (2011). Multispectral remote sensing from unmanned aircraft: Image processing workflows and applications for rangeland environments. Remote Sensing, 3(11): 2529–2551.
Li, J., and Tang, L. (2018). Crop recognition under weedy conditions based on 3D imaging for robotic weed control. J. Field Robot, 35: 596–611.
Lochan, K., Khan, A., Elsayed, I., Suthar, B., Seneviratne, L., and Hussain, I. (2024). Advancements in precision spraying of agricultural robots: A comprehensive Review. IEEE Access.
Lu, Y., and Young, S. (2020). A survey of public datasets for computer vision tasks in precision agriculture. Computers and Electronics in Agriculture, 178: 105760.
Machleb, J., Peteinatos, G. G., Kollenda, B. L., Andújar, D., and Gerhards, R. (2020). Sensor-based mechanical weed control: Present state and prospects. Computers and electronics in agriculture, 176: 105638.
Maillard, P., and Gomes, M. F. (2016). Detection and counting of orchard trees from vhr images using a geometrical-optical model and marked template matching. ISPRS Annals of the Photogrammetry, Remote Sensing and Spatial Information Sciences, 3: 75–82.
Majrashi, A., Okiely, S., and Alshammari, K. (2024). Power Bi-Based Analysis And Visualization Of Weed Plant Datasets Of Taif, Saudi Arabia. Applied Ecology & Environmental Research, 22(6).
Marble, S. C., Koeser, A. K., and Hasing, G. (2015). A review of weed control practices in landscape planting beds: Part II—Chemical weed control methods. HortScience, 50(6): 857–862.
Mazzia, V., Comba, L., Khaliq, A., Chiaberge, M., and Gay, P. (2020). UAV and machine learning based refinement of a satellite-driven vegetation index for precision agriculture. Sensors, 20(9): 2530.
Meshram, A. T., Vanalkar, A. V., Kalambe, K. B., and Badar, A. M. (2022). Pesticide spraying robot for precision agriculture: A categorical literature review and future trends. Journal of Field Robotics, 39(2): 153–171.
Mesnage, R., Székács, A., and Zaller, J. G. (2021). Herbicides: Brief history, agricultural use, and potential alternatives for weed control. pp. 1–20. *In:* Herbicides. Elsevier.

Monteiro, A., and Santos, S. (2022). Sustainable approach to weed management: The role of precision weed management. Agronomy, 12(1): 118.
Morra, M. J., Popova, I. E., and Boydston, R. A. (2018). Bioherbicidal activity of Sinapis alba seed meal extracts. Industrial Crops and Products, 115: 174–181.
Motmainna, M., Juraimi, A. S., Uddin, M. K., Asib, N. B., Islam, A. M., and Hasan, M. (2021). Bioherbicidal properties of *Parthenium hysterophorus*, *Cleome rutidosperma* and *Borreria alata* extracts on selected crop and weed species. Agronomy, 11(4): 643.
Motohka, T., Nasahara, K. N., Oguma, H., and Tsuchida, S. (2010). Applicability of green-red vegetation index for remote sensing of vegetation phenology. Remote Sensing, 2(10): 2369–2387.
Muchhadiya, R. M., Kumawat, P. D., Gohil, B. S., and Sakarvadia, H. L. (2024). Precision weed management using artificial intelligence tools and techniques for sustainable agriculture. Agriculture & Food e-Newsletter, 6(1).
Mulla, D. J. (2013). Twenty five years of remote sensing in precision agriculture: Key advances and remaining knowledge gaps. Biosystems Engineering, 114(4): 358–371.
Nguyen, T. T., Vandevoorde, K., Wouters, N., Kayacan, E., De Baerdemaeker, J. G., and Saeys, W. (2016). Detection of red and bicoloured apples on tree with an RGB-D camera. Biosystems Engineering, 146: 33–44.
Olorunmaiye, P. M., and Osunleti, S. O. (2025). Influence of plant population and different weed control methods on weed density and dominance in mango ginger (curcuma amada roxb.). International Journal of Pest Management, 1–11.
Pantović, J. G., and Sečanski, M. (2023). Weed Control in Organic Farming. Contemporary Agriculture, 72(1-2): 43–56.
Partel, V., Kakarla, S. C., and Ampatzidis, Y. (2019). Development and evaluation of a low-cost and smart technology for precision weed management utilizing artificial intelligence. Computers and Electronics in Agriculture, 157: 339–350.
Paturkar, A., Gupta, G. S., and Bailey, D. (2020). Non-destructive and cost-effective 3D plant growth monitoring system in outdoor conditions. Multimed. Tools Appl. 79: 34955–34971.
Pfanz, M., Schirrmann, M., and Nordmeyer, H. (2018). Drone based weed monitoring with an image feature classifier. Julius-Kühn-Archiv, 458: 379–384.
Potamitis, I., Rigakis, I., and Fysarakis, K. (2015). Insect biometrics: Optoacoustic signal processing and its applications to remote monitoring of McPhail type traps. PloS one, 10(11): e0140474.
Priyadarshini, A., Dash, S., Jena, J., Kusumavathi, K., Pattnaik, P., and Holderbaum, W. (2024). Review of the cutting edge technologies for weed control in field crops. International Journal of Agricultural and Biological Engineering, 17(5): 44–57.
Putra, B. T. W., and Soni, P. (2017). Evaluating NIR-Red and NIR-Red edge external filters with digital cameras for assessing vegetation indices under different illumination. Infrared Physics & Technology, 81: 148–156.
Putra, B. T. W., and Soni, P. (2020). Improving nitrogen assessment with an RGB camera across uncertain natural light from above-canopy measurements. Precision Agriculture, 21(1): 147–159.
Rai, N., Zhang, Y., Ram, B. G., Schumacher, L., Yellavajjala, R. K., Bajwa, S. et al. (2023). Applications of deep learning in precision weed management: A review. Computers and Electronics in Agriculture, 206: 107698.
Rai, P. K., Lee, S. S., Zhang, M., Tsang, Y. F., and Kim, K. H. (2019). Heavy metals in food crops: Health risks, fate, mechanisms, and management. Environment International, 125: 365–385.
Rakhmatulin, I., and Andreasen, C. (2020). A concept of a compact and inexpensive device for controlling weeds with laser beams. Agronomy, 10(10): 1616.
Rasmussen, J., Nielsen, J., Streibig, J. C., Jensen, J. E., Pedersen, K. S., and Olsen, S. I. (2019). Pre-harvest weed mapping of Cirsium arvense in wheat and barley with off-the-shelf UAVs. Precision Agriculture, 20: 983–999.
Riemens, M., Sønderskov, M., Moonen, A. C., Storkey, J., and Kudsk, P. (2022). An integrated weed management framework: a pan-European perspective. European Journal of Agronomy, 133: 126443.
Sa, I., Popović, M., Khanna, R., Chen, Z., Lottes, P., Liebisch, F. et al. (2018). WeedMap: a large-scale semantic weed mapping framework using aerial multispectral imaging and deep neural network for precision farming. Remote Sensing, 10(1): 1–25.

Sabzi, S., Abbaspour-Gilandeh, Y., and Arribas, J. I. (2020). An automatic visible-range video weed detection, segmentation and classification prototype in potato field. Heliyon, 6(5).

Saiz-Rubio, V., and Rovira-Más, F. (2020). From smart farming towards agriculture 5.0: A review on crop data management. Agronomy, 10(2): 207.

Sakamoto, T., Gitelson, A. A., Nguy-Robertson, A. L., Arkebauer, T. J., Wardlow, B. D., Suyker, A. E. et al. (2012). An alternative method using digital cameras for continuous monitoring of crop status. Agricultural and Forest Meteorology, 154: 113–126.

Scavo, A., and Mauromicale, G. (2020). Integrated weed management in herbaceous field crops. Agronomy, 10(4): 466.

Schmaltz, R., Coolidge, M., and Donald, G. (2022). U.S. Patent No. 11,386,378. Washington, DC: U.S. Patent and Trademark Office.

Shakoor, N., Lee, S., and Mockler, T. C. (2017). High throughput phenotyping to accelerate crop breeding and monitoring of diseases in the field. Current Opinion in Plant Biology, 38: 184–192.

Shaner, D. L., and Beckie, H. J. (2014). The future for weed control and technology. Pest management science, 70(9): 1329–1339.

Shanmugam, S., Assunção, E., Mesquita, R., Veiros, A., and Gaspar, P. D. (2020). Automated weed detection systems: A review. KnE Engineering, 271–284.

Shobila, P., and Mood, V. (2014). Automated irrigation system using robotics and sensors. Int. J. Sci. Eng. Res, 3(8): 9–13.

Singh, K. K. (2018, November). An artificial intelligence and cloud based collaborative platform for plant disease identification, tracking and forecasting for farmers. In 2018 IEEE international conference on cloud computing in emerging markets (CCEM) (pp. 49–56). IEEE.

Slaughter, D. C., Giles, D. K., and Downey, D. (2008). Autonomous robotic weed control systems: A review. Computers and Electronics in Agriculture, 61(1): 63–78.

Smith, L. N., Zhang, W., Hansen, M. F., Hales, I. J., and Smith, M. L. (2018). Innovative 3D and 2D machine vision methods for analysis of plants and crops in the field. Comput. Ind., 97: 122–131.

Su, W. H. (2020). Crop plant signaling for real-time plant identification in smart farm: A systematic review and new concept in artificial intelligence for automated weed control. Artificial Intelligence in Agriculture, 4: 262–271.

Su, W. H. (2020). Systemic crop signaling for automatic recognition of transplanted lettuce and tomato under different levels of sunlight for early season weed control. Challenges, 11(2): 23.

Su, W. H., Slaughter, D. C., and Fennimore, S. A. (2020). Non-destructive evaluation of photostability of crop signaling compounds and dose effects on celery vigor for precision plant identification using computer vision. Computers and Electronics in Agriculture, 168: 105155.

Subeesh, A., and Mehta, C. R. (2021). Automation and digitization of agriculture using artificial intelligence and internet of things. Artificial Intelligence in Agriculture, 5: 278–291.

Talaviya, T., Shah, D., Patel, N., Yagnik, H., and Shah, M. (2020). Implementation of artificial intelligence in agriculture for optimisation of irrigation and application of pesticides and herbicides. Artificial Intelligence in Agriculture, 4: 58–73.

Teal, S. L., Rudnicky, A. I. (1989). A performance model of system delay and user strategy selection. Conference on Human Factors in Computing Systems, California, USA. 1992 May; 3-7. 35. Washington R, Roth BH. Input Data Management in Real-Time AI System, 11th International Joint Conference on Artificial Intelligence, Michigan, USA, 1989, 20–25.

Torres-Sánchez, J., Peña, J. M., de Castro, A. I., and López-Granados, F. (2014). Multi-temporal mapping of the vegetation fraction in early-season wheat fields using images from UAV. Computers and Electronics in Agriculture, 103: 104–113.

Vasileiou, M., Kyrgiakos, L. S., Kleisiari, C., Kleftodimos, G., Vlontzos, G., Belhouchette, H. et al. (2024). Transforming weed management in sustainable agriculture with artificial intelligence: A systematic literature review towards weed identification and deep learning. Crop Protection, 176: 106522.

Verhoeven, G. (2008). Imaging the invisible using modified digital still cameras for straightforward and low-cost archaeological near-infrared photography. Journal of Archaeological Science, 35(12): 3087–3100.

Verma, J., Snehi, M., Kansal, I., Tiwari, R. G., and Prasad, D. (2023). 14 Real-time weed detection and classification using deep learning models and IoT-based edge computing for social learning applications. Augmented and Virtual Reality in Social Learning: Technological Impacts and Challenges, 3: 241.

Villa-Henriksen, A., Edwards, G. T., Pesonen, L. A., Green, O., and Sørensen, C. A. G. (2020). Internet of Things in arable farming: Implementation, applications, challenges and potential. Biosystems Engineering, 191: 60–84.

Wang, H. G. (2022). Smart phytoprotection and suggestions for its development. J. China Agric. Univ., 27: 1–21.

Wang, H., Cimen, E., Singh, N., and Buckler, E. (2020). Deep learning for plant genomics and crop improvement. Current Opinion in Plant Biology, 54: 34–41.

Wang, Y., Xie, L., Wang, H., Zeng, W., Ding, Y., Hu, T. et al. (2022). Intelligent spraying robot for building walls with mobility and perception. Automation in Construction, 139: 104270.

Warren, J., and Marz, N. (2015). Big Data: Principles and best practices of scalable realtime data systems. Simon and Schuster.

White, J. W., Hunt, L. A., Boote, K. J., Jones, J. W., Koo, J., Kim, S. et al. (2013). Integrated description of agricultural field experiments and production: The ICASA Version 2.0 data standards. Computers and Electronics in Agriculture, 96: 1–12.

Whitehead, K., and Hugenholtz, C. H. (2014). Remote sensing of the environment with small unmanned aircraft systems (UASs), part 1: A review of progress and challenges. Journal of Unmanned Vehicle Systems, 2(3): 69–85.

Williamson, H. F., Brettschneider, J., Caccamo, M., Davey, R. P., Goble, C., Kersey, P. J. et al. (2023). Data management challenges for artificial intelligence in plant and agricultural research. F1000 Research, 10: 324.

Wu, X., Aravecchia, S., Lottes, P., Stachniss. C., and Pradalier, C. (2020). Robotic weed control using automated weed and crop classification. J. Field Robot, 37: 322–340.

Xue, J., and Su, B. (2017). Significant remote sensing vegetation indices: a review of developments and applications. Journal of Sensors, 2017.

Yaoning, F. E. N. G., Liang, P., Xiao, C., Xiaobing, C., Yan, L., and Bin, C. (2021). Summary of the key technology of variable rate application. Journal of Chinese Agricultural Mechanization, 42(12): 65.

Zhang, Y., Teng, P., Shimizu, Y., Hosoi, F., and Omasa, K. (2016). Estimating 3D leaf and stem shape of nursery paprika plants by a novel multi-camera photography system. Sensors, 16: 874.

Zhang, X., Cao, Z., and Dong, W. (2020). Overview of edge computing in the agricultural internet of things: Key technologies, applications, challenges. Ieee Access, 8: 141748–141761.

Zheng, L. I. U., Chengqian, J. I. N., Yugang, F. E. N. G., and Tengxiang, Y. A. N. G. (2024). Research status and trends of intelligent technology in plant protection machinery. Journal of Intelligent Agricultural Mechanization, 5(1): 40–50.

Zhu, Y., Cao, P., Liu, S., Zheng, Y., and Huang, C. (2020). Development of a new method for turbidity measurement using two NIR digital cameras. Acs Omega, 5(10): 5421–5428.

Zimdahl, R. L., and Basinger, N. T. (2024). Fundamentals of weed science. Elsevier.

CHAPTER 14

Integration of AI with Precision Agriculture for Targeted Pest Control

Akhtar Hameed,[1,*] *Subhan Ali,*[1] *Hafiz Muhammad Usman Aslam,*[1,3] *Muhammad Waqar Alam,*[2] *Faizan Ali,*[4] *Rana Binyamin*[1] and *Muhammad Umair Rafiq*[1]

1. Introduction

Precision agriculture is a farming management concept that involves monitoring, quantifying and actuating on different crop attributes. This approach aims to achieve increased productivity in the agricultural industry by ensuring that operational decisions are data driven and resources are used in the correct area and at the correct time (Zhang et al., 2018; Raj et al., 2024). Recent studies have integrated the use of GPS, Geographical Information System (GIS) as well as remote sensing in practice of precision agriculture (Mani et al., 2021; Quamar et al., 2023). These tools help in identifying exact field conditions and make it easier for farmers to modify their operations in the best way possible concerning water, fertilizers and pesticides. Precision agriculture has the major objectives of enhancing efficiency, productivity and sustainable farming practices (Shafi et al., 2019). Thus, the farmers can avoid the unnecessary spendings on resources and bring more revenue by using precision

[1] Institute of Plant Protection, MNS University of Agriculture Multan 61000, Pakistan.
[2] Department of Plant Pathology, University of Okara 53600, Pakistan.
[3] Department of Plant Pathology, San Luis Valley Research Center, Colorado State University, Fort Collins, CO, United States.
[4] Department of Agricultural Science & Engineering, College of Agriculture, Tennessee State University, Nashville, TN 37209, USA.
* Corresponding author: akhtar.hameed@mnsuam.edu.pk

DOI: 10.1201/9781003546702-14

agriculture techniques. Research shows that precision approaches yields gains of about 500 kg/hectare on average, which equates to between 5–10 per cent yield improvement, at 15–30% less inputs (Mulla, 2013). New technologies are necessary to improve sustainable agriculture because this will help farmers to proceed with the better usage of inputs with environmental friendly practices.

1.1 Importance of Targeted Pest Control in Agriculture

Pests are a significant global constraint on crop production, causing a loss of 20–40% of global crop yields (Oerke, 2006). Pests not only decrease yields but also decrease the quality of produce, causing farmers to lose more of their earnings in selling their yield products. The current commonly used pest control methods, namely chemical pesticides come with limitations for instance; pest resistance to the chemicals, negative impacts on the bio- environment, and humans as well (Popp et al., 2013).

The importance of effective pest extension techniques is growing as primitive methods become less effective over time. Pest control in precision agriculture can be optimized by reducing the amount used by site-specific application that reduce chances of exposure to other forms (Pretty and Bharucha, 2015). The main idea is to concentrate on particular zones of pest attacks that help farmers to prevent crop losses, or strengthen their stand against a future infestation. The integration of AI with precision agriculture techniques is a comprehensive approach to pest management involving three interconnected stages namely; Monitoring, Management and Forecasting. During the Monitoring stage, technologies include AI, sensors, and mobile application in gathering information on pest activity. In the Forecasting stage,

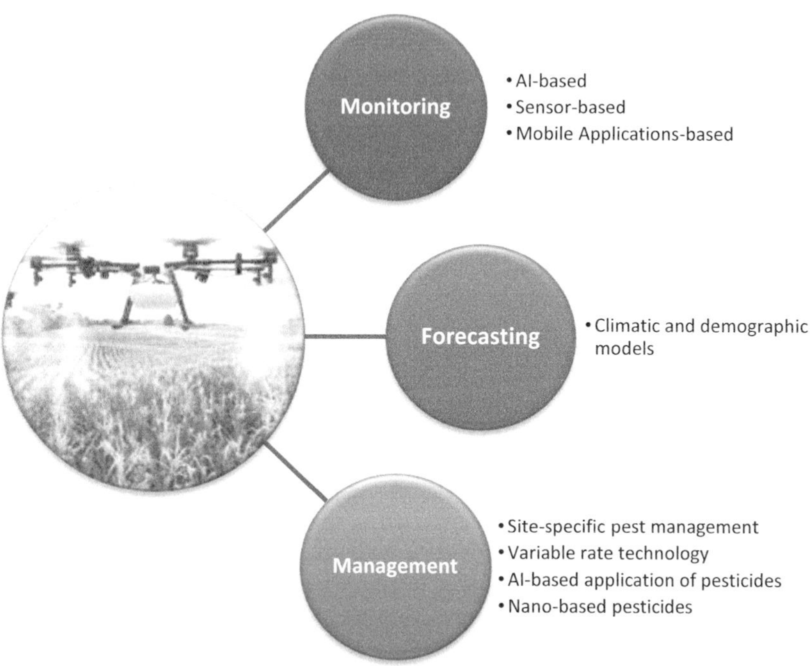

Figure 1. Integration of AI with Precision Agriculture Techniques for Pest Management.

the integrative models can be used such as climatic and demographic models to predict pest development and effects. Lastly, the Management stage entails specific pest control practices such as site-specific pest control, variable rate technology, artificial intelligence pesticide application and nano-pesticides respectively for effective and sustainable management of pests (Figure 1).

1.2 Role of AI in Optimizing Pest Control Strategies

AI has now started changing the pest control industry in agriculture by offering information to farmers that helps to make perfect decisions. These include machine learning modules that estimate pest occurrences to computer vision systems that identify the pests in real time (Kamilaris and Prenafeta-Boldú, 2018). The continuous advancement of AI technology offers novel points of view and approaches to the concerns with farming pest and disease concerns. As AI technology enables sharp identification and prognosis of pests and diseases, management becomes effective and advanced. Image recognition technology, data analysis, machine learning algorithms can help quick and more accurate identification of pests and diseases. The design and application of AI in early detection systems also improves the timeliness and flexibility of management. In addition, data-driven decision support system architectures, intelligent agriculture management platforms and auto control technologies offer innovative approaches to pest and disease control. As the application of AI technology strengthens and new technologies are constantly being developed, the use of AI in pest and diseases control in agriculture can give bigger outcomes and potential for sustainable agricultural production (Li and Wang, 2024; Yun et al., 2024; Li et al. 2023).

AI optimizes pest control strategies by enhancing detection and enabling predictive modeling, which forecasts pest outbreaks based on historical data and environmental conditions. Real-time monitoring through AI-driven Internet of Things (IoT) sensors and drones allows for immediate response to threats, while data-driven decision-making guides precise, targeted pesticide application, reducing chemical use and environmental impact. By integrating with robotics, AI automates pest control tasks across large fields, improving efficiency and reducing costs. Additionally, adaptive learning enables AI systems to refine strategies over time, continuously improving pest management effectiveness (shown in Figure 2).

AI-driven pest control offers several advantages over traditional methods. For instance, AI can process large amounts of data quickly, enabling farmers to respond to pest threats almost instantly, unlike conventional scouting methods that are labor-intensive and slower (Sharma, 2021). In addition, AI can detect pest activity characteristics of patterns that are not recognizable in most of the cases physically hence helping in management of pests in a better way. Due to its preservation of good decision making especially in pest control, AI plays its part in the goal of making precision agriculture sustainable. For instance, the application of predictive models enables one to forecast when and where pest invasions are most probable to happen to enable the farmer to prevent pest invaders and consequently reduce pesticide use (Shamshiri et al., 2018). This approach not only keeps the cost down, but also has

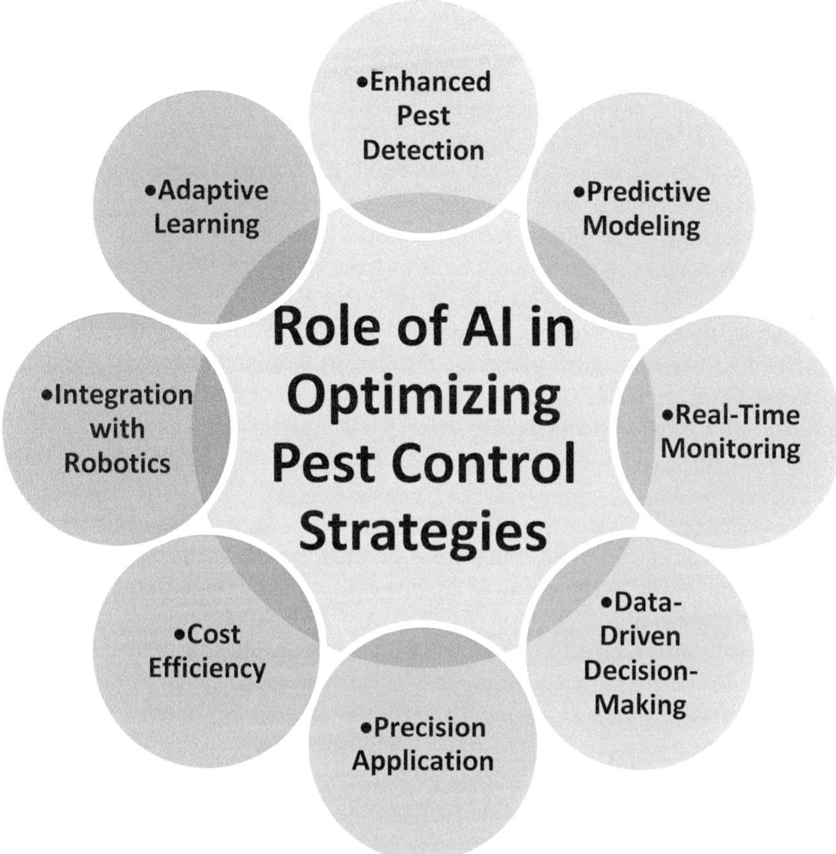

Figure 2. Optimizing Pest Control with AI for Enhancing Detection, Precision, and Efficiency.

less harm on the environment because the quantity of chemicals released into the environment decreases.

2. Understanding Pest Dynamics

2.1 Types of Pests and Their Impact on Crops

Agricultural pests can be classified into insects, mites, fungal, bacteria, viruses, nematodes and weeds. These pests can cause huge losses on the yields and quality of agricultural products. Insects as a group of pests are threats to the productivity of a farming enterprise more than any other group of pests. For example, the Cotton Bollworm (*Helicoverpa armigera*), which is a real threat to cotton and agricultural crops, such as tomatoes, corn and soybeans. Likewise, Japanese Beetle (*Popillia japonica*), feed on more than 300 species of plants and it is most damaging to fruits, vegetables and plants used for decoration with a desired on the leaves and flowers. Similarly, various fungal pathogens leading to crop spoilage and economic losses resulting from post-harvest decayers are able to effect diseases such as gray mold in many fruits and vegetables due to *B. cinerea*. (Romanazzi and Feliziani, 2014; Rhouma et al., 2023).

There is no doubt that pest infestations have its economic repercussions. It is globally a fact that pests reduces yields by at least 20–40% and the annual crop losses because of pests are in the hundreds of billions of dollars (Oerke, 2006). They grow side by side with crops for light, water and nutrients and some species such as *Amaranthus palmeri* has achieved resistance to several herbicides (Ward et al., 2013). Furthermore, some viruses that infect plant tissues distributed by insects for instance the tobacco mosaic virus are virulent hazards to crops as usually evident in cases of tobacco plants (Hanssen and Thomma, 2010). Also *Xanthomonas citri* pv. *citri*, the bacterium responsible for citrus canker, causes severe economic losses in various countries due to its damaging effects on citrus crops, leading to reduced fruit quality and yield (Ali et al., 2023; Ali et al., 2024). Similarly, Blackleg of potato, caused by the bacterium *Pectobacterium atrosepticum*, leads to significant crop losses globally by affecting tuber quality and yield, particularly in cool, wet climates (Hameed et al., 2024).

Other insects are the Two-Spotted Spider Mite (*Tetranychus urticae*) which particularly feeds on different crops such as strawberries, peppers, and tomatoes and causes the leaves to change color by transmitting the viruses or other microbes that affect the areas photosynthesis (Livinali et al., 2014). Flea beetles (*Phyllotreta* spp.) primarily attack cruciferous crops (e.g., cabbage, radish), puncturing leaves with small, distinct holes that depress plant growth and reduce yield (Metspalu et al., 2014). Green peach aphid (*Myzus persicae*) is a pest to a number of crops that include; peach, tomato and potatoes that transmits plant viruses as well as feed on the plant sap which in turn causes the plant's stunted growth and death of the plants (Van Emden et al., 1969). Some pests, such as the Fall Armyworm (*Spodoptera frugiperda*), are highly invasive and can affect a wide range of crops like corn, rice, and sugarcane. This pest migrates quickly and adapts to new environments, making it challenging to control (Srikanth et al., 2018). Other notorious pests include the Diamondback Moth (*Plutella xylostella*), which damages cabbage and other crucifers, and Cutworms, which attack seedlings at soil level, often cutting them off entirely (Furlong et al., 2013). The Brown Planthopper (*Nilaparvata lugens*) is a serious problem for rice plant suffering from its attacks by spreading plant viruses and feeding triggers loss in yield (Hu et al., 2014). Tobacco Whitefly (*Bemisia tabaci*) is another important pest that affects a number of crops, tobacco, cotton and vegetables and the viral diseases that it spreads causes significant damage to yield (Brown et al., 1995).

The ecological impact of pests also warrants consideration. Some effects of pest infestation are also destabilization of local ecosystems as the pests also distort the natural ratios and distribution of native species. For example, the underlisted invasive plant species can replace the native vegetation resulting in the displacement of native animal species like, *Parthenium hysterophorus*. Additionally, pests like nematodes penetrate and damage the crop roots apart from creating a relationship with soil microorganisms that are important for soil health (Nicol et al., 2011).

2.2 Factors Influencing Pest Outbreaks

Pests are sensitive to both endogenous and exogenous environmental factors, as well as those caused by humans. **Pest dynamics** depend on the environment such as climatic and ground conditions including the type of soil that preferred. For instance, by raising the temperatures and increasing the levels of humidity, pest life cycles are also increased hence more outbreaks occur that are usually severe (Deutsch et al., 2008). Soil health also directly impacts pest dynamics as young and well-fertilized plants can resist pest attacks better than plants grown in poor nutrient conditions, poor soil conditions stress plants (Altieri et al., 2012).

The **biological factors** are related with pest cycle and breeding schedules. Most pests have high reproduction rates and can reproduce rapidly when factors are ready for multiplication. Further, some pests have become acclimatized to natural enemies or harsh conditions by adapting to a perceived antagonistic environment (Gould, 1988). It is also important to note that because of certain flexibility which the pest can exercise in matters relating to climate, pest populations are capable of easily transiting from one region to another, therefore posing a higher risk to agriculture practices.

Another factor that is worth considering under **changing agricultural environment** is human induced factors in relation to pest outbreaks. Operation of monoculture is also disadvantageous as it involves farming a large area with a single crop species and this is likely to favors pest formations as they have a constant and unlimited source of food. Monoculture always have a negative effect on this aspect of the ecosystem and reduce the balance of predator and the prey. Furthermore, **chemical pest control** and the relentless utilization of chemical pesticides bring about growth of incompliant pest pests' populaces, making the control systems ineffective in the long run (Tabashnik et al., 2013). In addition, this characteristic also deters pest control since pests become resistant not only to the pesticide, but also to subsequent applications of a stronger or increased frequency of the pesticide.

2.3 Traditional Pest Control Methods and Limitations

Traditional pest control methods primarily rely on chemical, physical, and biological control strategies. **Chemical control** by means of pesticide application has always formed the bulk of the methods since they can instantly reduce pest numbers. However, this method has a numbers of disadvantages; the bugs and worms become resistant to the chemicals and other harmless forms of insects and other small animals such as bees and other insects that aid in pollination or pests predators are affected adversely by the chemicals (Georghiou, 1990). However, pesticide usage threatens soil and water quality and contributes to hazardous impacts on farm owners and purchasers (Aktar et al., 2009).

The **physical control methods** include traps and barriers and are options other than using chemicals when dealing with pests. For instance, pheromone traps may entice pests to move out of crops' vicinity while physical barriers may stop pests from accessing plants. With these methods, control can be effective in a small area, but they are not very efficient and can be impractical on a large scale (Pedigo and Rice, 2006). Further, while physical controls are sometimes used, it is often less successful in combating keenly mobile pests like flying insects.

Biological control are means and ways where pests are suppressed, managed or prevented by the help of natural enemies. This method takes little time, bearing in mind that it utilizes birds to control pests. And since birds are part of the environment, this makes it an environmental friendly practice (Van Driesche et al., 2009). However, biological control can be quite difficult to use especially due to the following reasons. For example, although the use of non-native predators can be sometimes harmless, those predators may at times pester indigenous species' equilibrium or over hunt other species that are not their target (Simberloff and Stiling, 1996). In addition, ability of biological control agent to control pests is usually slower than that of chemical pesticides which may be a disadvantage, particularly when there is a serious infestation that requires quick solutions. Even though conventional pest control methods have performed pivotal steps in the regulation of pest status, they are characterized by environmental aspects, cost and efficiency. These limitations justify the use of various methods of pest management in order to optimize the use of innovation technology like AI in pest control while reducing the impact of pests on ecosystems.

3. Integration of AI in Precision Agriculture

3.1 Introduction to AI Technologies in Agriculture

Artificial Intelligence (AI) in agriculture increases crop yields. AI systems predict things and that is done through algorithms and machine learning, which helps farmers in better decision making. They ensure crop performance and increased food availability by adjusting to various climatic conditions (Das et al., 2018). AI enhances the existing practices in farming and do not need spending a lot of on labour and also determining the status of the soil. AI robots can enable farmers to efficiently organize their time, plan, and make strategic decisions by autonomously performing chores (Eli-Chukwu, 2019). And you know what? Not only is it beneficial to the environment, but it is also economical as it reduces cost by streamlining and improving farming processes. Precision agriculture also includes cattle farming, viticulture and apiculture, which is a branch of the digital agriculture movement that uses large-scale data collection and computation to improve farm productivity. The advancement in farming technology over ages and an era starting from prehistoric period to the digital one is the testimony of the agricultural sector's resilience and its capacity to adapt with the climate change (Zha, 2020).

3.2 Application of AI in Pest Detection and Monitoring

Application of artificial intelligence in pest control involves all manners of systems and techniques in the identification, control and tracking of pests to minimize their impact on crop production. AI systems can use data derived from satellite images, aerial photography, and ground sensors, alongside others to provide early signs of insect intrusion (Javaid et al., 2023). These models employ the use of machine learning (ML) to spot slight trends and anamolies in vegetation, notify the users about pests or diseases before they manifest. ML techniques like as Decision Tree (DT), Decision Support System (DSS), Support Vector Machine (SVM), Naive Bayes (NB), and Random Forest (RF), as well as DL approaches like CNN, DCNN, Long Short-Term Memory (LSTM), and Deep Belief Network (DBN), are extensively used in farming for pest detection and crop protection (Jackulin and Murugavalli, 2022). Tamil Nadu Agricultural University (TNAU) recently created an Android smartphone application to detect fall armyworm infestations in maize crops as part of their integrated pest management (IPM) capsules. The software yields high levels of accuracy to the contaminated areas through integrating transfer learning and Deep Convolutional Neural Networks (DCNNs) with validation of 93.47 and 98.47 percent respectively (Table 1). It can also identify the period in which a particular crop is most vulnerable to fall armyworm infestations well before they require the combined pest suppression techniques and or pesticide application. Due to the advancement in technology, it has been made easy for farmers to use digital methods in the preservation of the harvested maize and livelihoods, this concludes that in the near future issues to do with pests affecting crops will be well handled through the use of these emerging technologies (Kariyanna and Sowjanya, 2024) (Figure 3).

AI-enabled drones equipped with high-definition cameras and sensors can carefully monitor forested or agricultural areas. Computer vision analysis applied to the data within images to identify regions that require attention, disease and yield assessments, and pest identification. The above information facilitates rationing of treatments and resources required in that particular specialism. AI is used to study crop status, pest locations and climate to minimize the application of pesticides and their effects on the environment (Buchelt et al., 2024). There are now cameras,

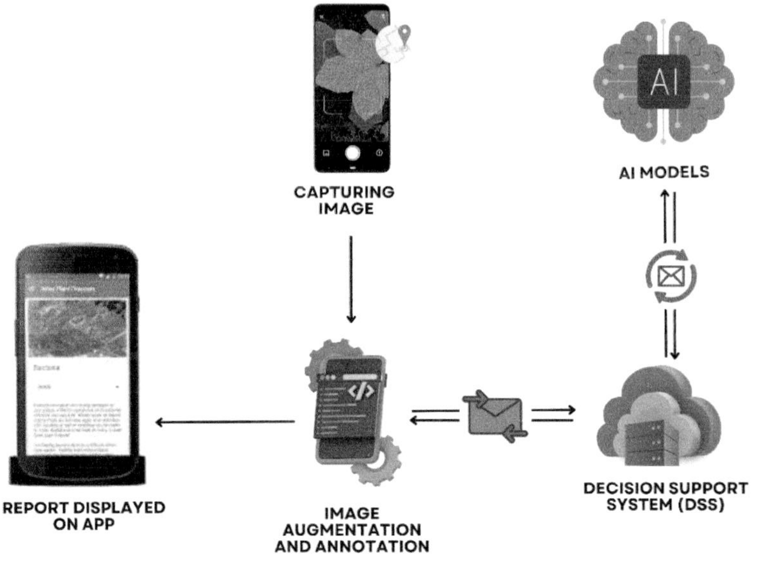

Figure 3. Schematic representation of the architecture for the AI based mobile application.

Table 1. AI-Based Applications for Pest Detection and Crop Health Monitoring in Agriculture.

AI-Based Application	Description	AI Techniques	Advantages	References
Trapview (Smart Pest Traps)	Smart traps for automatic pest detection and identification.	Machine learning, Computer vision	Provides real-time pest monitoring, reduces manual labor.	(Moura et al., 2023)
Xarvio Scouting App	AI-powered app for crop photos and pest/disease identification.	Deep learning, Image classification	Quick diagnosis and practical guidance.	(Calle and Böckmann, 2020a,b)
FarmSense Smart Insect Sensors	AI-enabled sensors for real-time pest detection through wingbeat analysis.	Machine learning, Pattern recognition	Immediate pest outbreak identification for tailored interventions.	(Burks, 2022)
Taranis	AI tool for analyzing drone photos to detect pests and diseases in crops.	AI-powered analytics, Computer vision	Enhanced crop health monitoring and high-resolution pest detection.	(Kariyanna and Sowjanya, 2024)
AgroScout	Evaluates satellite and drone imagery to detect pest-induced crop stress.	Deep learning, Satellite imagery analysis	Quick identification of affected areas and implementation of preventive measures.	(Uzhinskiy, 2023)
Precision Hawk	AI-powered drones and software for pest detection in large agricultural fields.	Data analytics, AI for remote sensing	Reduces pesticide usage by enabling focused treatment in specific areas.	(Puri et al., 2017)
DeepMind Pest Control (Research)	Experimental AI system using reinforcement learning for autonomous pest control in greenhouses.	Autonomous systems, Reinforcement learning	Encourages sustainable agriculture by reducing pesticide use.	(Chowdary, 2021)
Plantix	Mobile app using AI for diagnosing plant diseases and pests from photos taken by farmers.	AI-based image recognition	Provides accessible and immediate solutions for smallholder farmers.	(Samal et al., 2023)

sensors, and artificial intelligence that make the smart pest monitoring systems possible, and efficient as well. It is a challenge that insect pests like cotton bollworm, stress the need for an AI-based pest info system to operate pest smart traps and AI algorithms for effective sustainable cotton yield, least pesticide and highest output (Popescu et al., 2023).

These technologies raise agricultural output while reducing manual labor by using AI-powered devices to detect and identify pests. Automated robot vehicles equipped with sensors may perform mobile sensing across fields or woods in order to detect pest infested areas and make decisions about the best time to implement interventions (Khan et al., 2024).

3.3 Utilization of AI for Predictive Modeling of Pest Outbreaks

AI uses a range of facts to develop pest forecasting prediction models, including meteorological patterns, agricultural phenology and insect life cycles. These algorithms use past data as well as current feed to predict the occurrence of insect invasions. This allows the management strategies to be practical and the response to be immediate. The IoT camera and sensors captures insect populations, motions and behavior and the data collected is used for accurate and timely detection and tracking of pests and diseases. The rise of advanced sensors, IoT, and AI enables effective predictive modeling for pest control, helping achieve key Sustainable Development Goals (SDGs). AI × IoT domains in agriculture include robotic research for ultra-labor savings and outdoor and indoor smart farm operations (see in Figure 4). This can increase agricultural productivity to achieve the SDGs and effectively controls pest outbreaks (Al-Ahmed and Ahamed, 2024).

Farmers remain in a class of their own, with pest interaction, unstable climates, and overuse of pesticides (Markvic et al., 2023). To address these challenges, the study designed an IPM platform using artificial intelligence, various data types, and data analysis tools. This new tool identifies to complement pest control strategies in agriculture, and offer farmers with better as well as eco-friendly methods of pest control so as to reduce the harm on the environment, besides the formation

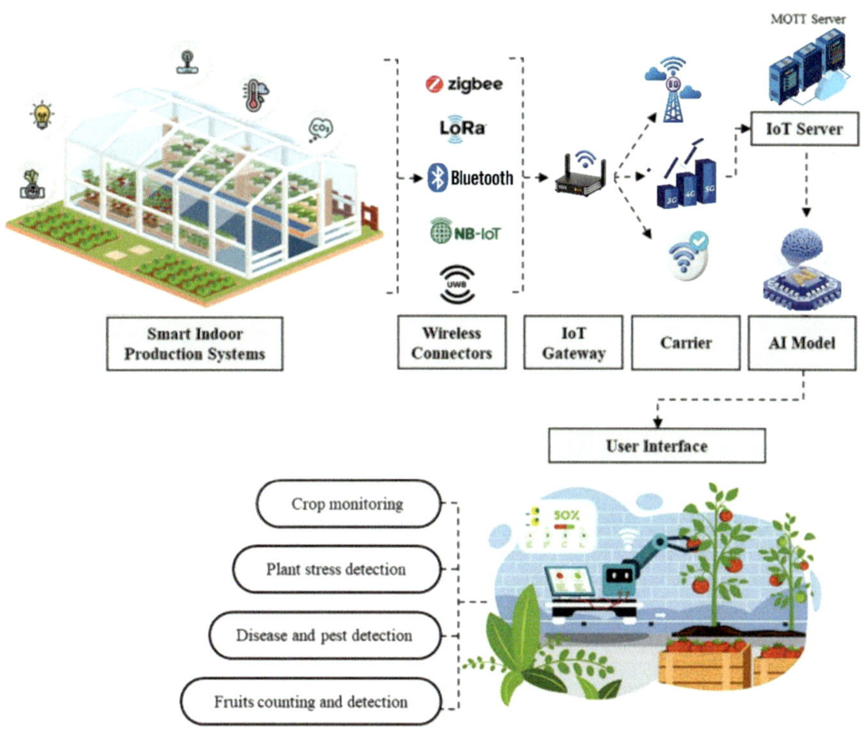

Figure 4. IoT-enabled smart indoor farming system with sensors, wireless connectors, and AI for monitoring crops, plant stress, diseases, and fruit counting via a user interface (Al-Ahmed and Ahamed, 2024).

of resistance. There are common AI techniques which are used to build predictive models for the pest management given below;

- **Time series analysis:** Time series analysis is used to analyze the historical pest data and their changing trends with the passage of time.
- **Machine Learning:** Predictive models are designed by using algorithms like support vector machines, random forests and gradient boosting, etc.
- **Deep Learning:** For the detection of signs of pest infestations, Neural networks particularly convolutional neural networks are effective to process image data (Balduque-Gil et al., 2023).

3.4 Integration of AI with Precision Agriculture Systems

Precision agriculture combines AI, GPS and data-driven insights to improve farming efficiency. This strategy increases agricultural production as farmers receive information on the soil, climate and aspects of their crops. Therefore, food production is enhanced, costs reduced and ever-increasing sustainable advances are achieved. Precision agriculture is the use of technology to optimize the farming methods that farmers use. It applies virtual reality, geography, precise navigation, and geolocation systems and remote sensing and data analysis to enhance efficiency and output to promote sustainable agriculture (Figure 5) (Khan et al., 2024).

Variable Rate Technology (VRT): VRT allows specific changes to input application rates including fertilizers, herbicides, and seeds at the field level to respond to variability in soil type to obtain the highest yield, minimized cost, and reduced environmental impact (Kushwaha et al., 2024).

Precision Irrigation Systems (PIS): PIS like drip irrigation with VRT, allow water delivery that responds to the characteristics of soil moisture and weather conditions.

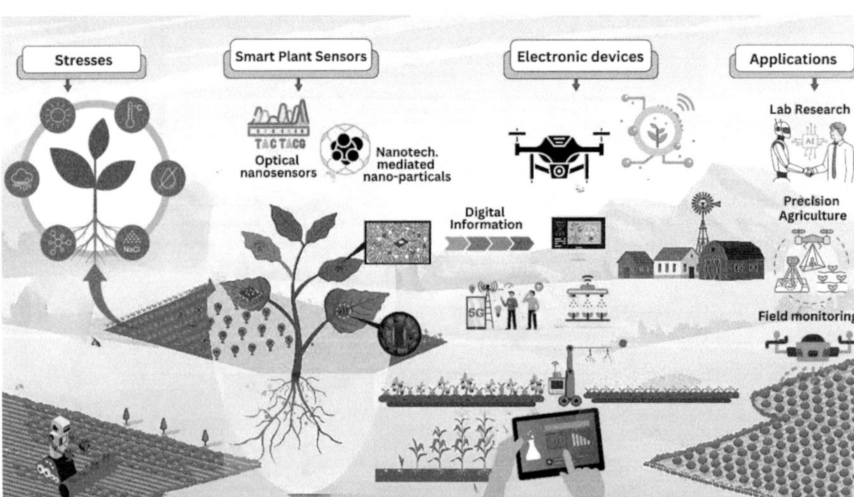

Figure 5. AI applications in precision agriculture for advancing towards sustainable crop production (Zain et al., 2023).

This enhances water saving and food security for cultivation (Papadopoulos et al., 2024).

Remote Sensing Technologies (RST): RST incorporating satellites and drones provide useful information on crop status, moisture and nutrient content. Some of the benefits include the following: This feature helps farmers to make correct choices to manage the crops properly (Hafeez et al., 2023).

Data Analysis and Decision Support Systems: Many data inputs come together to derive the best decisions on crop rain management, resource utilization, and risk management to enhance profitability and sustainability (Kumar et al., 2024).

Crop Monitoring and Management Software: It involves data from sensors, drones, and satellite photographing, to monitor crop progression. In addition, it defines pests and diseases and adjusts methods to allow for early intervention strategies (Puri et al., 2017).

AI and data driven approaches are reshaping farming as a business through precision agriculture. It means that farmers can leverage technology to raise production, impact the environment and improve farming operations. This gives an added advantage as robotics and AI techniques allow for crop monitoring and detection of soil and climatic conditions as they occur. The inclusion of analyzing technological factors within farming techniques, can prove to be efficient in use of resources necessary for farming, including water, fertilizers as well as pesticides and herbicides. This brings forth sustainability and productivity (Padhiary et al., 2024).

4. Targeted Pest Control Strategies

4.1 Precision Application of Pesticides using AI-driven Technologies

AI technologies have a major impact at the agricultural frontier because they are now being introduced to revolutionize the use of pesticides in real time. One of the most important practical use cases are **drones** that incorporate AI algorithms in surveillance together with application of targeted pesticides. These drones scan the crops' condition in real-time so that the pesticides can be applied in the site specific area. For instance, computer vision detects a localized pest invasion of crops, the drone that applies a pesticide sprays only that area and in a lesser quantity than in other healthier parts. This method minimizes the amount of pesticide used per acre by as much as 90% as compared to normal spraying (Sravani et al., e.d.). Another application unique approach is through **robotic systems** like Blue River Technology's "See & Spray" that incorporates camera with a high resolution, and artificial intelligence to crops from weed. This technology also enables spraying of herbicides to the areas specific to weeds without affecting the crops that are valuable. They also reduce the utilization of chemicals and also the total costs associated with a process and labor needed for that process (Sravani et al., e.d.). However, note that predictive analytics also is very useful in the management of pests as well. Since the extrinsic and intrinsic factors are determinable from historical information in combination with the present day's weather scenario and health of the soil, AI systems can predict pest arrivals in advance. Such detection is early enough to enable growers to apply prevention measures and minimize the application of pesticides, which are usually applied reactively. Programs like Plantix build up this asset by furnishing farmers with a library of plant diseases and pests which enables them to detect problems as they develop thus scholars optimize for interventions rather than spraying a large area with pesticide (Sravani et al., e.d.; Samal et al., 2023). Also, there are other sensing technologies embedded in AI to help farmers to understand the site-specific application of fertilizers or pesticides depending with the crop shown in Table 2.

Table 2. Overview of AI Models Applied in Precision Agriculture for Pest and Crop Management.

Model	Application	Efficiency and Accuracy	Main Focus	References
CNN	AI-driven drone	High (90%)	Pesticides recommendation and spraying using AI algorithms, achieving accuracy above 85%	Naseer et al., 2024
LSVM	Crop-row detection	High	The image collection was short and straightforward; it simply included pictures of sugarcane cultures	César Pereira et al. (2022)
SVM	Pesticide/ disease treatment	Precision: 85% Recall: 81% F1-score: 79%	NDVI problems can be resolved with a two-step method that combines semantic segmentation with pixel-wise NDVI computation	Tendolkar et al. (2021)
KNN3, KNN11	Crop-row detection	IOU: 0.76, F1: 0.86	Implemented algorithms that are the easiest to implement	César Pereira et al. (2020)
RF	Pesticide/ disease treatment	Recall of other weeds class: 45%, recall of saltbush class: 95%, recall of chamomile lass: 87%, and recall of sugar beet class: 78%. For anticipated items, the overall model accuracy was 86%.	The model can identify plants based only on their shape. Low recall and precision for plant identification in the "other weeds" class	Lottes et al. (2017)
		High (85%)	Integration of drone technology infarming, crop yield optimization, pest control	Chen and Patel (2018)
SVM	Fertilization	Dataset #1 (first configuration): [recall: 86%, accuracy: 72%, precision: 93%, accuracy: 78%] Dataset #2: [recall: 94%, accuracy: 73%, precision: 97%, accuracy: 83%]	Images captured by inexpensive cameras may be classified using a pattern recognition system.	Natividade et al. (2017)

Naseer et al. (2024) conducted a study that proposes a novel way of increasing the yields of crops in an environmentally friendly way by incorporating ML and drone technology. The use of the proposed method involves Convolutional Neural Networks (CNNs) that can be used to analyze multispectral images and different kind of data such as crop health status, weather conditions and fraction of soil color. This analysis is performable in real-time, the drone can give a recommendation on a suitable pesticide to use, quantity to apply and the time to apply it based on the pest and crop identified. Research findings show that the AI-powered drone boasts of an accuracy level of up to 90 % in the most infested pest areas hence reducing on unnecessary application of pesticide chemicals on other regions. This system improves pest efficiencies for pest management as well as encourages sustainable agriculture practices by giving information of crop status and pest variables over a period of time (Naseer et al., 2024).

4.2 Implementation of Biocontrol Methods Guided by AI

(1) **Data-driven methods:** Data-driven methods play a crucial role in biological control technologies. Biological control involves the use of biological products including biopesticides and biocontrol agents like predatory insects to control pests and diseases in the fields. Generally, biological control is more efficient than chemical pesticides and also has no harm to the environment. But it has drawbacks such as, application difficulty and mixed outcomes. Technological methods involve the use of computer programs to perform and coordinate the biological control mechanism for efficiency and efficacy. As one obtains vast field data and other important aspects in biological control, it enhances the decision making support to the agricultural producers in terms of application plans and management strategies. This in turn results into better pest management, higher production and better crop quality, thus promoting sustainable agriculture (Xiao et al., 2020) .

(2) **Machine learning for pest monitoring:** Machine learning is crucial in pest and disease monitoring. In machine learning analysis, it is now possible to match image data from the field and determine the presence, intensity and distribution of pests. Through a detailed analysis and simulation of these monitoring data, the trend analysis of pest outbreaks can be ascertained and the right control measures taken. For example, utilizing deep learning algorithms, many pest and disease images in fields are automatically recognized and classified, which enables agricultural producers to timely identify potential threats. It increases the effectiveness of monitoring the level of pest infestation in crops, detects the lowest values of crop yield and qualitative indicators, and offers decision support on crop production, based on accurate data obtained without any delay (Liao et al., 2023).

(3) **Application of intelligent agricultural robots.** Intelligent agricultural robots have enormous potential in biological control. Self-moving and having multiple sensors and/or actuators, such robots are capable of roaming fields independently and applying biological control strategies including, for instance, biopesticide spraying or the introduction of beneficial insects. Through integration of

machine learning algorithms and adaptive control technologies, these robots are capable of changing spraying quantities and methods according to the real field conditions resulting in improvement of biological control efficiency and accuracy. In this way, using large amounts of data collected, intelligent agricultural robots can fine-tune their choice and activities, and take a more accurate and efficient action in the fields. This reasonable biological control concept not only has a considerable effect on regulation of pests but also has less threats to the environment and enjoys several benefits to the development of sustainable agriculture (Li and Wang, 2024).

(4) Data analysis and optimization. Data analysis and optimization play a vital role in biological control. The results of comprehensive observations of the biological control process allow identifying trends in large amounts of data, which can be used to enhance the effectiveness of biological control technologies and their reliability. For instance, the amount and frequency of biopesticide application that machine learning algorithms are going to use for field pest control can be adjusted in real-time monitored fields regarding pest intensity and the existing field conditions to obtain the maximum control effect. This real-time optimization regularly enhances biological control technologies to be more intelligent and accurate to offer a better support to agricultural production (Zeng et al., 2024).

(5) Establishment of decision support systems. Establishing decision support systems is an important practice combining biological control technologies and AI algorithms. These systems give advice to producers of agricultural produces in relation to field management and control of diseases and pests. Several sources of field data can fit for decision support systems: pest risks in the location can be assessed, tendencies of pest outbreaks can be predicted, and suggested treatment and control measures provided. Compared with the traditional pest control methods that can only decrease the pest damage rate to crops, this system can not only decrease the pest damage rate to crops but also enhance the quality and quantity of agricultural production, which can provide more accurate and comprehensive information for agricultural producers' decisions and actions (Kang et al., 2023).

4.3 Use of Robotics and Drones for Targeted Pest Management

Drones and agricultural robots have become indispensable assets to modern precision agriculture especially when it comes to targeted pest control. Over a large area of farming, drones are used for doing field inspection, locating the areas that have been infested by pests and administering the pest control pesticides with accuracy. They are also fitted with modern imaging technologies such as multispectral or hyperspectral cameras to identify pest incidences at precocious stages. For instance, one flight of drone can capture images of up to 400 acre-farm age, which offers quality information to farmers on the areas that require attention (Zhang and Kovacs, 2012). This enables pest control to be done quickly and effectively saying no to the use of pesticides hence help to protect the environment.

Innovations in the field of agricultural robotics have gone even ahead to enhance the invention of machines that scan pest problems and fix them on their own. For instance, currently existent robots such Agrobot and Ecorobotix identify crops and weeds by using machine learning algorithms; the results benefit the process of applying herbicides. For instance, these robots can spray pests directly on the field –in which pesticide application is four times less than using conventional methods (Vasconcelos et al., 2023).

An example from a real world application includes the XAG R150 Unmanned Ground Vehicle which uses AI to search for pest in fields and treat them selectively. The R150 has been successfully used in rice fields where it cuts pesticide use by 50% with good pest control outcomes (Nahiyoon et al., 2024). Similarly, Blue River Technology's See & Spray robot uses AI to identify and treat weeds in real time, cutting herbicide use by ≈80% or more (Sravani et al., n.d.). The choice of these cases is aimed at elucidating the potential of robotics in pest control as a natural and cost-effective solution.

4.4 *Importance of Data-Driven Decision-Making in Pest Control*

Data-driven decision-making is integral to modern pest management, as it allows farmers to make informed choices based on real-time data and predictive analytics. AI solutions assist in processing the huge amounts of collected data using drones or other sensors to define patterns and possible invasions of pests. For instance, based on weather patterns, crop health and pest development, it can predict likely infestations in order to take appropriate action (Chen et al., 2022). There are lots of advantages in the utilization of data to control pests. Subsequently, employing deliberate data in managing farmers assist in preventing losses in crop through preventing environmental impacts. For instance, infield generated information can help adjust the frequency and quantity of pest control through pesticide usage and reduce on the usage of chemical and costs. Preliminary observations suggest that farms that use an analytic approach to pest management can achieve a 40% reduction in chemical application, improving crop yields and the quality of the land (Patrício and Rieder, 2018).

However, there are some problems associated with the installation of data-driven systems on farms. One of the biggest challenges is the high cost of acquiring the technology, which may be expensive for a small holder farmer. Furthermore, data issues and management remain a problem as poor data can result in a poor approach to pest control measures (Niloofar et al., 2021). There is also the question of additional education to ensure farmers are able to properly analyze the accumulation and application of new data to pest management. That being said, the nature of long-term advantages of the approach based on the analysis of big data speaks for its prospects in terms of non-trivial sustainable pest control.

4.5 *Best Techniques for Agricultural Problems*

The best solutions available at the moment for each issue together with the type of learning architecture that was employed is shown in Table 3. Overall, the deep learning and machine learning can produce respectable outcomes for a range of issues and occasionally fall inside the 80% accuracy range. There is, obviously, space for improvement in the majority of situations. YOLO-V5 specifically, was employed effectively by Jintasuttisak et al. (2022) to detect date palm trees in image data captured by UAV flying over farmlands in the Northern Emirates of UAE. A comparison of the detection results on date palm trees using YOLO-V5 and the other

Table 3. Best techniques results achieved in different agricultural problems.

Model	Type of Learning	Problem	Dataset	Best Results	References
YOLOv5	Supervised	Spatial segregation and segmentation	Date palm trees collected using a drone	mAP score of 92.34%	Jintasuttisak et al. (2022)
cGANs	Semi-supervised		5400 RGB images of pears and strawberries, 20% labeled	IoU score of 83.1% on mixed data	Fawakherji et al. (2020); Fawakherji et al. (2021)
ResNet18	Unsupervised		UAV images of spinach and bean fields	AUC: 91.7%	Bah et al. (2018)
DF-U-Net	Supervised	Pesticide/diseases treatment	Yangling UAV images	F1: 94.13%, Accuracy: 96.93%, Precision: 94.02%	Zhang et al. (2022)
Active learning: SVM	Semi-supervised		UAV images of Eucalyptus plantations	Accuracy: 98%, Recall: 97%	Coletta et al. (2020)
SGAN	Unsupervised		UAV images of strawberry and pea fields	Accuracy ~90%	Khan et al. (2021)
SVM	Supervised	Fertilization	UAV images of vineyards and forests	Accuracy: 83%, Precision: 97%, Recall: 94%	Natividade et al. (2017)
SSVT	Unsupervised		UAV images of a wheat field	96.5% on 384 × 384 pixel images	Zhang et al. (2022)
SVM	Supervised	Crop-row detection	Manually collected RGB images	88.01% F1-Score	Cesar Pereira et al. (2020)
FRCNN	Supervised	Tree/crop counting	Tree counting	IoU: 87.13% on palms, 49.41% on other trees	Ammar et al. (2021)
DENT	Semi-supervised		Yosemite tree dataset	MAE: 12.68%	Chen et al. (2022)
CNN	Supervised	Others	Various UAV images of land cover types	Accuracy: 96.3%	Aiger et al. (2017)

conventional CNN architectures, namely YOLOv3, YOLOv4, and SSD300 was made using the image acquisition from the drone which provided quantitative as well as qualitative comparisons. When it comes to accuracy, the training conducted herein established that the highest accuracy level recorded was 92.34% for the used training set applied in the YOLO-V5m (medium depth) model. In this way it became possible to actually locate and identify the date palm trees of different sizes in the places where date palm trees are not evenly distributed and in the areas with crowded overlapping surroundings.

Similarly, Bah et al. (2018) focused on detecting weeds from images acquired through UAV using unsupervised data labeling. There were two fields in the dataset: spinach and beans. Crop and weed were used as classification on the two datasets. Of the entire data, one-third was manually clustered and the rest were labeled in the supervised mode. In this paper, the ResNet18 model was used to perform the categorization. In bean field, ResNet18 was clearer better than SVM and RF techniques with average AUC 91.7 % for both supervised and unstructured labeled data while it was calculated 52.68% for SVM and 66.7% for Random forest technique. Nevertheless, RF generated a slightly superior average AUC% in the spinach field than ResNet18.

Zhang et al. (2022) used the dual-flow U-Net (DF-U-Net) to determine the level of vegetation of yellow rust in the farmland. The dataset is derived from the field experiment carried out in Yangling using a red-edge camera on a DJI M100 UAV with 1336 × 2991 sensor size. Accuracy, precision, and F1-score were therefore at 94.13%, 96.93%, and 94.02% respectively.

To identify new classes in an active learning fashion, Coletta et al. (2020) used a semi-supervised classification approach that might integrate data provided through clusters with that provided through a supervised method such as SVM. The authors state that this capability is uniquely valuable in volatile agricultural environments. The data was captured using a SenseFly eBee with an RGB camera. Two blocks made up the model: a contextual block (ConB) that gave more contextual information and a portion block (ClaB) that defined a 0.16m^2 area for classification. Two blocks create a circular pair which in turn generates feature vectors which are used for classification. To these vectors, one of three classifications was assigned based on researcher judgement. The classification uncertainty was measured next using a semi-supervised classifier and the importance of a classified feature vector was evaluated with a density measure. Some of the cases resulted in highly uncertain labels and were tagged as novelties to be learnt and, consequently, added to the training set for entropy- and density-based selection from which the files were tagged by an expert in the EDS domain. The findings showed a step wise increase in all class accuracy and recall values.

Pattern recognition system proposed by Natividade et al. (2017) provided NDVI scale to segment vegetation and classify vegetation into categories. This analysis used two datasets: A vineyard dataset that has three classes and a tree dataset with five classes were used in training an SVM model. On the two datasets, the best performing models attained an accuracy of about 72 percent.

Zhang et al. (2022) employed unmanned aerial vehicle (UAV) data to estimate crop nitrogen status employing a spectral spatial attention based transformer (SSVT).

The model is able to get spatial information from images and is a modification of the standard vision transformer (ViT). In view of this, the spectrum information that will comprise the majority of the properties in the agricultural applications may be estimated by the aforementioned model. Because of the SSL technology, which naturally allows models to learn from data that are not labeled, the model also solves the issue of computation originating from large pictures that ViT has. According to the data, the model with the accuracy of 96.2% was higher than the ViT model with the accuracy of 94.4%. However, this model required four million more parameters than the ViT model to achieve optimal performance.

The performance of many machine learning methods for the crop-row segmentation issue was compared by César Pereira et al. (2020). Their study involved just a single image of a sugar cane field and the results of the segmentation using different classifiers were compared to an image to which the authors had manually added annotations. Crop row and backdrop were the two groups the pixels of the manually segmented image were divided into. For textural characteristics, a gray-level co-occurrence matrix (GLCM) and a four-filter Gabor filter bank were used; for spectral characteristics, ExG and VI were used. SVM models were trained with RGB color characteristic and feature vectors. For the linear SVM model, the best feature combination was concluded to be RGB, EXG and Gabor filters. This combination produced an IoU percentage of 78.86% and an F1-score of 88.01%. RGB and GLCM were the worst feature combo. This combination produced an IoU percentage of 46.08% and an F1-score of 62.48%.

Ammar et al. (2021) proposed a new deep-learning method to detect the number of palms trees automatically from aerial images. They identified palm palms and other trees using a number of modern convolutional neural network models (faster RCNN, YOLOv3, YOLOv4, and EfficientDet) and performed the comparison analysis in the average precision and inference time. As for the two detectors, the speeds and accuracies were almost equally close to the optimum value – could be said that both YOLOv4 and EfficientDet-D5 (up to 99% mAP and 7.4 FPS) called the best among the variety of detectors.

Chen et al. (2022) proposed a new and efficient deep learning model called the density transformer (DENT) for tree counting from aerial images automatically. Four steps make up the model's architecture: a standard transformer encoder, a density map generator (DMG) to predict the density distribution over the input images, and a Multi-receptive field CNN (Multi-RF CNN) to calculate a feature map over the input images. Also, to the public they released the Yosemite tree dataset, a standard data set with aerial images for counting trees (Chen et al. (2022). With more than 17.3 and 22.6 with YOLOv3, the proposed model MAE of 10.7, and RMSE of 13.7 were higher than all the state-of-the-art techniques. It is important to note that the CANNet model (Liu et al., 2019) outperformed the DENT models in one of four locations and obtained the closest values, 10.8 and 13.8, respectively.

Aiger et al. (2017) presented a multi-view image projection-based image categorization technique. Their approach involved projecting many photos at various depth planes close to the surface that was rebuilt. This made it possible to classify objects with complicated reflection/light response qualities, such as water and trees, whose most obvious feature was their change in appearance under various perspectives. With their suggested 3D CNN, they achieved the highest accuracy of 96.3%.

5. Challenges and Future Directions

Ethical Considerations and Environmental Impact of AI-Driven Pest Control: The integration of AI in agriculture, particularly in pest control, raises several ethical considerations. One major concern is the idea that these new AI technologies that replace traditional agriculture works involved in pest monitoring and treatment (Susskind, 2015). However, these systems are closed structured, and sometimes the decision making process is unclear to the users, hence the issue of accountability arises. Should an AI system be incapable of accurately predicting or controlling a pest outbreak, it's unclear who is responsible: the farmer, the technology supplier, or the AI algorithm?

Then last but not the least, the negative effects of AI in pest control on the environment also deserves consideration. Although AI can reduce the wastage of pesticides and is hence more sustainable, AI innovation still depends on energy demanding data processing, especially by using more of the machine learning algorithms that require processing power (Rolnick et al., 2022). This is also valid to some extent due to potential contribution of the AI systems to reinforce some pest control strategies such as chemical pesticides. This could potentially harm non-target organisms and disrupt local ecosystems, underscoring the need for responsible AI implementation that balances effectiveness with ecological preservation.

Technical Challenges and Limitations of Current AI Technologies: Despite the promise of AI in precision agriculture, several technical challenges limit its effectiveness. The main challenge of using AI in pest control is real-time decisions; pest control frequently requires data at the exact moment of analysis. Nevertheless, dealing with large amounts of data in real time may take much time and current AI systems may face temporal and spatial limitations, let alone in remote areas (Moeuf et al., 2018).

Another important concern is the availability and quality of the information used in the present study. An effective AI model involves having a large data set to feed into the algorithm to cater for different conditions which may exist, unfortunately there is scarce data on pest which captures variations in climate, crop type or even pests behavior in many places. This means that with poor quality data, the AI system will be able to deliver poor predictions or recommendations to the system, thus making it less effective and reliable (Bronson and Knezevic, 2016). Also, there are challenges of infrastructure deployment and adoption using AI since most of the farms implement AI system do not have the technological framework that AI calls for such as the internet, sensors, and drones. Such hurdles put into focus the imperative of providing machinery and data capture systems in the rural areas so as to enhance AI for agriculture.

Potential Advancements and Future Trends in AI for Precision Agriculture: The future of AI in precision agriculture looks promising, with several emerging technologies set to enhance its capabilities. 5G technology is expected to improve connectivity in rural areas, enabling faster data transmission and supporting real-time AI applications in pest control (Tang et al., 2021; Mukherjee et al., 2021). With better internet connectivity will allow farmers to implement IoT sensors and drones that help in the collection of continuous data that can feed into the AI systems to assist in more timely and correct pest management decisions (Tang et al., 2021). Another innovation that can help avoid excessive latency is Edge computing, which pre-processes data near the source, in this case, on the farm itself rather than relying on widely located data centers (O'Grady et al., 2019). The AI systems will be able to perform optimally in situations where they can only get a limited or slow internet connection.

Decision-making autonomy is expected to keep rising alongside the growing sophistication of machine-learning and robotics systems. Self-driving drones and robots to detect, locate, and apply pest control without intervention are becoming more realistic as farmers have technology that can work around the clock (Inamdar et al., 2023). This additional freedom means AI could ease the operation of pest control systems since some of the farmers, especially smallholder farmers may not be able to effectively engage with the technologies (Ghosh et al., 2024).

Moreover, the application of AI technology in pest control is forecasted to increase in the future with advanced technique options like predictive analytics and integrated pest management (IPM) strategies that combine AI with biological control methods. It might possibly assist farmers to treat pest infestations even before they get out of hand and hence enhance productivity and efficiency. In addition, this report will also show that incorporating AI into IPM can reduce the dependency on chemicals through applying natural enemies approach in the management of pests (Demirel and Kumral, 2021). As these technologies evolve, they hold the potential to transform pest control into a proactive, precise, and environmentally responsible aspect of modern agriculture (Ghosh et al., 2024).

6. Conclusion

AI integration in precision agriculture offers a new level of accuracy and efficiency in pest control. By detecting, monitoring, and managing pests with precision, AI systems reduce chemical usage and promote sustainable farming practices. Predictive capabilities help mitigate potential outbreaks before they escalate, safeguarding crop yields and enhancing food security. AI with real-time processing of vast data reduces input costs and minimizes environmental impacts, allowing farmers to apply treatments more selectively. Moreover, AI is enhancing the adoption of Integrated Pest Management (IPM) strategies by combining biological controls with precise chemical applications. This technology improves productivity and resource management, benefiting farmers and the environment. However, challenges include high-quality data, real-time processing, and adequate infrastructure in rural areas. Ethical concerns include transparency and potential displacement of traditional agricultural roles. Moreover, AI could bring about reduced chemical inputs while advocating for the use of adequate technologies without which could upset an ecology if not well regulated. Towards the future, development of several technologies including, the fifth generation networking technology or 5G, or what is referred to as edge computing, is expected to boost the applicability of AI in agricultural practice. These advancements will further improve the consistency of and the speed of data transmission, paving the way for the application of AI-based pest control system at small holder farm, among others. As AI technology evolves, it is essential to balance innovation with sustainable practices that prioritize environmental health and resilience.

References

Aiger, D., Allen, B., and Golovinskiy, A. (2017). Large-scale 3d scene classification with multi-view volumetric cnn. arXiv preprint arXiv: 1712.09216.

Aktar, W., Sengupta, D., and Chowdhury, A. (2009). Impact of pesticides use in agriculture: their benefits and hazards. Interdisciplinary Toxicology, 2(1): 1–12.

Al-Ahmed, S. A., and Ahamed, T. (2024). AI× IoT: Increasing Agricultural Productivity of Crops, Orchards, and Livestock Management Using Smart Agricultural Space for Achieving SDGs. In IoT and AI in Agriculture: Smart Automation Systems for increasing Agricultural Productivity to Achieve SDGs and Society 5.0 (pp. 481–490). Singapore: Springer Nature Singapore.

Ali, S., Hameed, A., Binyamin, R., Alam, M. W., Aslam, H. M. U., Riaz, H. et al. (2024). Morphogenetic characterization of *Xanthomonas citri* pv. citri and its management. Journal of King Saud University-Science, 36(8): 103339.

Ali, S., Hameed, A., Muhae-Ud-Din, G., Ikhlaq, M., Ashfaq, M., Atiq, M. et al. (2023). Citrus canker: a persistent threat to the worldwide citrus industry—an analysis. Agronomy, 13(4): 1112.

Altieri, M. A., Ponti, L., and Nicholls, C. I. (2012). Soil fertility, biodiversity and pest management. Biodiversity and Insect Pests: Key Issues For Sustainable Management, 72–84.

Ammar, A., Koubaa, A., and Benjdira, B. (2021). Deep-Learning-Based Automated Palm Tree Counting and Geolocation in Large Farms from Aerial Geotagged Images. Agronomy, 11: 1458.

Bah, M.D., Hafiane, A., and Canals, R. (2018). Deep Learning with Unsupervised Data Labeling for Weed Detection in Line Crops in UAV Images. Remote Sens. 10: 1690

Balduque-Gil, J., Lacueva-Pérez, F. J., Labata-Lezaun, G., Del-Hoyo-Alonso, R., Ilarri, S., Sánchez-Hernández, E. et al. (2023). Big data and machine learning to improve european grapevine moth (*Lobesia botrana*) Predictions. Plants (Basel, Switzerland), 12(3): 633.

Bronson, K., and Knezevic, I. (2016). Big Data in food and agriculture. Big Data & Society, 3(1): 2053951716648174.

Brown, J. K., Frohlich, D. E., and Rosell, R. C. (1995). The sweetpotato or silverleaf whiteflies: biotypes of Bemisia tabaci or a species complex?.

Buchelt, A., Adrowitzer, A., Kieseberg, P., Gollob, C., Nothdurft, A., Eresheim, S. et al. 2024. Exploring artificial intelligence for applications of drones in forest ecology and management. For. Ecol. Manag. 551: 121530.

Burks, C. S. (2022). Comparison of navel orangeworm adults detected with optical sensors and captured with conventional sticky traps. AgriEngineering, 4(2): 523–532.

Calle, J. P. R., and Böckmann, E. (2020a). D3. 5: Report on case studies.

Calle, J. P. R., and Böckmann, E. (2020b). D3. 4: Reference document on SMART IPM techniques and methodologies.

César Pereira Júnior, P., Monteiro, A., Da Luz Ribeiro, R., Sobieranski, A. C., and Von Wangenheim, A. (2020). Comparison of supervised classifiers and image features for crop rows segmentation on aerial images. Applied Artificial Intelligence, 34(4): 271–291.

Chen, C. J., Li, Y. S., Tai, C. Y., Chen, Y. C., and Huang, Y. M. (2022). Pest incidence forecasting based on internet of things and long short-term memory network. Applied Soft Computing, 124: 108895.

Chen, G., Shang, Y. (2022). Transformer for Tree Counting in Aerial Images. Remote Sens. 14: 476.

Chowdary, V. (2021). Machine Learning Applications for Precision Agriculture: A Comprehensive Review.

Coletta, L. F. S., de Almeida, D. C., Souza, J. R., Manzione, R. L. (2022). Novelty detection in UAV images to identify emerging threats in eucalyptus crops. Comput. Electron. Agric., 196: 106901.

Das, S., Ghosh, I., Banerjee, G., and Sarkar, U. (2018). Artificial Intelligence in Agriculture: A Literature Survey.

Demirel, M., and Kumral, N. A. (2021). Artificial intelligence in integrated pest management. pp. 289–313. *In*: Artificial intelligence and IoT-based technologies for sustainable farming and smart agriculture. IGI Global.

Deutsch, C. A., Tewksbury, J. J., Huey, R. B., Sheldon, K. S., Ghalambor, C. K., Haak, D. C. et al. (2008). Impacts of climate warming on terrestrial ectotherms across latitude. Proceedings of the National Academy of Sciences, 105(18): 6668–6672.

Eli-Chukwu, N.C. 2019. Applications of Artificial Intelligence in Agriculture: A Review. 9.

Fawakherji, M., Potena, C., Prevedello, I., Pretto, A., Bloisi, D.D., Nardi, D. (2020). Data Augmentation Using GANs for Crop/Weed Segmentation in Precision Farming. pp. 279–284. *In*: Proceedings of the 2020 IEEE Conference on Control Technology and Applications (CCTA), Montreal, QC, Canada, 24–26 August. 135.

Fawakherji, M., Potena, C., Pretto, A., Bloisi, D.D., Nardi, D. (2021). Multi-Spectral Image Synthesis for Crop/Weed Segmentation in Precision Farming. Robot. Auton. Syst., 146: 103861.

Feng, H., Yao, Q., Hu, C., et al. (2023). Recent advances in intelligent techniques for monitoring and prediction of crop diseases and insect pests in China. Plant Protection, 49(5): 229–242.

Furlong, M. J., Wright, D. J., and Dosdall, L. M. (2013). Diamondback moth ecology and management: problems, progress, and prospects. Annual review of entomology, 58(1): 517–541.

Georghiou, G. P. (1990). Overview of insecticide resistance.

Ghosh, D., Siddique, M. A., and Pal, D. (2024). 6 AI-Driven Precision Agriculture Approach to. AI in Agriculture for Sustainable and Economic Management, 67.

Gould, F. (1988). Evolutionary biology and genetically engineered crops. BioScience, 38(1): 26–33.

Hafeez, A., Husain, M. A., Singh, S. P., Chauhan, A., Mohd.T., Khan, N. et al. 2023. Implementation of drone technology for farm monitoring & pesticide spraying: A review. Inf. Process. Agric. 10: 192–203.

Hameed, A., Zeeshan, M., Binyamin, R., Alam, M. W., Ali, S., Zaheer, M. S. et al. (2024). Molecular characterization of Pectobacterium atrosepticum infecting potato and its management through chemicals. PeerJ, 12: e17518.

Hanssen, I. M. and Thomma, B. P. (2010). Pepino mosaic virus: a successful pathogen that rapidly evolved from emerging to endemic in tomato crops. Molecular Plant Pathology, 11(2): 179–189.

Hanssen, I. M., and Lapidot, M. (2012). Major tomato viruses in the Mediterranean basin. pp. 31–66. *In*: Advances in virus research (Vol. 84). Academic Press.

Hu, G., Lu, M. H., Reynolds, D. R., Wang, H. K., Chen, X., Liu, W. C. et al. (2019). Long-term seasonal forecasting of a major migrant insect pest: the brown planthopper in the Lower Yangtze River Valley. Journal of Pest Science, 92: 417–428.

Inamdar, R., Bhalerao, S. R., Shinde, G. U., and Gupta, O. (2023). Precision Pest and Disease Detection System using Agricultural Robot.

Jackulin, C., and Murugavalli, S. (2022). A comprehensive review on detection of plant disease using machine learning and deep learning approaches. Meas. Sens., 24: 100441.

Javaid, M., Haleem, A., Khan, I. H., and Suman, R. (2023). Understanding the potential applications of Artificial Intelligence in Agriculture Sector. Advanced Agrochem., 2(1): 15–30.

Jintasuttisak, T., Edirisinghe, E., and Elbattay, A. (2022). Deep neural network based date palm tree detection in drone imagery. Comput. Electron. Agric., 192: 106560.

Kamilaris, A., and Prenafeta-Boldú, F. X. (2018). Deep learning in agriculture: A survey. Computers and Electronics in Agriculture, 147: 70–90.

Kang, M., Sun, H., and Wang, X. (2023). Systems agriculture: modeling and control based on social and economic attributes of agriculture. Chinese Journal of Intelligent Science and Technology, 5(1): 41–50.

Kariyanna, B., and Sowjanya, M. (2024). Unravelling the use of artificial intelligence in management of insect pests. Smart Agric. Technol., 8: 100517.

Kariyanna, B., and Sowjanya, M. (2024). Unravelling the use of Artificial Intelligence in Management of Insect Pests. Smart Agricultural Technology, 100517.

Khan, S., Tufail, M., Khan, M. T., Khan, Z. A., Iqbal, J., Alam, M. A. et al. (2021). novel semi-supervised framework for UAV based crop/weed classification. PLoS ONE, 16: e0251008.

Khan, S., Ullah, S., Ullah, K., Almutairi, S., and Aftan, S. (2024). Implementing Autonomous Control in the Digital-Twins-Based Internet of Robotic Things for Remote Patient Monitoring. Sensors, 24(17): 5840.

Kumar, V., Sharma, K. V., Kedam, N., Patel, A., Kate, T. R., and Rathnayake, U. (2024). A comprehensive review on smart and sustainable agriculture using IoT technologies. Smart Agric. Technol., 8: 100487.

Kushwaha, M., Singh, S., Singh, V., and Dwivedi, S. (2024). Precision Farming: A Review of Methods, Technologies, and Future Prospects. Int. J. Environ. Agric. Biotechnol., 9: 242–253.

Li, C., and Wang, M. (2024). Pest and disease management in agricultural production with artificial intelligence: Innovative applications and development trends. Advances in Resources Research, 4(3): 381–401.

LI, Z., LI, B. X., LI, Z. H., Zhan, Y. F., Wang, L. H., And Gong, Q. (2023). Research progress in crop disease and pest identification based on deep learning. Hubei Agricultural Sciences, 62(11): 165.

Liao, J., Tao, W., Zang, Y., Zeng, H., Wang, P., and Luo, X. (2023). Research progress and prospect of key technologies in crop disease and insect pest monitoring [J]. Transactions of the Chinese Society for Agricultural Machinery, 54(11): 1–19.

Liu, W., Salzmann, M., and Fua, P. (2019). Context-Aware Crowd Counting. arXiv.

Livinali, E., Sperotto, R. A., Ferla, N. J., and Souza, C. F. V. D. (2014). Physicochemical and nutritional alterations induced by two-spotted spider mite infestation on strawberry plants. Electronic Journal of Biotechnology, 17(5): 193–198.

Lottes, P., Khanna, R., Pfeifer, J., Siegwart, R., and Stachniss, C. (2017, May). UAV-based crop and weed classification for smart farming. pp. 3024–3031. *In*: 2017 IEEE international conference on robotics and automation (ICRA). IEEE.

Mani, P. K., Mandal, A., Biswas, S., Sarkar, B., Mitran, T., and Meena, R. S. (2021). Remote sensing and geographic information system: a tool for precision farming. Geospatial Technologies for Crops And Soils, 49–111.

Marković, S., Milovanović, T. P., Jelušić, A., Iličić, R., Medić, O., Berić, T. et al. (2023). Biological control of major pathogenic bacteria of potato by *Bacillus amyloliquefaciens* strains SS-12.6 and SS-38.4. Biological Control, 182: 105238.

Metcalf, R. L., and Metcalf, R. A. (1992). Destructive and useful insects. Their habits and control.

Metspalu, L., Kruus, E., Ploomi, A., Williams, I. H., Hiiesaar, K., Jõgar, K. et al. (2014). Flea beetle (Chrysomelidae: Alticinae) species composition and abundance in different cruciferous oilseed crops and the potential for a trap crop system. Acta Agriculturae Scandinavica, Section B—Soil & Plant Science, 64(7): 572–582.

Moeuf, A., Pellerin, R., Lamouri, S., Tamayo-Giraldo, S., and Barbaray, R. (2018). The industrial management of SMEs in the era of Industry 4.0. International journal of production research, 56(3): 1118–1136.

Moura, P., Pinheiro, I., Terra, F., Pinho, T., and Santos, F. (2023, November). Machine Vision for smart trap bandwidth optimization and new threat identification. In Biology and Life Sciences Forum (Vol. 27, No. 1, p. 42). MDPI.

Mukherjee, K., Mukhopadhyay, S., Roy, S., and Biswas, A. (2021). Application of IoT-enabled 5G network in the agricultural sector. Smart Agriculture Automation Using Advanced Technologies: Data Analytics and Machine Learning, Cloud Architecture, Automation and IoT, 151–164.

Mulla, D. J. (2013). Twenty five years of remote sensing in precision agriculture: Key advances and remaining knowledge gaps. Biosystems Engineering, 114(4): 358–371.

Nahiyoon, S. A., Ren, Z., Wei, P., Li, X., Li, X., Xu, J. et al. (2024). Recent development trends in plant protection UAVs: A Journey from Conventional Practices to Cutting-Edge Technologies—A Comprehensive Review. Drones, 8(9): 457.

Naseer, R., Sneha, M.K., Udayalaxmi., Usha, N.G., and Vismaya, M.D. (2024). AI-Based Drone for Pesticide Recommendation and Spraying for Precision Agriculture, International Journal Of Engineering Research & Technology (IJERT) Volume 13, Issue 04 (April 2024).

Natividade, J., Prado, J., and Marques, L. (2017, April). Low-cost multi-spectral vegetation classification using an Unmanned Aerial Vehicle. In 2017 IEEE International Conference on Autonomous Robot Systems and Competitions (ICARSC) (pp. 336–342). IEEE.
Nicol, J. M., Turner, S. J., Coyne, D. L., Nijs, L. D., Hockland, S., and Maafi, Z. T. (2011). Current nematode threats to world agriculture. Genomics and molecular genetics of plant-nematode interactions, 21–43.
Niloofar, P., Francis, D. P., Lazarova-Molnar, S., Vulpe, A., Vochin, M. C., Suciu, G. et al. (2021). Data-driven decision support in livestock farming for improved animal health, welfare and greenhouse gas emissions: Overview and challenges. Computers and Electronics in Agriculture, 190: 106406.
Oerke, E. C. (2006). Crop losses to pests. The Journal of Agricultural Science, 144(1): 31–43.
O'Grady, M. J., Langton, D., and O'Hare, G. M. P. (2019). Edge computing: A tractable model for smart agriculture?. Artificial Intelligence in Agriculture, 3: 42–51.
Padhiary, M., Saha, D., Kumar, R., Sethi, L.N., and Kumar, A. 2024. Enhancing precision agriculture: A comprehensive review of machine learning and AI vision applications in all-terrain vehicle for farm automation. Smart Agric. Technol., 8: 100483.
Papadopoulos, G., Arduini, S., Uyar, H., Psiroukis, V., Kasimati, A., and Fountas, S. 2024. Economic and environmental benefits of digital agricultural technologies in crop production: A review. Smart Agric. Technol., 8: 100441.
Patel, S. (2011). Harmful and beneficial aspects of *Parthenium hysterophorus*: an update. 3 Biotech, 1(1): 1–9.
Patrício, D. I., and Rieder, R. (2018). Computer vision and artificial intelligence in precision agriculture for grain crops: A systematic review. Computers and Electronics in Agriculture, 153: 69–81.
Pedigo, L. P., and Rice, M. E. (2006). Entomology and pest management. Prentice Hall.
Popescu, D., Dinca, A., Ichim, L., and Angelescu, N. (2023). New trends in detection of harmful insects and pests in modern agriculture using artificial neural networks. A review. Frontiers in Plant Science, 14: 1268167.
Popp, J., Pető, K., and Nagy, J. (2013). Pesticide productivity and food security. A review. Agronomy for Sustainable Development, 33: 243–255.
Pretty, J., and Pervez Bharucha, Z. (2015). Integrated pest management for sustainable intensification of agriculture in Asia and Africa. Insects, 6(1): 152–182.
Puri, V., Nayyar, A., and Raja, L. (2017). Agriculture drones: A modern breakthrough in precision agriculture. Journal of Statistics and Management Systems, 20(4): 507–518.
Quamar, M. M., Al-Ramadan, B., Khan, K., Shafiullah, M., and El Ferik, S. (2023). Advancements and applications of drone-integrated geographic information system technology—A review. Remote Sensing, 15(20): 5039.
Raj, E. F. I., Appadurai, M., Thiyaharajan, D., and Pushparaj, T. L. (2024). State-Of-The-Art Technologies for Crop Health Monitoring in the Modern Precision Agriculture. pp. 21–39. *In*: Precision Agriculture for Sustainability. Apple Academic Press.
Rhouma, A., Hajji-Hedfi, L., Kouadri, M. E., Atallaoui, K., Matrood, A. A., and Khrieba, M. I. (2023). Botrytis cinerea: The cause of tomatoes gray mold. Egyptian Journal of Phytopathology, 51(2): 68–75.
Rolnick, D., Donti, P. L., Kaack, L. H., Kochanski, K., Lacoste, A., Sankaran, K. et al. (2022). Tackling climate change with machine learning. ACM Computing Surveys (CSUR), 55(2): 1–96.
Romanazzi, G., and Feliziani, E. (2014). *Botrytis cinerea* (gray mold). pp. 131–146. *In*: Postharvest decay. Academic Press.
Samal, I., Bhoi, T. K., Pradhan, A. K., and Mahanta, D. K. (2023). Plantix app: A success story of artificial intelligence in plant protection. Van Sangyan, 24.
Shafi, U., Mumtaz, R., García-Nieto, J., Hassan, S. A., Zaidi, S. A. R., and Iqbal, N. (2019). Precision agriculture techniques and practices: From considerations to applications. Sensors, 19(17): 3796.
Shamshiri, R., Weltzien, C., Hameed, I. A., J Yule, I., E Grift, T., Balasundram, S. K. et al. (2018). Research and development in agricultural robotics: A perspective of digital farming.
Sharma, R. (2021, May). Artificial intelligence in agriculture: a review. pp. 937–942. *In*: 2021 5th international conference on intelligent computing and control systems (ICICCS). IEEE.
Simberloff, D., and Stiling, P. (1996). How risky is biological control?. Ecology, 77(7): 1965–1974.
Sravani, P., Jahnavi, K. V. S., and Anshuman, J. Precision Agriculture and the Role of AI and Robotics.

Srikanth, J., Geetha, N., Singaravelu, B., Ramasubramanian, T., Mahesh, P., Saravanan, L. et al. (2018). First report of occurrence of fall armyworm *Spodoptera frugiperda* in sugarcane from Tamil Nadu, India.

Susskind, R. (2015). The future of the professions: How technology will transform the work of human experts. Oxford University Press.

Tabashnik, B. E., Brévault, T., and Carrière, Y. (2013). Insect resistance to Bt crops: lessons from the first billion acres. Nature biotechnology, 31(6): 510–521.

Tang, Y., Dananjayan, S., Hou, C., Guo, Q., Luo, S., and He, Y. (2021). A survey on the 5G network and its impact on agriculture: Challenges and opportunities. Computers and Electronics in Agriculture, 180: 105895.

Tendolkar, A., Choraria, A., Manohara Pai, M. M., Girisha, S., Dsouza, G., Adithya, K. S. (2021). Modified crop health monitoring and pesticide spraying system using NDVI and Semantic Segmentation: An AGROCOPTER based approach. pp. 1–5. *In*: Proceedings of the 2021 IEEE International Conference on Autonomous Systems (ICAS), Montreal, QC, Canada, 11–13 August.

Tilman, D., Cassman, K. G., Matson, P. A., Naylor, R., and Polasky, S. (2002). Agricultural sustainability and intensive production practices. Nature, 418(6898): 671–677.

Uzhinskiy, A. (2023). Advanced Technologies and Artificial Intelligence in Agriculture. AppliedMath, 3(4): 799–813.

Van Driesche, R., Hoddle, M., and Center, T. (2009). Control of pests and weeds by natural enemies: an introduction to biological control. John Wiley & Sons.

Van Emden, H.F., Eastop, V.F., Hughes, R.D., and Way, M.G. (1969). The ecology of *Myzus persicae* (Sulzer). Annual Review of Entomology, 14: 197–270.

Vasconcelos, G. J. Q., Costa, G. S. R., Spina, T. V., and Pedrini, H. (2023). Low-cost robot for agricultural image data acquisition. Agriculture, 13(2): 413.

Ward, S. M., Webster, T. M., and Steckel, L. E. (2013). Palmer amaranth (*Amaranthus palmeri*): a review. Weed Technology, 27(1): 12–27.

Xiao, L. I. U., Yue, W. A. N. G., Kaiyun, M. A. O., Yuelei, F. A. N., Cheng, T. A. O., and Daming, C. H. E. N. (2020). Converge development of biotechnology and information technology. Bulletin of Chinese Academy of Sciences (Chinese Version), 35(1): 34–42.

Yun, D., Qiyao, F., Zhaowen, N., and Yanping, K. (2024). Fine grained classification of crop diseases and pests based on progressive growing of GANs. Journal of Chinese Agricultural Mechanization, 45(3): 156.

Zeng, X., Sun, Y., Wang, Q. et al. (2024). Research trends on soil microorganisms in paddy soil based on Web of Science. Heilongjiang Agricultural Sciences, 5: 114–118.

Zha, J. (2020). Artificial intelligence in agriculture. In Journal of Physics: Conference Series., 1693(1): 012058.

Zhang, C., and Kovacs, J. M. (2012). The application of small unmanned aerial systems for precision agriculture: a review. Precision Agriculture, 13: 693–712.

Zhang, T., Yang, Z., Xu, Z., and Li, J. (2022). Wheat yellow rust severity detection by efficient DF-UNet and UAV multispectral imagery. IEEE Sensors Journal, 22(9): 9057–9068.

Zhang, X., Han, L., Sobeih, T., Lappin, L., Lee, M., Howard, A., and Kisdi, A. (2021). The self-supervised spectral-spatial attention-based transformer network for automated, accurate prediction of crop nitrogen status from UAV imagery. arXiv preprint arXiv:2111.06839.

Zhang, Z., Wang, X., Lai, Q., and Zhang, Z. (2018). Review of variable-rate sprayer applications based on real-time sensortechnologies. Automation in Agriculture–Securing Food Supplies for Future Generations. Intech, 53–79.

Zain, M., Ma, H., Chaudhary, S., Nuruzaman, M., Azeem, I., Mehmood, F. et al. (2023). Nanotechnology in precision agriculture: Advancing towards sustainable crop production. Plant Physiology and Biochemistry, 108244.

CHAPTER 15

Artificial Intelligence-Powered Plant Protection Systems

Ethical Considerations, Challenges and Emerging Trends

Hafiz Muhammad Usman Aslam,[1,*] *Saba Aslam,*[2] *Akhtar Hameed,*[3] *Muhammad Zeeshan Mansha,*[4] *Hasan Riaz*[3] and *Faizan Ali*[5]

1. Introduction

As agriculture rapidly embraces technological advancement, Artificial Intelligence (AI) is now transforming indeed, a tool of life in plant protection. AI can provide precision, efficiency, and scalability in managing threats that crop production continues to experience such as climate change, pests, and pathogens. These systems observe and respond to plant health threats in real-time, changing the concept of plant protection fundamentally. The ability to detect and respond to such changes through the use of predictive analytics, alarms, and self-actuating countermeasures places AI at the center of the next revolution in crop preservation.

The application of Artificial Intelligence in agriculture can be generally categorized across various fields such as machine learning, computer vision, and

[1] Department of Plant Pathology, San Luis Valley Research Center, Colorado State University, USA.
[2] Department of Plant Pathology, University of Agriculture, Faisalabad, Pakistan.
[3] Department of Plant Pathology, Institute of Plant Protection, MNS-University of Agriculture, Multan, Pakistan.
[4] Department of Plant Pathology, Faculty of Agricultural Sciences and Technology, University of Layyah, Pakistan.
[5] Department of Agricultural Science and Engineering College of Agriculture, Tennessee State University, Nashville, TN, USA.
* Corresponding author: usman.aslam@colostate.edu

DOI: 10.1201/9781003546702-15

robotics used in improving crop management as well as resource utilization. The principal functions of AI in the field of plant protection are the identification of symptoms, prediction of disease occurrence, and optimization of pesticide use. By processing big data sets allied to weather, soil health, and pest population, AI provides comprehensive solutions to tackle the most important challenges that rely on a higher amount of chemicals and increased resistance of crops (Hassan et al., 2022).

AI is especially important in plant protection due to the constantly rising world population and unfavorable conditions for agriculture. The performance of these systems could substantially decrease losses resulting from pests and diseases, which act as triggers for instabilities in the economies of various countries and global food insecurity. The advancement of AI accordingly averts potential threats, and optimizes both the yield and minimal negative impact on the environment so it is possible to achieve the goals of radical farming (Mahlein et al., 2024).

This chapter discusses the ethical considerations, challenges, and new trends, of AI-powered plant protection systems. AI has amazing potential, but it also raises ethical questions around data privacy, labor impacts, and the risk of excessive use of tech. In addition, to achieve equitable and accountable deployment data quality, model bias, and technology access must be addressed. This chapter will discuss these dimensions and articulate the ethical and practical landscape related to the use of AI in plant protection.

2. Role of AI in Modern Plant Protection

Advances in AI are increasingly transforming agriculture by providing farmers with enabling technologies that increase the precision of their activity, optimize the use of specific resources, and protect crops from insects and diseases. Plant protection requires the modernization of traditional farming practices using AI, switching from reactive, broad approaches to targeted, more proactive use cases (Balaska et al., 2023). This section uncovers the center stage that AI plays in precision agriculture, especially in combating pests and diseases, in building early detection systems, and finally its contribution to data-driven decision-making. Real-world case studies featuring the impact of AI-powered plant protection systems will enhance the discussion.

Precision Agriculture and AI Integration

Precision agriculture utilizes tools and techniques to apply water, fertilizers, and pest control chemicals at the exact right time and in the right amount. While the old way of applying inputs across an entire field is a blanket approach, precision agriculture targets a specific area for the application of an input at the right time and in the right amount to give maximum yields and optimum use of resources while checking the environmental degradation.

AI plays a pivotal role in precision agriculture by:

1. Streamlining Data Collection: Real-time data on soil health, crop situations, and climate are captured with the help of sensors, drones, and satellites backed up by

artificial intelligence. According to Shaikh et al. (2022), such valuable systems integrate, analyze, and provide information for farmers.

2. Optimizing Resource Use: AI can be used to monitor crops, deciding when and how much water, pesticides, and fertilizers to apply, yet effectively avoiding wastage, cost, and pollution.
3. Enhancing Yield Predictions: By using machine learning models over historical data, yield forecasts can be made to help farmers make more accurate planning on when to harvest and what resources to allocate (Hassan et al., 2022).

Thus, by applying AI to precision agriculture, farmers reach a level of accuracy and results providing plant protection through early detection and intervention while sparing resources.

Key Applications of AI in Plant Disease Detection and Pest Management

AI's ability to process large amounts of data and learn from patterns has revolutionized the detection and management of plant diseases and pests. Some of the key applications include:

- **Image Recognition for Disease Identification:** AI systems, using computer vision and machine learning, analyze images of crops to detect diseases. By comparing images of healthy and infected plants, AI can identify specific symptoms such as leaf discoloration, lesions, or mold growth. Farmers can capture these images using smartphones or drones, enabling rapid and remote disease detection (Li and Wang, 2024).
- **Pest Monitoring and Control:** AI-powered sensors can detect pests in the field, and AI algorithms can predict pest outbreaks by analyzing environmental conditions and historical pest activity. Some AI-driven systems also integrate with autonomous robots to apply pesticides or physically remove pests without damaging crops (Figure 1).
- **AI-Powered Drones and Robots:** Drones equipped with AI technology survey fields and detect signs of pest infestations or plant diseases. Autonomous robots can then be deployed to intervene with targeted spraying or mechanical removal of affected plants or pests (Adetunji et al., 2023).
- **Disease Risk Prediction Models:** AI models predict the likelihood of disease outbreaks by analyzing a variety of factors such as humidity, temperature, wind patterns, and soil moisture. This predictive power allows farmers to take preventive measures before diseases manifest, reducing crop loss.

AI in Early Detection Systems for Crop Health

In plant protection, one of the most useful features of using AI is that it can usually identify risks that are still small and have not yet developed into full-blown problems. Real-time detection systems that are driven by AI models can detect signs of illness

Figure 1. AI detection techniques that improve crop growing and monitoring include the assessment of soil health, determination of yields, and detection of pests or diseases. It makes it possible for the farmers to make decision which is most appropriate for the management of agriculture.

or stress at an early stage, or even before they appear, and then initiate remedial action. Key systems include:

1. **AI for Remote Sensing:** Crop spraying drones and satellites; ground-level sensors keep track of crop health by measuring deviations in plant metabolism that infer stress from pests or diseases. For example, infrared imaging can identify these changes before visible symptoms appear. Gupta et al. (2021) highlights that early detection allows for timely action to prevent disease spread.
2. **AI-Powered Sensor Networks:** Field sensors monitor external environmental conditions such as soil moisture levels, nutrient content of the soil, and growth status of the plants and AI comprises a consistent check of field indicators in search of such signs as nutrient deficiencies, fungal attack, and drought stress among the plants (Figure 1).
3. **AI-Driven Irrigation and Fertilization Systems:** In these systems, real-time crop data is fed into the system, and the water and fertilizers are applied automatically. As Ramdinthara et al. (2022) mentioned that efficient irrigation can be decreased in the areas affected by disease, and immunity in plants can be improved due to the optimization of the fertilization process with AI help.
4. **Genomic and Molecular Data for Early Detection:** In genomics, the markers of diseases in crops can be identified using artificial intelligence. Other applications of AI include gene analysis for the selection of disease-resistant forms of crops and the identification of unfavorable microorganisms and parasites before their detrimental impact is noticeable.

Data-Driven Decision-Making for Disease and Pest Control

The use of AI will no longer be based on speculation but on real and predicted data, which will influence the decisions regarding plant protection. There are several critical steps in the process, such as:

- **Data Collection:** AI systems collect data from a wide range of sources, including weather data, soil sensors, drone imagery, and satellite data. These inputs give a holistic picture of the farm's health and help us in precisely monitor crop conditions.
- **Predictive Analytics:** Using apparent and historical data AI algorithms predict the chances of a pest infestation or disease outbreak. With AI, farmers can begin to understand and see patterns, and even correlations, so that they can start to anticipate potential problems and take preventative measures.
- **Automated Decision Support:** AI systems can identify threats and provide suggestions or actions to reduce their impact. Artificial intelligence (AI)-powered recommendations for targeted pesticide applications or irrigation schedule adjustments according to soil moisture levels are examples of possible implementations.
- **Continual Learning:** As AI systems learn from fresh data and feedback, they improve their models and eventually become even more accurate in disease and pest identification. This improvement happens over time. The system can be upgraded to adapt to changing environmental circumstances and dangers through the establishment of continuous improvement.

Case Studies of AI-powered Plant Protection Systems

1. **The Banana Xanthomonas Wilt (BXW) Project:** AI systems were implemented in East Africa to counter the widespread Banana Xanthomonas Wilt, a deadly bacterial disease. AI algorithms, by analyzing satellite and drone photographs of the disease-affected areas, spotted early indicators of the disease and thus could target interventions. It did not only lower pesticide use but also lowered the disease's transmission.
2. **Purdue University's AI-Enhanced Pest Monitoring System:** Purdue University researchers created an AI system that tracks pest population numbers in agricultural fields. Predicts pest outbreaks with this system that employs AI-driven sensors to watch for insect activity and environmental conditions. Farmers are alerted in real-time and given recommendations for how to manage pests, cutting agricultural losses sharply (Gardezi et al., 2022).
3. **FarmWise's AI-Driven Weeding Robots:** A U.S.-based startup called FarmWise has designed autonomous robots that are powered by AI to differentiate a crop from weeds. Designed to weed crops precisely without the use of chemicals, these robots boost crop yield and also reduce environmental harm. According to Mathushika et al. (2022), the robots continuously adapt to changing conditions by assessing soil characteristics and adjusting their weeding strategies accordingly.

4. **Google's DeepMind in Plant Health:** In partnership with agricultural experts, Google's AI division, DeepMind, has created models to detect precursors to diseases in tomato crops. This technique uses picture recognition and machine learning algorithms to effectively reduce crop loss by diagnosing diseases and recommending targeted remedies.

AI is now a useful tool which aids with plant protection, evolving into how it's done for diseases and pests. In other words, with precision agriculture, early detection systems, and data-driven decision-making, AI supports better, more sustainable, and more scalable agriculture. These advancements lead to timely response, resource utilization efficiency, and a significant decrease in environmentally borne disturbances. Nonetheless, like every progressive technology, the integration of AI is composed of challenges and ethical dilemmas which are going to be discussed in the subsequent segment of this chapter.

3. Emerging AI Technologies in Plant Protection

The increase in AI within agriculture, especially plant protection, has led to phenomenal progress in terms of efficiency and accuracy. Emerging technologies in the field of AI are slowly transforming conventional approaches to farming by providing real-time tools to estimate, detect, and address plant health challenges (Mesías-Ruiz et al., 2023). The leading AI technologies transforming plant protection today include robot technology, big data technology, remote sensing technology, aerial robots, auto image identification technology, and data mining technology. These innovations are resulting in a new age of enhanced intelligent farming (Table 1).

Table 1. Emerging Technologies in AI-Powered Plant Protection.

Technology	Description	Applications
Machine Learning	Algorithms that utilize data to generate predictions.	Anticipating diseases and forecasting pests.
Deep Learning	Robust neural networks for complex data analysis.	Image recognition explicitly for identifying pests and diseases.
Remote Sensing	Utilization of satellite or drone photography for the assessment of crop health.	Thorough field monitoring and evaluation.
IoT (Internet of Things)	A network of sensors collecting real-time agricultural information.	Ecological monitoring as well as settings and intensity of agroactivities.
Robotics	Self-operating devices intended for performing tasks including weeding.	Pesticide applicator accuracy and efficient labor operations.

Machine Learning and Deep Learning for Predictive Analytics

Machine learning and deep learning (DL) are two of the most powerful artificial intelligence (AI) methods driving predictive analytics in agriculture. These techniques enable systematic analysis of large datasets, pattern recognition, and prediction of potential threats to plant health.

- **i.** The applications of ML algorithms involve using historical information including climate, soil type, and history of pests and diseases to develop models that predict the likely times that diseases and pests are most likely to appear. From these predictions, farmers can take course preventive action or, if necessary, improve pesticide application (Brown et al., 2022).
- **ii.** It is also noteworthy that DL models are more suitable than other models for accurate yield forecasts because they can process large amounts of data. Such models consider both the type of crop planted, climate information, and other practices to make sound estimates of harvest possible in order to help farmers improve the allocation of limited resources.
- **iii.** Real-time Data Processing: Both ML and DL systems perform well in handling data in real-time. It enables farmers to respond to threats when the threats appear unlike in other systems where farmers wait for damage to occur before initiating action, this is due to the fact that they receive data from field sensors and adjust forecasting and recommended actions correspondingly.
- **iv.** Anomaly Detection: Artificial intelligence mechanisms can also find abnormal conditions in crops with reference to the pattern of healthy and unhealthy crop conditions. This early detection of disease or pest attacks is very important in reducing crop yield losses through early actions.

Remote Sensing and Drones for Field Monitoring

Remote sensing using advanced AI and drones helps farmers supervise every aspect of crop growing from above (Mohd Noor et al., 2018). These systems provide fundamental information relating to the health of a plant, the status of the terrain, and threats.

1. **Drones powered by artificial intelligence** and fitted with multispectral and hyperspectral cameras monitor the health status of plants by featuring detailed photographs of fields. The conventional AI algorithms on these images are based on factors such as chlorophyll level, moisture in leaves, and temperature of the canopy to infer the plant's status. Remote sensing, therefore, means that farmers can control diseases and pest attacks before they manifest themselves on plants (Johansen et al., 2014).
2. **Field Mapping and Analysis:** They produce accurate field plans; by producing these plans, one is likely to recognize the disparity in the field. AI models of this data generate visual representations of productivity with high and low zones known as 'heat maps' to assist farmers in focusing their irrigation systems, fertilization parts, and pest control parts of the farm.
3. **Crop Growth Monitoring:** AI-powered drones can monitor crop growth over time, enabling farmers to track progress and detect any abnormalities. This continuous monitoring helps optimize crop cycle management, ultimately leading to improved yields.

4. **Pest and Disease Surveillance:** Drones can also be employed in identifying disease and pest spread in large fields. A fully automated AI system scans for signs of fully analyzing images of stress indicators such as scorched or wilted leaves. This makes it possible to prescribe the treatments based on the areas that are most affected, hence minimizing the rates of chemical use and thus adverse impacts on the physical environment (Maslekar et al., 2020).

AI-based Image Recognition for Disease and Pest Identification

Most of the pest and disease identification tools used in real-time have been enhanced with artificial intelligence-driven image recognition technology. They identify diseases or pest infections in the crops they map by analyzing crop images through computer vision and machine-learning algorithms (Tirkey et al., 2023).

1. **Immediate Disease Detection**: Instead of people physically measuring crop damage, they can use their smartphone, drone, or even a stationary camera. For instance, these pictures undergo recognition with AI-enabled image analysis tools to check for diseases like spots, lesions, or wilting. Real-time analysis allows acting promptly, which helps to avoid the spread of disease to other plants (Figure 2).
2. **Pest Detection:** Pattern recognition systems can recognize individual pests by evaluating images of a pest's appearance. This is especially desirable when the observer seeks to find small and relatively fast-moving insects that may be very hard to spot with the naked eye. Thus, once the potential pest is identified, the system offers specific control measures excluding the usage of broad-spectrum insecticides.

Figure 2. The use of sensors on drones improves agricultural surveillance since it can measure the soil moisture, humidity as well as nutrient content. It assists the farmers explain pest or disease risks, and the development of a map that helps one explain extra water or fertilizer needed. Different drones powered by artificial intelligence parse through this massive volume of data and then offer farmers advice regarding the best course of action.

3. **Scalable and Cost-Effective Monitoring:** In contrast to the field-based surveys that are prone to human error and can only be conducted for a limited number of animals at a time, using an image recognition system based on artificial intelligence is more feasible to cover a population of animals. Farming has been made very efficient and natural through the use of REMS so that farmers are not required to be on-site all the time to supervise expansive farmlands.
4. **Machine Learning for Enhanced Accuracy:** These algorithms of image recognition get refined in their accuracy with the use of a database of big images. The more images of healthy and diseased plants undergo interpretation, the efficiency of the system improves and even when there is misdiagnosis there cannot be a high possibility of false positive or false negative results.

Autonomous Robotic Systems for Weed and Pest Control

Autonomous robotic systems, equipped with AI, are transforming how farmers address weeds and pests, offering a more sustainable alternative to traditional chemical-based methods. These robots can navigate fields independently, identify unwanted plants or pests, and take action with precision.

1. **AI-powered Weeding Robots:** One of the key applications of autonomous robots in plant protection is weed management. AI-driven weeding robots use computer vision to distinguish between crops and weeds. These robots can physically remove weeds by cutting them or applying localized herbicides, drastically reducing the need for blanket herbicide spraying and minimizing chemical runoff (Anand and Kumar, 2024).
2. **Precision Pest Control:** Insect tracking systems are some of the things that can be found on some self-directed robots that are used for removing pests/diseases from crops. Using AI, the automation system for these robots will locate pests and diseases, identify the best intervention approach, and either target pesticides to the pests or physically remove the pests. This method restores ecological harmony in the field while cutting down on pesticide usage.
3. **Automated Monitoring and Intervention:** In addition to controlling weeds and pests, robots can also monitor crops all the time. The good news is that these systems can detect early signs of stress or disease and prevent the problem from escalating so much that treatment becomes necessary. Combining machine learning with robotics gives farmers a high level of automation and precision when it comes to plant protection tasks.
4. **Efficiency and Labor Savings:** AI-powered robots offer significant efficiency benefits by operating continuously, even in harsh weather conditions or at night. This automation reduces the need for human labor in repetitive tasks like weeding or pest monitoring, allowing farmers to focus on more complex aspects of farm management (Slaughter et al., 2008).

Use of Big Data, IoT, and AI for Comprehensive Ecosystem Monitoring

Big Data, IoT, and AI have already become the solid foundation for monitoring all aspects of agricultural ecosystems. The nature of this scheme allows continuous monitoring and creating a complete picture of crop ground on the side of farmers and helps them make the most appropriate decisions.

1. **IoT-Enabled Sensors for Continuous Monitoring:** IoT sensors placed throughout agricultural fields collect data on soil moisture, temperature, humidity, and plant health. These sensors communicate with AI systems, which analyze the data to identify patterns, predict future conditions, and recommend actions. For example, AI can determine the optimal time for irrigation by analyzing real-time moisture levels, preventing both over-watering and under-watering (Bibri et al., 2023).
2. **Big Data Analytics for Ecosystem Insights:** The vast amounts of data gathered from IoT devices, satellite imagery, and drones are processed by AI algorithms to generate a comprehensive understanding of the agricultural ecosystem. AI can analyze trends in soil health, weather patterns, and crop growth to develop strategies that improve overall agricultural performance, including pest control and soil management.
3. **Predictive Maintenance for Farm Equipment:** The integration of IoT and AI allows the monitoring of real-time health and status of agricultural machinery. Some of the applications of AI in farming include the diagnosis of tractor and harvester sensors as well as indicators of irrigation systems. This approach greatly prevents the occurrence of downtime and guarantees that any machinery will be running to the fullest capacity.
4. **Sustainability and Environmental Impact Monitoring:** By fusing AI with IoT and Big Data, farmers can closely trace the impact of their practices through the environmental impact. AI tracks farms, which monitor chemical runoff, carbon emissions, and water usage, providing insights into which more sustainable methods to adopt. It helps to make an agricultural ecosystem more eco-friendly without having an adverse environmental imprint when it comes to farming.

Plant protection was also able to embrace new approaches like machine learning, remote sensing, image recognition, and robots. It allows farmers to utilize data flow and expert system analysis to lower the cost, increase scale, and use complex future technology in plant health risk control. Combining AI with IoT and Big Data creates an ecological system of monitoring that can improve decision-making and minimize environmental Preinjury. Yet, these advancements are accompanied by potential problems and ethical issues, which will be discussed in the subsequent part of the chapter.

4. Challenges in Implementing AI-powered Plant Protection Systems

Self-driving machinery in plant protection incorporates attributes that can transform the agricultural field by presenting different techniques of pest and disease control that are reliable, flexible, and effective. However, some challenges prevent the widespread application and high effectiveness of such technologies. This chapter explores the key obstacles that have to be addressed to achieve operationality in different environments of the agriculture sector supported by AI solutions.

Data Collection and Quality Issues

Data is a crucial factor for the capability of AI systems and agricultural environments come with unique challenges regarding data quality and quantity.

1. **Inconsistent Data Collection:** Most AI models on plant protection rely on big data generated by sensors, drones, or satellite imagery. However, the units of agricultural production are located in different locations, and methods of data collection are not standardized. Variations in sensor placement, calibration, and environmental conditions during data gathering can lead to inconsistent or incomplete datasets, thereby limiting the AI system's ability to provide reliable and consistent predictions (Qazi et al., 2022).
2. **Limited Data for Specific Crops or Regions:** AI needs vast historical data for proper working, and it gets saturated within a few months. For new areas of farming or crops that are less tested in digital farming technologies, or even in some regions of the world, there is not enough data to train AI systems. This lack of data may result in inferior performance or biased AI solutions in a given field of work, which is not an ideal result.
3. **Data Quality and Noise:** Data quality is affected by noise, including erroneous signals from a sensor or the presence of unrelated information. Decision-making that uses AI is likewise prone to false positive or negative results in diseases or pests if it relies on low-quality data. The accuracy of data plays an important role in determining AI system performance, but the challenge lies in the ability to maintain such standards across different sources (Nagaty, 2023).
4. **Data Privacy and Ownership Issues:** One of the biggest issues with data acquisition is the matter of ownership of the collected data. Agricultural data may not be easily shared with other entities because of issues of privacy, patent rights, and abuse by other entities. Ensuring the proper ethical, secure, and transparent method of data management is essential to gaining the trust of farmers so they can integrate AI technologies into their work on a large scale.

Technical Challenges: Accuracy, Scalability, and Adaptability

Technical constraints that exist in current plant protection AI techniques and apparatuses are also major impediments to their expansion.

1. **Accuracy of Predictions:** Even though AI models can analyze various pieces of information to predict the conditions of the crop's health and the possible occurrence of pests, it is still difficult to reach a high degree of accuracy. Poor predictions may cause wrong diagnoses of diseases or pests and thereby the farmer may either intervene unnecessarily or fail to intervene when he/she should do so, and the output is loss-making (Murthy and Bobba, 2021).
2. **Scalability of Solutions:** A lot of AI technologies are effective in small-scale or limited settings like in test farms, or for small-scale research. However, when applying these systems to large commercial farms or various regions with different climatic conditions the challenges are enormous. To address the complexity of crop production, AI models applied to agriculture must be proficient for a broad range of ecosystems from plant species to the climate (Lwakatare et al., 2020).
3. **Adaptability to Changing Conditions:** Agricultural environments are unsteady, and by the time a researcher starts studying a given environment, the conditions such as weather, quality of soil, and density of pests change. On the other hand, AI systems have to learn new conditions and have to adapt to those changing conditions. Nevertheless, a large number of existing AI models lack flexibility and often need to be rewired or updated periodically. This process can be time-consuming, labor-intensive, and demanding in terms of data input, making it challenging and burdensome for farmers.
4. **Interoperability with Other Systems:** Irrigation is one of the essential fad management tools employed by farmers while others use sophisticated farm managing applications. The challenge comes when implementing AI-powered plant protection systems with these tools, where systems are often not developed in a way that complements others. One of the key findings is that AI systems need to be integrated with other systems as this is vital for mass deployment.

Integration with Traditional Farming Practices

Implementing AI technologies within the sphere of agriculture is both a cultural and practical issue, especially if we are talking about countries where agricultural practices have been inherited from generation to generation.

1. **Resistance to Technological Change:** In most farming regions of the world, conventional crop care practices are cultural practices, making farmers reluctant to transform to AI-driven methods. Farmers may have a lack of trust in new technologies of this type, considering them too complicated to be applied to farms or as unnecessary additions to traditional farming proceedings. This resistance can hinder the implementation of AI systems, especially if the public has little or no confidence in modern agricultural technology.
2. **Balancing AI with Human Expertise:** AI systems are very useful in analyzing big data, but they cannot replace the human point of view and experience. Farmers use their contextual knowledge about the crops and the farms that grow them, and it can be hard to translate this to an AI. The correct blend of using

AI in decision-making and conventional farming techniques that have been in use for centuries ought to be maintained to ensure that the farmers are able to make their decisions based on what AI is recommending and not solely on it (Holzinger et al., 2024).

3. **Compatibility with Low-tech Farming Practices:** In most regions, especially in the third world, farmers use basic or even draft techniques in farming or managing their production. Unfortunately, many AI systems cannot be regarded as easily applicable to work under these practices since they need such equipment as drones, sensors, or smartphones. Integrating AI into existing low-tech settings requires greater effort to ensure such advancement is available and feasible for a broad range of farmers.

Cost and Accessibility of AI Technologies for Smallholder Farmers

The cost and availability of AI technology remain a major problem since a high percentage of farmers are smallholders.

1. **High Initial Costs:** The use of such systems entails the purchase of capital equipment that is usually expensive and software costs amounting to many thousands of US dollars, costs that are usually beyond the reach of smallholder farmers. Acquiring drones, sensors or even AI-based management systems could be beyond the resources of many due to high costs, poor investors' support, or low access to credit facilities.
2. **Limited Access to Technological Infrastructure:** Farmers in rural areas and smallholders in developing countries often face challenges in adopting AI technologies due to unreliable internet connectivity and limited access to electricity. In contrast, large-scale farmers with sufficient capital and better infrastructure can afford and maintain these advanced systems. This disparity creates a technological divide, leaving smallholder farmers at a disadvantage and limiting their ability to benefit from AI-driven agricultural innovations.
3. **Need for Affordable and Scalable Solutions:** AI systems should also be as low a cost as possible to efficiently accommodate the constraints of small-holder farmers. This may range from creating AI systems with non-expensive hardware and lower-cost structures or making certain access to AI-based platforms cheaper by offering subscription services that clients can afford as opposed to paying one lump sum.
4. **Digital Literacy and Training:** For farmers to be able to optimally make use of the AI systems, they have to be quite computer literate. However, smallholder farmers may not possess the skills required for the optimum usage of devices that are based on artificial intelligence. However, for enhanced uptake and utilization of these technologies, capacity authorities within the farming fraternity should undertake training and availing mechanisms through which the technologies can be manned appropriately.

Dependency on Technology and Human Expertise

While artificial intelligence systems do provide useful capabilities for plant protection, there are also issues and limitations with how they use information and technology.

- **Overreliance on AI:** One of the challenges that arise as AI systems continue to be adopted within the domain of agriculture is that they become more or less relied on. Artificial intelligence (AI) recommendations may replace farmers' observations and even conventional wisdom. This could lead to its drawbacks, for example when algorithms do not take into consideration other peculiarities of the local environment or improper crop conduct.
- **Technological Failures:** Any technology, including AI systems, can be exposed to certain technical problems. System failures, including hardware and software, together with connectivity problems, may hamper the functionality of the tools at the disposal of the farmers leading to loss-making due to a lack of the right time to intervene or the wrong advice. Therefore, there is a need to develop contingency measures and safety nets so that farmers can exhaust other ways when technologies do not work.
- **Need for Human Expertise:** It is appreciated, however, that current endorsements of AI systems do not imply the total replacement of human judgment and expertise. Farmers and agronomists interact with the data generated by AI and make decisions based on that data while taking into account technology affordances as well as contextual factors. AI systems should not be thought of as substitutes for human knowledge to avoid the zero-sum model of AI systems' usage.

Despite the higher capabilities offered by AI-based plant protection systems, a few difficult situations need to be addressed before transforming AI methods extensively for diverse plants. Concerning data quality, technical problems, combined with conventional approaches, accessibility, and reliance on technology, the possible problems are critical. Addressing these issues will require the collective effort of technology implementers and farmers, policymakers and academics, and researchers to build AI systems that are robust, sustainable, and equitable so that progressive innovation in farming is made available to as many farmers as possible.

5. Ethical Considerations in AI-driven Plant Protection

Besides technical and operational problems, there are several ethically challenging themes related to AI as one of the elements of contemporary plant protection with its extensive use. (Ali et al., 2024). Concerns begin to emerge as AI uses become more deeply embedded in farming and could pose both benefits and drawbacks. This chapter seeks to identify the main ethical questions of plant protection technologies based on AI, which include unemployment or underemployment, privacy or lack thereof, opaqueness, ecological responsibility as well as revenue generation in plant protection technologies. Particular emphasis is placed on how these technologies alter the environment of agriculture and the social impacts thereof (Table 2).

Table 2. Ethical Considerations in AI-Driven Plant Protection.

Ethical Consideration	Description	Implications
Employment Impact	Automation may diminish employment prospects for agricultural laborers.	Requirement for retraining initiatives and the establishment of new employment opportunities.
Data Privacy	Concerns around the ownership and utilization of data gathered from agricultural operations.	Establish explicit protocols around data utilization and proprietorship.
Algorithmic Bias	AI algorithms may unintentionally favor specific crops or practices, resulting in disparities.	Regular updates and checks are required for AI systems.
Environmental Impact	The implementation of AI must be harmonized with ecological issues to prevent adverse consequences.	Highlight sustainable methodologies in the deployment of artificial intelligence.
Profit vs. Responsibility	Corporations must reconcile business objectives with social and environmental obligations.	Formulate ethical standards for AI engineers.

Impact on Employment and Farmer Autonomy

1. **Loss of Employment and Shifts in Labor Demand:** Probably the most concerning when it comes to ethical issues is the impact when artificial intelligence-powered systems take human jobs, primarily physical, low-skill jobs like pest control, field monitoring, crop planning, and many others. It threatens job displacement in regions where agriculture comprises a large part of the population, including self-driving drones and robots and AI-based decision support systems. In the opinion of Liu et al. (2023), this shift can lower demand for low-skilled employees and thus cause increased rates of unemployment as well as inequality, further eradicating it in rural regions.
2. **Loss of Farmer Autonomy:** AI when used to support decision-making in agriculture is beneficial but, on the downside, it has demerits in that it can bring out the inequality of farmers. AI-based technologies automatically offer recommendations on pest control, water, and pesticide uses respectively. Eventually, these hints make farmers fully dependent on them, and farmers will doubt their abilities or knowledge. This may lead to farmers using AI guidance without further reflection and decision making thus reducing the level of decision-making autonomy of the farming businesses (Altayeb et al., 2024).
3. **The Digital Divide and Skills Gap:** As the use of AI escalates in the sector, this article provides a reality check on how the trend runs the danger of further widening the digital divide. Farmers incorporating the latest AI applications as a tool can get an advantage over competitors, with inadequate funds or an inability to use such systems. To fill this void, it has been important to ensure the smallholder and low-technology farmers acquire the knowledge required to fully implement the AI systems hence closing the gap between technological advancement and those who do not benefit from it.

Privacy and Data Ownership Issues

1. **Farmers' Data as Intellectual Property:** AI systems are said to need large data sets to be sort of run as is expected by many businesses. This data in agriculture may contain important information about the status of the soil and crops, climate conditions, and pest status from the farmer's field. There is one predominant ethical concern, which embodies the question of ownership of this data. Over 90% of AI applications are established by private sector entities that collect and process agricultural data to provide recommendations or information. However, where there are no well-defined ownership rights of the data and this may become a problem with the farmers losing ownership authority over their data, raising issues on privacy and property.
2. **Risk of Data Exploitation:** Agricultural data is a resource that is owned and can be sold to various companies, mostly to the disadvantage of the farmer. The big farming companies and technology firms that gather this data may develop their models for their benefit or sell the information to other third parties without paying farmers. This dynamic gives rise to kinds of dependency, such that farmers depend on some external corporations for access to information that could be derived from the data they input into the system (Ali et al., 2024).
3. **Data Security and Misuse:** Apart from the most common and potential problems regarding ownership issues, many risks directly connected with system security can be highlighted. Weather and soil information, farming practices, and yields, market secrets, and current trends in the market can potentially be accessed by a third party. It can cause various data breaches with the potential to endanger farmers with financial shocks or place them at a competitive disadvantage. The use of agricultural information in the alteration of prices or the unfair exploitation of the markets has relevant ethical questions for which proper data protection should be implemented.

Algorithmic Transparency and Bias in AI Models

1. **Opacity in AI Decision-Making:** One of the major ethical challenges in AI is the lack of transparency in how AI systems make decisions, often referred to as the "black box" problem. Most AI models, especially those that are under the group of deep learning do not have processes that make it easy to understand the output even for the programmers. In agriculture, this opacity leads to a lack of confidence from farmers who, for instance, may not grasp how the recommendation on fertilization they receive as a suggestion from an AI, or why this specific decision is being proposed. There is a risk that farmers themselves will not be able to critically think about AI-based interventions and object to them, due to lack of information about them (Daneshjou et al., 2021).
2. **Bias in AI Models:** This means that AI systems are only as good as the data that has been fed into them and get distorted or misrepresented by data containing bias or only part of the information. In plant protection, AI models may be designed to learn data from particular regions and particular kinds of crops, or

particular farming techniques, in which case there can be a scenario designed to believe some kinds of farms but not others. For instance, AI systems that are developed based on large commercial farming data may not function well in smallholder or organic farming settings. Elimination of bias in AI models is highly important to guarantee that the effectiveness of the technology in as many ways as possible can be utilized in agriculture (Mensah, 2023).

3. **Fairness and Accountability:** Ensuring fairness in AI systems involves making sure that the benefits of AI are distributed equitably across different types of farmers, regions, and agricultural practices. Companies and researchers developing AI-powered plant protection systems must be held accountable for the performance and fairness of their models, ensuring that they do not disproportionately benefit large-scale, high-tech operations at the expense of smaller, less-resourced farms.

Ecological Impact and Sustainability Concerns

1. **Overuse of Pesticides and Chemicals:** Machine learning algorithms, in particular aids in focusing the pesticide on the areas that require it to eliminate pests while sparing the rest of the crop. However, there is a high risk that the AI systems will consistently recommend chemicals be used in agriculture. If the AI models are completely focused on growing the rate of return today, then they may promote chemical sales rather than lending for homegrown farming. This could lead to some or all of the following effects: soil erosion, water pollution and negative effects on useful insects and animals (Mutsuddi, 2024).
2. **Ecosystem Disruption:** The urgent utilization of such AI technologies as drones and robots for pest and weed control can disrupt ecosystems at the local level. For instance, pest-controlled robots can be negative to many other forms of life in the environment apart from the pests that they are aimed at destroying. Reducing the impact of artificial intelligence on ecosystems and establishing sustainable plant protection measures is one of the ethical challenges (Galaz et al., 2021).
3. **Balancing Long-Term Sustainability with Short-Term Gains:** One of the greatest ethical issues that have emerged in the agricultural sector relates to short-term results and profit. Quite a few AI systems are built to optimize effectiveness in executions. This is a prescription for reinventing practices that derogate the environment in the long run. Those concerned with Artificial Intelligence plant protection technologies should be in a position to provide plant protection solutions that do not compound the situation but rather make it better.

Balancing Profit with Environmental and Social Responsibility

1. **Profit-driven AI Solutions:** Most advanced applied artificial intelligence technologies in agriculture are created by private firms that operate for profit. Such a lure for the shareholders' profits might contribute to designing technologies that remain beyond the reach of small farmers or technologies that are more

beneficial to their investors financially in the short term but devastating to the environment and communities in the long term. Ethical approaches towards the development of AI need to consider not only the question of profitability of the developed technologies but also other potential consequences for society and the environment.

2. **Ethical Business Models:** This means there is a need to develop ethical business models that will enhance the use of AI technologies in a way that benefits farmers, consumers as well as the environment. Such models could be tools in the form of open-source AI for smallholder farmers, cost models within which farmers in low-income categories can afford the services that AVs offer. Further, there is a need to encourage the creation of AI models that popularize sustainable approaches in farming rather than those aimed at generating high revenues (Mutsuddi, 2024).
3. **Corporate Responsibility and Transparency:** Companies developing AI technologies for plant protection must ensure these tools are used ethically and responsibly. In cases of misuse or rule violations, they should take accountability and prevent further exploitation. Transparency is essential, firms should clearly disclose how their AI models are developed, the types of data utilized, and the basis for their recommendations. Moreover, corporations must ensure that they do not exploit farmers' data solely for profit but instead share the benefits by providing value and support back to the farming community..

The ethics involved when using AI in plant protection systems are quite numerous and present significant issues. The steps that current AI technologies undertake need to address employment, privacy, transparency, ecological status, and the multiple costs and benefits within determination and multiple social responsibilities in assisting farmers. In this way, the proposed ethical challenges will help the agricultural industry to navigate the creation of improved and not only technologically valuable but ethically sustainable AI systems for the future.

6. Regulatory and Policy Challenges

The application of AI technology to plant protection systems is a good chance to increase production in agriculture, stability, and security of food in society. Even when there has been such fast growth in the sector, there have not been so many developed rules and structures that could serve as the opposite regulation of the field. AI tools require proper regulation in a way that ensures the application of the software is safe and efficient. This chapter discusses the legal and regulatory barriers to the use of AI for plant protection and control: as legal situation now, harmonization, issues regarding IP, and the overall policy environment.

Current Regulations on AI in Agriculture

1. **Fragmented Regulatory Landscape:** At this time, the regulation of AI in agriculture is in its infancy, with different degrees of surveillance depending on the area. Different countries have established governing bodies that direct

the use of AI, but these are often focused on a wide range of AI applications, without specific consideration to plant protection systems (Dara et al., 2022). In agriculture, the current regulations are more generally focused on the safety of pesticides, chemicals, and genetically modified crops, rather than the technologies or algorithms used to manage plant pests or diseases.

In the European Union, tight guidelines govern pesticide usage under the Common Agricultural Policy (CAP) and several directives about environmental sustainability; nevertheless, there is minimal particular regulation of artificial intelligence in plant protection. In contrast, countries like the United States have regulatory bodies like the USDA and FDA, but AI regulation remains a challenge as AI technologies are treated more like software tools rather than agricultural inputs. This fragmented landscape creates inconsistencies and gaps in oversight.

2. **Lack of AI-specific Guidelines:** Presently, most of the software, inclusive of the AI tools utilized in plant protection, is governed by general software regulations that lack a specialized approach for the agriculture industry. Agriculture involves a fluctuating environment, and most decision-making processes must occur simultaneously; therefore, AI regulation should not be as rigid as that implemented in industries or hospitals. For instance, AI for pest outbreak detection or early disease surveillance needs continuous monitoring and operational response, which again, most existing regulations do not consider or fail to address the risk of AI system failure when deployed in such contexts.
3. **Risk of Delayed Oversight:** The major drawback of AI advancement in the agriculture sector is that someday regulatory authorities may be outcompeted in terms of the number of innovations to approve. Outdated or delayed regulations could result in the use of unproven or dangerous AI tools that can damage crops, the environment, and even the farmer. This lack of connection also allows room for the market to raise AI marketing products without verifying or testing them along state-of-the-art methods.

Need for Standardization and Certification of AI Tools

1. **Importance of Standardization in AI Development:** The absence of a clear regulatory framework gives rise to standardization. The term AI plant protection system refers to the creation of common technical specifications for the development, launching, and assessing of AI-based plant protection systems. Standardization proclaims to the world that these AI technologies meet certain accuracy, safety, reliability, and ethical use standards. Detecting diseases and pests with the aid of AI is particularly important since the wrong classification means we can overuse pesticides or fail to take the right action to protect the environment, and it affects yields and environmental health. The big lack in industry standards at the moment is in how AI algorithms are designed, trained, and tested so farmers don't know which tools are reliable and usable.
2. **Certification of AI Tools:** Certification is essential for establishing trust in AI technologies. The certification procedure will entail third-party assessments

verifying that AI-powered tools adhere to regulatory and performance guidelines. AI models used in plant protection may be accredited according to their proficiency in reliably identifying particular diseases, pests, or other indications of crop health. Such certificates would instill confidence in farmers and agribusinesses regarding the safety, efficacy, and dependability of the AI tools they employ. Establishing certification methods for agricultural AI tools is challenging owing to the diverse environmental circumstances, crop species, and dynamic characteristics of pests and diseases. Certification systems must exhibit sufficient flexibility to accommodate this variability while upholding stringent testing and validation standards.

3. **Data Quality Standards:** The quality of data inputs for the AI algorithms has a verbatim effect on performance system execution. Erroneous or biased information affects the development of proper recommendations and successful solutions. Hence, they have high requirements for the quality of AI system data input since the data has to be correct and credible. As this data is used in training AI models for disease prediction, pests management, and environmental monitoring, these standards should guide how data is collected, labeled, and used in training AI models.

Intellectual Property and Licensing Challenges

1. **Ownership of AI-generated Models:** One of the primary intellectual property challenges in AI-driven plant protection systems involves the ownership of AI-generated models and insights. When AI algorithms are trained on data collected from farms, questions arise about who owns the resulting models, and the knowledge derived from them. Should ownership belong to the technology developers, or should farmers who provide the data retain some form of control or rights over the AI models that benefit from their input?

   These IP issues are further complicated by the fact that AI systems often evolve, continuously learning from new data inputs. Determining ownership of these adaptive models requires careful consideration, as farmers might feel exploited if their data is used to create proprietary models that they cannot access or afford to use.

2. **Licensing and Access to AI Tools:** Adapting the AI plant protection systems for licensing is another challenge. Almost all AI tools are created by private entities that market them as commercial products. Smallholder farmers may not be able to afford technologies such as GPS or NT passports because of high licensing costs or restrictive usage contracts. This could potentially result in a phenomenon of accumulated AI technologies in the hands of the giant, affluent agricultural firms, enhancing the already existing divide between resource base and productivity of giant and smallholder operations.

   Possibilities to make certain AI models open-source, or provide for more permissive licensing agreements, are the ways to prevent primary plant protection technologies based on AI from becoming a realm of only well-funded farmers. A possible solution to this problem is to make those data and tools as openly available as possible or to offer them at affordable subscription prices to allow as many users as possible to access AI solutions.

3. Patentability of AI Algorithms: Another IP issue concerns the patentability of AI algorithms used in plant protection. Although software patents are widespread, AI algorithms provide distinct problems due to their adaptable and self-learning features. Securing a patent for an AI program that always adapts to fresh data inputs can present legal challenges. Major ethical issues arise due to the patents that have been granted on algorithms produced in a way that uses publicly accessible data in agriculture or data provided by crowd-sourcing. The public authorities must therefore navigate between the protection of innovations on the one hand and the maintenance of the availability of the tools which are useful for the farmers and necessary for agriculture on the other hand.

Policy Frameworks to Support AI in Plant Protection

1. **Government Support and Incentives for AI Adoption:** Governments need to develop legal bases that suggest regulatory oversight and financial backing to ensure the wider use of AI-based plant protection systems. This may include grants or tax credits for the farmers implementing the AI, especially the smallholder farmers who would regard them as financially unfathomable. Moreover, the options of public funding for AI research and development which are conducted within agriculture may spur innovation and uncover novel ways through which precise solutions can be implemented in many places and agricultural systems (Mahlein et al., 2024).

   Policymakers can undertake programs that promote technology transfer, which ensures that the advancements made in the field of AI research are translated into tangible solutions that can be availed by farmers. To push the envelope of AI-based plant protection solutions' development and deployment, governments may involve research institutions, universities, and private entities in their cooperation.

2. **Public-Private Partnerships (PPPs):** Public-private partnership is a necessity for advancing AI in agriculture. The collaborations help bridge the gap between the companies that market AI technologies and the research organizations that make them. These can be backed by governments, such as funding for new institutions, for innovation hubs, as well as using the public sector, academic, and industrial domains.

   AI technologies can be more effectively scaled through public-private partnerships that make them available to farmers all over the world, and adaptable to local conditions. These partnerships also ensure that smallholder farmers can afford and access AI technologies so that AI benefits don't accrue to agribusinesses.

3. **Sustainability-Oriented Policies:** Policymakers must also assess the long-term viability of AI-driven agricultural protection systems. This involves formulating regulations that foster the utilization of AI by environmental sustainability objectives. AI technologies that enhance pesticide application or improve water

management can assist more sustainable agricultural practices. Policymakers can facilitate the advancement and implementation of AI systems that demonstrate verifiable good environmental effects while dissuading the utilization of technologies that highlight immediate gains at the expense of ecological well-being.

4. International Collaboration and Standardization of Regulations: For AI and plant protection, since the agricultural sector is worldwide in nature, international cooperation is required to control such technologies. Agricultural trade can be difficult around the world because of inconsistent regulations from different countries. Different countries can approve technology for use and ban it in another, causing trade barriers, legal disputes all the way (Eruaga, 2024).

The United Nations Food and Agriculture Organization (FAO), and the World Trade Organization (Batson et al.) can not only help harmonize regulations but can also set up global standards for AI in agriculture. Unifying the regulation will assure global benefit from AI in agriculture while at the same time assuring food safety and sustainability standards.

The legislative and policy landscape for AI-driven plant protection systems is still in flux and still has many challenges. Since there is no turning back from the importance of AI in modern agriculture, governments, industry pinch, and industry stakeholders as well as researchers have to come together to develop a comprehensive regulatory framework that ensures safety, equity, and sustainability. To advance innovation and allow for the responsible use of AI in agriculture, standardization, certification, intellectual property considerations, and supportive policy frameworks will be important.

7. Emerging Trends in AI-powered Plant Protection

The digital revolution does not spare the agricultural industry and AI plant protection as one of the prospects for the development of this field is actively emerging. However, this movement also highlights what AI can possibly to do for the crop protection sector while stressing the importance of and for sustainable agriculture. The primary focus of this chapter is to explore the revolutionary AI-based plant protection systems, such as developing disease-resistant crops, integrating AI into the supply chain for better traceability, AI managing climate-smart agriculture, and stakeholders, and possible role of AI in further sustainability of the agriculture system.

AI-Driven Plant Breeding for Disease Resistance

1. **The Role of AI in Advancing Plant Breeding:** Developing disease-resistant crops with traditional plant breeding methods is typically a time-intensive and laborious process. Rapid genetic data analysis, determining useful traits like disease resistance, and predicting ideal crossbreeding combinations is possible today thanks to AI (Farooq et al., 2024). Using machine learning algorithms, we

can process vast databases of crop genomes, disease patterns, and environmental variables to rapidly select disease-resistant plant varieties.

The genetic markers that give resistance to specific plant diseases can be detected using AI algorithms. AI improves breeding methods by incorporating data from past breeding trials, DNA sequences, and specific environmental conditions to turn out crops that are more resistant to diseases, pests, and climate change. A key capability in an era where increasing numbers of novel plant diseases are arising as a result of climate change and global trade.

2. **Genomic Selection and Predictive Breeding:** Genomic selection enabled by AI uses ML algorithms to predict a plant's phenotype (observable traits), given its genotype (genetic makeup). Using this predictive approach, breeders can focus on specific genetic modifications that provide disease resistance, rather than needing to spend time in field trials and making the breeding process more efficient. Combining AI with genomic selection helps bring about crops with greater disease resistance faster, cultivating a more productive footprint overall and needing fewer chemical pesticides (Fang, 2024).

Integration of AI with Blockchain for Supply Chain Transparency

1. **Enhancing Traceability and Food Safety:** The integration of AI with blockchain technology is appearing as a powerful tool for advancing transparency and traceability in the agricultural supply chain. Blockchain is a decentralized ledger system that works to securely record and track transactions (Patil, 2023). By working with AI-enabled plant protection systems, blockchain can provide up-to-date insights into crop health, disease management, and the use of pesticides throughout the supply chain.

   With AI, we can assess crops, to detect signs of diseases or pest infestation at an early stage, and with blockchain, ensure that every bit of this data is safely recorded and uninterruptedly traced, from farm to market. This approach makes sure that only crops without harmful substances or diseases can be supplied to the consumers thus ensuring food safety. It addresses apprehensions about the ethical and sustainable use of AI in agriculture which, in turn, enables it to provide consumers and other stakeholders with verifiable information about the method of farming.

2. **Supply Chain Efficiency and Farmer Accountability:** Agricultural input and output monitoring can be automated by linking AI-driven plant protection systems with blockchain technology to lengthen the supply chain. Blockchain can record data in a secure system that complies with sustainability and AI can monitor pesticide use, crop health, and water consumption in real time. Such accountability comes from this incorporation, which facilitates transparency in the use of chemical inputs (pesticides, fertilizers etc.) (Mana et al., 2024).

Additionally, blockchain streamlines product certification related to organic, non-GM, and disease-free labeling thus enabling farmers to receive higher market pricing while sating the consumer demand for transparency with respect to their product. AI teamed up with the blockchain will aid in increasing customer trust and sustainable practices.

AI in Climate-Resilient Agriculture

1. **Adapting to Climate Change:** Extreme weather, shifting growing seasons, and new plant diseases and pests are just some of the many challenges created by climate change, which is today making agriculture more vulnerable than ever. The role of AI-driven plant protection systems is becoming a necessary tool to aid farmers in coping with these changes. Through studying climate data, soil health, and crop performance, AI is capable of creating predictive models to enable farmers to make well-informed decisions regarding climate-resilient agriculture.

   Some of the examples that AI can predict include farmers adjusting planting schedules or using preventive measures such as drought, heat waves, or periods of heavy rainfall. AI systems also capture data on crop conditions to help farmers detect problems such as temperature elevation and water shortages and offer ways to minimize these consequences. Climate change conditions will have to be addressed early enough to ensure that crop production is not affected to mean that there will be low crop yields and food security in future (Khan et al., 2022).
2. **Breeding Climate-Resilient Crops:** Plant breeding is one application of AI that generates varieties resistant to diseases and adaptable even to a more extreme climate. AI examines data on plant traits, environmental conditions, and genetic deviations to define attributes that lead to greater crop resilience to drought, heat, and other climate stressors. By taking a data-driven approach to crop improvement, farmers can grow hardier crops, rely less on chemicals and more on soils to protect yields from climate change impacts (Harfouche et al., 2019).

Collaborations Between Tech Companies, Researchers, and Farmers

1. **Bridging the Gap Between Research and Practical Application:** One of the dominant tendencies of AI's further application in plant protection is the expansion of cooperation between tech companies, researchers, and farmers. Such collaborations are important in farming as it facilitates the uptake of higher-level artificial intelligence research into usable technology. For example, technology companies are collaborating with agricultural organizations to build AI models for disease diagnosis while researchers engage farmers to use AI-based models that have been developed in real-life environments.

   These collaborations make sure that AI-based solutions are developed in a way to cater to the requirements of a farmer as it depends on local environmental conditions, crop types, and farming practices. Training and support are also offered to farmers by researchers and technology companies in order for AI solutions to be easily integrated into their operations.
2. **Building Trust and Supporting Adoption:** A critical factor to the success of AI-driven plant protection systems is stakeholder engagement, namely enabling appropriate support to foster trust and encourage take-up. Sometimes,

new technologies have to cost a lot and may need you to be really techy, and farmers are very slow to embrace new technologies. For this reason, technology providers and researchers must work together to eliminate farmers' fears and illustrate the value of AI in crop protection. It also directly addresses the need to ensure that even the best skilled farmers understand and implement these new sophisticated solutions without feeling overwhelmed by their complexity.

Future of AI and Plant Protection: Towards Sustainable Agriculture

1. **Sustainability as a Core Focus:** The future of AI in plant protection is closely tied to the broader goal of achieving sustainable agriculture. AI-driven plant protection systems have the potential to reduce the environmental impact of farming by minimizing the use of chemical pesticides, optimizing water and nutrient use, and enhancing crop resilience to diseases and pests. By integrating AI with sustainability principles, the agricultural industry can move towards a future where food production is not only more efficient but also more environmentally friendly (Vasileiou et al., 2023).
2. **AI as a Catalyst for Precision Agriculture:** Precision agriculture will be the future of plant protection, and AI will be the main playmaker in that process. Livestock farming employs data derived from drones, satellites, sensors, and AI models, among others to help farmers to assess the health of crops, quantity and quality of soils, and weather conditions (Figure 3). By integrating these two

Figure 3. AI increases the efficiency of precision farming since data from sensors and satellite imaging is averted to improve crop yield, predict yields, and minimize resource use. It assists farmers in overall decision-making, enhances production and sustainable practices to raise the yield level of agriculture with reducing the environmental costs.

pieces of data, AI can tell farmers the best time to plant, irrigate, use fertilizers, or control pests. It may prevent the use of excessive resource input which is costly and may also reduce injury to the environment, thus making the agriculture sector more sustainable (Shankar et al., 2020).

3. **Addressing Global Food Security Challenges:** The power of AI in plant protection systems is expected to help address some of today's most urgent challenges in food security. With an increasing global population and increasing demand for agricultural productivity from limited resources, we are increasingly turning towards crop and livestock genetic engineering as a way to cope with pressures alone. Through AI-driven applications, farmers can increase their production efficiency, prevent crop losses due to pest and disease infestation, and raise overall production-critical factors for feeding a growing population. AI in plant protection provides viable solutions that ensure a correlation between agricultural production growth and residents' population growth with minimum negative effects on the environment based on leveraging advanced technologies. The integration of innovation in this manner means that future food demands would be met in a balanced manner (Gupta et al., 2024).
4. **Balancing Technology with Ecological Stewardship:** Despite the advantages of using AI in plant protection, its future should be environmentally sustainable. This means that AI solutions for agriculture should not only serve the purpose of high yields or even high incomes, etc. but also the sustainability of ecosystems. The principles underlying the design of AI tools have to reflect such regenerative farming techniques that would not require synthetic inputs and enhance biodiversity. This balance will be important in order to guarantee that AI-powered plant protection systems will help to promote sustainable, healthy, and ethical agriculture (Kumari et al., 2024).

Emerging technologies and collaborative workloads have the potential to fuel the future of AI-powered plant protection. The next generation of farming practices will be informed by AI's use of disease-resistant plant breeding, blockchain integration for transparency, climate resilience, and sustainable agriculture. As such trends continue to change, they not only help protect crops but also make for a more sustainable, resilient, and transparent agricultural system. The integration of AI in plant protection is poised to revolutionize the way we produce food, ensuring that agriculture remains productive and sustainable in the face of global challenges.

8. Recommendations for Ethical AI Implementation in Agriculture

With the rising concern in plant protection and agriculture, the adoption of AI should not go astray without taking into consideration some of the ethical concerns. The following considerations focus on the need to enhance inclusiveness, improve access to resources, development of a transparent AI framework, farmers' awareness raising, and policies to facilitate the usage of AI in the agricultural field in appropriate and sustainable manners. These guidelines MUST be applied and followed while using

Table 3. Recommendations for Ethical AI Implementation.

Recommendation	Description	Expected Outcomes
Promoting Inclusivity	Involve many types of stakeholders when developing artificial intelligence.	AI tools rank lower in power dynamics and are more oriented towards the user.
Ensuring Fair Access	Support AI as a service to the smallholder farmers.	Increased adoption of AI technologies by every farmer.
Building Transparency	AI systems must disclose what types of information they use to draw their conclusions and where they get their data.	Improved user confidence in dealing with AI technologies.
Education and Training	Design proficiency development seminars to give agriculturalists knowledge of artificial intelligence systems.	The efficacy of AI technologies relies on their optimization and the successful realization of advantages from their deployment.
Sustainable Use Policies	Regulations should be developed which will urge farmers to use artificial intelligence in a proper way.	Favor human beings and other living organisms within the community.

AI technology to effectively make success and impact cut across smallholder farmers to large corporations in agriculture without causing ethical and social concerns (Table 3).

Promoting Inclusivity in AI Tool Development

1. **Incorporating Diverse Perspectives in AI Design:** To ensure the accessibility and fairness of AI tools in agriculture, their development must involve a diverse range of stakeholders, including farmers from various geographical, social, and economic backgrounds, agronomists, policymakers, and IT developers. By integrating feedback from these groups, AI solutions can become more inclusive, adaptable, and beneficial, particularly for agricultural communities in developing regions or those on the margins.

   AI tools must address the varied needs and resource constraints of their users. For developing countries where small-scale farmers are so often a significant proportion of the population, advanced mechanized equipment may well prove prohibitively expensive. There are considerable benefits to AI-driven solutions tuned to a particular condition. These innovations boost the practical applicability of these tools if they make them more representative of AI tools. Such attempts are representative of the goals of modern agricultural technologies: to cultivate farming without relegating workers to the sideline if they cannot navigate complicated machinery.

2. **Culturally Sensitive AI Implementation:** It is important to apply culture and tradition to solve ethical AI problems in agriculture. The majority of farming communities depend on the era's agricultural knowledge and techniques of ancestry or culture used in agriculture. Therefore, the AI systems that have respect and incorporation of these methods are likely to be accepted and function well.

It mainly focuses on developing AI tools that can work as an addition to traditional practices in order to combine progressive advancements with experience. In this way, AI can contribute to the process of creating a more effective and effective sustainable agriculture as well as protecting cultural identity.

Ensuring Fair Access to AI Technologies

1. **Addressing the Digital Divide:** One of the major ethical considerations in the launch of AI-based plant protection systems is that its outcome may deepen the digital divide. The availability of AI technologies is commonly centralized in more developed areas or among affluent growers, hence marginalizing smallholders and resource-constrained farmers. Appropriate measures need to be taken to make resource-friendly options available for farmers of different backgrounds to prevent any kind of prejudice against the tools available.

   To ensure that relevant advances in technology benefit people living in developing and rural areas, international technology specialists, civil society, and government policymakers need to work together to develop the necessary physical infrastructure such as reliable internet connections and cheap smart devices. The availability of AI technology to all farmers, particularly smallholder farmers, is commendable to forward inclusive agricultural transformation.
2. **Financial Assistance for Smallholder Farmers:** Policies that offer financial aid, such subsidies or grants, to smallholder farmers can facilitate their access to AI technologies. The need for the acquisition and maintenance of the systems is a major cost factor by far in most cases, as is the case with AI-driven systems in this case. It is possible to decrease these investments through specific governmental and international organizations' funding activities. Moreover, the adoption of AI tools can be facilitated where the farmers cannot afford to pay for the technology through public-private partnerships.

Building Transparent and Accountable AI Systems

i. **Algorithmic Transparency:** The ethics that govern the usage of AI systems in the farming industry include assertion of transparency. This requires farmers and other stakeholders to know how the used AI algorithm arrives at certain conclusions when it is, for example, diagnosing crop infections, suggesting pest control measures, or controlling production tactics. Transparent process is always a measure of trust when it comes to AI and the farmer is able to implement an AI recommendation without doubt.

   Currently, there is a lack of practical recommendations for developers of AI systems regarding how the algorithms are trained, the sources of data, potential bias, and steps to mitigate such bias. In addition, farmers should also have some level of control to question or validate the decision made by artificial intelligence in a product in order to ensure its compliance with the productive demands of the clients.

ii. **Addressing Bias in AI Models:** In agriculture, the quality and diversity of the datasets used for developing AI systems are of critical importance to the effectiveness of those systems. Lack of, or selective as well as limited data can produce inaccurate or unfair results. Thus, to prevent this, developers need to gather detailed, assorted datasets covering various agricultural situations and agriculture types.

Pre-scheduled and periodic reviews are important to establish the absence or presence of prejudices in the algorithms used in agriculture. Through such analyses, biases are uncovered, and continual enhancement can be made to ensure that the required positive changes in the outlook of the long-term disabled are attained. The approach mentioned is proactive in preventing the AI systems from getting out of control and becoming beneficial to some farmers more than others. The ideas of constant auditing and enhancement ensure integrity throughout contributing to an improved standard of living for all members involved through harnessing the power of AI in agriculture.

Strengthening Farmer Education and Training on AI Tools

1. **Empowering Farmers with Knowledge:** The application of ethical AI in agriculture requires providing farmers with awareness and skills to utilize AI tools efficiently. However, such programs have to be created that teach farmers how to work with AI-driven systems, interpret data, and application recommendations so that AI technology is adopted successfully (Singh et al., 2024).

   It is essential for these programs to be specified by farmers' educational requirements of varying degrees of technical proficiency. Farmers can have more confidence in effectively using AI systems if they have practical workshops, user-friendly manuals, and support in local languages. Building such prototypes, however, has spread its importance, something not many farmers observe presently.

2. **Cultivating Regional Knowledge:** It is imperative to invest in the professional growth of farmers in order to advance their understanding of digital agriculture and AI. The technique of both building a network of local technicians, agronomists, and AI professionals and using third-party reporting to quickly send growers actionable, accurate advice makes the process timely and valuable. As Bhat et al. (2024) highlight that on-site extension workers are very much needed to demystify AI tools for farmers as they adapt those tools to their specific microclimates and local crops.

Educational institutes and governmental bodies should endorse programs that emphasize the application of AI in agriculture to cultivate the next generation of agricultural specialists. Maintaining local expertise in AI-driven plant protection and other farming practices will be crucial for advancing sustainable and efficient agricultural systems.

Policies to Encourage Sustainable and Responsible AI Use

- **Creating a Supportive Policy Environment:** Policymakers, especially at the national level, have an important duty of uplifting legislation that would promote the usage of ethical AI in farming. Policies have to encourage the development of AI solutions that are effective as well as environmentally friendly and that support sustainable farming. For instance, rules that promote the development of AI tools, which reduce or optimize the usage of pesticides or water, shall ensure that technologies being developed support environmental priorities (Zhao and Gómez Fariñas, 2023).

In particular, regulatory bodies should set the rules for AI application in agriculture to guarantee AI tools meet safety, environmental, and ethical requirements. Such guidelines should encompass risks such as fairness, privacy, and transparency in artificial intelligence systems with concern to farmers and consumers.

- Supporting Public-Private Partnerships: Public-private partnerships (PPPs) can drive innovation in AI-powered plant protection by pooling resources and expertise from both sectors. The governments should engage technology companies, research institutions, and agricultural organizations to get funding to support research, create new and improved AI models, and the deployment of AI in farming villages. Such collaborations may guarantee that advancements in AI technologies are realized not only for profitability, but for sustainability in the entire agriculture value chain.
- Long-Term Ethical Monitoring: Last but not least, policymakers need to identify how the ethical monitoring of the long-standing AI in agriculture systems will be sustained. This concerns yearly reviews of the social and environmental repercussions of the AI tools in use: whether the tools are still consonant with ethical norms. The flow of rules should also involve farmers, researchers, and environmentalists in order to get their input towards the change of rules as appropriate. This persistence and flexibility in policy checking will further guarantee that AI remains a tool for responsible and sustainable agriculture (Wang et al., 2020).

The implementation of AI in agriculture has great potential to enhance plant protection through the delivery of appropriate solutions at the right time, raise yields, and encourage sustainable farming practices. However, to attain these benefits, it is important to ensure that AI tools are developed and deployed ethically, inclusively, and transparently. Through advocating for inclusiveness, providing fair access, developing transparent systems, educating farmers, and building supportive policy frameworks, we can direct the ethical upscaling of AI-powered plant protection systems. As a roadmap for promoting responsible AI innovation that serves farmers, consumers, and the environment, these recommendations were developed.

9. Conclusion

It is important to address how artificial intelligence will affect plant protection as it continues to be incorporated into agriculture. This concluding chapter reviews the main ideas presented in the book while also considering potential trends in artificial intelligence in the future. In the concluding part of the chapter, we contemplate creating innovative responsibility and hospitable strategies for the world.

Summary of Key Points

1. **The Role of AI in Plant Protection:** The application of artificial intelligence in agriculture is revolutionizing conventional practices, especially in plant protection. From planting to lethal pest detection apps, artificial intelligence presents new ways to think about pest and disease control. Other key applications described in prior chapters include the use of artificial intelligence to diagnose diseases, control pests, and make decisions, all of which can prove invaluable in improving crop production.
2. **Emerging Technologies:** Technological enhancements in the fields of machine learning, deep learning, remote sensing, and robotics have provided impetus to the integrated approach in plant protection systems. Drone, IoT, as well as big data analysis technologies, help farmers to have a comprehensive and timely assessment of field threats. This technological change not only optimizes the established breakthrough but also addresses specific concerns about environmental effects.
3. **Challenges in Implementation:** There are major challenges to the adoption of AI in plant protection. These deep issues are with data quality, integration of AI in traditional farming practices, and accessibility for smallholder farmers. In addition, technology dependency can expose farmers to vulnerabilities when they do not have the necessary knowledge, or when technical breakdowns occur.
4. **Ethical Considerations:** The uses of AI in agriculture have broad and diverse ethical concerns including employment, privacy concerns, algorithmic bias, and sustainability. Approaching these issues with great care is necessary for AI technologies to be capable of addressing and resolving deepening inequalities or, on the contrary, for them to aggravate the worst manifestations of those issues.
5. **Regulatory and Policy Frameworks:** Local laws concerning AI application in agriculture change constantly. Creating guidelines for AI tools, regulating IP matters, and making policies that will create a healthy landscape for advanced technologies are key aspects that need to be put in place to guarantee that technological goals are achieved without wronging consumers and farmers.
6. **Emerging Trends:** The future of AI in plant protection is one filled with promising prospects such as the explicit combination of blockchain with AI for transparency in the supply chain, further developments in AI technology that are set to bring revolutionary change in plant breeding for disease resistance and AI in partnership with tech firms, researchers as well as farmers. These trends show that AI can enhance sustainable and climate-smart agriculture.

Future Directions for AI in Plant Protection

1. **Enhanced Data Utilization:** In the future, the ability to extract and use meaning from agricultural data will be increasingly important as these data continue to grow. Future AI systems should be designed to take in large datasets across different sources such as environmental sensors satellite imagery and agricultural inputs to gain actionable insights. Improved prediction in terms of pest infestations and disease outbreaks can be expected as we continue to see advancements in data analytics.
2. **Focus on Sustainability:** The application of AI technology in plant protection should be made sustainable. Subsequent advancements should therefore focus on developing AI systems that increase crop production whilst at the same time lowering chemical usage and consequently the effects of AI on the environment. AI can improve practices like precision fertilization and integrated pest management as a way of improving sustainable agriculture.
3. **Interdisciplinary Collaborations:** Despite the challenges, agriculture today requires interdisciplinary collaboration. Partnerships between agronomists, ecologists, data scientists, and AI specialists will lead to future AI applications. Such collaborations can create innovation to solve both the ecological and social aspects of agricultural systems.
4. **Improved Accessibility:** Equitable agricultural progress requires that AI technologies are accessible to smallholder farmers. In the future, the focus must be on developing AI solutions that are the most affordable and user-friendly for use in the farmers who are least resourced. For instance, it might entail developing mobile applications offering real-time data and recommendations to farmers resulting from AI analysis with which farmers can be able to make decisions.
5. **Evolving Regulatory Frameworks:** The legal structures need to change frequently as the technology of artificial intelligence progresses. Regulating the use of AI, therefore, requires the involvement of governments in associations with farmers and developers to determine the right code of ethics for AI applications. The above framework must be in a position to factor critical areas like, who is to blame in case for arising complications in the AI systems, who owns the data, and how the data should be protected to ensure that all parties have confidence in the systems at hand.

Final Thoughts on Balancing Innovation with Ethical Responsibility

The integration of AI in plant protection systems holds immense promise for enhancing agricultural yields and promoting sustainability. However, the rapid pace of technological advancement underscores the need for a well-structured implementation plan that balances innovation with ethical responsibility. Collaboration among all stakeholders in agriculture is essential to achieve this balance. The development and deployment of AI must prioritize societal and environmental well-being. By embedding principles of inclusion, transparency, and accountability into AI systems, we can create solutions that positively transform the agricultural landscape. The goal is not merely to implement AI but to use it as a catalyst for developing sustainable agricultural practices that benefit farmers, consumers, and the environment alike.

Ethical AI transcends tools and techniques and embodies a forward-thinking approach that anticipates possibilities and addresses probable challenges. This vision will shape the future of agriculture, ensuring a resilient food supply while safeguarding the environment. Embracing this responsibility unlocks AI's full potential, enabling transformative advancements in agricultural systems while preserving the foundational principles of sustainability.

Acknowledgment

The corresponding authors sincerely thank all co-authors for their significant contributions to the book chapter. We have purchased the pictures from Shutterstock with the following details:

shutterstock.com/Pand P Studio/1969644328

shutterstock.com/earth phakphum/2430033455

shutterstock.com/Leonid Sorokin/1915344580.

References

Adetunji, C. O., Olaniyan, O. T., Anani, O. A., Inobeme, A., Osemwegie, O. O., Hefft, D. et al. (2023). Artificial Intelligence and Automation for Precision Pest Management. In Sensing and Artificial Intelligence Solutions for Food Manufacturing (pp. 49–70). CRC Press.

Ali, G., Mijwil, M. M., Buruga, B. A., Abotaleb, M., and Adamopoulos, I. (2024). A Survey on Artificial Intelligence in Cybersecurity for Smart Agriculture: State-of-the-Art, Cyber Threats, Artificial Intelligence Applications, and Ethical Concerns. Mesopotamian Journal of Computer Science, 2024: 53–103.

Altayeb, J. M., Eleyan, H., Wishah, N. D., Elmahmoum, A. E., Khalil, A. J., Abu-Nasser, B. S. et al. (2024). AI-Driven Innovations in Agriculture: Transforming Farming Practices and Outcomes. International Journal of Academic Applied Research (Ijaar) 8(9): 1–6.

Anand, S., and Kumar, C. A. (2024). Ai-Powered Weed Detection System For Precision Farming. Journal of Basic Science and Engineering, 21(1): 1844–1867.

Balaska, V., Adamidou, Z., Vryzas, Z., and Gasteratos, A. (2023). Sustainable crop protection via robotics and artificial intelligence solutions. Machines, 11(8): 774.

Batson, A., Spawton, K., Katz, R., and Du Toit, L. (2022). Cladosporium leaf spot caused by *Cladosporium variabile* in winter high tunnel production of spinach (*Spinacia oleracea*) in Maine, United States. Plant Disease, 106(8): 2260.

Bhat, P. P., Prasad, R., Anil, K., Jadhav, A., Manohar, K., Rajesh, C. et al. (2024). The role of information and communication technology in enhancing the effectiveness of agricultural extension programs worldwide: A review. Journal of Scientific Research and Reports, 30(7): 963–976.

Bibri, S. E., Alexandre, A., Sharifi, A., and Krogstie, J. (2023). Environmentally sustainable smart cities and their converging AI, IoT, and big data technologies and solutions: an integrated approach to an extensive literature review. Energy Informatics, 6(1): 9.

Brown, M. E., Mugo, S., Petersen, S., and Klauser, D. (2022). Designing a pest and disease outbreak warning system for farmers, agronomists and agricultural input distributors in East Africa. Insects, 13(3): 232.

Daneshjou, R., Smith, M. P., Sun, M. D., Rotemberg, V., and Zou, J. (2021). Lack of transparency and potential bias in artificial intelligence data sets and algorithms: a scoping review. JAMA dermatology, 157(11): 1362–1369.

Dara, R., Hazrati Fard, S. M., and Kaur, J. (2022). Recommendations for ethical and responsible use of artificial intelligence in digital agriculture. Frontiers in Artificial Intelligence, 5: 884192.

Eruaga, M. (2024). Enhancing global food safety standards through international collaboration and policy harmonization. Int. J. Sch. Res. Multidiscip. Stud, 4: 20–32.

Fang, J. (2024). Breeding 5.0: AI-Driven Revolution in Designed Plant Breeding. Molecular Plant Breeding, 15(1): 27–33.

Farooq, M. A., Gao, S., Hassan, M. A., Huang, Z., Rasheed, A., Hearne, S. et al. (2024). Artificial intelligence in plant breeding. Trends in Genetics, 40(10): 891–908.

Galaz, V., Centeno, M. A., Callahan, P. W., Causevic, A., Patterson, T., Brass, I. et al. (2021). Artificial intelligence, systemic risks, and sustainability. Technology in Society, 67: 101741.

Gardezi, M., Adereti, D. T., Stock, R., and Ogunyiola, A. (2022). In pursuit of responsible innovation for precision agriculture technologies. Journal of Responsible Innovation, 9(2): 224–247.

Gupta, D. K., Pagani, A., Zamboni, P., and Singh, A. K. (2024). AI-powered revolution in plant sciences: advancements, applications, and challenges for sustainable agriculture and food security. Exploration of Foods and Foodomics, 2(5): 443–459.

Gupta, M., Kumar, R., and Singh, D. (2021). Crop Diseases Detection and Prevention Using AI and Machine Learning Techniques. pp. 61–82. *In*: Digital Twin Technology. CRC Press.

Harfouche, A. L., Jacobson, D. A., Kainer, D., Romero, J. C., Harfouche, A. H., Mugnozza, G. S. et al. (2019). Accelerating climate resilient plant breeding by applying next-generation artificial intelligence. Trends in biotechnology, 37(11): 1217–1235.

Hassan, M., Malhotra, K., and Firdaus, M. (2022). Application of artificial intelligence in IoT security for crop yield prediction. ResearchBerg Review of Science and Technology, 2(1): 136–157.

Holzinger, A., Fister Jr, I., Fister, I., Kaul, H.-P., and Asseng, S. (2024). Human-centered ai in smart farming: Towards agriculture 5.0. Ieee Access, 12: 62199–62214.

Johansen, K., Sohlbach, M., Sullivan, B., Stringer, S., Peasley, D., and Phinn, S. (2014). Mapping banana plants from high spatial resolution orthophotos to facilitate plant health assessment. Remote Sensing, 6(9): 8261–8286.

Khan, M. H. U., Wang, S., Wang, J., Ahmar, S., Saeed, S., Khan, S. U. et al. (2022). Applications of artificial intelligence in climate-resilient smart-crop breeding. International Journal of Molecular Sciences, 23(19): 11156.

Kumari, N., Pandey, A. K., Singh, A. K., and Singh, A. (2024). Sustainable Agriculture: Balancing Productivity and Environmental Stewardship for Future Generations. Journal of Scientific Research and Reports, 30(8): 629–639.

Li, C., and Wang, M. (2024). Pest and disease management in agricultural production with artificial intelligence: Innovative applications and development trends. Advances in Resources Research, 4(3): 381–401.

Liu, Y., Meng, X., and Li, A. (2023). AI's Ethical Implications: Job Displacement. Advances in Computer and Communication, 4(3): 138–142.

Lwakatare, L. E., Raj, A., Crnkovic, I., Bosch, J., and Olsson, H. H. (2020). Large-scale machine learning systems in real-world industrial settings: A review of challenges and solutions. Information and software technology, 127: 106368.

Mahlein, A.-K., Arnal Barbedo, J. G., Chiang, K.-S., Del Ponte, E. M., and Bock, C. H. (2024). From Detection to Protection: The Role of Optical Sensors, Robots, and Artificial Intelligence in Modern Plant Disease Management. Phytopathology®, 114(8): 1733–1741.

Mana, A., Allouhi, A., Hamrani, A., Rahman, S., el Jamaoui, I., and Jayachandran, K. (2024). Sustainable AI-Based Production Agriculture: Exploring AI Applications and Implications in Agricultural Practices. Smart Agricultural Technology, 100416.

Maslekar, N., Kulkarni, K. P., and Chakravarthy, A. K. (2020). Application of unmanned aerial vehicles (UAVs) for pest surveillance, monitoring and management. Innovative Pest Management Approaches for the 21st Century: Harnessing Automated Unmanned Technologies, 27–45.

Mathushika, M., Vinushayini, R., and Gomes, C. (2022). Smart Farming Using Artificial Intelligence, the Internet of Things, and Robotics: A Comprehensive Review. Artificial Intelligence and Smart Agriculture Technology, 1–19.

Mensah, G. B. (2023). Artificial intelligence and ethics: a comprehensive review of bias mitigation, transparency, and accountability in AI Systems. Preprint, November, 10.

Mesías-Ruiz, G. A., Pérez-Ortiz, M., Dorado, J., De Castro, A. I., and Peña, J. M. (2023). Boosting precision crop protection towards agriculture 5.0 via machine learning and emerging technologies: A contextual review. Frontiers in Plant Science, 14: 1143326.

Mohd Noor, N., Abdullah, A., and Hashim, M. (2018). Remote sensing UAV/drones and its applications for urban areas: A review. IOP conference series: Earth and Environmental Science. 169: 012003.

Murthy, P., and Bobba, S. (2021). AI-powered predictive scaling in cloud computing: Enhancing efficiency through real-time workload forecasting. ICONIC Research and Engineering Journals, 5(4): 143.

Mutsuddi, I. (2024). Ethical Considerations in the Intersection of AI and Environmental Science. pp. 283–310. *In*: Maintaining a Sustainable World in the Nexus of Environmental Science and AI. IGI Global.

Nagaty, K. A. (2023). IoT commercial and industrial applications and AI-powered IoT. pp. 465–500. *In*: Frontiers of Quality Electronic Design (QED) AI, IoT and Hardware Security. Springer.

Patil, A. (2023). Use of Artificial Intelligence to Hasten Progress in Plant Genetics. Intelligence (AI), 1, 3.

Qazi, S., Khawaja, B. A., and Farooq, Q. U. (2022). IoT-equipped and AI-enabled next generation smart agriculture: A critical review, current challenges and future trends. IEEE Access, 10: 21219–21235.

Ramdinthara, I. Z., Bala, P. S., and Gowri, A. (2022). AI-based yield prediction and smart irrigation. Internet of Things and Analytics for Agriculture, 3: 113–140.

Shaikh, T. A., Rasool, T., and Lone, F. R. (2022). Towards leveraging the role of machine learning and artificial intelligence in precision agriculture and smart farming. Computers and Electronics in Agriculture, 198: 107119.

Shankar, P., Werner, N., Selinger, S., and Janssen, O. (2020). Artificial intelligence-driven crop protection optimization for sustainable agriculture. 2020 IEEE/ITU International Conference on Artificial Intelligence for Good (AI4G).

Singh, P., Kansal, M., Tyagi, G., Tanwar, K., Singh, A. P., and Yadav, R. (2024). Empowering Farmers: An AI-Based Solution for Agricultural Challenges. pp. 401–417. *In*: Computational Intelligence in Internet of Agricultural Things. Springer.

Slaughter, D. C., Giles, D., and Downey, D. (2008). Autonomous robotic weed control systems: A review. Computers and Electronics in Agriculture, 61(1): 63–78.

Tirkey, D., Singh, K. K., and Tripathi, S. (2023). Performance analysis of AI-based solutions for crop disease identification, detection, and classification. Smart Agricultural Technology, 5: 100238.

Vasileiou, M., Kyriakos, L. S., Kleisiari, C., Kleftodimos, G., Vlontzos, G., Belhouchette, H. et al. (2023). Transforming weed management in sustainable agriculture with artificial intelligence: A systematic literature review towards weed identification and deep learning. Crop Protection, 106522.

Wang, Y., Xiong, M., and Olya, H. (2020). Toward an understanding of responsible artificial intelligence practices. Proceedings of the 53rd Hawaii international conference on system sciences. Hawaii International Conference on System Sciences (HICSS 2020), Maui, Hawaii, USA. pp. 4962–4971. ISBN 9780998133133.

Zhao, J., and Gómez Fariñas, B. (2023). Artificial intelligence and sustainable decisions. European Business Organization Law Review, 24(1): 1–39.

CHAPTER 16

AI-Powered Innovations in Plant Virology

Prabha K.

1. Introduction

Artificial intelligence (AI) is intelligence created artificially. AI is a branch of computer science dedicated to creating systems with human-like intelligence. In the past few decades, algorithmic and computational advances in machine learning (ML) have led to many breakthroughs in natural language processing, image recognition, robotics, data processing, and many other fields. The advances in AI can contribute significantly to plant pathology, specifically, plant virology. The revolution in AI and ML is making rapid advances in medical virology, virus evolutionary study, identification, etc. This chapter discusses the advancement in the application of AI in plant virus diagnostics, epidemiology, and management.

1.1 AI and Plant Pathology

Artificial intelligence has revolutionized biotic stress phenotyping, sensor-based biomarkers of diseases and automated interventions for plant disease management. Automated polyhouses with integrated nutrient, pest and disease management strategies have become a common sight today. Hyperspectral data-based stress phenotyping is widely adopted in plant disease diagnostics. The detection of variations in the optical properties of a plant results from the infections along with other sensors detecting environmental features like moisture, plant surface temperature, etc., which helps in non-invasive disease diagnostics (Figure 1). Artificial Neural Networks (ANN), Convolution Neural Network (CNN), Probabilistic Neural Network (PNN), Random Forest and Support Vector Machines (SVM), are the dominant machine

ICAR-Directorate of Floricultural Research, Pune, India, 411028.
Email: prabhaicardfr@gmail.com

DOI: 10.1201/9781003546702-16

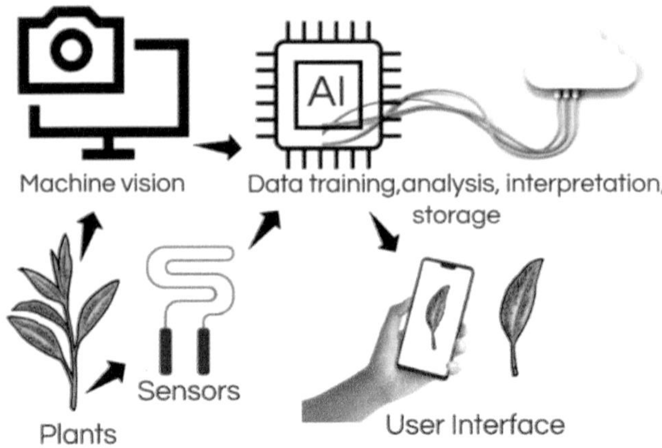

Figure 1. AI-Assisted Plant Disease Diagnostics.

learning techniques used for image classification in disease identification. Unmanned aerial vehicles (UAVs) deployed with disease identification algorithms and models are used for automated disease identification and management interventions in orchards and playhouses. Penetration of smartphone technology into large populations enabled on-field detections of diseases at fingertips through various mobile applications. Pocket Agronomist, Plantix, etc., are a few among them. Like any other plant disease diagnostic techniques, the technique employed, method of sampling, size of the sample, time taken for sample processing, prevailing weather conditions and microclimate, growth stage of the crop, other background noise, etc., determines the accuracy of diagnostics. Similar to other diagnostic techniques, machine learning-based detection is much more difficult for plant viruses compared to other phytopathogens like fungi and bacteria, where these pathogen infections result in more visible impacts on the plants.

2. Detection and Diagnostics of Plant Viruses

Timely detection and right identification are the keys to managing any virus infection of plants. Plant viruses are difficult to detect. In most cases, visual detection is confusing for the farmer. In contrast, other detection methods are disruptive and involve state-of-the-art facilities and skilled personnel and are costly. Developments in plant phenotyping incorporated with AI and machine vision are being employed in disease detection. Dataset of images used for classification of diseased and healthy plants has to be verified and large enough to develop models for accurate disease diagnosis (Prabha et al., 2020). A list of major reports of diagnostics of plant virus infections with machine learning is listed in Table 1.

To distinguish cassava diseases like cassava brown streak disease (CBSD), and cassava mosaic disease (CMD) (or lack thereof), a transfer learning based on a deep convolutional neural network was employed with an accuracy of 98 and 96% respectively along with other cassava diseases and pests such as brown leaf spot (BLS), red mite damage (RMD) and green mite damage (GMD) (Ramcharan et al., 2017). The best-trained model accuracies were 98% for cassava brown streak

Table 1. Applications of AI in plant virus diagnostics

Sl. No.	Virus and Host	Model	References
1.	Cassava mosaic disease (CMD), Cassava brown streak disease (CBSD)	Deep convolutional neural network -Deep CNN	Ramcharan et al., 2017
2.	*Vigna mungo*	VirLeafNet - CNN	Joshi et al., 2021
3.	Zucchini yellow mosaic virus, Melon yellow spot virus	CNN	Kawasaki et al., 2015
4.	Potato virus Y	Support Vector Machines (SVM)	Griffel et al., 2018
5.	*Grapevine leafroll-associated virus 3* (GLRaV-3)	Hyperspectral imaging	Gao et al., 2020
6.	Tospovirus	Generative Adversarial Nets (GAN)	Wang et al., 2019
7.	Ground nut bud necrosis virus	SVM	Kadam, 2020
8.	Tobacco -*Tomato leaf curl Gujarat virus* (ToLCGV) ,*Tomato leaf curl New Delhi virus* (ToLCNDV)	CNN	Chakraborty et al., 2021
9.	Geminivirus in Chilli	Squeeze-Net training architecture	Kanaparthi and Ilango et al., 2023
10.	Alfalfa Mosaic Virus (AMV) in tomato	Electric signal patterns -Random Forest, SVM, K-Nearest Neighbours	Ghasemi et al., 2024

disease (CBSD), and 96% for cassava mosaic disease (CMD). Further, an attention-based approach is integrated into pre-trained CNN-based Efficient Net models to improve disease detection in cassava (Ravi et al., 2022). Automated diagnosis of viral infection in the *Vigna mungo* using a CNN model, VirLeafNet integrated with drones for wider crop area analysis was successfully demonstrated (Joshi et al., 2021). By training a total of 800 cucumber leaf images, CNN automatically acquired the requisite features for classification and achieved high classification performance with an average accuracy of 94.9% in classifying cucumber diseases, Zucchini yellow mosaic virus and Melon yellow spot virus (Kawasaki et al., 2015).

Spectral reflectance pattern of PVY-infected and healthy seed potato plants was classified by an SVM classifier for Potato virus Y detection and the task was achieved with an accuracy of 89.8%.

This is an example of the integration of remote sensing and machine learning for plant virus detection where the alteration in reflectance of electromagnetic energy from potato plants during virus infection (Griffel et al., 2018). The potential of the Hyperspectral imaging technique in non-destructively detecting virus-infection in plants at an early asymptomatic stage has been employed in grapevines to detect *Grapevine leafroll-associated virus 3* (GLRaV-3) (Gao et al., 2020). Another hyperspectral proximal sensing method based on generative adversarial nets (GAN) named as outlier removal auxiliary classifier generative adversarial nets (OR-AC-GAN) was used as an improvement on classic spectrum band selection methods with a plant-level classification accuracy of 96.25% at asymptomatic stages of a Tospovirus infected plant (Wang et al., 2019). Ground nut bud necrosis virus has been

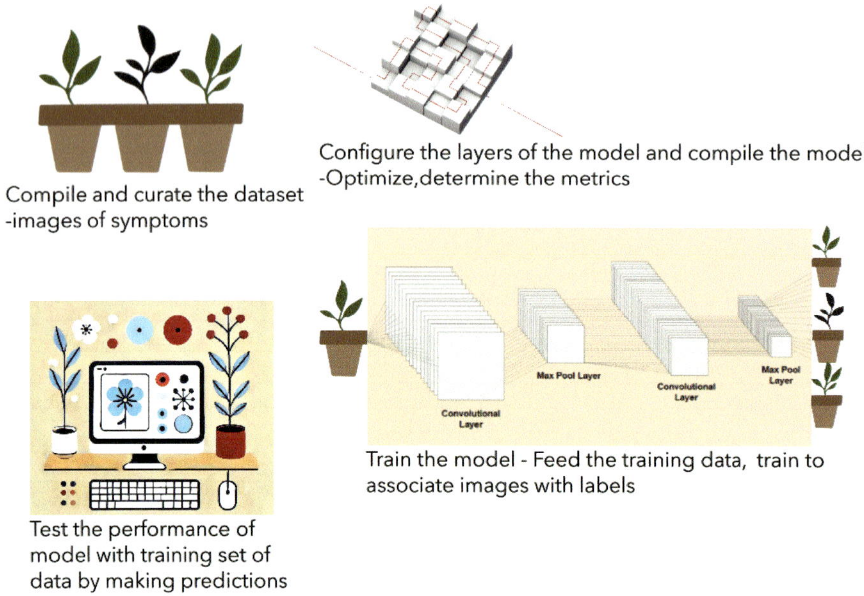

Figure 2. A neural network model for image detection and prediction.

detected in Tomato plants deploying Support Vector Machine based segmentation algorithm (Kadam et al., 2020). Two begomoviruses, *Tomato leaf curl New Delhi virus* (ToLCNDV) and *Tomato leaf curl Gujarat virus* (ToLCGV) were detected, with an accuracy of 97.211% from inoculated tobacco plants using a deep CNN known as Visual Geometry Group 16 or VGG-16(Chakraborty et al., 2021). A study was conducted to evaluate the efficacy of the Squeeze-Net training architecture in Gemini virus detection in Mosaic-infected chilli leaves using 160 images obtained an accuracy varying from 50% to 100% based on CNN optimizers deployed (Kanaparthi and Ilango, 2023). Among the 12 pre-trained DL models evaluated for detecting chili viruses the highest accuracy of 83.54% was obtained for VGG19 (Naik et al., 2022). The accuracy was enhanced to 99.28% by designing a squeeze-and-excitation-based convolutional neural network (SECNN) model. A mulitmodel framework fusing digital photographic dataset and multispectral datasets was developed for Cotton leaf curl Multan Virus infection in cotton. These dataset were trained to four machine learning models (random forest (RF), simple logistics (SL), sequential minimal optimization (SMO), multilayer perceptron (MLP)) separately and the fusion model turned out to be more precise in virus detection 96% accuracy (Ahmad et al., 2024).

If we analyse the reports on the diagnostics studies, almost of the diagnostics methods are machine vision-based. This includes the following steps: image analysis, segmentation, training, and prediction of test images. The details of a neural network model for image detection and prediction are given in Figure 2. The efficacy of this application in real field conditions needs to be looked into. In real field conditions, isolated occurrences of diseases will be rare, and often nutritional deficiency and virus infections are confused. Thus, training the models with images of real field

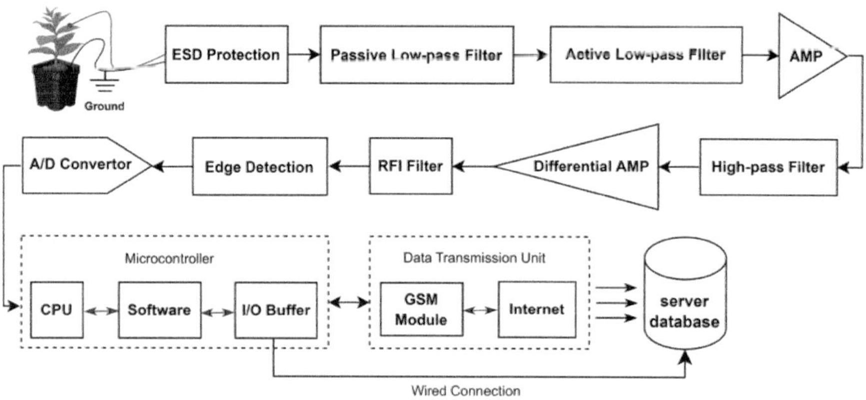

Figure 3. AI-based metagenomic mining of viral sequence.

conditions is a major challenge in machine vision-based diagnostics. Machine vision combined with other sensors of metabolites, insect vectors, environmental conditions, and the package of practices will deliver a more reliable diagnostic tool for virus infection detection in real-time. In crops like apple and citrus, advanced and modified algorithms like YOLO, and Deep Forest, in combination with unmanned aerial vehicles (UAVs) are employed in the detection of trees, fruits, and diseases of fungal and bacterial origin in real-time (Jemaa et al., 2023; da Silva et al., 2023). Another non-destructive method of detection of virus infections is through biomarkers for infestation from host plants including metabolites and volatiles, where sensor-based detection can be done. Integration of both machine vision and other metabolite biomarkers, environmental conditions and cultivation practices can increase the accuracy of AI-based disease detection. A new machine learning technology where post infection electrical signal patterns have been interpreted for prediction of virus infection has been developed with an accuracy of 97% in tobacco plants infected with Alfalfa Mosaic Virus (AMV) (Ghasemi et al., 2024). The signal recording and interpretation system is given in the Figure 3. In this method, minimal input data is required compared to other techniques.

3. Genomes and Viromics

High throughput sequencing is widely employed in viral genomics and the use of generative AI in bioinformatic tools is on the rise. There are many pipelines incorporated with AI available for analysis of next-generation sequencing (NGS) data and interpretation of viromes. Machine learning method using *k*-mer frequency based contig identification of virus genomes are widely used for deciphering viral genomes from prokaryotes in metagenomic data. The VirFinder programme is first of this kind (Ren et al., 2017). V-PIPE is such a computational pipeline used to assess genetic diversity in virome from the NGS data which is freely available at https://github.com/cbg-ethz/V-pipe (Posada-Céspedes et al., 2021). To measure the intra-host virus diversity in Zika and West Nile virus , amplicon sequencing of RNA viruses directly from clinical samples, Primal Scheme (multiplex primer design tool) coupled to PrimalSeq, a laboratory protocol has been used and sequencing results were analyzed through iVar pipeline (Grubaugh et al. 2019). This technique may be manipulated in plant viruses as well. VirSorter (Roux et al., 2015), VirSorter 2 (Guo et al., 2021), DeepVirFinder (Ren et al., 2020), MARVEL (Amgarten et al., 2018), and VIBRANT (Kieft et al., 2020) are similar tools for identifying microbial viruses from genome sequences. Fangorn Forest (F2) is a machine-learning approach utilizing genomic sequence information and has been successfully employed in classifying genes and genera of family Geminiviridae (Silva et al., 2017). The optimization of neural network-based models that predict probable point mutations in the RNA genome of human and animal viruses may be optimized for the prediction of mutations in plant viral genome (Ghosh et al., 2022). Rapid analysis of large-scale viral sequence data and reconstruction of consensus genomes is possible through ViReflow. This is a user-friendly pipeline where an AWS and the Reflow system are employed (Moshiri et al., 2022). Sequence calling of viral genomes from read alignment data piped directly from a read mapper can be achieved through a model called ViralConsensus. (Moshiri, 2023).

A deep learning classifier, CHEER (hierarchical taxonomic classification for viral mEtagEnomic data via deep learning) is a hierarchical model, in which species masking features can be trained for identifying novel RNA viruses' taxonomy and classification (Shang and Sun, 2020). Verifier is another tool for mining viral sequences from metagenomic data (Yan et al., 2021). Recurrent Neural Network (RNN)-VirSeeker is another viral sequence mining tool in a fasta file of metagenomic sequences (Liu et al., 2022). VirHunter is a deep learning tool executed to identify viruses in plant virome sequencing datasets. There are 8 trained models of VirHunter: *carrot*, *grapevine*, *lettuce*, *peach*, *rice*, *sugar beet*, *tomato*, and *generalistic* where a generalistic model is trained with a mixture of plants from *dicots* and *monocots* clades. This tool enables the identification of new RNA viruses from plant genomic sequences (Sukhorukov et al., 2022).

While all the above models are efficient in mining small sequences and longer sequences are divided into non-overlapping short sequences, they perform poorly in identifying long viral sequences. Thus, the existing methods have been further improved for the identification of longer sequences and some composite models have emerged to improve the efficiency of virome mining in metagenomics data.

By constructing relationships of short subsequences from long ones and long viral sequences are deciphered by Graph Convolutional Network (GCN) based VirGrapher, trained by the Graph Self-Attention (GSA) mechanism (Miao et al., 2024).

Composite model that incorporates multi-head attention mechanisms with graph isomorphic neural networks are the latest advancement in virome mining tools. TransGINmer is a tool that uses a transformation method based on k-mer frequencies, embedding nucleotide sequences of varying lengths directly into feature matrices for virus identification (Wang et al., 2024). AI-based metagenomic mining for finding virus genomes has revolutionized the genomics of viruses along with protein structure prediction algorithms. LucaProt is a DL algorithm integrates the discovery of highly divergent RNA-dependent RNA polymerase (RdRP). This has been used topredict 161,979 potential RNA virus species and 180 RNA virus supergroups from 10,487 metatranscriptomes of diverse global ecosystems including plants (Hou et

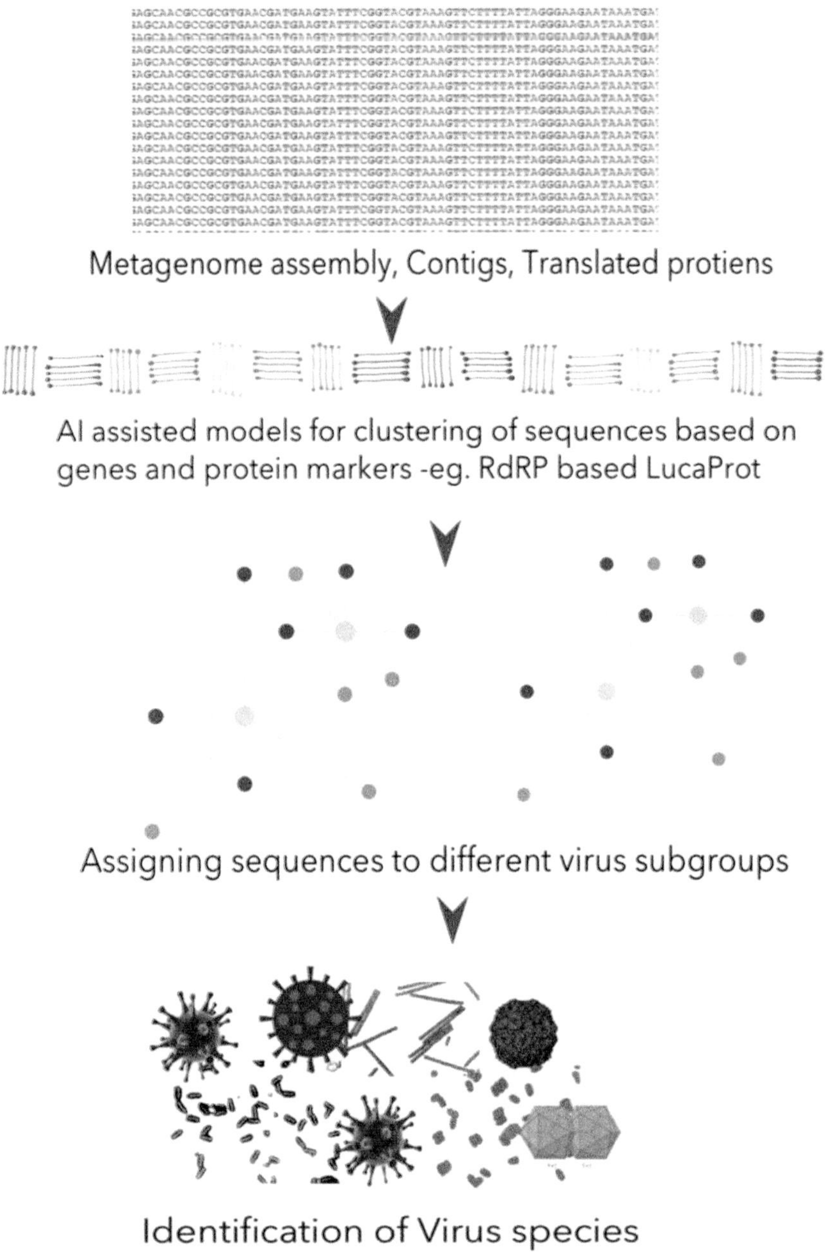

Figure 4. Block diagram of the signal recording system for capturing electrical signals from plants. ESD electrostatic discharge, amplification (AMP), radio frequency interference (RFI), analog-to-digital (A/D). (Directly adopted from Ghasemi et al., 2024, Under a Creative Commons license).

al., 2024). In the parameters such as precision, recall, and long sequence processing, this algorithm outperformed the RNA virus discovery tools, CHEER, VirHunter, Virtifier, and VirSeeker.

Though AI has achieved the discovery of viruses at the genome level, Virus classification at a deep evolutionary scale is a complex and challenging task in the present day. Due to greater genetic divergence in viruses, neither sequence nor structural homology will be able to uncover the true evolutionary journey of the viruses. Another challenge is, though viruses are identified in silico, real-time isolation of the virus from biological samples remains a challenge.

4. Host-virus Interactions

The accumulation of transcriptomic data enables generative AI models to predict and analyze the interactions between plant viruses and their host plants at the molecular level. AI has been successfully manipulated in unravelling the gene regulatory networks from RNA-sequencing data from a single cell or nuclei and this can be compared at the population level. Single-Cell Oriented Reconstruction of PANDA Individually Optimized gene regulatory Networks (SCORPION) is an R package that reconstruct gene regulatory networks that are comparable from RNA seq data (Osorio et al., 2024). BEELINE is another such Benchmarking algorithm for inferring gene regulatory networks (Pratapa et al., 2020). *In silico* prediction of interaction between CLCuMuB-βC1 and a host protein in a cotton plant was successfully validated by wet lab experiments indicating the potential of computational technologies in predicting host-virus interactions (Kamal et al., 2019).

In 2020, Hassabis and John developed an AI model called AlphaFold2, capable of virtually predicting the structure of 200 million proteins from protein sequence data (Callaway, 2022). ESMFold is a language model-based sequence-to-structure predictor that has characterized more than 772 million structures of proteins for the first time, resulting in an open atlas of predicted proteins called the ESM Metagenomic Atlas (Lin et al., 2023). AI-enabled predictive modelling of the structure and function of proteins has wide application in predicting virus proteins and their interaction with host proteins. AlphaFold2 was employed in the prediction of the structure of a protein in Rice black-streaked dwarf virus (RBSDV) (Liu et al., 2022) for studying the interaction of P10 protein with lipids. Similarity of 30K Movement Proetin's (MPs) core structural domain to the jelly-roll domain of the capsid proteins (CPs) was unravelled in the analysis by RoseTTA Fold of MPs of Plant RNA and DNA viruses (Butkovic et al., 2023). Predictive modelling of proteins of grapevine fanleaf virus (GFLV) two genomic RNAs, and their satellite RNAs validated their multifunctionality with functions as RNA silencing suppressor, protease, viral genome-linked protein, movement protein, coat protein, helicase, RNA-dependent RNA polymerase, etc. (Roy et al., 2024).

A variety of computational tools has been employed for analysis of the interaction of the host protein (Papaya Plant) with Papaya Leaf Curl Virus (PaLCuV) and its beta satellite (PaLCuB) protein. Where Phyre2 (Protein Homology/analogy Recognition Engine v.2.0), was used for prediction and analysis of the structure of PaLCuV proteins. Homology modelling was achieved with SWISS-MODEL, and structure validation of viral and plant protein models was done from the PDB sum server to predict interaction sites of viral proteins in papaya plants. SPPIDER (Solvent accessibility-based Protein-Protein Interface iDEntification and Recognition) was used, and Docking analysis of PaLCuV and plant proteins was performed using the HADDOCK v2.4 (High Ambiguity Driven Protein-Protein Docking program) (Srivastava et al., 2024).

A variety of computational tools has been employed for analysis of the interaction of the host protein (Papaya Plant) with Papaya Leaf Curl Virus (PaLCuV) and its beta satellite (PaLCuB) protein. Where Phyre2 (Protein Homology/analogy Recognition Engine v.2.0), was used for prediction and analysis of the structure of PaLCuV proteins. Homology modelling was achieved with SWISS-MODEL, and structure validation of viral and plant protein models was done from the PDB sum server to predict interaction sites of viral proteins in papaya plants. SPPIDER (Solvent accessibility-based Protein-Protein Interface iDEntification and Recognition) was used, and Docking analysis of PaLCuV and plant proteins was performed using the HADDOCK v2.4 (High Ambiguity Driven Protein-Protein Docking program) (Srivastava et al., 2024).

Most of the AI-based protein structure prediction models are based on amino acid sequences. A challenge for the present-day algorithms is, in the case of plant-host interaction and defense signalling, is post-translational protein modifications (PTM). PTM play an important role and is unable to predict with the models based on amino acid sequences. Attempts are on the way to rectify this caveat as well. A protein prediction suite, Rosetta has been incorporated with trained artificial neural networks (ANNs) to predict PTMs, like protein glycosylation, phosphorylation, methylation, and deamidation among the eighteen modifications trained to the model (Ertelt et al., 2024). A prompt-based interpretable protein language model, PTMGPT2, is a trained transformer model used to predict PTMs (Shrestha et al., 2024).

AI tools have been employed further to identify the localization of viral and host proteins in host cells successfully. LOCALIZER is an online ML model that predicts sub-cellular localization of effector proteins and host factors from amino acid sequences as inputs (Sperschneider et al., 2017). Plant-mSubP is another AI-integrated protein localization prediction tool in plants (Sahu et al., 2019). Ensemble Multi-Localized Attention (E-MuLA) is a computational model used in the accurate prediction of subcellular localization of viral proteins with an accuracy of 94.87% (Kissanga et al., 2024). Knowledge on RNA silencing and virus suppressors of gene silencing is important in the context of host-virus interaction. a random forest based prediction tool, *plant VsupPred* helps in predicting such viral suppressors (Jagga and Gupta, 2014).

5. Viral Epidemiology and Management

AI has enabled the prediction of the incidence, spread, and severity of diseases based on environmental and historical data. This technology has been widely employed in the prediction of pandemics in human and animal diseases and may be extrapolated to plant virus infection as well. Predictive modeling of the incidence of insect vectors can be used as an indirect way to determine the occurrence of virus diseases in protective cultivation where clean certified planting materials are used. Smart monitoring and management of insect vectors will in turn manage the incidence of viruses in plants.

AI-enabled smart traps are widely used to monitor insect pests in crop fields where pheromone traps are used to attract pests, insects are photographed by a camera inside the trap and the images are searched against the image database and the insect pests are identified. Once the insect is recognized as a pest, an alarm is sent to the farmer. Trapview Vertical SC trap successfully monitors the American grapevine leafhopper (*Scaphoideus titanus*), the vector of Flavescence dorée phytoplasma, and enables early detection and control of insect nymphs (Trapview, 2024). MaxEnt is a software for modeling species niches and geographical distributions by applying a machine-learning technique where environmental and climate data is employed (Phillips et al., 2017). This can be used to predict the spread and impact of a new plant virus species, introduced into a new area by overlapping the insect vector distribution. This method is promising in quarantine and pest risk analysis. A MaxEnt model was used to see the distribution of *Impatiens necrotic spot tospovirus* and its vector *Frankliniella occidentalis* in the current climate and projection has been made to visualize the dynamics of both INSV and its vector in the future of the changing climate (Prabha et al., 2020).

Genome editing has found wide application in managing virus diseases through manipulating gene/genes/non coding region in the viral genome, host factors required by virus for its propagation, RNA silencing mechanisms as well as developing crop varieties resistant to plant viruses. Precise deployment of substitutions, insertions, and deletions in genomes of living systems can be achieved through prime editing. AI-enabled prime editing software can be deployed for engineering directed mutations in genes and guide RNAs for the management of disorders and imparting disease resistance in plants (Gould et al., 2024). A python based PEGG - Prime Editing Guide Generator is such an algorithm for precision genome editing through high-throughput prime editing sensor libraries (https://pegg.readthedocs.io/en/latest/). The prime editor has successfully engineered virus-like particles (PE-eVLPs) as transient ribonucleoprotein complexes to deliver proteins and guide RNAs in human cells (An et al., 2024).

Management of virus diseases begins from the selection of planting material and use of certified virus-free planting material. Management of vectors in the field is the key to managing virus diseases in the field to reduce economic loss. To achieve these objectives, the right diagnostics is important. In diagnostics, the advances in AI are making positive impacts. The source of all information about the virus is from its genome and AI-enabled viral genomics and proteomics are heading to multitudes. Thus, the current scenario of applications of AI in plant virus management is promising.

6. Plant Virology Academics

Generative AI, based on Large Language models (LLM) is now widely researched and applied in research and academics. These large language models (LLMs) are engineered to understand and produce human language by combining memory and deep learning techniques. As a result, they generate meaningful responses to specific questions. This has enabled the analysis and interpretation of large amounts of information in a short time, making literature review and summarization a cakewalk. When they are fine-tuned and trained to perform specific tasks, it helps in improving scientific writing. AI can assist in writing and editing manuscripts, grant proposals, proofreading, and improving the quality of written content. Multilingual support makes communicating science in regional languages hassle-free. LLMs help in creating educational materials, tutorials, interactive learning modules, etc. These algorithms help in creating engaging presentations and visual representations of the scientific results after interpreting the data. Though GenAI has promising applications, the threat to academic integrity becomes diluted in many instincts like non-invigilated assessment tasks and presentation of machine-generated outputs (Perkins, 2023). The introduction of GenAI tools into research and academics is inevitable. Thus, laying out proper guidelines to check the aberrations is the only option left for streamlining such interventions.

7. Conclusion

Viruses are sub-microscopic organisms that cannot be cultured in vitro. This makes their study difficult compared to other plant microbes. Generative AI models like Generative Adversarial Networks (GANs) can relieve the current bottleneck of data scarcity by creating synthetic data that mimics real plant virus genomes (Bhugra et al., 2024). Virtual simulations and *in silico* analysis are enabling the studies on plant viruses further handy. The computational field has been revolutionized after the entry of AI algorithms and its impressions are visible in plant virus studies as well. Applications of AI have cleared many grey areas of plant virology while the vast area is yet to be explored. With their minimalistic genomes, virus jumble with the host genome machineries for their reign over the host. Despite virtual advancement in plant virology, the impact on real-field situations is still meagre. *In situ* management of the virus infections needs more uplifting. To conclude, Generative AI is transforming plant virology research by providing innovative tools and insights that were previously unattainable.

References

Ahmad, N., Qadri, S., and Akhtar, N. (2024). A multimodal fusion framework to diagnose cotton leaf curl virus using machine vision techniques. Cogent Food & Agriculture, 10(1): 2339572. https://doi.org/10.1080/23311932.2024.2339572.

Amgarten, D., Braga, L. P. P., da Silva, A. M., and Setubal, J. C. (2018). MARVEL, a tool for the prediction of bacteriophage sequences in metagenomic bins. Frontiers in Genetics, 9: 304. https://doi.org/10.3389/fgene.2018.00304.

An, M., Raguram, A., Du, S. W., Banskota, S., Davis, J. R., Newby, G. A. et al. (2024). Engineered virus-like particles for transient delivery of prime editor ribonucleoprotein complexes in vivo. Nature Biotechnology, 42: 1526–1537. https://doi.org/10.1038/s41587-023-02078-y.

Bakanina Kissanga, G.-M., Zulfiqar, H., Gao, S., Yussif, S. B., Momanyi, B. M., Ning, L. et al. (2024). E-MuLA: An ensemble multi-localized attention feature extraction network for viral protein subcellular localization. Information, 15(3): 163. https://doi.org/10.3390/info15030163.

Bhugra, S., Srivastava, S., Kaushik, V., Mukherjee, P., and Lall, B. (2024). Plant data generation with generative AI: An application to plant phenotyping. *In*: Lyu, Z. (Ed.). Applications of Generative AI (pp. 26). Springer, Cham. https://doi.org/10.1007/978-3-031-46238-2_26.

Butkovic, A., Dolja, V. V., Koonin, E. V., and Krupovic, M. (2023). Plant virus movement proteins originated from jelly-roll capsid proteins. PLoS Biology, 21(6): e3002157. https://doi.org/10.1371/journal.pbio.3002157.

Callaway, E. (2022). The entire protein universe: AI predicts shape of nearly every known protein. Nature, 608: 15–16. https://doi.org/10.1038/d41586-022-02083-2.

Chakraborty, S., Kodamana, H., and Chakraborty, S. (2021). Deep learning aided automatic and reliable detection of tomato begomovirus infections in plants. Journal of Plant Biochemistry and Biotechnology, 66: 1–8.

Da Silva, J. C. F., Silva, M. C., Luz, E. J. S., Delabrida, S., and Oliveira, R. A. R. (2023). Using mobile edge AI to detect and map diseases in citrus orchards. Sensors, 23: 2165. https://doi.org/10.3390/s23042165.

Ertelt, M., Mulligan, V. K., Maguire, J. B., Lyskov, S., Moretti, R., Schiffner, T. et al. (2024). Combining machine learning with structure-based protein design to predict and engineer post-translational modifications of proteins. PLoS Computational Biology, 20(3), e1011939. https://doi.org/10.1371/journal.pcbi.1011939.

Gao, Z., Khot, L. R., Naidu, R. A., and Zhang, Q. (2020). Early detection of grapevine leafroll disease in a red-berried wine grape cultivar using hyperspectral imaging. Computers and Electronics in Agriculture, 179: 105807.

Ghasemi, E., Ebrahimie, E., and Niazi, A. (2024). Machine learning for early detection of plant viruses: Analyzing post-infection electrical signal patterns. Smart Agricultural Technology, 9: 100668. https://doi.org/10.1016/j.atech.2024.100668.

Ghosh, D., Chakraborty, S., Kodamana, H., and Chakraborty, S. (2022). Application of machine learning in understanding plant virus pathogenesis: Trends and perspectives on emergence, diagnosis, host-virus interplay and management. Virology Journal, 19: 42. https://doi.org/10.1186/s12985-022-01767-5.

Gould, S. I., Wuest, A. N., Dong, K., Johnson, G. A., Hsu, A., Narendra, V. K. et al. (2024). High-throughput evaluation of genetic variants with prime editing sensor libraries. Nature Biotechnology. https://doi.org/10.1038/s41587-024-02172-9.

Griffel, L., Delparte, D., and Edwards, J. (2018). Using support vector machines classification to differentiate spectral signatures of potato plants infected with Potato Virus Y. Computers and Electronics in Agriculture, 153: 318–324.

Grubaugh, N. D., Gangavarapu, K., Quick, J., Matteson, N. L., de Jesus, J. G., Main, B. J., et al. (2019). An amplicon-based sequencing framework for accurately measuring intrahost virus diversity using PrimalSeq and iVar. Genome Biology, 20, 8. https://doi.org/10.1186/s13059-018-1618-7.

Guo, J., Bolduc, B., Zayed, A. A., Varsani, A., Dominguez-Huerta, G., Delmont, T. O. et al. (2021). VirSorter2: A multi-classifier, expert-guided approach to detect diverse DNA and RNA viruses. Microbiome, 9(1): 37. https://doi.org/10.1186/s40168-020-00990-y.

Hou, X., He, Y., Fang, P., Mei, S.-Q., Xu, Z., Wu, W.-C., et al., (2024). Using artificial intelligence to document the hidden RNA virosphere. Cell. https://doi.org/10.1016/j.cell.2024.09.02.

Jagga, Z., and Gupta, D. (2014). Supervised learning classification models for prediction of plant virus encoded RNA silencing suppressors. PLoS One, 9(5): e97446. https://doi.org/10.1371/journal.pone.0097446.

Jemaa, H., Bouachir, W., Leblon, B., LaRocque, A., Haddadi, A., and Bouguila, N. (2023). UAV-based computer vision system for orchard apple tree detection and health assessment. Remote Sensing, 15: 3558. https://doi.org/10.3390/rs15143558.

Joshi, R. C., Kaushik, M., Dutta, M. K., Srivastava, A., and Choudhary, N. (2021). VirLeafNet: Automatic analysis and viral disease diagnosis using deep-learning in Vigna mungo plant. Ecological Informatics, 61: 101197. https://doi.org/10.1016/j.ecoinf.2020.101197.

Kadam, K. U. (2020). Identification of groundnut bud necrosis virus on tomato fruits using machine learning-based segmentation algorithm. International Journal of Future Generation Communication and Networking, 13: 259–264.

Kamal, H., Minhas, F. A., Farooq, M., Tripathi, D., Rana, M. H., Mustafa, R. et al. (2019). *In silico* prediction and validations of domains involved in *Gossypium hirsutum* SnRK1 protein interaction with cotton leaf curl Multan betasatellite encoded βC1. Frontiers in Plant Science, 10, 656. https://doi.org/10.3389/fpls.2019.00656.

Kanaparthi, K. R., and Ilango, S. S. (2023). A survey on training issues in chili leaf diseases identification using deep learning techniques. Procedia Computer Science, 218: 2123–2132. https://doi.org/10.1016/j.procs.2023.01.188.

Kawasaki, Y., Uga, H., Kagiwada, S., and Iyatomi, H. (2015). Basic study of automated diagnosis of viral plant diseases using convolutional neural networks. pp. 638–645. *In:* International Symposium on Visual Computing. Springer.

Kieft, K., Zhou, Z., and Anantharaman, K. (2020). VIBRANT: Automated recovery, annotation and curation of microbial viruses, and evaluation of viral community function from genomic sequences. Microbiome, 8: 90. https://doi.org/10.1186/s40168-020-00867-0.

Kissanga, G.-M. B., Zulfiqar, H., Gao, S., Yussif, S. B., Momanyi, B. M., Ning, L. et al. (2024). E-MuLA: An Ensemble Multi-Localized Attention Feature Extraction Network for Viral Protein Subcellular Localization. Information, 15(3): 163. https://doi.org/10.3390/info15030163.

Lin, Z., Akin, H., Rao, R., Hie, B., Zhu, Z., Lu, W., et al. (2023). Evolutionary-scale prediction of atomic-level protein structure with a language model. Science, 379(6637): 1123–1130. https://doi.org/10.1126/science.ade2574.

Liu, F., Miao, Y., Liu, Y., and Hou, T. (2022). RNN-VirSeeker: A deep learning method for identification of short viral sequences from metagenomes. IEEE/ACM Transactions on Computational Biology and Bioinformatics, 19(3): 1840–1849. https://doi.org/10.1109/tcbb.2020.3044575.

Liu, H., Peck, X. Y., Choong, Y. K., Ng, W. S., Engl, W., Raghuvamsi, P. V. et al. (2022). Identification of putative binding interface of PI(3,5)P2 lipid on rice black-streaked dwarf virus (RBSDV) P10 protein. Virology, 570: 81–95. https://doi.org/10.1016/j.virol.2022.03.010.

Miao, Y., Sun, Z., Ma, C., Lin, C., Wang, G., and Yang, C. (2024). VirGrapher: A graph-based viral identifier for long sequences from metagenomes. Briefings in Bioinformatics, 25(1): bbae036. https://doi.org/10.1093/bib/bbae036.

Moshiri, N., Fisch, K. M., Birmingham, A., DeHoff, P., Yeo, G. W., Jepsen, K. et al. (2022). The ViReflow pipeline enables user friendly large scale viral consensus genome reconstruction. Scientific Reports, 12(1): 5077. https://doi.org/10.1038/s41598-022-09035-w.

Moshiri, N. (2023). ViralConsensus: A fast and memory-efficient tool for calling viral consensus genome sequences directly from read alignment data. Bioinformatics, 39(5), btad317. https://doi.org/10.1093/bioinformatics/btad317.https://github.com/niemasd/ViralConsensus.

Naik, N., Malmathanraj, R., and Palanisamy, P. (2022). Detection and classification of chili leaf disease using a squeeze-and-excitation-based CNN model. Ecological Informatics, 69: 101663. https://doi.org/10.1016/j.ecoinf.2022.101663.

Osorio, D., Capasso, A., Eckhardt, S. G., Giri, U., Somma, A., Pitts, T. M., et al. (2024). Population-level comparisons of gene regulatory networks modeled on high-throughput single-cell transcriptomics data. Nature Computational Science, 4: 237–250. https://doi.org/10.1038/s43588-024-00597-5.

Peng, Y., Dallas, M. M., Ascencio-Ibáñez, J. T., Hoyer, J. S., Legg, J., Hanley-Bowdoin, L. et al. (2022). Early detection of plant virus infection using multispectral imaging and spatial–spectral machine learning. Scientific Reports, 12: 3113. https://doi.org/10.1038/s41598-022-06372-8.

Perkins, M. (2023). Academic integrity considerations of AI large language models in the post-pandemic era: ChatGPT and beyond. Journal of University Teaching and Learning Practice, 20(2).

Phillips, S. J., Anderson, R. P., Dudík, M., Schapire, R. E., and Blair, M. (2017). Opening the black box: An open-source release of Maxent. Ecography. 40: 887–893. https://doi.org/10.1111/ecog.03049.

Posada-Céspedes, S., Seifert, D., Topolsky, I., Jablonski, K. P., Metzner, K. J., and Beerenwinkel, N. (2021). V-pipe: A computational pipeline for assessing viral genetic diversity from high-throughput data. Bioinformatics, 37(12): 1673–1680. https://doi.org/10.1093/bioinformatics/btab015.

Prabha, K. (2020). Disease sniffing robots to apps fixing plant diseases: Applications of artificial intelligence in plant pathology—a mini review. Indian Phytopathology. https://doi.org/10.1007/s42360-020-00290-3.

Prabha, K., Sivaraj, N., Holajjer, P., Gaikwad, H. D., Gupta, N., Girish, K. S. et al. (2020). Modeling of current and future distribution of Impatiens necrotic spot tospovirus and its vector Frankliniella occidentalis under climate change scenario in India. In International Conference VIROCON 2020 on the theme "Evolution of Viruses and Viral Diseases", February 18–20, 2020.

Pratapa, A., Jalihal, A. P., Law, J. N., Bharadwaj, A., and Murali, T. M. (2020). Benchmarking algorithms for gene regulatory network inference from single-cell transcriptomic data. Nature Methods, 17: 147–154. https://doi.org/10.1038/s41592-019-0690-6.

Ramcharan, A., Baranowski, K., McCloskey, P., Ahmed, B., Legg, J., and Hughes, D. P. (2017). Deep learning for image-based cassava disease detection. Frontiers in Plant Science, 8: 1852.

Ravi, V., Acharya, V., and Pham, T. D. (2022). Attention deep learning-based large-scale learning classifier for Cassava leaf disease classification. Expert Systems, 39(2): e12862. https://doi.org/10.1111/exsy.12862.

Ren, J., Song, K., Deng, C., Ahlgren, N. A., Fuhrman, J. A., Li, Y. et al. (2020). Identifying viruses from metagenomic data using deep learning. Quantitative Biology, 8: 64–77. https://doi.org/10.1007/s40484-019-0187-4.

Ren, J., Song, K., Deng, C., Ahlgren, N. A., Fuhrman, J. A., Li, Y., et al. (2020). Identifying viruses from metagenomic data using deep learning. Quantitative Biology, 8: 64–77. https://doi.org/10.1007/s40484-019-0187-4.

Roux, S., Enault, F., Hurwitz, B. L., and Sullivan, M. B. (2015). VirSorter: Mining viral signal from microbial genomic data. PeerJ, 3, e985. https://doi.org/10.7717/peerj.985.

Roy, B. G., Choi, J., and Fuchs, M. F. (2024). Predictive modeling of proteins encoded by a plant virus sheds a new light on their structure and inherent multifunctionality. Biomolecules, 14(1): 62. https://doi.org/10.3390/biom14010062.

Sahu, S. S., Loaiza, C. D., and Kaundal, R. (2019). Plant-mSubP: A computational framework for the prediction of single- and multi-target protein subcellular localization using integrated machine-learning approaches. AoB Plants, 12(3): plz068. https://doi.org/10.1093/aobpla/plz068.

Shang, J., and Sun, Y. (2020). CHEER: Hierarchical taxonomic classification for viral metagenomic data via deep learning. bioRxiv. https://doi.org/10.1101/2020.03.26.009001.

Shrestha, P., Kandel, J., Tayara, H., and Chong, K. T. (2024). Post-translational modification prediction via prompt-based fine-tuning of a GPT-2 model. Nature Communications, 15: 6699. https://doi.org/10.1038/s41467-024-51071-9.

Silva, J. C. F., Carvalho, T. F. M., Fontes, E. P. B., and Cerqueira, F. R. (2017). Fangorn Forest (F2): A machine learning approach to classify genes and genera in the family Geminiviridae. BMC Bioinformatics, 18: 431.

Sperschneider, J., Catanzariti, A. M., DeBoer, K., Petre, B., Gardiner, D. M., Singh, K. B. et al. (2017). LOCALIZER: subcellular localization prediction of both plant and effector proteins in the plant cell. Scientific Reports, 7: 44598. https://doi.org/10.1038/srep44598.

Srivastava, A., Pandey, V., Marwal, A., Ali, A., and Gaur, R. K. (2024). *In-silico* prediction and validation of Carica papaya protein domains interaction with the Papaya leaf curl virus and associated betasatellite encoded protein. Discover Applied Sciences, 6: 287. https://doi.org/10.1007/s42452-024-05961-8.

Sukhorukov, G., Khalili, M., Gascuel, O., Candresse, T., Marais-Colombel, A., and Nikolski, M. (2022). VirHunter: A deep learning-based method for detection of novel RNA viruses in plant sequencing data. Frontiers in Bioinformatics, 2: 867111. https://doi.org/10.3389/fbinf.2022.867111.

Trapview. (2024). Enhancing vineyard health: Successful Scaphoideus titanus monitoring with Trapview solution. Retrieved September 10: 2024, from https://trapview.com/enhancing-vineyard-health-successful-scaphoideus-titanus-monitoring-with-trapview-solution/.

Wang, D., Vinson, R., Holmes, M., Seibel, G., Bechar, A., Nof, S., et al. (2019). Early detection of tomato spotted wilt virus by hyperspectral imaging and outlier removal auxiliary classifier generative adversarial nets (OR-AC-GAN). Scientific Reports, 9: 4377. https://doi.org/10.1038/s41598-019-40066-y.

Wang, J., Sun, Z., Wang, G., and Yan, M. (2024). TransGINmer: Identifying viral sequences from metagenomes with self-attention and graph isomorphism network. Future Generation Computer Systems, 161. https://doi.org/10.1016/j.future.2024.07.025.

Yan, M., Sun, Z., Ma, C., Lin, C., Wang, G., and Yang, C. (2024). VirGrapher: A graph-based viral identifier for long sequences from metagenomes. Briefings in Bioinformatics, 25(2). https://doi.org/10.1093/bib/bbae036.

Yan, M., Liu, F., Hou, T., and Liu, Y. (2021). Virtifier: A deep learning-based identifier for viral sequences from metagenomes. Bioinformatics, 38. https://doi.org/10.1093/bioinformatics/btab845.

Index

N

O

P

R

S

T

V

W

About the Editors

Kamel A. Abd-Elsalam, Ph.D., is currently a Research Professor and Deputy Director for Research at the Plant Pathology Research Institute, Agricultural Research Center, Giza, Egypt. Dr. Kamel's research interests include developing and improving plant biosecurity diagnostic tools, understanding fungal pathogen genomes, and creating eco-friendly hybrid nanomaterials for controlling toxigenic fungi and plant diseases as well as applications in agroecosystems. Kamel earned his Ph.D. in Molecular Plant Pathology from Christian Albrechts University of Kiel (Germany) and Suez Canal University (Egypt). In 2008, he was awarded a postdoc-toral fellowship from the same institution. Dr. Kamel served as a Visiting Associate Professor at Mae Fah Luang University in Thailand, the Institute of Microbiology at TUM in Germany, the Laboratory of Phytopathology at Wageningen University in the Netherlands, and the Plant Protection Department at Sassari University in Italy.

He has published 41 books related to nanobiotechnology applications in agriculture and plant protection with major publishing houses such as Springer, Taylor & Francis, and Elsevier. Since 2019, he has served as the Editor-in-Chief of the Elsevier book series Nanobiotechnology for Plant Protection and as the Series Editor for Applications of Genome Modified Plants and Microbes in Food and Agriculture. Additionally, he has been an active member of the Elsevier Advisory Panel since 2020, providing feedback and suggestions to improve Elsevier's products and services. He has published more than 328 scientific research articles in international and regional specialized journals with high impact factors, and has an H-index of 55, an i-10 index of 173, and over +11802 citations. He served as a Guest Editor for the Journal of Fungi, Plants, and Microorganisms, and as a Review Editor for Frontiers in Genomic Assay Technology, Agrochemicals, and Plant Nano Biology. He was ranked in the top 2% of the most influential scientists in the world in nanobiotechnology for 2021, 2022, 2023, 2024 and 2025 by Elsevier and Stanford University. In 2014, he was awarded the Federation of Arab Scientific Research Councils Prize for excellent scientific research in biotechnology (fungal genomics), ranking first. Additionally, he received the Egyptian State Award for Agricultural Scientific Excellence from ASRT, Egypt, in 2022 with a silver medal.

Aftab Ahmad, Ph.D., is currently working as an Associate Professor at the Natural and Medical Sciences Research Center (NMSRC), University of Nizwa, Oman. Dr. Aftab's research interests include developing biotic stress-tolerant gene-edited plants using CRISPR/Cas technology. He has published four international books on CRISPR and more than 40 research articles in international peer reviewed journals including Frontiers in Plant Science, Journal of Biomedical Sciences, Scientific Reports-Nature, PLOS One, Journal of Translational Medicine, Plant Molecular Biology Reporter, Molecules, Journal of Cereal Science, and Genetic Resources and Crop Evolution. Dr. Aftab pursued his Ph.D. and postdocs in Plant Molecular Biology from the University of Shizuoka, Japan, under the Center of Excellence (COE, 21st Century) program. He works as a guest editor in Frontiers in Genome Editing and Frontiers in Plant Sciences. Dr. Aftab also served as a visiting scientist in the Department of Plant Sciences, University of California, Davis, USA. Dr. Aftab has more than ten years of teaching experience in biochemistry and molecular biology, and his research focuses on CRISPR/Cas-based genome editing in plants.

For Product Safety Concerns and Information please contact our
EU representative GPSR@taylorandfrancis.com Taylor & Francis
Verlag GmbH, Kaufingerstraße 24, 80331 München, Germany